FUNCTION AND BIOSYNTHESIS OF LIPIDS

ADVANCES IN EXPERIMENTAL MEDICINE AND BIOLOGY

Recent Volumes in this Series

Volume 82
ATHEROSCLEROSIS: Metabolic, Morphologic, and Clinical Aspects
Edited by George W. Manning and M. Daria Haust • 1977

Volume 83
FUNCTION AND BIOSYNTHESIS OF LIPIDS
Edited by Nicolás G. Bazán, Rodolfo R. Brenner, and Norma M. Giusto • 1977

Volume 84
MEMBRANE TOXICITY
Edited by Morton W. Miller and Adil E. Shamoo • 1977

Volume 85A
ALCOHOL INTOXICATION AND WITHDRAWAL – IIIa: Biological Aspects of Ethanol
Edited by Milton M. Gross • 1977

Volume 85B
ALCOHOL INTOXICATION AND WITHDRAWAL – IIIb: Studies in Alcohol Dependence
Edited by Milton M. Gross • 1977

Volume 86A
PROTEIN CROSSLINKING: Biochemical and Molecular Aspects
Edited by Mendel Friedman • 1977

Volume 86B
PROTEIN CROSSLINKING: Nutritional and Medical Consequences
Edited by Mendel Friedman • 1977

Volume 87
HYPOTHALAMIC PEPTIDE HORMONES AND PITUITARY REGULATION
Edited by John C. Porter • 1977

Volume 88
AVIAN IMMUNOLOGY
Edited by Albert A. Benedict • 1977

Volume 89
MUCUS IN HEALTH AND DISEASE
Edited by Max Elstein and Dennis V. Parke • 1977

Volume 90
PARKINSON'S DISEASE: Neurophysiological, Clinical, and Related Aspects
Edited by Fathy S. Messiha and Alexander D. Kenny • 1977

FUNCTION AND BIOSYNTHESIS OF LIPIDS

Edited by

Nicolás G. Bazán

Universidad Nacional del Sur – CONICET
Bahía Blanca, Argentina

Rodolfo R. Brenner

Universidad Nacional de La Plata
La Plata, Argentina

and

Norma M. Giusto

Universidad Nacional del Sur – CONICET
Bahía Blanca, Argentina

PLENUM PRESS • NEW YORK AND LONDON

Library of Congress Cataloging in Publication Data

International Symposium on Function and Biosynthesis of Lipids, Sierra de la Ventana, Argentine Republic, 1976.
Function and biosynthesis of lipids.

(Advances in experimental medicine and biology; v. 83)
Includes index.
1. Lipids–Congresses. I. Bazán, Nicolás. II. Brenner, Rodolfo Roberto, 1922-
III. Giusto, Norma M. IV. Title. V. Series. [DNLM: 1. Lipids–Biosynthesis–Congresses. 2. Lipids–Metabolism–Congresses. W1 AD559 v. 83 1976/QU85 I598f 1976]
QP751.I637 1975 599'.01'9247 77-6831
ISBN 978-1-4684-3278-7 ISBN 978-1-4684-3276-3 (eBook)
DOI 10.1007/978-1-4684-3276-3

Proceedings of the International Symposium on Function and Biosynthesis of Lipids held at Sierra de la Ventana, Tornquist, Province of Buenos Aires, Argentina, November, 1976 (Symposium of the International Union of Biochemistry, No. 79)

Softcover reprint of the hardcover 1st edition 1977
A Division of Plenum Publishing Corporation
227 West 17th Street, New York, N.Y. 10011

PREFACE

This volume contains the proceedings of the International Symposium on Function and Biosynthesis of Lipids held November 1976 at Sierra de la Ventana, Argentina. The conference was organized by the Argentine Biochemical Society and co-sponsored by the International Union of Biochemistry and the Pan American Association of Biochemical Societies.

The Symposium's great success is owed to all the participants, most of whom were actively involved in the interesting and stimulating discussions that followed every single lecture and communication delivered. Many thanks also to Drs. R. Anderson, S. Bonting, K. K. Carroll, B. Cecarelli, M. A. Crawford, R. T. Holman, J. Mead, R. H. Michell, G. Porcellati, W. Stoffel, W. Thompson, and H. Wiegandt for chairing Symposium sessions.

The contributions are organized in four sections, preceded by the opening lecture on the role of dolichol phosphate in glycoprotein biosynthesis delivered by Dr. Luis F. Leloir. The first section, on aspects of lipid involvement in the biogenesis of membranes, covers the biosynthesis of saturated and unsaturated fatty acids, including enzymology, biosynthetic pathways, and regulation and modifications during development, in the fetus and in tumor cells. Papers on the composition and assembly of membrane components, lipid interactions with proteins and enzymes, developmental changes of lipids, and newer approaches to survey the organization and biogenesis of cellular membranes complete this section.

New information as well as reviews of current concepts about structural specificity, biosynthesis, and function of glycosphingolipids are included in the second section.

The third section is devoted to various aspects of biosynthesis, turnover, and function of lipids in the central nervous system, such as metabolism of polyenoic fatty acids, base-exchange reactions, biosynthesis of complex lipids, and the latest progress in phosphatidylinositol biochemistry. In addition papers on the implications of lipids in transmitter release and in receptors are included.

The fourth section is composed of several contributions on essential fatty acids in human nutrition, diet, brain, developing retina, and testes, as well as on enzymological studies.

We are grateful to Aerolíneas Argentinas; Bodegas y Viñedos Edmundo Norton; Comisión de Investigaciones Científicas of the Province of Buenos Aires; Consejo Nacional de Investigaciones Científicas y Técnicas, Argentina; the Embassy of the German Federal Republic in Buenos Aires; FIDIA Research Laboratories, Abano Terme, Padua, Italy; Gobernación of the Province of Buenos Aires; International Union of Biochemistry; Municipalidad of Tornquist; Secretaría de Estado de Salud Pública; Ministerio de Bienestar Social of Argentina; Secretaría de Estado de Ciencia y Tecnología; Laboratorios Promeco; Ministerio de Cultura y Educación of Argentina; The British Council; and Universidad Nacional del Sur for their financial support of the Symposium.

We should like to express our gratitude to the staff of the Instituto de Investigaciones Bioquímicas, Universidad Nacional del Sur-Consejo Nacional de Investigaciones Científicas y Técnicas, for its valuable help both in the organization of the Symposium and in the preparation of this book. Thanks are also due to Plenum Publishing Corporation for its support.

We hope that this volume will stimulate further productive research on the function of lipids in the cell.

Nicolás G. Bazán

Rodolfo R. Brenner

Norma M. Giusto

Bahía Blanca, January 1977

CONTENTS

OPENING LECTURE

I. LIPID INVOLVEMENT IN THE BIOGENESIS OF MEMBRANES

(A) Biosynthesis of fatty acids

(B) Biogenesis and organization of cellular membranes

II. GLYCOSPHINGOLIPIDS—GANGLIOSIDES

III. LIPIDS IN NEURAL TISSUE

(A) Biosynthesis and turnover

(B) Functions

IV. ESSENTIAL FATTY ACIDS IN NUTRITION

IN MEMORIAM

PROF. WERNER SEUBERT

In life it is impossible to have only happiness. When we were organizing this Symposium we invited Prof. Werner Seubert to give a lecture on his specialty, "The elongation reaction of fatty acids." However, destiny was against our will since his sudden death in 1975 silenced his experienced voice. This was a real loss for international lipid biochemistry, although his knowledge will remain in his collaborators and publications.

Werner Seubert was born in Munich on February 9th, 1928. He obtained his degree in chemistry at the University of Munich in 1952, and worked on his doctoral thesis in Professor Lynen's laboratory. Subsequently he worked for several years in Lynen's group at the Max Planck Institut für Zellchemie in Munich on the mechanism of fatty acid synthesis.

In 1957 he was invited by Prof. Earl Stadtman to go to the Laboratory of Biochemistry at the National Heart Institute, Bethesda, Maryland, USA. He stayed there until 1959, studying the mechanism of biological catabolism of branched carbon chains on models such as farnesole and geraniole. When he came back to Munich, he continued these studies on the metabolism of terpenes.

In 1962 he accepted an invitation of Professor Heinz to go to the Institute of Physiological Chemistry at the University of Frankfurt where he started to build up a department of enzymology. In 1966 he was appointed Professor of Physiological Chemistry at Frankfurt. During 1967 he was offered chairs at Münster, Heidelberg, Regensburg, Frankfurt, and Göttingen, and in 1968 he became Head of the Institute of Physiological Chemistry at the University of Göttingen, where he worked until his untimely death.

In 1969 he received the Heinrich Wieland award for his work on fatty acid chain elongation in mitochondria, and in 1970 he was elected to membership in the Academy of Science in Göttingen.

His main interest in the last ten years was in the field of the regulation of gluconeogenesis and ketogenesis and the mechanism and biological significance of fatty acid elongation. With respect to gluconeogenesis he worked on the problem of compartmentation and control of gluconeogenic enzymes such as pyruvate carboxylase and PEP-carboxykinase, studies which led to the problem of mitochondrial heterogeneity and new aspects of subunit structure of pyruvate carboxylase. His investigations on ketogenesis led to the characterization of the thiolase reaction as a control point in ketone body formation and the isolation and characterization of different mitochondrial acetoacetyl-CoA thiolases.

His studies and interest in fatty acid chain elongation started at Prof. Lynen's laboratory during the fifties. At that time he showed that fatty acids can be synthesized from acetyl-CoA in the presence of NADH and NADPH by reversal of fatty acid oxidation in an *in vitro* system containing thiolase, β-hydroxyacl-CoA dehydrogenase, crotonase, and a newly found "reducing enzyme" (enoyl-CoA reductase) from pig liver and yeast.

Some years later, he came back to this line of work, purified the enoyl-CoA reductase from rat liver, and began systematically to investigate malonyl-CoA-*dependent* and -*independent* fatty acid elongation systems. Together with Dr. Podack he separated a mitochondrial and a microsomal enoyl-CoA reductase. The investigation of the malonyl-CoA-dependent chain elongation led to the purification and characterization of the microsomal fatty acid chain elongation system from beef adrenal cortex and rat liver, and Seubert suggested a multienzyme complex catalyzing the entire reaction sequence.

Studies of the malonyl-CoA-*independent* chain elongation in collaboration with Dr. Hinsch led to further characterization of the mitochondrial chain elongating system of rat liver and pig kidney cortex. Studies on the properties of this system in various tissues led to Dr. Seubert's proposal of two types of mitochondrial chain elongation: a "heart type," needing only NADH as nucleotide, important in the conservation of reducing equivalents or acetate units in the anaerobic state, and a "liver type," playing a role in the transfer of hydrogen from NADPH to the respiratory chain. But these studies were interrupted by Dr. Seubert's sudden death on September 8th, 1975. We thank him for his contribution to knowledge.

Rodolfo R. Brenner

A SHORT HISTORY OF THE BIOCHEMISTRY OF LIPIDS IN ARGENTINA (OPENING ADDRESS)

Rodolfo R. Brenner

Cátedra de Bioquímica, Instituto de Fisiología
Facultad de Ciencias Médicas, La Plata, Argentina

On behalf of the Scientific Committee of the Symposium I will say a few words to open the meeting and welcome our distinguished guests.

I will not refer to the contributions each of our guests has made to science since that would take many hours. I will only say that their contributions to the methodology of lipid biochemistry and to basic and applied research are outstanding.

I think that it is interesting for you to know the historical background of the various lipid biochemistry research groups in our country.

The pioneer of lipid biochemistry in our country is Prof. Pedro Cattáneo, who, following the Hildich school, began in 1940 a series of studies (that still go on) on the fatty acid composition of Argentine plants. He is unquestionably the grandfather of our lipid family. Curiously, he began his research line studying the fatty acid composition of the ceibo (<u>Citrina cristagalli</u>), a small tree whose beautiful red blossom is our national flower. His work has been done in the Faculty of Ciencias Exactas at Buenos Aires University. I personally had the pleasure of working in his laboratory and there began in 1946 several studies on lipids. First we studied fish lipids and waxes. Then we started a series of research projects on polyunsaturated fatty acid biosynthesis in different species of the animal kingdom as well as on the lipid metabolism and the effect of diabetes on tumor cells. Our group, the oldest in continuous research in lipid biochemistry, developed in the Cátedra de Bioquímica of the Instituto de Fisiología of the Faculty of Medicine of La Plata University. To this group belong such senior investigators as R. O. Peluffo, O. Mercuri, M. E. De Tomás, A. Nervi, A. Catalá, M. J. T. de Alaniz, I. N. T. de Gómez Dumm, J. C. Castuma, S. Ayala, and others. I have at times been considered the father of the lipid family.

Dr. Luis F. Leloir, of course, is also a member of this family. He showed years ago, in 1943, his interest in fatty acid β oxidation. He is an uncle of the family. He showed in that year that β oxidation of fatty acids could be produced by a cell-free system derived from mitochondria. Much later, in 1965, he came back to the lipid field, beginning a propitious research in polyprenols in collaboration with N. Behrens, M. A. Dankert, and A. Parodi. R. Staneloni, H. Carminatti, E. Belocopitow, and others joined their efforts. This research is done in Campomar Institute in Buenos Aires.

In 1963 our other uncle, Dr. R. Caputto, came back from the United States and made his home in Córdoba. In the Department of Biological Chemistry of Córdoba University he began to develop an active and well oiled group. His line is the biosynthesis and function of gangliosides. To this group belong H. Maccioni, A. Arce, J. Curtino, C. Landa, and others. Dr. F. A. Cumar also works there in lipids.

Dr. E. De Robertis is an uncle-in-law of the family since he also showed some interest in lipids to resolve his own synaptic problems at the Faculty of Medicine of Buenos Aires studying proteolipids. From his laboratory stems our wandering son, E. G. Lapetina, who -- rather alone -- works on phosphatidylinositol.

Dr. J. C. Gómez and Dr. E. F. Soto are other uncles, who work on the lipids of nervous tissues in collaboration with J. M. Pasquini at the Faculty of Biochemistry of Buenos Aires.

A new figure in the field is R. N. Farías, one of the sons of the family. At Tucumán University he is developing an active group that studies the structure of cellular membranes in essential fatty acid deficiency. This work was initiated in Buenos Aires by Dr. Trucco in 1971 and, after traveling to Córdoba, reached and stayed in Tucumán at the Institute of Biological Chemistry.

Another son of the family is the secretary of this Symposium, Dr. N. G. Bazán. He came back from Canada in 1970 and, pushing impetuously, organized the Institute of Biochemistry in Bahía Blanca University. He is surrounded by a number of young collaborators, a very active group consisting of N. M. Giusto, M. I. A. de Caldironi, E. B. R. de Turco, H. E. P. de Bazán, A. M. P. de D'Angelo, H. A. Caldironi, G. D. Cascone, M. F. Pediconi, M. G. I. de Boschero, I. C. B. de Romanelli, and T. S. Alonso, who constitute the local organizing committee of the Symposium. His main work is related to lipid biochemistry of excitable and developing membranes.

At the Institute of Biological Chemistry of the Faculty of Medicine in Buenos Aires University, Prof. A. Stoppani has been working for a long time in enzymatic problems of mitochondria. Diverging from this line, J. C. Vidal entered the lipid family, at first reluctantly, purifying the phospholipase of "yararâ" venom (Bothrops atrox) (1965-1969) and afterwards in 1972 more decisively. At the moment he is studying the interactions between phospholipids and proteins in the 3-hydroxybutyrate dehydrogenase of mitochondrial membrane. He is undoubtedly our adoptive son.

The result of the existence of this family, which has grown very rapidly in recent years, is a newborn. The newborn is this Symposium. It took a long time to be delivered. It was originally planned for 1974, but many reasons delayed it until this year, when we finally closed our eyes and jumped into a void without really knowing whether we would land on solid ground or in a pit.

Due mainly to the extraordinary activity of our secretary, Dr. N. G. Bazán, many of the difficulties were overcome. Vital also was the extraordinary help we received from government and the national and international private organizations mentioned in our program. We thank them heartily. To them belongs the honor of today's reality.

Now comes the other part, the scientific part. Here we are optimistic because of the unexpected response we had from all over the world, as a result of which we have here representatives of the science of America, Asia, and Europe. We are especially optimistic because of the quality and extraordinary scientific level of our foreign colleagues, whom we welcome heartily. You are at home because our home is yours.

OPENING LECTURE

THE ROLE OF DOLICHOL IN PROTEIN GLYCOSYLATION

L. F. Leloir

Instituto de Investigaciones Bioquímicas "Fundación Campomar" y Facultad de Ciencias Exactas y Naturales, Obligado 2490, Buenos Aires, Argentina

Lipid intermediates were first detected in studies on the biosynthesis of lipopolysaccharide and of proteoglycan in bacteria. These intermediates turned out to be sugar derivatives of poly-isoprenoid alcohols. A series of these compounds has been isolated from various organisms. Their general formula is shown as follows:

$$H(CH_2-\overset{\displaystyle CH_3}{\overset{|}{C}}=CH-CH_2)_nOH$$

The substance involved in sugar transfer reactions in bacteria is undecaprenol which has n=11 and 2 trans double bonds. A similar compound ficaprenol is present in plants. It has n=11 and 3 trans double bonds. It was isolated as the free alcohol after treatment of leaves with strong alkali and at present there is no evidence that its derivatives act as intermediates in sugar transfer.

Another compound of interest is dolichol which was first isolated from liver and is now known to be present in many other animal tissues. It is a long chain molecule (n=17-21), its first isoprene residue is saturated and it has 2 trans double bonds. It is found in liver part as the free alcohol and another part esterified with fatty acids. We now know that a small amount is present as the phosphate or pyrophosphate and may have various sugars joined to it. These dolichol derivatives are involved in the glycosylation of proteins. An excellent review on polyprenols has been written by Hemming (1974).

Sugar-aminoacid linkages. Various sugar-aminoacid linkages have been found in glycoproteins, some are formed directly by transfer from sugar nucleotides, others from dolichol derivatives and information is lacking for several of them.

Serine-threonine. One of the common sugar-aminoacid linkages is that involving the hydroxy aminoacids serine or threonine. This linkage is found in several glycoproteins for instance in chondroitin sulfates and similar substances (usually referred to as proteoglycans). The polypeptide backbone carries many saccharide side chains in which the sequence of the innermost part usually called the linkage region is: GlUA-Gal-Gal-Xyl-Ser (Thre). In chondroitin acetylgalactosamine is joined to the glucuronic acid residue and then follows another acetylgalactosamine and another glucuronic acid residue and so on. Sulfate residues may also be present. The different transfer reactions involved in the biosynthesis have been fairly well studied and there is no evidence that lipid intermediates are involved.

Another type of linkage involving serine-threonine is found in submaxillary mucins. The innermost sugar is N-acetylgalactosamine which may have galactose and neuraminic acid residues joined to it. Here again the glycosylation occurs presumably by direct transfer from the sugar nucleotides. The antifreeze protein of antartic fish has the sequence Gal-β—1.4-Gal-NAc-Ser (Thre). Another glycoprotein containing serine-threonine linked sugars is yeast mannan which has side chains of two to four mannose residues. Besides these it also has some asparagine linked residues.

Galactose linked to serine or threonine has been found and in earthworm cuticula collagen. Futhermore a compound containing the disaccharide glucosyl-β-1.3 fucose joined to threonine has been isolated from urine.

Hydroxylysine. Another glycoprotein which has been studied from the point of view of structure and biosynthesis is collagen. The sugar residue of collagen and of kidney glomerular basement membrane is joined to hydroxylysine as follows: Glc-α.1.2-Gal-β-Hyl. Studies on the biosynthesis have shown that the transfer is direct from the corresponding uridine nucleotides.

Hydroxyproline. Glycoproteins containing sugars linked to hydroxyproline comprise those in which the sugar is arabinose as in plant extensin or galactose as in a protein found in the alga chlamidomonas.

Cystein. Di-galactosylcystein has been detected in a urine glycopeptide and another compound containing cystein and three glucose residues was found in red blood cells. It is not known whether these are degradation products of a glycoprotein.

Asparagine. Glycoproteins containing asparagine linked oligosaccharides are numerous. They comprise all the serum proteins except albumin and many membrane linked glycoproteins.

The sugars involved are acetylglucosamine, mannose, neuraminic acid, fucose and galactose. The oligosaccharides may be of two types: the lactosamine or complex type and the oligomannosidic or high mannose type.

In both types the core that is the innermost sugars has the following structure:

```
Man
   α 1·3
         Man-β-1·4 GlcNAc-β-1·4 GlcNAc-Asn
   α 1·6
Man
```

This common core is found in many glycoproteins such as ovalbumin, immunoglobulins, thyroglobulin, some plant glycoproteins, etc.

In the lactosamine type one to four lactosamine residues (Gal-β-1·4 GlcNAc) may be joined to the mannoses and also some fucose and neuraminic acid residues.

In the high mannose type several chains of mannose residues are joined to the core. Other oligosaccharides have been described that are not included in this classification may have the mannose intercalated in the chitobiose unit or have N-acetyl-glucosamine residues intercalated in the mannose chains.

Mixed type of glycoproteins. Many glycoproteins have oligosaccharides linked to different aminoacids. For instance glycophorin, the main glycoprotein of red blood cells has about 60% of carbohydrate. Each molecule has 15 serine/threonine linked oligosaccharides and one of the asparagine type. Another example of the mixed type is the blood protein fetuin which has both serine/threonine and asparagine linked oligosaccharides. For reviews on glycoproteins see Gottschalk (1972), Spiro (1973), Marshall & Neuberger (1970), Montreuil (1975) and Kornfeld & Kornfeld (1976).

Initial experiments with dolichol phosphate and UDP-glucose. In 1967 our colleague Marcelo Dankert returned from a stay in Boston where he worked with Dr. Robbins' group on the identification of undecaprenol pyrophosphate as an intermediate in the biosynthesis of lipopolysaccharide in Salmonella. He was so enthusiastic with polyprenols that he infected several of us. With Dr. Behrens we mixed a crude enzyme from rat liver which was used as glycogen synthetase with radioactive UDP-glucose and after incubation measured the label extracted by organic solvents. A very small difference with the control was observed. The difference could be increased by optimizing the conditions and it was found that addition of a lipid extract of liver produced a large increase in the formation of glucose bound to lipid. Since within limits the radioactivity soluble in organic solvents was proportional to the amount of lipid extract added we had a method of measuring the active substance. After purification and comparison with a specimen obtained by chemical phosphorylation of dolichol we became convinced (Behrens & Leloir, 1970) that the substance was dolichol monophosphate and that the reaction we were measuring was as follows:

$$\text{Dol-P} + \text{UDP-Glc} \longrightarrow \text{Dol-P-Glc} + (\text{UDP}) \qquad (1)$$

The procedure which we used for measuring DolMP is still the only one we have available for small amounts. It is not quite specific, thus ficaprenol phosphate, undecaprenol phosphate and other polyprenol phosphates also act as acceptors when tested with the liver enzyme. This method of estimation is quite useful since dolichol phosphate is not easy to identify in crude extracts.

Incubation of Dol-P-(^{14}C)-Glc with liver microsomes was found to lead to the transfer of the label to an endogenous acceptor which on the basis of somewhat indirect evidence was assumed to be a dolichol pyrophosphate oligosaccharide (Behrens et al., 1971).

Acid hydrolysis was found to give rise to a product behaving like Dol-P when tested with the enzyme system catalyzing the reaction (1). The presence of a pyrophosphate was deduced from the behaviour of the compound in DEAE cellulose columns and towards alkaline hydrolysis. As to the water soluble moiety which is released by mild acid treatment it was found to give the pattern of an oligosaccharide when acetolyzed and chromatographed before or after deacetylation. Measurements of the molecular weight with Sephadex columns and peptides as standards gave a value of 3500. When run on paper in butanol-pyridine-water 4:3:4 it behaved like a maltooligosaccharide of 17-20 units. Treatment with alkali of the oligosaccharide obtained by acid methanolysis gave products which had one or two negative charges and that became neutral by N-acetylation or by treatment with nitrous acid. It was concluded that the products obtained with alkali corresponded to substances in which one or two N-acetyl residues were deacetylated that is as if the oligosaccharide contained two hexosamine residues (Behrens et al., 1971a; Parodi et al., 1973).

The substance is present in very small amounts in tissues but a large scale preparation (Tábora, thesis, Buenos Aires University, 1976) allowed an analysis by gas liquid chromatography. The results approximated mannose 12, glucose 4, glucosamine 2. The free oligosaccharide was reduced with tritiated sodium borohydride and then hydrolyzed completely and run on paper. The labelled product behaved like glucosaminitol thus showing that acetylglucosamine is present at the reducing end. It is interesting to point out that the glucose containing oligosaccharide has been found to be formed in all the different tissues tested, rat and pig liver, hen oviduct, rat kidney, rat brain and lymphocytes. It has recently been found also in bakers' yeast by Parodi and Barrientos. Such a wide distribution indicates that it must have some important physiological role.

Transfer from UDP-acetylglucosamine. Incubation of liver microsomes with Dol-P and labelled UDP-acetylglucosamine was found to give a product soluble in organic solvents which is now known to be Dol-P-P-N-acetylglucosamine (Behrens et al., 1971b). After mild acid hydrolysis and paper chromatography the radioactivity was found in a product which behaved like N-acetylglucosamine. A small peak of a slower moving substance was also detected and its amount could be increased by further incubation with unlabelled UDP-N-acetylglucosamine (Leloir et al., 1973).

The reactions were written as follows:

$$\text{UDP-GlcNAc} + \text{Dol-P} \rightarrow \text{Dol-P-P-GlcNAc}$$

$$\text{UDP-GlcNAc} + \text{Dol-P-P-GlcNAc} \rightarrow \text{Dol-P-P-diacetylchitobiose} + \text{UDP}$$

The di,N-acetylchitobiose obtained by mild acid hydrolysis was identified by paper chromatography under various conditions. The distribution of radioactivity in the monosaccharide residues within the di,N-acetylchitobiose after reduction with borohydride was consistent with the reactions as formulated and excluded Dol-P-P-acetylglucosamine as donor in the second reaction.

Transfer from GDP-mannose. The transfer of mannose according to the following equation:

$$\text{GDP-Man} + \text{DolMP} \rightarrow \text{Dol-P-Man} + \text{GDP}$$

is catalyzed by an enzyme which is more active than those which transfer glucose or N-acetylglucosamine. Transfer from UDP-N-acetylgalactosamine or UDP-galactose does not seem to occur (Behrens et al., 1971b) although some workers have reported transfer of galactose.

Besides Dol-P-Man which is the major reaction product in the incubations transfer occurs to acceptors present in the enzyme yielding lipid linked oligosaccharides. Their yield can be greatly increased if exogenous acceptors are added. These have been obtained by chromatography crude extracts of liver on DEAE-cellulose and selecting the fractions where Dol-P-P-oligosaccharides are known to emerge (Behrens et al., 1973).

The Dol-P-P-oligosaccharides which become labelled are of several sizes and run on paper with butanol pyridine water 4:3:4 as solvent like maltooligosaccharides of 3 to 18 or more units.

The smallest oligosaccharide labelled in the mannose was formed by incubation of GDP-Man with Dol-P-P-di'N,acetylchitobiose and is presumably the trisaccharide mannosyl diacetylchitobiose (Levy et al., 1974). It should be noted that in many oligosaccharides the mannose joined to the diacetylchitobiose is β whereas the mannoses in the chains attached to it in the 3 and 6 position are all α. There is evidence suggesting that these two different types of bonds may arise through transfer reactions involving different sugar donors. In the first case it may be GDP-Man which has an α configuration and in the latter Dol-P-Man which is β. An inversion of the configuration would occur then in each transfer reaction.

The pattern of the oligosaccharides which become labelled with GDP-Man is strikingly different from that obtained with UDP-Glc. In the first case they are very polydisperse whereas in the latter they appear after paper chromatography as a definite peak with the same mobility as a maltooligosaccharide of 17 to 20 units.

Transfer to protein. The first dolichol diphosphate oligosaccharide which was found to act as donor of oligosaccharide to protein was the glucose labelled compound (Parodi et al., 1972). The enzyme as well as the acceptor protein was found in the microsomal fraction of liver. The reaction requires manganese ions.

The glycoproteins formed appear to be membrane bound. Dr. Parodi investigated the point using rabbit reticulocytes. He reasoned that since red blood cell glycoproteins have been more studied than any other, it appeared convenient to use this material for glycosylation experiments. However since red blood cells did not yield an active enzyme preparation he used reticulocytes. The proteins which became labelled on incubation with dolichol-diphosphate-(^{14}C)-glucose oligosaccharide were separated by polyacrylamide electrophoresis and it was found that there was no coincidence between the radioactivity and any of the glycoproteins detectable with Schiff's reagent. The label appeared not in one band but in many small peaks.

One may assume that the protein might be degraded by proteases either before or after glycosylation or that the acceptor protein might be glycosylated in different stages of completion while still bound to ribosomes. Carminatti and Idoyaga have observed that the smooth microsomes from rat liver that do not have ribosomes are fully active in transfering the glucose containing oligosaccharide to protein. The rough microsomes after removal of the ribosomes were also active. As to the action of proteases Parodi observed that the pattern of proteins and glycoproteins in rabbit reticulocytes was not affected during the period of incubation required for the "in vitro" glycosylation.

Another possibility is that the glycoproteins in question are many and are present in very small amounts.

On the other hand glucose is not a usual component of the asparagine type of glycoproteins. One possibility that can be considered is that the glucose is added to the oligosaccharide as a signal for transfer to protein and then removed by the action of a glycosidase. The presence of glucose in several glycoproteins has been described but it is not quite certain that glucose was not present as a contaminant. The detailed structure of no oligosaccharide containing glucose, mannose and acetylglucosamine has as yet been described.

Transfer from dolichol diphosphate-mannose labelled oligosaccharide to protein has been studied with enzymes from liver (Behrens et al., 1973), hen oviduct (Lucas et al., 1975) and myeloma cells (Hsu et al., 1974). The results were similar to those described before in that not one but many proteins seem to become labelled. In the case of hen oviduct which produces mainly ovalbumin only less than 10% of the label was recovered in ovalbumin (Lucas et al., 1975). For reviews on the subject see Behrens (1974), Lennarz (1975), Parodi & Leloir (1975) and Waechter & Lennarz (1976).

A general scheme. The facts which we have described can be made to fit into the general scheme shown in Fig. 1. Dolichol phosphate shown in the middle may react with UDP-acetylglucosamine to yield first dolicholdiphosphate acetylglucosamine and then dolicholdiphosphate di'N,acetylchitobiose. After addition of mannose from UDP-mannose or dolichol phosphate mannose the resulting compound may either be transferred to protein as such or after accepting glucose from dolichol phosphate glucose.

Transfer reactions from sugar nucleotides to the protein-linked oligosaccharide would lead to the formation of a more complex molecule which might contain neuraminic acid, galactose and glucose besides acetylglucosamine and mannose.

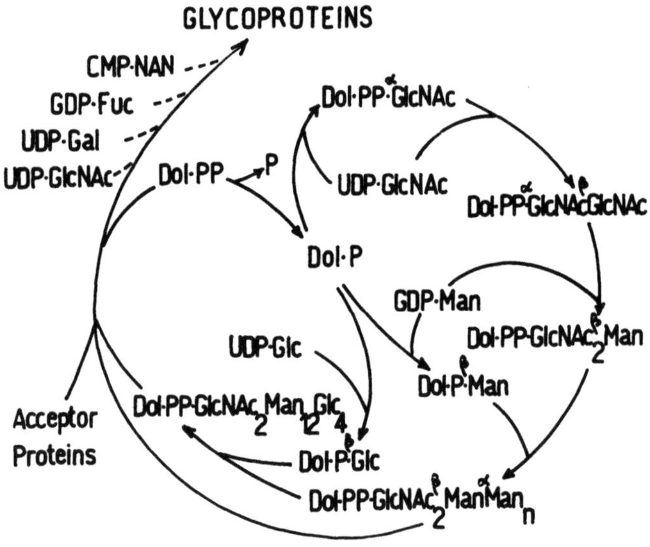

Fig. 1. A scheme for protein glycosylation (Behrens, unpublished)

The scheme describes asparagine linked oligosaccharide formation in liver but it also seems to be valid for other animal tissues (Adamany & Spiro, 1975) and for insects (Quesada Allue, 1976) and probably plants.

The scheme shown in Fig. 1 does not show all the reactions in which dolichol phosphate has been found to be involved. Thus yeast enzymes have been found to transfer mannose from dolichol-phosphate-mannose to serine or threonine residues (Babczinki & Tanner, 1973). Furthermore addition of one or more mannoses from GDP-mannose would give rise to the formation of the short mannose chains found in yeast mannan. The other sugar protein linkage of mannan, which involves di'N,acetylchitobiose is presumably formed through the dolichol containing intermediates.

Some of the reactions shown in the scheme have also been detected in plants (Pont Lezica et al., 1975; Lehle et al., 1976; Brett & Leloir, 1976). The polyprenol involved has been shown, by chromatography on Sephadex in deoxycholate solution to be smaller than liver in size, it has about 17 or 18 isoprene residues as compared to liver in which the compound with 19 residues is the most abundant. The compound is quite different from ficaprenol which is the major polyprenol found in plants and has 11 isoprene residues. The reactions detected in plants apparently include the formation of Dol-P-Man, Dol-P-Glc, Dol-P-P-acetylglucosamine and also some Dol-P-P-oligosaccharides. Since plants have glycoproteins of the same type found in animals it seems very likely that the scheme is also valid for them.

The role of glycoproteins in cell physiology is being actively studied and important facts have been uncovered. One might mention the specificity of blood groups, the importance of sugars in determining liver uptake of proteins, glycoproteins as specific cell receptors or in regulating the viscosity or freezing temperature of proteins solutions and many other important functions. There is no doubt that the field will yield important results in the future.

ACKNOWLEDGMENTS

The work carried out in our laboratory was supported by Grant number GM 19808-03 from the National Institutes of Health, the Argentine Research Council and the Organization of American States.

REFERENCES

The selection of references is strongly biased in favour of our laboratory and I apologize for this. Other workers whose names do not appear have also made important contributions.

ADAMANY A. M. & SPIRO R. G. (1975) J. Biol. Chem. 250, 2842-2854
BABCZINKI P & TANNER W. (1973) Biochem. Biophys. Res. Commun. 54, 1119-1124
BEHRENS N. H. & LELOIR L. F. (1970) Proc. Natl. Acad. Sci. U.S.A. 66, 153-159
BEHRENS N. H., PARODI A. J. & LELOIR L. F. (1971a) Proc. Natl. Acad. Sci. U.S.A. 68, 2857-2860
BEHRENS N. H., PARODI A. J., LELOIR L. F. & KRISMAN C. R. (1971b) Arch. Biochem. Biophys. 143, 375-383
BEHRENS N. H., CARMINATTI H., STANELONI R., LELOIR L. F. & CANTARELLA A. I. (1973) Proc. Natl. Acad. Sci. U.S.A. 70, 3390-3394
BEHRENS N. H. (1974) Biology and Chemistry of Eucaryotic Cell Surfaces (Lee, E.Y.C. and Smith E.E. eds.), Vol.7, pp. 159-178, Academic Press, New York.
BRETT C. & LELOIR L. F. (1976) Biochem. J. (in press)
GOTTSCHALK A., ed. (1972) Glycoproteins, Elsevier, Amsterdam
HEMMING F. W. (1974) in Biochemistry of Lipids (Goodwin T. W., ed.), Vol. 4 pp. 39-97, Butterworths and University Park Press, London and Baltimore
HSU A. F., BAYNES J. W. & HEATH E. C. (1974) Proc. Natl. Acad. Sci. U.S.A. 71, 2391-2395
KORNFELD R. & KORNFELD S. (1976) Annu. Rev. Biochem. 45, 217-237
LEHLE L., FARTACZEK F., TANNER W. & KAUSS H. (1976) Arch. Biochem. Biophys. 175, 419-426
LELOIR L. F., STANELONI R. J., CARMINATTI H. & BEHRENS N. H. (1973) Biochem. Biophys. Res. Commun. 52, 1285-1292
LENNARZ W. J. (1975) Science 188, 986-991
LEVY J. A., CARMINATTI H., CANTARELLA A. I., BEHRENS N. H., LELOIR L. F. & TABORA E. (1974) Biochem. Biophys. Res. Commun. 60, 118-125
LUCAS J. J., WAECHTER C. J. & LENNARZ W. J. (1975) J. Biol. Chem. 250, 1992-2002
MARSHALL R. D. & NEUBERGER A. (1970) in Advances of Carbohydrate Chemistry and Biochemistry (Stuart Tipson, R. and Horton, D., eds), Vol. 25, pp. 407-478. Academic Press
MONTREUIL J. (1975) Pure and Applied Chem. 42, 431-477
PARODI A. J., BEHRENS N. H., LELOIR L. F. & CARMINATTI H. (1972) Proc. Natl. Acad. Sci. U.S.A. 69, 3268-3272.
PARODI A. J., STANELONI R., CANTARELLA A. I., LELOIR L. F., BEHRENS N. H., CARMINATTI H. & LEVY J. A. (1973) Carbohyd. Res. 26, 393-400
PARODI A. J. & LELOIR L. F. (1975) Trends in Biochemical Sciences 1, 58-59
PONT-LEZICA R., BRETT C., ROMERO P. & DANKERT M. (1975) Biochim. Biophys. Res. Commun. 66, 980-987

QUESADA ALLUE L. A., MARECHAL L. R. & BELOCOPITOW E. (1976) FEBS Lett. 67, 243-247

SPIRO R. G. (1973) in Advances in Protein Chemistry (Anfinsen C.B., Edsall J.T. and Richards F.M. eds) Vol. 27, pp. 349-467, Academic Press, New York

WAECHTER C. J. & LENNARZ W. J. (1976) Annu. Rev. Biochem. 45, 95-112

I. LIPID INVOLVEMENT IN THE BIOGENESIS OF MEMBRANES

(A) BIOSYNTHESIS OF FATTY ACIDS

THE ROLE OF LIPID IN STEARYL CoA DESATURATION

P.W. Holloway, M. Roseman, and A. Calabro

Department of Biochemistry, University of Virginia

Charlottesville, Virginia 22901 U.S.A.

INTRODUCTION

The fatty acid desaturase system was one of the first microsomal enzymes for which a lipid requirement was established. Using Fleischer's procedure of aqueous acetone extraction (Lester & Fleischer, 1961) we were able to reduce the desaturase activity of hen liver microsomes to very low levels and restore activity completely with a mixture of lipids isolated from the microsomes (Holloway & Peluffo, 1964; Jones at al., 1969). Shortly after this observation cytochrome b_5 was implicated in the desaturation process (Oshino et al., 1966). Progress in the isolation and purification of the stearyl CoA desaturase has been slow and it is only in the last two years that the complete resolution and reconstitution of the stearyl CoA desaturase has been achieved by Strittmatter and co-workers (Strittmatter et al., 1974).

Milestones along the way have, I think, been the initial postulation by Sato and co-workers of a role for cytochrome b_5 (Oshino et al., 1966) and later their spectral studies showing the acceleration of cytochrome b_5 re-oxidation by added stearyl CoA (Oshino et al., 1971).

RESULTS AND DISCUSSION

Our approach from the beginning was to use detergents to resolve the desaturase system. Although we never pushed these techniques to the point of obtaining homogeneous desaturase protein, we were able to use this approach to remove cytochrome b_5 from the majority of the

Table 1. - Regeneration of Stearyl CoA Desaturase Activity

Components[a]	Stearyl CoA Desaturase[b]
	nmoles oleate formed
Experiment 1	
P3 + Fp	0
P3 + Fp + d-b_5	0.56
P3 + Fp + PC-0	0
P3 + Fp + PC-0 + d-b_5	1.29
P3 + Fp + PC-0 + d-b_5 + DOC	4.17
P3 + Fp + PC-0 + DOC	0.26
P3 + Fp + d-b_5 + DOC	1.07
p3 + PC-0 + d-b_5+ DOC	1.07
Fp + PC-0 + d-b_5 + DOC	0
Microsomes	2.18
Experiment 2	
P3 + Fp + PC vesicles + d-b_5 + DOC	5.09
P3 + Fp + PC vesicles + d-b_5	0.90
P3 + Fp + PC vesicles + t-b_5 + DOC	0

[a] The components added to the standard assay system as indicated were : P3 and microsomes 0.1 mg protein, 3 µg NADH-cytochrome b_5 reductase (Fp), 0.12 mg phosphatidylcholine-oleic acid (4:1 w/w) vesicles (PC-0), 5 µg d-b_5, 4 µg t-b_5 0.3 mg sodium deoxycholate (DOC),or 0.10 mg phosphatidylcholine vesicles (PC vesicles).

[b] The standard assay system contained 60 mM potassium phosphate (pH 7.2), 100 µM NADH, 20 µM $|1\text{-}^{14}C|$-stearyl CoA, protein and other components, and water to a final volume of 0.5 ml. Incubations were at 37°C for 15 min in air. The product $|1\text{-}^{14}C|$-oleate was separated as the methyl ester, after saponification and methylation, by thin layer chromatography on $AgNO_3$-impregnated Silica Gel H (Jones et al., 1969).

microsomal protein and hence, show a dependence of desaturation upon cytochrome b_5 (Holloway, 1971; Holloway & Katz, 1972). This useful procedure is outlined below.

A microsomal subfraction was dissolved in deoxycholate and resolved by gel filtration into two protein fractions (Holloway & Katz, 1972). The first fraction designated P_3 was almost devoid of lipid and cytochrome b_5, which were eluted later in the second fraction. Neither fraction alone had any desaturase activity but a combination of the two had activity. The second fraction could be replaced by cytochrome b_5 isolated by detergent procedure, plus lipid vesicles. As shown in Table 1 cytochrome b_5 isolated by a trypsin procedure was unable to replace the cytochrome b_5 isolated by the detergent procedure. Desaturation was also stimulated by NADH-cytochrome b_5 reductase, in agreement with our earlier demonstration that this protein was essential for desaturase activity (Holloway & Wakil, 1970). Similar requirements were established by Sato and co-workers at the same time, although, they used different resolution procedures (Shimakata et al., 1972).

Again all these studies, and of course the more recent ones, pointed to a requirement for lipid as seen in the original microsomes. We examined the role of lipid over the years and succeeded in eliminating two postulated roles; that lipid was required to provide the true substrate for the reaction--it does not. Stearyl CoA is the true substrate and oleyl CoA the product (Holloway & Holloway, 1974). A second postulate, that lipid altered the redox behaviour of cytochrome b_5 , was also shown to be incorrect (Sullivan & Holloway, 1973). Indeed, the fact that we saw no alteration in redox behaviour, spectral properties or ease of reduction by NADH-cytochrome b_5 reductase when detergent-isolated cytochrome b_5 was bound to phosphatidylcholine vesicle was the first direct proof that the catalytic portion of the molecule was remote from the lipid bilayer and was in an aqueous environment (Sullivan & Holloway, 1973).

All these studies suggested that the lipid must be playing some kind of structure role in the desaturation process and this supposition has been amply supported by subsequent work from several groups. In common with several other groups (Dufourcq et al., 1975; Faucon et al., 1976; Rogers & Strittmatter, 1975; Robinson & Tanford, 1975), we have chosen to examine the cytochrome b_5 - phosphatidylcholine vesicle system. We would like to use this system both to understand the desaturase and also to examine lipid-protein interaction in general. Two specific aspects of our studies I would like to discuss were initiated to characterize the mechanism of binding of cytochrome b_5 to lipid vesicles.

Cytochrome b_5 as isolated from microsomal membranes by detergent solubilization (d-b_5) has been reported to exist as an octomer, presumably held together via non-polar interactions of the hydrophobic tails. It seemed that the actual binding of d-b_5 to phosphatidylcholine vesicles would more likely involve prior dissociation of octomer to monomer and we initiated studies to look for the presence of d-b_5 monomer. In initial studies we used gel filtration on Sephadex G200 to detect the presence of monomer; monomer MW 16,700 should readily separate from octomer MW 135,000. As shown in Fig. 1 gel filtration of d-b_5 gave two peaks suggesting that two species of d-b_5, with different Stokes radius, exist and that these species can be separated by gel filtration (Calabro et al., 1976).

We determined the Stokes radius of the second peak, by calibration of the column with standard proteins and also by another technique of equilibrium saturation gel filtration which proved very

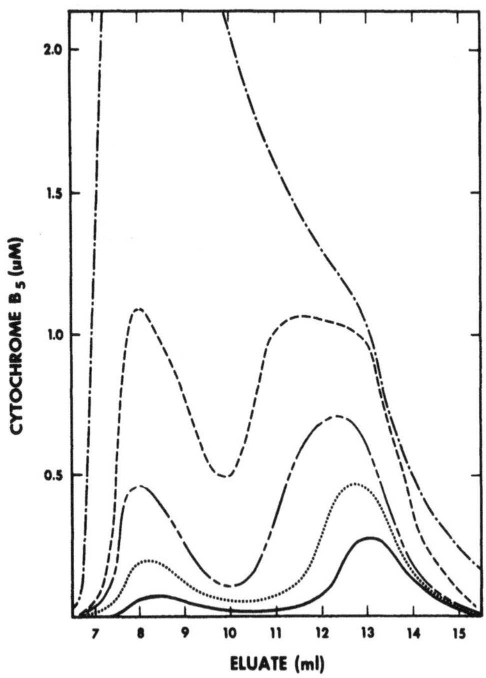

Fig. 1. Effect of loading concentration on the elution properties of d-b_5. Samples of d-b_5 (0.12 ml) in 10 mM Tris acetate/0.1 mM EDTA/0.4 M sucrose (pH 8.15 at 20°) were chromatographed on a column of Sephadex G-200 (Sperfine) at 20°equilibrated in 10 mM Tris acetate/0.1 mM EDTA (pH 8.15 at 20°). The initial loadings of d-b_5 were 0.834 nmol, 1.64 nmol, 3.41 nmol, 6.80 nmol, and 12.54 nmol. (Calabro et al., 1976).

convenient. This second technique involved the direct observation of the d-b_5 in a small Sephadex column by use of a dual wavelength spectrophotometer. The absorbances of the sample, a standard protein excluded from the gel, and a pigment completely included in the gel are measured in the column and later, at the same concentration in a 1 cm cuvette. The partition coefficients (σ) calculated from these data can be used to determine the Stokes radius (Fig. 2). These results were confirmed by experiments with the analytical ultracentrifuge which showed that solutions of d-b_5 were composed of a mixture of monomer and octomer. The thermodynamics of the system are such that the monomer concentration remains fairly constant once a certain total concentration of d-b_5 is reached (Table 2).

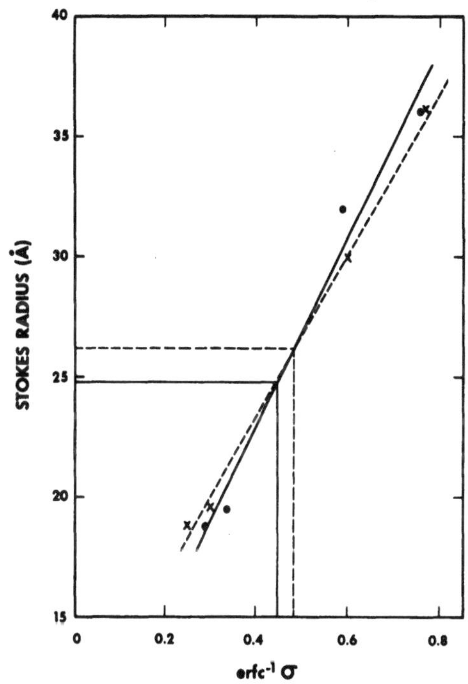

Fig. 2. The Stokes radius of d-b_5 monomer as determined by both small zone and equilibrium saturation gel filtration. The five standard proteins used to calibrate the small zone (- - -) and equilibrium saturation (——) gel filtration experiments were (in order of decreasing Stokes radius) bovine serum albumin, hemoglobin, ovalbumin, cytochrome b_5 isolated by a trypsin procedure (t-b_5), and myoglobin. From the elution volumes in the small zone experiments and the partition cross sections in the equilibrium saturation experiments, the partition coefficients (σ) were calculated. The values of erfc $^{-1}\sigma$ for d-b_5 monomer in 10mM Tris acetate/0.1 mM EDTA (pH 8.15 at 20°) in the small zone and equilibrium saturation gel filtration experiments were 0.58 and 0.45 respectively which corresponds to a Stokes radius of 26.4 (- - -) or 25.4 (——) Å (Calabro et al., 1976).

Table 2. - Analysis of Analytical Ultracentrifugation Data

D-b_5 at an initial concentration of 2 μM in 10 mM Tris acetate, 0.1 mM EDTA pH 8.15 was subjected to equilibrium ultracentrifugation. The absorbance of the sample at 410 nm was measured across the cell and the data was analysed to give the dependence of apparent weight average molecular weight upon total d-b_5 concentration (Calabro et al., 1976). These data, in turn, were analysed to give the dependence of weight fraction of monomer upon the total d-b_5 concentration, assuming a monomer-octomer system (Calabro et al., 1976).

TOTAL d-b_5 (μM)	0.80	1.20	1.60	2.00	2.40	2.80	3.20	3.60
d-b_5 MONOMER (μM)	0.64	0.86	1.02	1.16	1.19	1.22	1.27	1.25

This behaviour is of course similar to that seen with detergents above their critical micelle concentration and indeed the d-b_5 octomer may be thought of as a super-micelle with a critical micelle concentration of approximately 1.2 μM in low salt. In keeping with this analogy d-b_5 in both monomeric and octomeric forms can be seen to have rather atypical dimensions. The monomer as we saw in Fig. 2 has a Stokes radius of 26 Å whereas the octomer, by gel filtration on Sepharose, has a Stokes radius of 65 Å (Holloway & Katz, 1975). Both these values are larger than one would predict for typical globular proteins of 16,700 and 135,000 molecular weights respectively. More acceptable values would be 19 Å and 45 Å. This would indicate the monomer is rather asymmetric, as suggested by Visser et al. (1975), and the octomer is either asymmetric or, perhaps, has a less compact interior than a normal globular protein.

Our published data, and those experiments of others where d-b_5 binding to microsomes was investigated, suggest a rapid binding of d-b_5 to membranes does occur and we would like to extend these studies to investigate the topography of the complex either by chemical modification or by spectral procedures. We were concerned that results of these types of experiments could be invalidated if the binding of d-b_5 to lipid vesicles was reversible and d-b_5 spent a considerable portion of its time completely separated from the vesicle. Our recent experiments have been directed towards this latter possibility. Specifically, we asked: does d-b_5 exchange between lipid vesicles?

Our previous report on the d-b_5 lipid system used analytical

Table 3 .- Formation of a Single Vesicle Population From Two Vesicle Populations

In the first incubation PC vesicles (5mM) were incubated either alone or in the presence of d-b_5 at 37°. After 20 min additional PC vesicles were added to experiments 3 and 5 and a second 20 min incubation at 37° was performed. The five samples were analyzed immediately following the second incubation in the analytical ultracentrifuge and in each ultracentrifuge run a single symmetrical schlieren peak was observed. The s values of these peaks are shown.

	FIRST INCUBATION			SECOND INCUBATION		
Expt.	PC	d-b_5	AVG.d-b_5/ VESICLE	PC	AVG.d-b_5/ VESICLE	$s_{20,w}$
1.	5 mM	--	--	--	--	2.3
2.	5 mM	25 µM	12	--	--	5.6
3.	5 mM	25 µM	12	5 mM	6	3.4
4.	5 mM	10 µM	5	--	--	3.7
5.	5 mM	10 µM	5	5 mM	2.5	3.0

ultracentrifugation to characterize the complex (Holloway & Katz, 1975). These data showed that the complex formed at a level of 2.5 d-b_5 per vesicle was as homogenous as the original vesicle preparation. We were originally gratified at this result and took it to mean the vesicle was not excessively perturbed by the addition of d-b_5. The increase in sedimentation coefficient was compatible with the increased density of the particle. On reconsideration, it is apparent that these results are not expected if d-b_5 is randomly distributed among the vesicles. According to the Poisson distribution we would expect a great heterogeneity in population. The existence of a symmetrical peak with the theoretical sedimentation coefficient could only have arisen by exchange of d-b_5 between vesicles or if the d-b_5 was distributed uniformly.

To test the possibility of exchange we again resorted to the analytical ultracentrifuge. The complex formed at a given ratio of d-b_5 per vesicle was incubated a second time alone or with added lipid vesicles. In all cases a single symmetrical schlieren peak resulted. These experiments showed that after the second incubation no pure lipid vesicles (s_{20}= 2.3) remained and that the new complex has a sedimentation coefficient intermediate between that of the original complex and the vesicles alone (Table 3). We propose that this new complex is formed by exchange of d-b_5 between the two vesicle populations. It is, however conceivable that the new complex could result from fusion or aggregation of the two original populations. These alternatives, although extremely unlikely in view of the low s values recorded in Table 3, cannot be rigorously excluded by ultracentrifugation theory.

We chose to test more directly, the possibility that the new species we detected in the ultracentrifuge was formed by aggregation or fusion. In this discussion we would like to define fusion as the interaction between two or more vesicles leading to the formation of a single larger unilamellar vesicle. This process leads to mixing of the internal contents of the two vesicles. Aggregation is defined as the interaction of two or more vesicles leading to the formation of a complex comprised of discrete vesicle subunits. Light scattering measurements, which would monitor both fusion and aggregation, gave no indication of these processes. We also monitored for fusion directly by using two vesicle populations where one population had its aqueous compartment loaded with ferricyanide, and the other population had its aqueous compartment loaded with ascorbate. Fusion of the vesicles would be evidenced by the loss of absorbance at 420 nm as the ferricyanide was reduced by the ascorbate. By the experimental design it was possible to also monitor ferricyanide leakage from inside the ferricyanide-loaded vesicles, ascorbate leakage from inside the ascorbate-loaded vesicles, total ferricyanide trapped inside the ferricyanide loaded vesicles at the end of the incubation and total ascorbate trapped inside the ascorbate-loaded vesicles at the end of the incubation. As shown in Table 4, we were not able to distinguish fusion from vesicle leakage but as this latter rate could be measured independently at another time during the experiment we can see that in no instance was "FUSION + LEAKAGE" significantly greater than "LEAKAGE". In all experiments large amounts of ferricyanide and ascorbate remained trapped inside even after 45 minute incubation at 37°.

From the analytical ultracentrifugation, light scattering and optical experiments just described, it is apparent that d-b_5 can exchange between lipid vesicles. The mechanism of exchange has not yet been determined but could involve either the transfer of d-b_5 upon collision of two vesicles or depend upon the transfer of free d-b_5 through the aqueous phase. Although we intuitively think intrinsic membrane proteins are firmly attached to the membrane, Dehlinger & Schimke (1971) have suggested, from protein turnover experiments, that membrane proteins are degraded only when they are in the cytoplasm. This implies that a membrane protein is in equilibrium between membrane bound and cytoplasmic states. Perhaps, the exchange of d-b_5 proceeds through free d-b_5 and is the first *in vitro* demonstration of the reversibility of binding of intrinsic membrane proteins.

Table 4.- Absence of Fusion between Two Vesicle Populations

PC vesicles were preparated at an initial concentration of 30 mM by sonication in either 0.5 M $K_3Fe(CN)_6$ or 0.5 M sodium ascorbate (pH 6.5). The two populations of vesicles were purified by centrifugation and gel filtration on Sepharose 4B (Holloway & Katz, 1975). For each experiment shown ferricyanide (FeCN)-loaded vesicles or ascorbate (ASC)-loaded vesicles were added with d-b_5 to 10 mM Tris acetate/0.1 mM EDTA pH 8.0 in the orders indicated. The final concentration of d-b_5 in all but experiment 4 was 1 μM and the final concentration of each species of vesicle was 1 mM. The mixtures were incubated at 37° under argon and the rate of FeCN reduction by ASC was monitored at 420 nm.

EXPT[a]	FeCN LEAK ALONE	FeCN LEAK + b_5	FeCN LEAK +FUSION	FeCN INSIDE AT END	TIME WITH b_5	ASC LEAK ALONE	ASC LEAK + b_5	ASC LEAK + FUSION	ASC INSIDE AT END	TIME WITH b_5
	nmole /min	nmole /min	nmole /min	nmole	min	nmole /min	nmole /min	nmole /min	nmole /min	min
1						0	2		142	12
2	0	3	4	136	40			4	148	23
3	0		3	183	23	0		1	152	23
4	0	5	3	134	45			2	165	28
5			5	213	17	0	0	0	132	20

[a]The orders of addition of the three components in each experiment were:
Expt. 1. ASC-loaded vesicles, d-b_5 . Expt. 2. FeCN-loaded vesicles, d-b_5, ASC-loaded vesicles. Expt. 3. ASC-loaded vesicles, FeCN-loaded vesicles, d-b_5. Exp. 4. FeCN-loaded vesicles, d-b_5, ASC-loaded vesicles, d-b_5. Expt. 5. ASC-loaded vesicles, d-b_5, FeCN-loaded vesicles.

CONCLUSIONS

The data presented on the stearyl CoA desaturase system demonstrate requirements for NADH-cytochrome b_5 reductase, cytochrome b_5, a third protein fraction, lipid and detergent for regeneration of stearyl CoA desaturation in agreement with studies by others (Strittmatter et al., 1974; Shimakata et al., 1972). Two roles for lipid in the system were excluded and it appears that the lipid is playing a structural role rather than supplying a substrate-precursor or altering the redox potential of the cytochrome b_5. The further delineation of the role which lipid plays in the desaturase system may come from the investigation of the interaction of d-b_5 with lipid vesicles. The complete understanding of this simpler model system is now much more feasible with our two new observations on d-b_5.

The existence of a d-b_5 monormer-octomer system and the demonstration of exchange of d-b_5 between vesicles may now enable the thermodynamics of the lipid-d-b_5 interaction to be evaluated. These observations do, however, introduce complications into any topographical experiments one would want to perform.

Speculations on the repercussions of d-$b_5$exchange could include the possibility that d-b_5 in moving from membrane to membrane functions as a phospholipid carrier protein or as a carrier of electrons to the terminal desaturase proteins. If exchange were fast enough the d-b_5 could interact with the other components of the desaturase even in the absence of a fluid mosaic membrane.

ACKNOWLEDGEMENTS

This work was supported by the American Cancer Society (Grant BC-71,D,E) and by the USPHS (Grant GM 18970).

REFERENCES

CALABRO, M. A., KATZ, J. T. & HOLLOWAY, P. W. (1976) J. Biol. Chem. 251, 2113-2118.

DEHLINGER, P. J. & SCHIMKE, R. T. (1971) J. Biol. Chem. 246, 2574-2583.

DUFOURCQ, J., FAUCON, J. F., LUSSAN, C. & BERNON, R. (1975) FEBS Letters 57, 112-116.

FAUCON, J. F., DUFOURCQ, J., LUSSAN, C. & BERNON, R. (1976) Biochim. Biophys. Acta 436, 283-294.

HOLLOWAY, P. W. (1971) Biochemistry 10, 1556-1560.

HOLLOWAY, C. T. & HALLOWAY, P. W. (1974) Lipids 9, 196-200.

HOLLOWAY, P. W. & KATZ, J. T. (1972) Biochemistry 11, 3689-3696

HOLLOWAY, P. W. & KATZ, J. T. (1975) J. Biol. Chem. 250, 9002-9007.
HOLLOWAY, P. W. & PELUFFO, R. O. (1964) Abstracts Sixth International Congress of Biochemistry, IUB Vol. 32, p.577, VII-68.
HOLLOWAY, P. W. & WAKIL, S. J. (1970) J. Biol. Chem. 245, 1862-1865.
JONES, P. D., HOLLOWAY, P. W., PELUFFO, R. O. & WAKIL, S. J. (1969) J. Biol. Chem. 244, 744-754.
LESTER, R. L. & FLEISCHER, S. (1961) Biochim. Biophys. Acta 47, 358-377.
OSHINO, N., IMAI, Y. & SATO, R. (1966) Biochim. Biophys. Acta 128, 13-28.
OSHINO, N., IMAI, Y. & SATO, R. (1971) J. Biochem. (Tokyo) 69, 155-167.
ROBINSON, N. C. & TANFORD, C. (1975) Biochemistry 14. 369-378.
ROGERS, M. J. & STRITTMATTER, P. (1975) J. Biol. Chem. 250, 5713-5718.
SHIMAKATA, T., MIHARA, K. & SATO, R. (1972) J. Biochem. (Tokyo) 72, 1163-1174.
STRITTMATTER, P., SPATZ, L., CORCORAN, D., ROGERS, M. J., SETLOW, B. & REDLINE, R. (1974) Proc. Natl. Acad. Sci. U.S.A. 71, 4565-4569.
SULLIVAN, M. R. & HOLLOWAY, P. W. (1973) Biochem. Biophys. Res. Commun. 54, 808-815.
VISSER, L., ROBINSON, N. C. & TANFORD, C. (1975) Biochemistry 14, 1194-1199.

BIOSYNTHETIC PATHWAYS OF POLYUNSATURATED FATTY ACIDS

Howard Sprecher

Dept. Physiological Chemistry, The Ohio State University

333 W. 10th Avenue, Columbus, Ohio 43210

INTRODUCTION

The type and amount of any unsaturated fatty acid found in lipids must be regulated by integrated control mechanisms directed both towards unsaturated fatty acid and lipid biosynthesis. The types and amounts of unsaturated fatty acids produced must be determined both by the number of enzymes involved in this process and the substrate specificity of these enzymes. In addition those factors which influence enzyme activity must also regulate the rate at which unsaturated fatty acids are produced. The results presented below summarize the studies carried out in our laboratory in an attempt to further determine what regulates unsaturated fatty acid biosynthesis.

INTERRELATIONSHIPS BETWEEN UNSATURATED FATTY ACID BIOSYNTHESIS AND LIVER LIPID FATTY ACID COMPOSITION

Metabolic Conversions of Acids in the Linoleate Sequence

Under normal dietary conditions liver lipids contain high levels of linoleate and arachidonate but only trace or negligible amounts of other acids in this metabolic sequence. We have measured rates of both desaturation and chain elongation reactions in the linoleate sequence to determine what relationship exists between rates of conversion and the type of fatty acid found in liver lipids (Bernert & Sprecher, 1975). Dietary linoleate is either incorporated into liver lipids, or as shown in Figure 1, it is desaturated to 6,9,12-18:3 at a rate of 1.0 nmole/min/mg microsomal protein. Liver lipids never contain measureable amounts of 6,9,12-18:3 even when this acid is added directly to the diet

11,14-20:2 $\xrightarrow{0.3}$ 5,11,14-20:3　　10,13,16-22:3

↑ 0.4　　↑ 0.2

9,12-18:2 $\xrightarrow{1.0}$ 6,9,12-18:3 $\xrightarrow{4.4}$ 8,11,14-20:3 $\xrightarrow{0.8}$

5,8,11,14-20:4 $\xrightarrow{1.2}$ 7,10,13,16-22:4 $\xrightarrow{0.1}$ 4,7,10,13,16-22:5

FIGURE 1. Rates of desaturation and chain elongation for acids in the linoleate sequence. Rates expressed as nmoles product/min/mg microsomal protein. Each incubation for desaturation contained 10 μmoles ATP, 2 μmoles NADH, 0.3 μmole CoA, 150 μmoles potassium phosphate buffer, pH 7.4, 150 ηmoles radioactive fatty acid, 5 mg liver microsomal protein from rats raised on a fat-free diet and 5 mg bovine serum albumin. Chain elongation conditions were the same except 2.0 μmoles of NADPH were used instead of NADH and 0.3 μmole of malonyl-CoA was added. All incubations were run in 1.5 ml for 3 minutes at 37°C. (Bernert & Sprecher, 1975)

(Sprecher, 1974b). The absence of 6,9,12-18:3 in liver lipids may possibly be attributed to the rapid rate at which it is removed by chain elongation to 8,11,14-20:3. The 8,11,14-20:3 may then be chain elongated to 10,13,16-22:3, desaturated to arachidonate, incorporated into lipids or serve as a substrate for prostaglandin biosynthesis. The 8,11,14-20:3 was chain elongated to 10,13,16-22:3 at a rate of only 0.2 nmole/min/mg microsomal protein. If this slow reaction did take place *in vivo* it is questionable if it is of any physiological significance since 10,13,16-22:3 is rapidly converted back to 8,11,14-20:3 by retroconversion (Sprecher, 1968a). Retroconversion is a partial degradation process taking place in the mitochondria (Stoffel *et al.*, 1970; Kunau & Bartnik, 1974). The principal role of 8,11,14-20:3 must be to serve as a precursor for arachidonate. However, the rate of desaturation of 8,11,14-20:3 to arachidonate is not as great as the rate of chain elongation of arachidonate to 7,10,13,16-22:4 and yet liver lipids contain high levels of arachidonate but low levels of both 7,10,13,16-22:4 and 4,7,10,13,16-22:5. Feeding studies have shown that both 7,10,13,16-22:4 (Sprecher, 1967) and 4,7,10,13,16-22:5 (Verdino *et al.*, 1964) are preferentially converted back to arachidonate by retroconversion rather than being incorporated directly into liver lipids. The termination of polyunsaturated fatty acid biosynthesis may thus be regulated by a balance between retroconversion and chain elongation. Removal of 7,10,13,16-22:4 and 4,7,10,13,16-22:5 by retroconversion would terminate polyunsaturated fatty acid biosynthesis by removing the substrates required for continuation of this process.

Finally, the absence of large amounts of 8,11,14-20:3 in liver lipids can not be attributed to a lack of enzyme specificity for incorporating this acid into liver lipids. Both arachidonate and 8,11,14-20:3 were acylated at the 2-position of 1-acyl-sn-glycero-3-phosphorylcholine at about the same rate (Hill & Lands, 1968).

Although the above studies fail to explain why 8,11,14-20:3 is not a major liver lipid component, the factors which regulate the metabolism of this acid are most important since 8,11,14-20:3 and arachidonate are both substrates for prostaglandin biosynthesis. The results in Figure 2 were obtained when seven different methyl branched isomers of 8,11,14-20:3 were used to study both the substrate specificity of rat liver microsome 5-desaturase (Do & Sprecher, 1975) and prostaglandin synthetase in bovine vesicular gland microsomes (Do & Sprecher, 1976). The rate of desaturation at the 5-position was inhibited most markedly when the methyl group was located near the site of double bond insertion. The rate of desaturation was not reduced as markedly when the methyl branch was located towards the terminal end of the substrate. In agreement with more extensive studies on how substrate modification influences prostaglandin biosynthesis (van Dorp & Christ, 1975) we also found that the rate of prostaglandin biosynthesis was most markedly depressed when the methyl group was located near the terminal end of the substrate. Two microsomal enzymes, using the

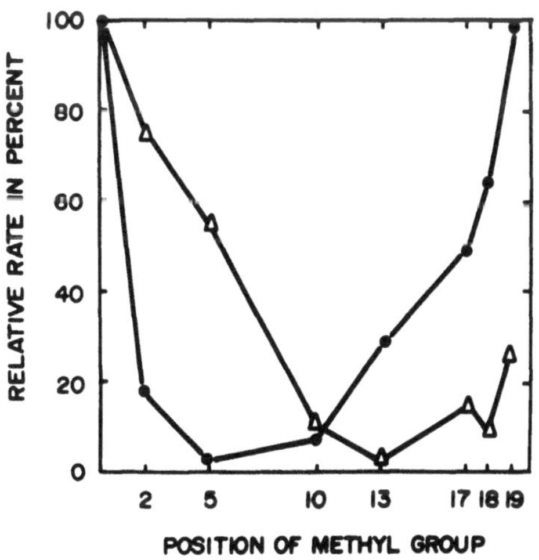

FIGURE 2. Rates of desaturation (•) and conversion into prostaglandins (Δ) using methyl branched isomers of 8,11,14-eicosatrienoic acid. (Do & Sprecher, 1975; Do & Sprecher, 1976)

same substrates, thus exhibit somewhat opposite substrate specificities.

Previously it has been suggested that arachidonate could also be produced via the following optional pathway: 9,12-18:2 → 11,14-20:2 → 8,11,12-20:3 → 5,8,11,14-20:4 (Stoffel, 1963). In liver we feel that this pathway is inoperative since as shown in Figure 1 linoleate is chain elongated to 11,14-20:2 at a slow rate. When 11,14-20:2 is produced it may be converted back to linoleate by retroconversion (Stearns et al., 1967) or desaturated to 5,11,14-20:3 rather than to 8,11,14-20:3 (Ullman & Sprecher, 1971b). Any 5,11,14-20:3 which is produced is not further desaturated to 5,8,11,14-20:4 (Schlenk et al., 1970). Additional studies have shown that 11-20:1, 11,14-19:2, 11,14-21:2 and 11,14,17-20:3 were all desaturated at the 5- rather than the 8-position thus prompting us to suggest that liver microsomes do not contain an 8-desaturase (Ullman & Sprecher, 1971b; Sprecher & Lee, 1975).

Metabolic Conversions of Acids in the Oleate Sequence

The tissue lipids of animals raised on a fat free diet contain high levels of 5,8,11-20:3. This acid is not normally detected in lipids of animals raised on a balanced diet. Competitive feeding experiments (Holman, 1964) and enzymatic experiments (Brenner & Peluffo, 1968; Brenner, 1974) suggest that liver microsomes contain a common 6-desaturase that acts on oleate, linoleate and linolenate. When linoleate and linolenate are dietary components these acids serve as the preferred substrates for the 6-desaturase and thus prevent desaturation of oleate to 6,9-18:2. Rates of reactions in the oleate sequence, as shown in Figure 3, demonstrate that desaturation of oleate to 6,9-18:2 is the rate limiting step in this sequence. Selective desaturation of linoleate or linolenate would thus be an effective control to minimize 5,8,11-20:3 production. A possible optional pathway bypassing this slow step would be 9-18:1 → 11-20:1 → 8,11-20:2 → 5,8,11-20:3. This pathway is of no significant importance since 9-18:1 is chain elongated to 11-20:1 at a slow rate - i.e. Figure 3. The 11-20:1 is rapidly

11-20:1
↑ 0.3 (from 9-18:1)

18:0 $\xrightarrow{3.3}$ 9-18:1 $\xrightarrow{0.2}$ 6,9-18:2 $\xrightarrow{3.0}$ 8,11-20:2 $\xrightarrow{0.8}$ 5,8,11-20:3

FIGURE 3. Rates of desaturation and chain elongation for acids in the oleate sequence. Rates expressed as nmoles product/min/mg microsomal protein. Incubation conditions were as described in Figure 1. (Bernert & Sprecher, 1975)

converted to 9-18:1 by retroconversion (Sprecher, 1972) and any 11-20:1 which is desaturated is converted to 5,11-20:2 rather than to 8,11-20:2 (Ullman & Sprecher, 1971b).

In the oleate sequence the conversion of 6,9-18:2 to 8,11-20:2 and the desaturation of 8,11-20:2 to 5,8,11-20:3 proceed at about the same rate as the corresponding chain elongation of 6,9,12-18:3 to 8,11,14-20:3 and desaturation of 8,11,14-20:3 to 5,8,11,14-20:4 in the linoleate sequence. These findings suggest that rates of conversion, beyond the rate limiting step in the oleate sequence, are rapid enough so that any 6,9-18:2 which is produced should be converted to 5,8,11-20:3 and be made available for incorporation into lipids. Feeding studies, in which various acids of the linoleate sequence were fed to rats raised on a fat free diet, demonstrate that 6,9,12-18:3, 8,11,14-20:3 and arachidonate are all more effective in depressing the level of 5,8,11-20:3 in liver lipids than was observed when the same amount of linoleate was fed - i.e. Table 1 (Sprecher, 1974b). Since there is a similarity in structure between 6,9,12-18:3 and 6,9-18:2 a common microsomal chain elongating enzyme may act on both substrates. In a similar way a common 5-desaturase may act on both 8,11-20:2 and 8,11,14-20:3. Competitive enzyme studies showed that 6,9,12-18:3 did inhibit the microsomal chain elongation of 6,9-18:2 to 8,11-20:2 (Sprecher, 1974a) and in a similar way 8,11,14-20:3 inhibited the desaturation of 8,11-20:2 to 5,8,11-20:3 (Ullman & Sprecher, 1971a). These findings suggest that these substrates were acted on by common enzymes and that this type of competition may represent a secondary control mechanism to further prevent synthesis of 5,8,11-20:3. If this is indeed true then it

TABLE 1

Percent of 5,8,11-Eicosatrienoic Acid and of Arachidonic Acid in Total Rat Liver Lipids After Feeding Acids of the Linoleate Sequence to Rats Raised on a Fat-Free Diet

Dietary Components*	Percent in Liver Lipids 5,8,11-20:3	5,8,11,14-20:4
Fat Free Controls	15	6
9,12-18:2	7	13
6,9,12-18:3	4	20
8,11,14-20:3	4	22
5,8,11,14-20:4	4	25

*Fed at the rate of 80 mg/day for a period of 9 days. (Sprecher, 1974b)

should be possible to verify this hypothesis by competitive feeding experiments.

The compositional data in Table 2 was obtained when rats, which were raised on a fat free diet, were all fed a constant level of 6,9,12-18:3 and various levels of 6,9-18:2 (Sprecher, 1974b). Neither of the dietary fatty acids was directly incorporated into liver lipids in measureable amounts. As expected the level of 5,8,11-20:3 in liver lipids was depressed when rats received only 6,9,12-18:3. When rats received a constant amount of 6,9,12-18:3 and increasing levels of 6,9-18:2 the level of arachidonate in liver lipids never dropped below 70 percent of that found in those rats receiving only 6,9,12-18:3. Conversely, even when rats received 1.5 times as much 6,9-18:2 and 6,9,12-18:3 the level of 5,8,11-20:3 was only 64 percent of that found in the fat free controls. Similar results were obtained when identical competitive feeding experiments were carried out with 8,11-20:2 and 8,11,4-20:3 and also with 5,8,11-20:3 and 5,8,11,14-20:4. The compositional data in Table 3 is typical of the results obtained in this study and shows that the amount of 5,8,11-20:3 and arachidonate in liver lipids is the same when rats recieved equal amounts of different metabolites from both the oleate and linoleate biosynthetic sequences. The amount of 5,8,11-20:3 and arachidonate in liver lipids is independent of whether these acids themselves were fed or whether precursors to these acids were fed.

TABLE 2

Percent of 5,8,11-Eicosatrienoic Acid and of Arachidonate Acid in Total Liver lipids After Feeding a Constant Level of 6,9,12-Octadecatrienoate and an Increasing Level of 6,9-Octadecadienoate to Rats Raised on a Fat-Free Diet

Dietary Components*	Percent in Liver Lipids 5,8,11-20:3	5,8,11,14-20:4
Fat-Free Controls	14	6
80 mg 6,9,12-18:3	4	20
80 mg 6,9,12-18:3 + 40 mg 6,9-18:2	5	19
80 mg 6,9,12-18:3 + 80 mg 6,9-18:2	8	16
80 mg 6,9,12-18:3 + 120 mg 6,9-18:2	9	18

*Indicates amounts fed on a daily basis for a period of 9 days. (Sprecher, 1974b)

TABLE 3

Percent of 5,8,11-Eicosatrienoic Acid and Arachidonic Acid in Total Liver Lipids After Feeding a Mixture of 6,9-Octadecadienoate and 6,9,12-Octadecatrienoate to Rats Raised on a Fat-Free Diet

Dietary Components*	Percent in Liver Lipids 5,8,11-20:3	5,8,11,14-20:4
Fat-Free Controls	14	6
80 mg 6,9,12-18:3 + 80 mg 6,9-18:2	8	16
80 mg 8,11,14-20:3 + 80 mg 8,11-20:2	7	19
80 mg 5,8,11-20:3 + 80 mg 5,8,11,14-20:4	7	20

*Indicates amount fed on a daily basis for a period of 9 days. (Sprecher, 1974b)

It would thus appear that once the rate limiting step in the oleate sequence is bypassed that the type of unsaturated fatty acid found in liver lipids is dictated by specificities for incorporation rather than for regulation at the site of polyunsaturated fatty acid biosynthesis.

Metabolic Conversions of Acids in the Palmitoleate Sequence

Even when rats are raised on a fat free diet the liver lipids do not contain high levels of polyunsaturated fatty acids derived from palmitoleate. On the basis of structure this is somewhat surprising since the metabolites from 18:2 to 20:4 in the palmitoleate sequence differ from those in the linoleate sequence only in that the double bond is shifted one position closer to the carboxyl carbon. The rates of desaturation and chain elongation of acids in the palmitoleate sequence are shown in Figure 4. Failure to find major amounts of 4,7,10,13-20:4 in liver lipids may be attributed to the following three slow desaturation steps: 9-16:1 → 6,9-16:2; 8,11-18:2 → 5,8,11-18:3 and 7,10,13-20:3 → 4,7,10,13-20:4. The compositional data shown in Table 4 supports this hypothesis. The amount of 4,7,10,13-20:4 in liver lipids was the same when equal amounts of 5,8,11-18:3 or 7,10,13-20:3 were fed to rats raised on a fat free diet (Sprecher, 1971). These findings suggest that the only regulatory point in the reaction sequence 5,8,11-18:3 → 7,10,13-20:3 → 4,7,10,13-20:4 is the slow desaturation step. Failure to find elevated levels of 4,7,10,13-

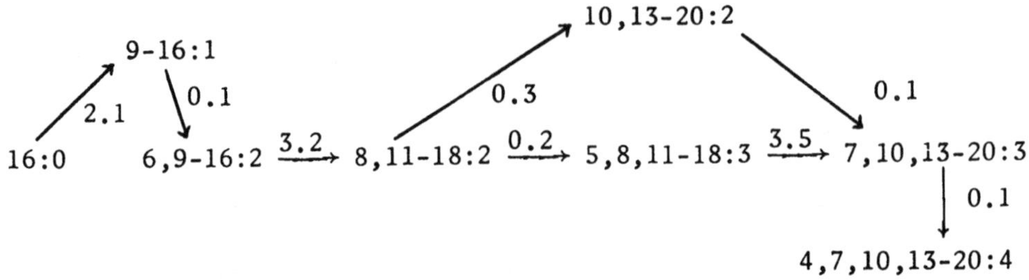

FIGURE 4. Rates of desaturation and chain elongation for acids in the palmitoleate biosynthetic sequence. Rates expressed as nmoles product/min/mg microsomal protein. Incubation conditions were as described in Figure 1. (Bernert & Sprecher, 1975)

TABLE 4

Composition of 20:4 from Total Rat Liver Lipids After Feeding 5,8,11-Octadecatrienoate or 7,10,13-Eicosatrienoate to Rats Raised on a Fat-Free Diet

Dietary Component*	Percent in Total Liver Lipids 4,7,10,13-20:4	5,8,11,14-20:4
Fat-Free Controls	0.7	4.2
5,8,11-18:3	5.5	4.2
7,10,13-20:3	5.5	4.1

*Fed at the rate of 150 mg/day for 12 days. (Sprecher, 1970)

20:4 in liver lipids, when 8,11-18:2 was fed to rats raised on a fat free diet (Klenk, 1965) may be attributed to the necessary utilization of two slow desaturation steps. An optional pathway avoiding a desaturation of 8,11-18:2 to 5,8,11-18:3 would be as follows: 8,11-18:2 → 10,13-20:2 → 7,10,13-20:3. This pathway is probably of no major significance since, as shown in Figure 4, both reactions in this sequence proceed very slowly. In addition, 10,13-20:3 is preferentially converted back to 8,11-18:2 by retroconversion rather than serving as a substrate for desaturation (Budny & Sprecher, 1971).

If 4,7,10,13-20:4 is produced it is incorporated into lipids. The compositional data in Table 5 was obtained when either arachidonate or 6,9,12,15-22:4 were fed to rats which had been raised on a fat free diet (Sprecher, 1968a). When this data is compared with that in Table 4 it suggests that 4,7,10,13-20:4 was produced more rapidly by retroconversion from 6,9,12,15-22:4 than by direct synthesis from either 5,8,11-18:3 or 7,10,13-20:3. The total amount of 20:4 in liver lipids was about the same when equal amounts of arachidonate and 6,9,12,15-22:4 were fed to rats - i.e. Table 5. Failure to find large amounts of unsaturated acids of the palmitoleic acid sequence in liver lipids may thus most likely be attributed to the slow desaturation steps in this sequence.

The Microsomal Chain Elongation System

There is increasing evidence to show that rat liver microsomes contain different desaturases for introducing double bonds at positions 5, 6 and 9 (Brenner, 1974; Gurr, 1974; Mead & Fulco, 1976). In contrast little is known about what regulates the microsomal chain elongation system or how many different chain elongation systems there are in microsomes. Recent studies suggest that the conversion of palmitate to lignocerate, in mouse brain microsomes, may involve three different chain elongation systems (Bourre et al., 1973; Goldberg et al., 1973). When rats were fasted the rates of chain elongation of palmitic acid, 6,9-18:2 and 6,9,12-18:3 in liver microsomes were all depressed to about the same extent (Sprecher, 1974a). When rats, which had been fasted, were refed the rates of chain elongation of 6,9-18:2 and 6,9,12-18:3 returned only to the level found in non-fasted controls. Conversely, the rate of chain elongation of palmitate to stearate increased above that found for the non-fasted controls. These studies, coupled with competitive enzyme studies, suggest that liver microsomes contain different chain elongation systems for saturated versus unsaturated fatty acids (Sprecher, 1974a).

In 1965 it was shown that it was possible to measure individual reactions in the overall malonyl-CoA dependent chain elongation reaction (Nugteren, 1965). We have studied intermediate

TABLE 5

Composition of 20:3 and 20:4 in Total Rat Liver Lipids After Feeding Arachidonate or 6,9,12,15-Docosatetraenoate to Rats Raised on a Fat-Free Diet

Dietary Component*	Percent in Total Liver Lipids			
	5,8,11-20:3	7,10,13-20:3	4,7,10,13-20:4	5,8,11,14-20:4
Fat-Free Controls	11.9	0.9	1.1	3.0
6,9,12,15-22:4	8.2	0.9	10.8	3.5
5,8,11,14-20:4	2.9	0.3	0.5	17.0

*Fed at the rate of 50 mg/day for 14 days. (Sprecher, 1968))

reactions in the microsomal chain elongation pathway to determine what regulates this metabolic process and also to determine if the microsome contains more than one chain elongating system (Bernert & Sprecher, unpublished results). Initial experiments were carried out to determine if the condensation reaction between palmitoyl-CoA (16:0 CoA), 6,9-octadecadienoyl-CoA (6,9-18:2 CoA) and 6,9,12-octadecatrienoyl-CoA (6,9,12-18:3 CoA) was inhibited to different degrees by N-ethylmaleimide. When the pseudo first order rate constants obtained with various concentrations of N-ethylmaleimide were plotted versus the indicated inhibitor concentration, the plots shown in Figure 5 were obtained. The apparent second order rate constants obtained through regression analysis of the data in Figure 5 were respectively 2.43, 3.75 and 4.20 L mol^{-1} min^{-1} for 16:0 CoA, 6,9-18:2 CoA and 6,9,12-18:3 CoA. Statistical analysis of the data used in generating the second-order rate plots indicates that the difference between the slope obtained with 16:0 CoA as substrate versus that obtained with 6,9-18:2 CoA and 6,9,12-18:3 CoA was highly significant ($p < .001$ in both cases). The difference

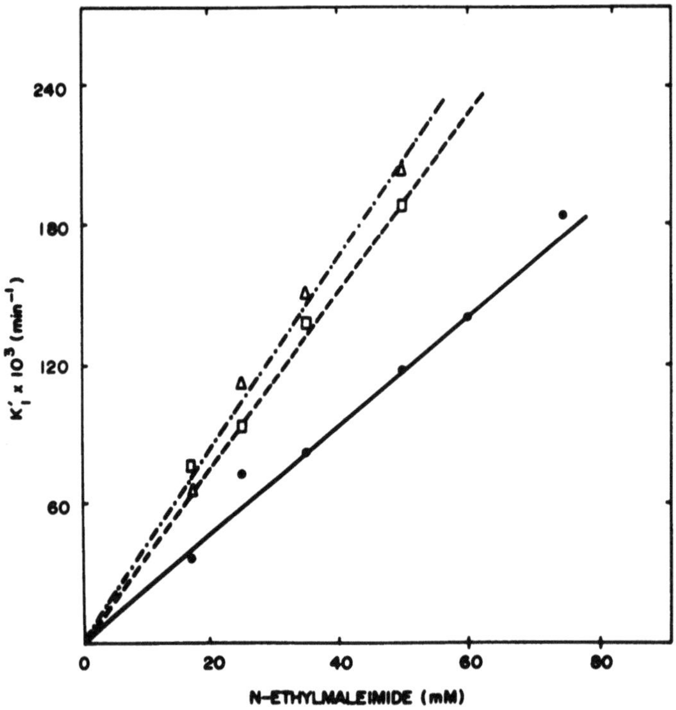

FIGURE 5. Plots of the apparent pseudo first-order rate constants for the N-ethylmaleimide inhibition of the condensation reaction using 16:0 CoA (•), 6,9-18:2 CoA (◦), and 6,9,12-18:3 CoA (Δ) as substrates.

between the slopes for 6,9-18:2 CoA and 6,9,12-18:3 CoA was not significant ($0.1 < p < 0.2$). These data suggest that the condensation reaction with 16:0 CoA involves at least one different enzyme than utilized for either 6,9-18:2 CoA or 6,9,12-18:3 CoA. The results are consistent with a common condensation pathway for both 6,9-18:2 CoA and 6,9,12-18:3 CoA.

The results in Table 6 were obtained when both the overall rates of chain elongation and the condensation reaction were measured with 16:0 CoA, 6,9-18:2 CoA and 6,9,12-18:3 CoA as substrates. The β-hydroxy acyl-CoA dehydrase reaction was assayed only with the CoA derivatives of DL-β-hydroxy-stearic acid (β-OH-18:0 CoA) and DL-β-hydroxy-8,11-eicosadienoic acid (β-OH-8,11-20:2 CoA). The 2-trans-enoyl-CoA reductase reaction was assayed with the CoA derivatives of 2-trans-octadecenoic acid (2-trans-18:1 CoA) and 2-trans-8,11-eicosatrienoic acid (2-trans-8,11-20:3 CoA). All reactions were assayed with the microsomes from rats raised on both a chow diet and a fat-free diet. Both the rates of the condensation reaction and the overall chain elongation reaction were depressed when rats were fed a chow diet. These depressed rates of conversion were more pronounced when 16:0 CoA was used as substrate than when 6,9-18:2 CoA or 6,9,12-18:3 CoA were used as substrates. These results support our N-ethylmaleimide inhibition studies and are consistent with the concept that rat liver microsomes contain at least two different condensing enzymes. One condensing enzyme would preferentially utilize saturated substrates while another would act on unsaturated substrates.

The rates of the β-hydroxy-acyl-CoA dehydrase reaction were more rapid than found for the condensation or overall chain elongation reactions - i.e. Table 6. The rate of the β-hydroxy-acyl-CoA dehydrase reaction was the same when β-OH-18:0 CoA or β-OH-8,11-20:2 CoA were used as substrates. The rate of this reaction, with both substrates, was the same with microsomes from rats raised on either a chow or a fat-free diet. In a similar way the rates of the 2-trans-enoyl-CoA reductase reaction were the same for both substrates and the rates of this reaction with both substrates were not altered by dietary modification. Again the rate of this reaction with both substrates was much more rapid than found for either the condensation reaction or the overall chain elongation reaction.

The results in Table 6, with all three substrates, show that in every instance the rate of condensation was the same as the overall rate for chain elongation. In order to determine whether there is a general correlation between the rate of condensation with that of overall chain elongation we measured the rates of these two reactions with eight isomers of octadecadienoic acid (Ludwig & Sprecher, unpublished results). As shown in Figure 6 in every instance there was good agreement between the rates for

TABLE 6

Rates of Component Reactions in the Microsomal Chain Elongation Pathway*

Initial Substrate	16:0 CoA			6,9-18:2 CoA			6,9,12-18:3 CoA		
Dietary Condition	Normal	Fat-Free	[N/FF]±	Normal	Fat-Free	[N/FF]±	Normal	Fat-Free	[N/FF]±
Chain-Elongation	0.7	2.9	[0.25]	2.8	4.0	[0.70]	3.5	6.0	[0.59]
Condensation	0.9	3.0	[0.30]	3.0	4.1	[0.73]	4.3	6.8	[0.63]
β-Hydroxy-Acyl-CoA Dehydrase	45.4	49.5	[0.92]	45.7	49.3	[0.93]			
2-*trans*-Enoyl-CoA Reductase	99.6	101.8	[0.98]	99.0	104.2	[0.95]			

*All results expressed as nmoles/min/mg microsomal protein.

±Ratio of specific activity with microsome from rats raised on a normal versus a fat-free diet.

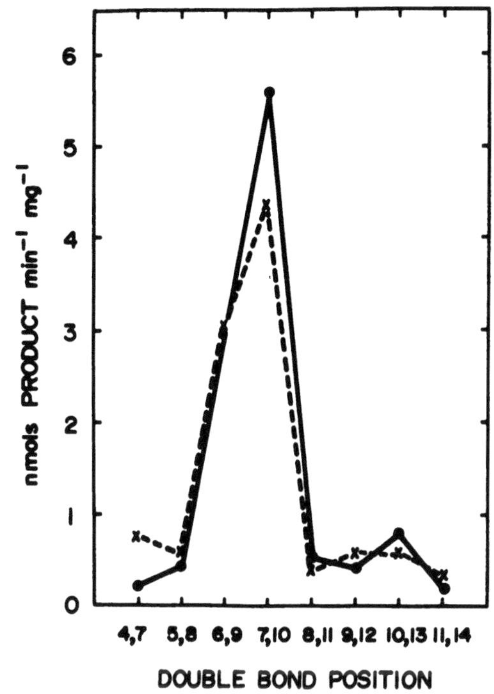

FIGURE 6. Rates of chain elongation (•) and condensation (x) with various octadecadienoic acid isomers.

condensation and overall chain elongation. It was somewhat surprising that 7,10-18:2 was the best substrate for these reactions. In the oleate, palmitoleate, linoleate and linolenate biosynthetic sequences 6,9-16:2, 6,9-18:2, 6,9,12-18:3 and 6,9,12,15-18:4 are respectively the natural substrates for chain elongation. It might thus be predicted that 6,9-18:2 would be the best substrate for condensation and chain elongation. Feeding studies, as shown in Table 7, demonstrate that the amount of linoleate and arachidonate in total liver lipids was about the same when equal amounts of 7,10-16:2 or linoleate were fed to rats raised on a fat-free diet (Sprecher, 1968b). These feeding studies show that an acid with the first double bond at position 7 was readily chain elongated *in vivo*

We have also measured the rates of the β-hydroxyacyl-CoA dehydrase and 2-*trans*-enoyl-CoA reductase reactions using the CoA derivatives of the appropriate intermediates produced during chain elongation of 5,8-18:2, 7,10-18:2 and 8,11-18:2 (Ludwig & Sprecher, unpublished results). The β-OH-7,10-20:2 CoA, β-OH-9,12-20:2 CoA and β-OH-10,13-20:2 CoA were all dehydrated at about the same rate as was found for β-OH-8,11-20:2 CoA and β-OH-18:1 CoA - i.e.

TABLE 7

Percent of Linoleic Acid and of Arachidonic Acid in Total Rat Liver Lipids After Feeding 7,10-Hexadecadienoate and Linoleate to Rats Raised on a Fat-Free Diet

Dietary Components*	Percent in Liver Lipids 9,12-18:2	5,8,11,14-20:4
Fat-Free Controls	1	3
7,10-16:2	5	12
9,12-18:2	7	14

*Fed at the rate of 100 mg/day for a period of 16 days. (Sprecher, 1968))

Table 6. In a similar way 2-trans-7,10-20:3 CoA, 2-trans-9,12-20:3 CoA and 2-trans-10,13-20:3 CoA were reduced at about the same rate as found for 2-trans-18:1 CoA and 2-trans-8,11-20:3 CoA - i.e. Table 6. The β-hydroxyacyl-CoA dehydrase and 2-trans-enoyl-CoA reductase reactions are thus not highly substrate specific. Since both of these reactions proceed much more readily than the condensation reaction it might suggest that two or more independent condensing enzymes contribute there products to a common set of enzymes for subsequent reactions. Failure of dietary modification to alter the rates of these two reactions supports this hypothesis. Moreover, we have also shown that the CoA derivative of β-keto-stearic acid was converted to stearic acid at least four times as rapidly as β-ketostearic acid was produced from 16:0 CoA and malonyl-CoA. The β-keto reductase reaction would thus not be rate limiting in this metabolic process. It thus appears that the rate at which an acid is chain elongated is dictated by the rate of the condensation reaction and that the rate of this reaction is controlled by structural features which are an inherent part of the substrate.

CONCLUSIONS

These studies show that in some cases the types of unsaturated fatty acids found in liver lipids are determined primarily by regulation of polyunsaturated fatty acid biosynthesis. In other cases, rates of conversion must be correlated with other controls such as competitive interactions, retroconversion, and specificities for incorporating given acids into lipids in order to explain the types of unsaturated fatty acids found in liver lipids.

Within any given unsaturated fatty acid biosynthetic sequence the rates of chain elongation generally proceed more rapidly than the desaturation reactions. The rate of Chain elongation of an acid is determined by the rate of the initial condensation step and the rate of the condensation reaction is regulated by structural features inherent in the substrate. Several lines of evidence strongly suggest that rat liver microsomes contain a different malonyl-CoA dependent condensing enzyme for saturated acyl-CoA substrates than is used for unsaturated acyl-CoA derivatives.

ACKNOWLEDGEMENTS

These studies were in part supported by grants AM-09758 and AM-18844 from the United States Public Health Service.

REFERENCES

BERNERT, J. T. & SPRECHER, H. (1975) Biochim. Biophys. Acta 398, 354-363.
BOURRE, J. M., POLLET, S., CHAIX, G., DAUDU, O. & BAUMAN, N. (1973) Biochimie 55, 1473-1479.
BRENNER, R. R. & PELUFFO, R. O. (1966) J. Biol. Chem. 241, 5213-5219.
BRENNER, R. R. (1974) Mol. Cell Biochem. 3, 41-52.
BUDNY, J. & SPRECHER, H. (1971) Biochim. Biophys. Acta 239, 190-207.
DO, U. H. & SPRECHER, H. (1975) Arch. Biochem. Biophys. 171, 597-603.
DO, U. H. & SPRECHER, H. (1976) J. Lipid Res. 17, 424-430.
GOLDBERG, I., SHECHTER, I. & BLOCH, K. (1973) Science 182, 497-499.
GURR, M. I. (1974) in Biochemistry of Lipids (Goodwin, T. W., ed.) Vol. 4, pp. 181-236, University Park Press, Baltimore.
HILL, E. E. & LANDS, W. E. M. (1968) Biochim. Biophys. Acta 152, 645-648.
HOLMAN, R. T. (1964) Federation Proc. 23, 1062-1067.
KLENK, E. (1965) in Advances in Lipid Research (Paoletti, R. & Kritchevsky, D. eds.) Vol. 3, pp. 1-23, Academic Press, London & New York.
KUNAU, W. H. & BARTNIK, F. (1974) Eur. J. Biochem. 48, 311-318.
MEAD, J. F. & FULCO, A. J. (1976) in The Unsaturated and Polyun saturated Fatty Acids in Health and Disease (Kugelmass, I. N. ed.) pp. 52-103, Charles C. Thomas, Springfield, Illinois.
NUGTEREN, D. H. (1965) Biochim. Biophys. Acta 106, 280-290.
SCHLENK, H., SAND, D. M. & GELLERMAN, J. L. (1970) Lipids 5, 575-577.
SPRECHER, H. (1967) Biochim. Biophys. Acta 144, 296-303.
SPRECHER, H. (1968a) Biochim. Biophys. Acta 152, 519-530.
SPRECHER, H. (1968b) Lipids 3, 14-20.
SPRECHER, H. (1971) Biochim Biophys. Acta 231, 122-130.
SPRECHER, H. (1972) Federation Proc. 31, 1451-1457.
SPRECHER, H. (1974a) Biochim. Biophys. Acta 360, 113-123.
SPRECHER, H. (1974b) Biochim. Biophys. Acta 369, 34-44.
SPRECHER, H. & LEE, C. (1975) Biochim. Biophys. Acta 388, 113-129.

STEARNS, E. M., RYSAVY, J. A. & PRIVETT, O. S. (1967) J. Nutr. 93, 485-490.
STOFFEL, W. (1963) Z. Physiol. Chem. 333, 71-88.
STOFFEL, W. ECKER, W., ASSAD, H. & SPRECHER, H. (1970) Z. Physiol. Chem. 351, 1545-1554.
ULLMAN, D. & SPRECHER, H. (1971a) Biochim. Biophys. Acta 248, 61-70.
ULLMAN, D. & SPRECHER, H. (1971b) Biochim. Biophys. Acta 248, 186-197.
VAN DORP, D. A. & CHRIST, E. J. (1975) Rec. Trav. Chim. 94, 247-276.
VERDINO, B., BLANK, M. L., PRIVETT, O. S. & LUNDBERG, W. O. (1964) J. Nutr. 83, 234-238.

THE SPECIFICITY OF MAMMALIAN DESATURASES

A.T. James

Biosciences Division, Unilever Research Laboratory
Colworth/Welwyn
Colworth House, Sharnbrook, Bedford, England

INTRODUCTION

Studies of the metabolism of unsaturated fatty acids by mammals, particularly the rat, have been done over many years both by feeding unlabelled fatty acids followed by analytical and structural studies of the resultant tissue fatty acids and by using labelled substrates in whole animals, individual organs and tissue homogenates. Many pathways for individual fatty acids have been elucidated particularly in the essential fatty acid series, but until recently there has been much confusion in the literature as to the number and type of desaturase enzymes. Only in the last few years has there been any attempt at systematic investigation of enzyme-substrate specificity since such work is hampered by the failure to isolate single enzymes. All the desaturases so far studied appear to be particulate of very high molecular weight and cannot be purified by conventional means.

Our own studies using very crude preparations, from a variety of sources, of enzymes converting saturated to monoenoic and monoenoic to dienoic acids have shed some light on enzyme-substrate relationships. With this as a basis and using the data and ideas from many workers, especially Klenk, Holman, van Dorp, Mead, Stoffel, Sprecher, and Brenner and their colleagues (for references see later) this paper makes an attempt at interpretation of the pathways that leads to an apparent simplification.

Fatty acid ⟶ acyl-S-CoA ⟶

- (1) Complete β-oxidation to CO_2
- (2) Retroconversion (loss of 2C atoms
- (3) Dehydrogenation to a variety of unsaturated acids
- (4) Chain elongation, by stages of 2C atoms
- (5) Acyl lipid formation

Fig. 1. Competing fates of fatty acids.

DISCUSSION

1. General

Any tissue fatty acid after activation to the acyl-S-CoA or ACP thiol ester is open to a large range of possible fates (Fig. 1). Its distribution between the competing pathways will be a function of (a) the particular tissue and the previous feeding regime insofar as they govern the relative amounts of the individual enzymes, (b) the relative amounts of competing substrates and (c) the relative rates of the competing processes.

After reactions (2), (3) and (4) the products are subject to the same choices so that definition of the relative reaction velocities of any stages by study of complete systems is doomed to failure. In addition, reaction (5) may remove the substrate completely or partially. Nevertheless, sufficient data have now been accumulated to allow one to make some proposals as to forbidden and acceptable conversions.

2. Enzyme-Substrate Specificity

(a) The Δ9 desaturase. Studies of specificity of the enzymes forming Δ9 monoenoic fatty acids, are summarised in Table 1. The following conclusions are suggested by the results :-

(i) The double bond is always introduced at the 9-10C atoms specified solely with respect to the carboxyl group.

(ii) Two enzymes appear to exist, one having maximum conversion for an 18C chain and other maximum conversion for an 14C chain. (Gurr et al., 1972).

Table 1. Fatty acids shown to be dehydrogenated at the Δ9-10 position

Saturated acids	Branched chain acids	Unsaturated acids	
10:0, 11:0, 12:0		Δ18 - nonadecenoic acid	(Brett et al., 1971)
13:0, 14:0, 15:0		cis Δ12 - octadecenoic acid	(Gurr et al., 1972)
16:0, 17:0, 18:0			
19:0			
	2-, 3-, 4-, 16-, 17-methyl stearic acids		(Howling et al., 1972)
		trans Δ5 - hexadecenoic acid	Pollard et al. (1976)
		trans Δ5 - heptadecenoic acid	Pollard et al. (1976)
		trans Δ5 - octadecenoic acid	Pollard et al. (1976)
		trans Δ12 - heptadecenoic acid	Pollard et al. (1976)
		trans Δ13 - octadecenoic acid	Pollard et al. (1976)
		trans Δ7 - hexadecenoic acid	Pollard et al. (1976)

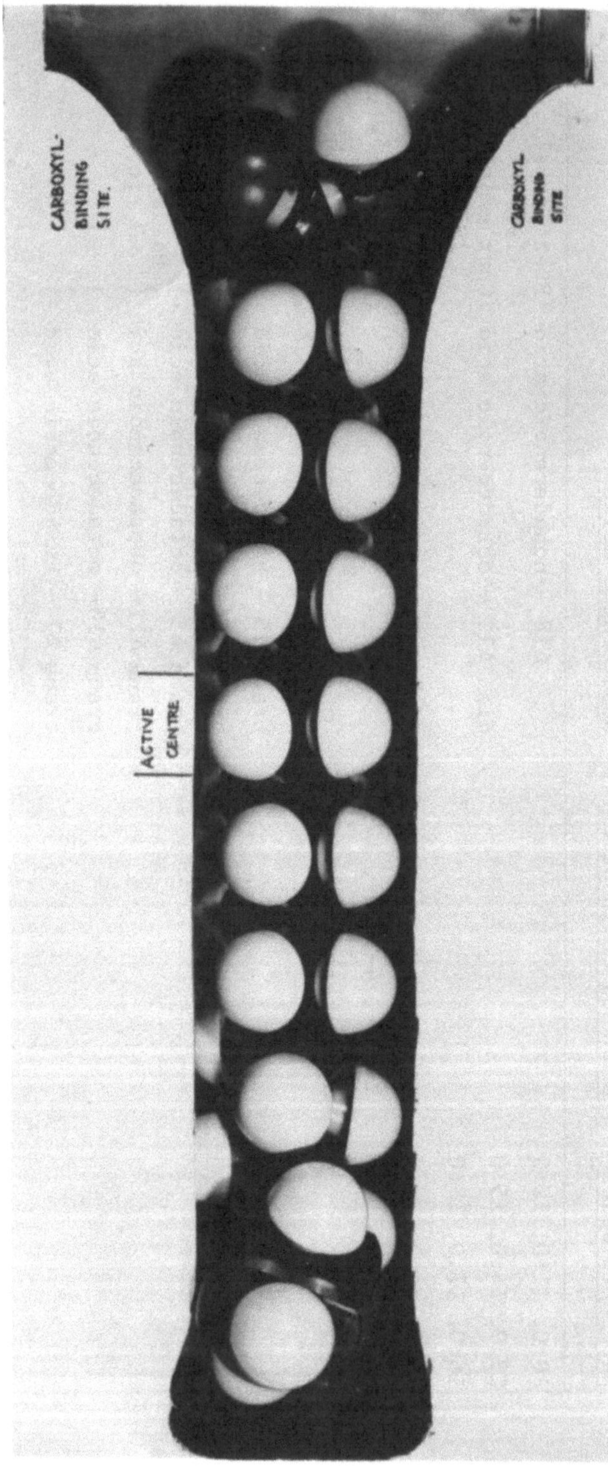

Fig. 2. Suggested cleft structure capable of accepting the substrates of the enzyme having maximal activity for a C_{18} chain.

(iii) Substrates already containing a double bond are acceptable to the enzymes provided the double bond (if cis) is upchain towards the methyl group, and if trans in configuration it can be at Δ5, Δ12 , or Δ13 (Pollard et al., 1976). For example, the stearic acid desaturase which showed low conversion of a saturated C_{19} acid, had a high conversion rate of a C_{19} acid with a double bond between C atoms 18 and 19. A C_{18} unsaturated acid with a *cis* double bond in the 12-13 position was readily converted to a linoleic acid (Δ9:12 octadecadienoic acid) by those crude enzyme systems apparently containing the desaturase with maximal activity with a 14C chain (hen liver) but not by systems containing only the desaturase with maximal activity at the 18C chain (rat liver) (Table 1).

Investigation of the acceptability of a range of methyl branched stearic acids showed that a methyl group anywhere on C atoms 5 to 15 prevented dehydrogenation. This indicated that this length of the methylene chain was rigidly held by the enzyme and hence the enzyme could be portrayed as having a cleft equal in length to the extended staggered stearic acid molecule and close to the width of the methylene chain except for the region of C1 to C4 and C16 to C18(Fig. 2). The high stereospecificity of the dehydrogenation (only the D-H atoms being removed) also argues for a rigid holding of the fatty acid chain around the active centre which must itself be located in a fixed position with respect to the carboxyl group of the substrate.

A fatty acid whose chain length is greater than that of the cleft could be accommodated only by adopting a configuration which lifts the methyl end away from the cleft, this would account for the lower conversions of such substrates.

A *cis* unsaturated acid, however, would lift the distal part of the chain out of the cleft if the extended zig-zag configuration were adopted. This would have the effect of shortening the length of methylene chain in close contact with the enzyme and could be regarded as mimicking the configuration of a shorter chain saturated acid. This would account for the introduction of a 9:10 double bond into Δ12-octadecenoic acid by those crude systems containing the enzyme capable of rapidly converting tetradecanoic acid to Δ9-tetradecenoic acid. The Δ12-octadecenoic acid in the extended chain configuration could, therefore, be claimed to possess an "effective" chain length of about 14C atoms (Fig. 3). More cis double bonds in the molecule further upchain would still further prevent any steric interaction of the distal parts of the chain with the enzyme and make it easier for a highly unsaturated acid to "mimic" a short chain saturated acid.

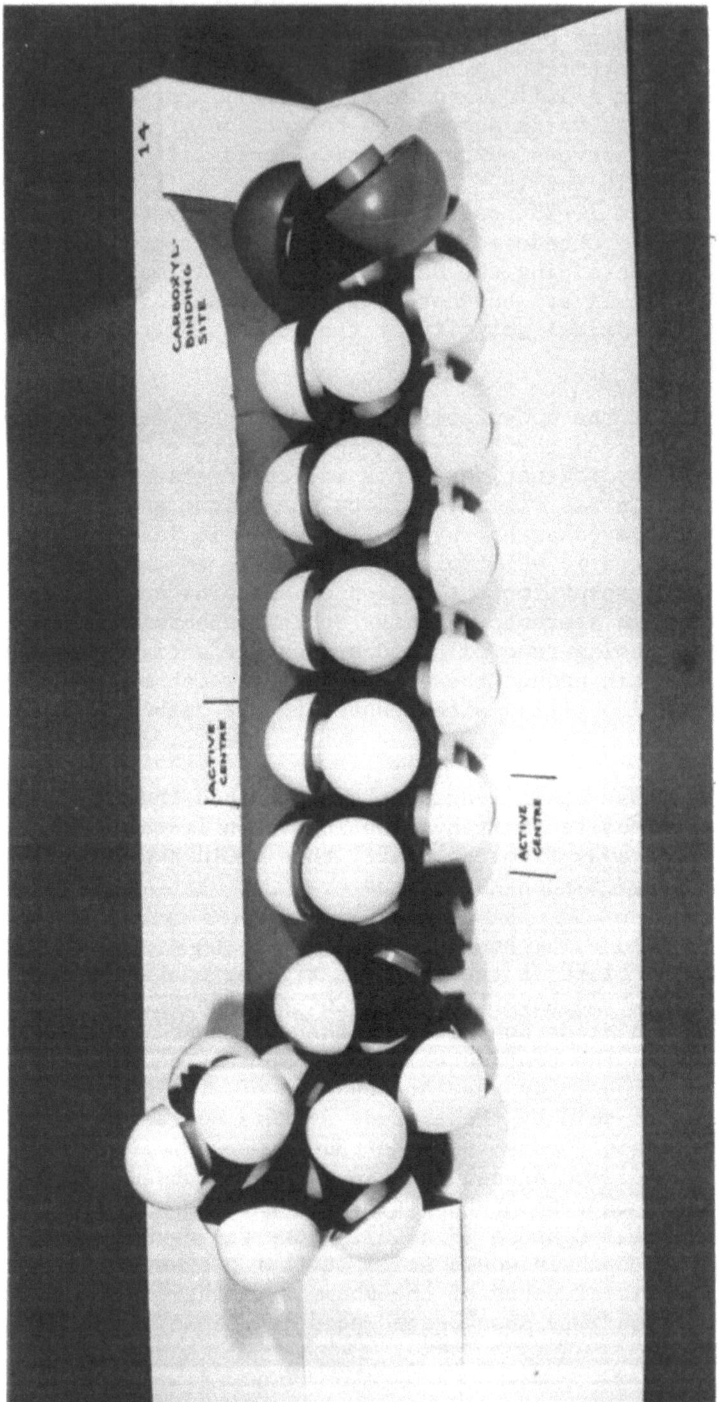

Fig. 3. Model of Δ 12 octadecenoic acid fitted into a slot of length equal to that of a saturated C_{14} chain.

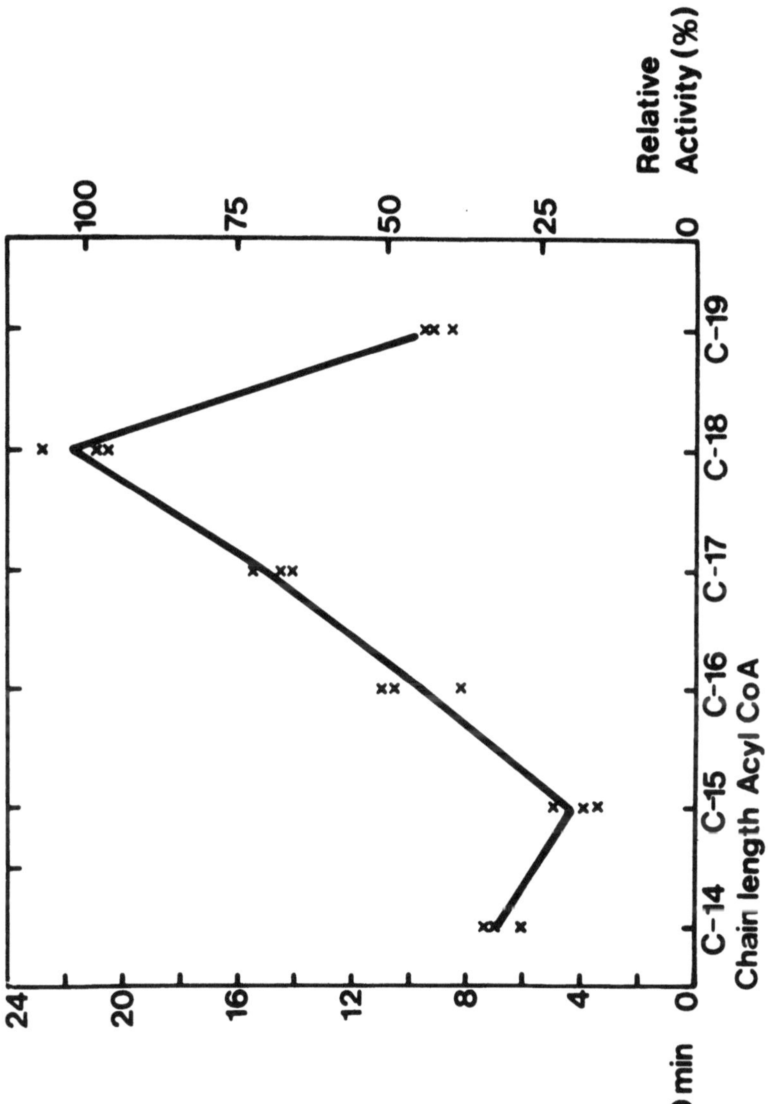

Fig. 4. Δ9 Desaturation of C_{14}-C_{19} Acyl CoAS at 30°pH 7.4

The existing double bond would need to be separated from the reaction centre upchain by at least one methylene group, to prevent any alteration of the electron distribution at the reacting C atoms. Any cis double bond between the reaction centre and the carboxyl group would lift the chain away from the cleft and so prevent the dehydrogenation from occurring. This would apply only to those enzymes which introduce double bonds specified with respect to the carboxyl group.

Our most recent results (Pollard et al., 1976) have shown that Δ5 trans monoenoic acids (see Table 1) are as acceptable substrates as are the corresponding saturated acids. Trans fatty acids with the double bonds upchain from the 9-10C atoms are also acceptable. Since trans double bonds do not perturb the normal extended chain configuration, our original views on the laying down on the enzyme of the two parts of the molecule on either side of the 9-10C atoms needs no modification.

Earlier work on the specificity of Δ9 desaturases was based on the use of free fatty acids and hence was subject to the criticism that ease of formation of the corresponding acyl CoA thiol ester was an undetermined factor in the results.

Recently we have demonstrated (Jeffcoat et al., 1976) that the apparent chain length specificity demonstrated by adding free fatty acids to a crude enzyme system containing the acyl-CoA actuating system, is identical to that obtained by using chemically synthesized acyl-CoA thioesters. (Fig. 4). The initial approach we and others have used is, therefore, vindicated.

The discussion so far has introduced the "rules" of (a) carboxyl end controlled positional specificity of the dehydrogenation and (b) mimicking of a short chain saturated acid by cis-mono, or polyunsaturated fatty acids with double bonds distal by at least one methylene group from the reaction centre.

(b) Δ6 desaturase. In Table 2 are shown all those substrates demonstrated to be dehydrogenated at the Δ6-7 positions. That the enzyme(s) is (are) carboxyl end controlled is shown by the conversion of both odd and even numbered fatty acids whether mono or dienoic. Specifying double bond position with respect to the methyl group would argue for three enzymes since the new double bond would be introduced at (n-10) for the C_{16} acid , (n-11) for the C_{17} acid and (n-12) for the C_{18} acid. On the basis that the simplest explanation is the correct one, this would suggest a single Δ6 desaturase enzyme. Brenner (1971) has already published convincing evidence for such a single Δ6 desaturase capable of accepting Δ9, Δ9:12 and Δ9:12:15 unsaturated acids, based on (a) Kinetic data with competing substrates and (b) inhibition by competing substrates.

Table 2.- Structure of fatty acids shown to be dehydrogenated at the Δ6-7 positions

Δ9 monoenoic acids	Δ9:12 dienoic acids	Δ9:12:15 trienoic acids
C_{15} Pollard et al. (1976)		
C_{16} Fulco & Mead (1959)	C_{16} Holman (1970)	
C_{17} Pollard et al. (1976)	C_{17} Schlenk, Sand & Sen (1964)	
C_{18} Mead & Slaton (1956) Fulco & Mead (1959)	C_{18} Brenner & Peluffo (1966)	C_{18} Brenner & Peluffo (1966)
Other monoenoic acids:		
C_{17} Δ12 trans Pollard et al. (1976)	C_{17} Δ9 cis 12 trans Pollard et al. (1976)	
C_{16} Δ8 trans Pollard et al. (1976)	C_{18} Δ9 cis 13 trans Pollard et al. (1976)	

Brenner (1971) suggested that the increase in conversion rate in the series Δ9 18:1, Δ9:12 18:2 and Δ9:12:15 18:3 is due to better fit of the double bonds and hence the shape of the distal parts of the chain to the enzyme. An alternative suggestion is that as the number of *cis* double bonds increases, the more is the distal part of the chain lifted away from the enzyme, so decreasing steric hindrance. This would assume that the enzyme cleft has a length equal to an 11 or 12 saturated C atom chain, into which a Δ9 unsaturated acid would fit less easily than a Δ9:12 acid, and this less easily than a Δ9:12:15 acid. In this case short chain saturated acids should be good substrates for the enzyme. Experiments to test this with both normal and starved and refed rats gave no formation of Δ6 monoenoic acids from C_9, C_{10}, C_{12} or C_{14} fatty acids whilst under the same conditions linoleic acid was readily converted to Δ6:9:12 octadecatrienoic acid (Hall & James, 1976). This demonstrates that interaction of substrate with enzyme involves other groups in addition to the carboxyl to 6-7C atoms part of the molecule, i.e. that Brenner's explanation is basically correct.

The necessary features of the substrate seem to be the following :- (1) a double bond of cis configuration at the 9-10 position (Δ9 trans acids do not act as substrates, Pollard et al., 1976); and (2) a cis or trans Δ12 or Δ13 double bond; the existence of a cis Δ15 double bond further increases reaction rate (there is no data available for a trans Δ15 double bond). The trans Δ12 double bond exerts a greater effect than a cis Δ9 double bond since 17:1 trans Δ12 is converted to 17:2 cis Δ6 trans Δ12 in 15% yield whilst 17:1 cis Δ9 is converted to 17:2 cis Δ6 cis Δ9 in only 9% yield.

The shape of the enzyme cleft is clearly more complex than for the Δ9 desaturase and involves polar sites capable of weakly interacting with existing double bonds in the substrate at positions 9-10, 12-13, 13-14 and 15-16. A greater range of possible substrates needs to be tested before the shape and dimensions of the cleft can be more accurately defined.

(c) The Δ5 desaturase. In Table 3 are brought together all those acids known to be converted to a Δ5 acid. No published data exits for the conversion of Δ8 monoenoic acids to Δ5:8 dienoic acids; our experiments suggest that cis Δ8 16:1 is a very poor substrate (Pollard et al., 1976). The C_{16} and C_{18} Δ8:11 dienoic acid conversions are, however, well documented but Rahm & Holman (1964) were unable to demonstrate any conversion of an unlabelled Δ8:11 C_{17} dienoic acid. This is puzzling and does not fit the concept of a single enzyme with a wide chain length specificity, the experiment needs repeating with labelled substrate. The Δ8:11:14 trienoic acid conversions of both odd an even number acids are well documented as is the conversion of the Δ8:11:14:17 C_{19} tetraenoic acid. With one exception therefore the published data do fit the concept of a single enzyme capable of accepting acids with a double bond at the closest position of Δ8 and introducing a double bond at the Δ5:6 position. That the Δ8 double bond is not absolutely necessary was shown by Ullman & Sprecher (1971) who demonstrated the conversions Δ11, 20:1 to Δ5:11, 20:2 - 11:14, to Δ5:11:14, 20:3, as well as our results with Δ9 trans acids.

The major specificity seems to reside in the Δ11 and Δ14 double bonds since (in terms of percentage conversion) Δ11c:14c, 20:2 is as good a substrate as Δ8c: 11c:14c, 20:3. The Δ9 trans 17:1 and 18:1 acids are much poorer substrates as is the Δ11 cis 18:1. We may, therefore, conclude that the enzyme has interacting sites at a distance corresponding to Δ11-12 and Δ14-15 from the carboxyl group, interactions with double bonds at other positions are much less critical but overall chain length may play some role as well.

Table 3.- Structure and chain length of fatty acids shown to be dehydrogenated at the Δ5-6 positions

Monoenoic acids	Δ8:11 Dienoic acids	Δ8:11:14 trienoic acids	Δ8:11:14:17 tetraenoic acids
	C_{16} Holman (1970)		
C_{20} Δ11 Ullman & Sprecher (1971)	C_{18} Budney & Sprecher (1971) Holman (1970)	C_{18} Stoffel (1973) Schlenk et al. (1964) Beerthuis et al. (1968)	
C_{17} Δ9 trans Pollard et al.,(1976)			
C_{18} Δ9 trans ibid.	C_{19} Klenk & Tschope (1963)	C_{19} Schlenk et al. (1964) Beerthuis et al. (1968)	C_{19} Schlenk et al. (1964)
	C_{20} Stoffel & Scheid (1967)	iso-C_{20} Stoffel & Scheid (1967) C_{20} Holman, R. T. (1970) C_{21} Beerthuis et al. (1968)	

Table 4.- Chain lengths of fatty acids shown to be dehydrogenated at the Δ4-5 position

Δ7:10 dienoic acids	Δ7:10:13 trienoic acids	Δ7:10:13:16 tetraenoic acids
	C_{16} Klenk (1965)	
C_{18} Holman (1970)		
		C_{19} Beerthuis et al. (1968)
	C_{20} Beerthuis et al. (1968) Budney & Sprecher (1971)	
		C_{21} Schlenk et al. (1964)
		C_{22} Sprecher (1967a, 1967b)

(d) A Δ4 desaturase. In Table 4 is collected all the published data demonstrating the dehydrogenation of fatty acids to Δ4 acids. The chain length and position of double bonds in the substrates also fits that expected for a single carboxyl end determined Δ4 desaturase. No data bearing on the possibilities of conversion of a Δ7 monoenoic acid have yet been published. This also needs experimental verification. Little can there be said as to the nature of the enzyme-substrate complex.

DISCUSSION

All the experimental evidence available supports the contention that in mammalian systems there are only four desaturases, each of which introduces double bonds in positions defined exclusively with respect to the carboxyl group, the positions being Δ4, Δ5, Δ6, and Δ9; each enzyme having different substrate specifications. No desaturases seem to exist capable of inserting double bonds at the Δ7, Δ8 or Δ11 positions, double bonds in these positions arise only by chain elongation of Δ5, Δ6 or Δ9 acids. This proposition is in line with the definitive earlier work (Klenk 1965) showing that fatty acids of the oleic series and α-linolenic series are unable to enter the linoleic series at any stage.

Table 5.- Polyunsaturated fatty acids shown to be chain elongated

Δ4:7:10	(16:3)	Δ6:9:12	(18:3)	Stoffel (1966)
Δ4:7:10:13	(16:4)	Δ6:9:12:15	(18:4)	Klenk (1963)
Δ4:7:10:13	(19:4)	Δ6:9:12:15	(21:4)	Nugteren (1965)
Δ5:8	(14:2)	Δ7:10	(16:2)	Sprecher (1968) Stoffel (1966)
Δ5:8:11	(20:3)	Δ7:10:13	(22:3)	Stoffel (1966)
Δ5:8:11:14	(20:4)	Δ7:10:13:16	(22:4)	Stoffel (1966)
Δ6:9	(18:2)	Δ8:11	(20:2)	Stoffel (1966)
Δ6:9:12	(18:3)	Δ8:11:14	(20:3)	Stoffel (1966)
Δ6:9:12:15	(18:4)	Δ8:11:14:17	(20:4)	Stoffel (1966)
Δ7:10	(16:2)	Δ9:12	(18:2)	Stoffel (1966)
Δ7:10	(18:2)	Δ9:12	(20:2)	Holman (1970)
Δ8:11	(16:2)	Δ10:13	(18:2)	Holman (1970)
Δ8:11	(17:2)	Δ10:13	(19:2)	Rahm & Holman (1964) Budney & Sprecher (1971)
Δ8:11	(18:2)	Δ10:13	(20:2)	Holman (1970)
Δ8:11:14	(18:3)	Δ10:13:16	(20:3)	Stoffel (1966)
Δ8:11:14	(20:3)	Δ10:13:16	(22:3)	Stoffel (1966)
Δ9	(18:1)	Δ11	(20:1)	Stoffel (1966)
Δ9:12	(14:2)	Δ11:14	(16:2)	Holman (1970)
Δ9:12	(17:2)	Δ11:14	(19:2)	Schlenk, Sand & Gellerman (1969)
Δ9:12	(18:2)	Δ11:14	(20:2)	Stoffel (1966)
Δ9:12	(19:2)	Δ11:14	(21:2)	Holman (1970)
Δ9:12:15	(18:3)	Δ11:14:17	(20:3)	Stoffel (1966)
Δ10:13	(18:2)	Δ12:15	(20:2)	Holman (1970)
Δ10:13	(19:2)	Δ12:15	(21:2)	Holman (1970)
Δ11	(20:1)	Δ13	(22:1)	Stoffel (1966)
Δ11:14	(19:2)	Δ13:16	(21:2)	Holman (1970)
Δ11:14	(20:2)	Δ13:16	(22:2)	Holman (1970)

Table 6.- Unsaturated fatty acids shown to be chain shortened

Precursor	Product	Reference
Δ9 (18:1)	Δ7 (16:1)	Chang & Holman (1972)
Δ10:13 (20:2)	Δ8:11 (18:2)	Budney & Sprecher (1971)
Δ11:14 (20:2)	Δ9:12 (18:2)	Stearns et al. (1967)
Δ6:9:12:15 (21:4)	Δ4:7:10:13 (19:4)	Schlenk et al. (1969)
Δ7:10:13:16 (22:4)	Δ5:8:11:14 (20:4)	Sprecher (1967b) Verdino et al. (1964)
Δ4:7:10:13:16 (22:5)	Δ5:8:11:14 (20:4)	Schlenk et al. (1969)
Δ4:7:10:13:16:19(22:6)	Δ5:8:11:14:17 (20:5)	ibid(1969)

The transformations that also affect the fate of an ingested fatty acid are chain elongation and chain shortening, either process could change an "unacceptable" fatty acid to one capable of further desaturation and on the other hand, change an acceptable to an unacceptable substrate. Chain elongation apppears to have less substrate specificity (unless one involves a large number of different enzymes) than desaturation, as in shown by the long list of defined elongations in Table 5.

Chain shortening is well defined for a number of polyunsaturated acids (Table 6) ranging in chain lenght from C_{20} to C_{22}. Some acids appear in both tables as either precursor or product and hence one can consider chain shortening and elongation as reversible processes (but not in the sense that a single enzyme is concerned).

Thus Δ 8:11 (18:2) $\rightleftharpoons$ Δ 10:13 (20:2)

Once, however, that the double bond is moved up chain towards the methyl end by the addition of a C_2 unit, the product may be then unacceptable to the desaturases. Such products can thus be regarded as "stop points" so far as desaturation is concerned. They can be further metabolised either by complete oxidation to CO_2 or desaturated after chain shortening. The metabolism of the Δ8:11

(18:2) acid could be represented as follows.

Δ8:11 (18:2) ⇌ Δ10:13 (20:2) STOP POINT
↓
Δ5:8:11 (18:3) ⇌ Δ 7:10:13 (20:3)
↓
Δ 4:7:10:13 (20:4) ⇌ Δ6:9:12:15 (22:4)

The horizontal changes are reversible but the vertical changes of desaturation are in general irreversible. However, Δ4:7:10:13:16 (22:5) has been shown to be converted to Δ5:8:11:14 (20:4) (See Table 6) and this involves the reduction of a double bond as well as the loss of 2C atoms (Schlenk et al., 1969). The reason for this is not hard to see since on chain shortening a Δ4 double bond becomes Δ2, such acids are readily hydrogenated. Such a sequence allows us to add another feasible pathway to those shown above, viz.

Δ5:8:11 (18:3) ⇌ Δ7:10:13 (20:3)
↑ ↓
Δ2:5:8:11 (18:4) ⇌ Δ4:7:10:13 (20:4)

Such pathways enable the drawing of reaction schemes which fit all the known data and can be applied to any fatty acid - one such is shown in Fig. 5, for Δ9:12 (17:2) such a scheme could be called the "linoleic" series. Similar sequences are shown in Fig. 6 for the "oleic" series of Δ9 monoenoic acids and in Fig. 7 for Δ9:12:15 acids of the "linolenic" series.

By analogy with the range of reaction velocities :

Δ9 (18:1) < Δ9:12 (18:2) < Δ9:12:15 (18:3)

established for the Δ6 desaturase by Brenner (1971) one would expect for the other desaturases a similar sequence.

Thus : <u>Δ5 desaturase</u>
Δ8 monoenes < Δ8:11 dienes < Δ8:11:14 trienes < Δ8:11:14:17 tetraenes

<u>Δ4 desaturase</u>
Δ7 monoenes < Δ7:10 dienes < Δ7:10:13 trienes < Δ7:10:13:16 tetraenes

This could explain Klenk's results (1965), that Δ8:11 (18:2) showed no apparent further metabolism whereas Δ8:11:14 (18:3) gave rise to highly labelled (but small amounts) of C_{20} polyenoic acids. However, Budny & Sprecher (1971) demonstrated that Δ8:11 (18:2) is indeed converted to Δ5:8:11 (18:3) in appreciable amount.

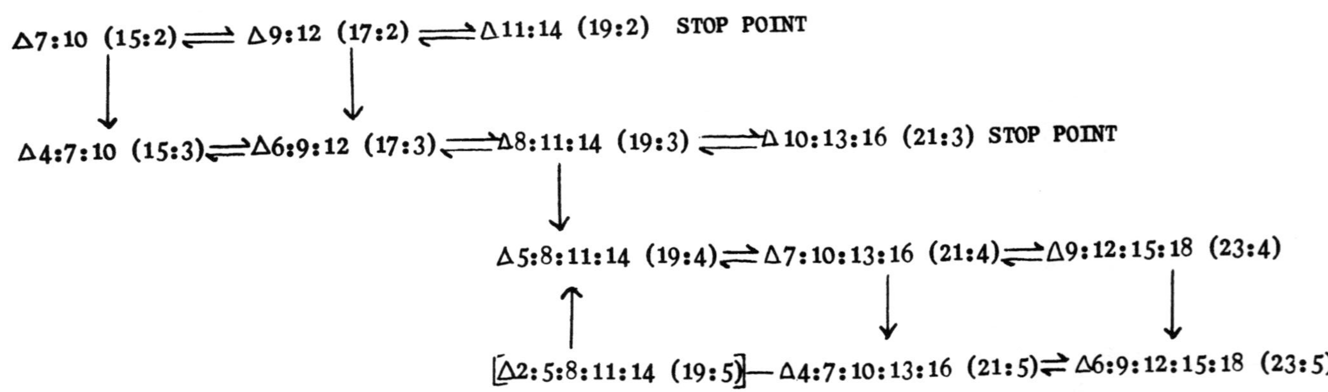

Fig. 5. Metabolic sequences for fatty acids having Δ9:12 double bonds (linoleic series) using Δ9:12 (17:2) as an example.

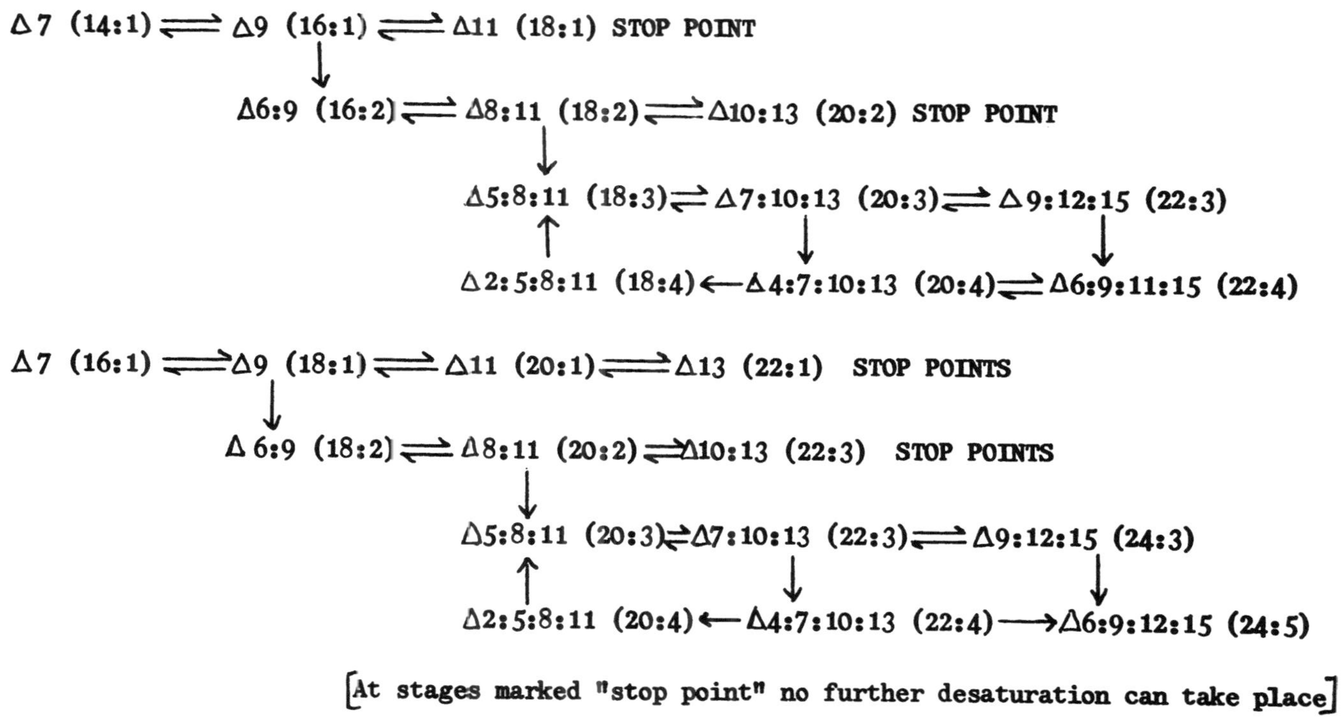

Fig. 6. Metabolic sequences for fatty acids having cis 7 or 9 double bonds (oleic series)

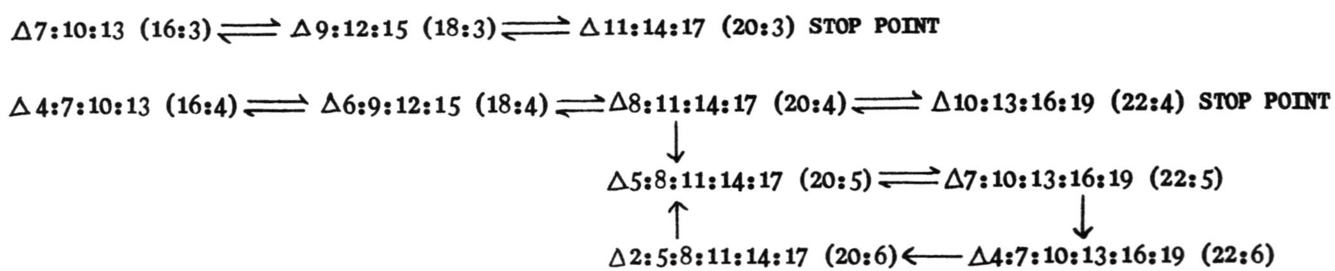

Fig. 7. Metabolic sequences for fatty acids having Δ9:12:15 double bonds (α - linolenic series)

The proposition of absence of any Δ12 or Δ10 desaturases is supported by the failure of Sprecher (1967a) to obtain any conversion of Δ9:15 (18:2) or Δ7:13 (20:2). Δ12:15 (18:2) was also found not to be desaturated in the rat by the same author, this agrees with our failure to demonstrate desaturation of Δ12 (18:1) in this animal (Gurr et al., 1972). However, other animal systems capable of converting Δ12 (18:1) to Δ9:12 (18:2) i.e. hen, goat mammary gland microsomes, etc. (Gurr et al., 1972), might be expected to convert Δ12:15 (18:2) into Δ9:12:15 (18:3).

The pathways put forward here should enable the prediction of the products from any ingested fatty acid. Such predictions can be evaluated for those acids shown to be "essential" to the developing rat. This was done by van Dorp and his colleagues (Beerthuis et al., 1968) following on their studies of prostaglandin synthetase. They demonstrated that although the synthetase would accept a variety of polyunsaturated fatty acids as substrates, physiologically active prostaglandins were produced only from fatty acids that could, in the intact animal, give rise to fatty acids in the sequences leading to Δ5:8:11:14 polyenoic acids of chain lengths C_{19}, C_{20} and C_{21}. Comparison of such acids with their essential fatty acid effects in deficient rats gives a virtually complete correlation.

The proposals made in this paper on allowable enzymic transformations also fit their results. In table 7 are listed all those fatty acids tested for EFA activity and it will be seen that only those acids capable of being converted (in the rat) to the Δ5:8:11:14 C_{19}, C_{20} or C_{21} tetraenes show any activity. Thus Δ12 (18:1) cannot be converted to Δ9:12 (18:2) by the rat, and is EFA inactive, however see Jacob & Grimmer (1971). Whilst Δ8:11:14 (18:3) can produce a Δ5:8:11:14 tetraene, it has the wrong chain length (C_{18}) for prostaglandin synthesis. Δ10:13 (19:2) after chain shortening can give rise to only a Δ4:7:10:13-C_{19} tetraenoic acid as with Δ7:10:13 (19:3). Δ8:11:14 (19:3) will give rise to some Δ5:8:11:14 (19:4) hence its appreciable EFA activity. Δ11:14 (20:2) cannot be directly converted to Δ5:8:11:14 (20:4) because of the absence of a Δ8 desaturase. Its activity is therefore probably due to chain shortening to Δ9:12 (18:2) which then enters the normal linoleic acid sequence. The reasons for activity or inactivity of the remaining acids in the Table is obvious. The last acid Δ4:7:10:13:16 (22:5) is, however, of some interest since it presumably acts as a reservoir for formation of Δ5:8:11:14 (20:4) as shown in Fig. 5. Indeed the same would be expected to be true for Δ7:10:13:16 (22:4), Δ9:12:15:18 (24:4) and Δ6:9:12:15:18 (24:5). Such a concept of "reservoirs" of potential arachidonic acid might partially explain the difficulty of depleting the adult animal of essential fatty acids.

Table 7.- Essential Fatty Acid Activities in the Rat

Structure		Activity in Units/g	Reference	
Δ12	(18:1)	0	Thomasson	(1953)
Δ9:12	(17:2)	100	Schlenk & Sand	(1967)
Δ9:12	(18:2)	100	Deuell & Greenberg	(1950)
Δ6:9:12	(18:3)	115	Thomasson	(1962)
Δ8:11:14	(18:3)	0	Beerthuis et al.	(1968)
Δ10:13	(19:2)	9	Thomasson	(1967)
Δ7:10:13	(19:3)	6	Thomasson Rahm & Holman	(1967) (1964)
Δ8:11:14	(19:3)	22 to 33	Beerthuis et al.	(1968)
Δ5:8:11:14	(19:4)	49	Beerthuis et al.	(1968)
Δ11:14	(20:2)	46	Thomasson	(1962)
Δ8:11:14	(20:3)	100	Thomasson	(1962)
Δ7:10:13	(20:3)	0	Beerthuis et al.	(1968)
Δ8:11:14	(21:3)	56	Beerthuis et al.	(1968)
Δ5:8:11:14	(21:4)	62 to 78	Beerthuis et al.	(1968)
Δ8:11:14	(22:3)	0	Beerthuis et al.	(1968)
Δ4:7:10:13:16	(22:5)	139	Verdino et al. De Iong & Thomasson	(1964) (1956)

On this basis one can now list all these acids expected to show some EFA activity (acids with a first double bond at the Δ2 or Δ3 positions are left out since they are likely to be hydrogenated). Such a list is shown in Table 8, the upper part consisting of only the linoleic series, this gives a total of 33 fatty acids. This is likely to be the limit for the rat but for those animals (see Gurr et al., 1922) capable of inserting a Δ9 double bond in Δ12 acids, the list can be extended by a further 21 acids giving a total of 54 essential fatty acids.

These considerations render 3 (n-x) classification of double bond position in fatty acids introduced by Thomasson (1953) less useful since new double bonds seem to be introduced in animals only with respect to the carboxyl end rather than the methyl end of the molecule. I would therefore suggest that this nomenclature be dropped since although it identifies fatty acids in their chain elongation or shortening sequences it conceals the important controlling effects of the carboxyl (Δ) specification of desaturation.

Table 8.- List of fatty acids expected to show some E.F.A. activity if this is due entirely to formation of active prostaglandins

(a) In all animals

Δ5:8 of chain length 13,14,15	Δ4:7:10 of chain length 15,16,17	Δ5:8:11:14 of chain length 19,20,21	
Δ7:10 of chain length 15,16,17	Δ6:9:12 of chain length 17,18,19		
Δ9:12 of chain length 17,18,19	Δ8:11:14 of chain length 19,20,21	Δ7:10:13:16 of chain length 21,22,23	Δ4:7:10:13:16 of chain length 21,22,23
Δ11:14 of chain length 19,20,21	Δ10:13:16 of chain length 21,22,23		

33 acids in all species

(b) In some species excluding the rat

	chain length
Δ4	9, 10, 11
Δ6	11, 12, 13
Δ8	13, 14, 15
Δ10	15, 16, 17
Δ12	17, 18, 19
Δ14	19, 20, 21

51 acids in total (for some species only)

CONCLUSIONS

1. Using the concepts of enzyme-substrate interaction already developed for the enzyme converting stearic to oleic acid, almost all published data on transformation of fatty acids can be accounted for by the action of the following enzymes.

   (a) A Δ6 desaturase introducing a double bond solely in position 6-7 with respect to the carboxyl group.

   (b) A Δ5 desaturase introducing a double bond solely in positions 5-6 with respect to the carboxyl group.

   (c) A Δ4 desaturase introducing a double bond solely in positions 4-5 with respect to the carboxyl group.

   (d) Two Δ9 desaturases introducing a double bond solely in positions 9-10 with respect to the carboxyl group, but each having a different chain length specificity and distribution.

   (e) A low-specificity chain extending enzyme or enzymes.

   (f) A low specificity chain shortening enzyme or enzymes.

2. The specificity of each desaturase enzyme is examined and preliminary conclusions drawn as to the nature of the enzyme-substrate complex.

3. It is suggested that the (n-x) classification of double bond position be dropped since while it shows elongation relationships, it often conceals the real relationships between precursors and products, the older Δ or carboxyl end classification is proposed as the replacement.

REFERENCES

BEERTHUIS, R. K., NUGTEREN, D. H., PABON, H. J. J. & van DORP, D. A. (1968) Recuit 87, 461.
BRENNER, R. R. & PELUFFO, R. O. (1966), J. Biol. Chem. 241, 5213.
BRENNER, R. R. (1971) Lipids 6, 567.
BRETT, D., HOWLING, D., MORRIS, L. J. & JAMES, A. T. (1971), Arch. Biochem. Biophys. 143, 535.
BUDNY, J. & SPRECHER, H. (1971), Biochim. Biophys. Acta 239, 190.
CHANG HUEI-CHE & HOLMAN, R. T. (1970) Biochem. Biophys. Acta 280, 17.
DE IONG, H. & THOMASSON, H. J. (1956) Nature 175, 1051.
DEUELL, H. J. Jr. & GREENBERG, S. M. (1950) Fortschr. Chem. Org. Naturstoffe 6, 1 Wien Springen Verlag.
FULCO, A. J. & MEAD, J. F. (1959) J. Biol. Chem. 234, 1411.
GURR, M. I., ROBINSON, M. P., JAMES, A. T., MORRIS, L. J. & HOWLING, D. (1972) Biochim. Biophys. Acta 280, 415.
HOLMAN, R. T. (1970) Progress Chem. Fats 9, 611.
HOWLING, D., MORRIS, L. J., GURR, M. I. & JAMES, A. T. (1972) Biochim. Biophys. Acta 260, 10.
HALL, S. & JAMES, A. T. (1976) (Unpublished observations)
JAMES, A. T. (1973) : in "Current Trends in the Biochemistry of Lipids", edited by J. Ganguly & R.M.S. Smellie and published by Academic Press, p. 49.
JACOB, J. & GRIMMER, G. (1971) Hoppe Seyler's Z.Physiol. Chem. 352, 1445.
KLENK, E. & TSCHOPE, G. (1963) Z. Physiol. Chem. 334, 193.
KLENK, E. (1965) Adv. in Lipid Res. 3, 1.
KLENK, E. (1963) Z. Physiol. Chem. 331, 50.
MEAD, J. F. & SLATON, W. H. Jr. (1956) J. Biol. Chem. 219, 705.
MEAD, J. F. & HOWTON, D. R. (1957) J. Biol. Chem. 229, 575.
NUGTEREN, D. H. (1965) Biochim. Biophys. Acta 106, 280.
POLLARD, M., GUNSTONE, F., MORRIS, L. J. & JAMES, A. T. (1976) (In preparation).
RAHM, J. J. & HOLMAN, R. T. (1964) J. Lipid Res. 5, 169.
SCHLENK, H., GELLERMAN, J. L. & SAND, D. M. (1967) Biochem. Biophys. Acta 137, 420.
SCHLENK, H., SAND, D. M. & SEN, N. (1964) Biochim. Biophys. Acta 87, 361.
SCHLENK, H., GERSON, T. & SAND, D. M. (1969) Biochim. Biophys. Acta 176, 740.
SCHLENK, H., SAND, D. M. & GELLERMAN, J. L. (1969) Biochim. Biophys. Acta 187, 201.
SCHLENK, H. & SAND, D. M. (1967) Biochim. Biophys. Acta 144, 305.
SPRECHER, H. (1967a) Lipids 2, 122.
SPRECHER, H. (1967b) Biochim. Biophys. Acta 144, 296.
SPRECHER, H. (1968) Lipids 3, 14.
STEARNS, E. M., RYSARY, J. A. & PRIVETT, O.S. (1967) J. Nutr. 93, 485.

STOFFEL, W. (1966) Naturwiss 24, 621.
STOFFEL, W. & SCHEID, A. (1967) Hoppe Seyler's Z. Physiol. Chem. 348, 205.
STOFFEL, W. (1973) Z. Physiol. Chem. 333, 71.
THOMASSON, H. J. (1953) Intern. Z. Vitaminforsch 25, 62.
THOMASSON, H. J. (1962) Nature 194, 973.
ULLMAN, D. & SPRECHER, H. (1971) Biochim. Biophys. Acta 248, 186.
VERDINO, B., BLANK, M. L., PRIVETT, O. S. & LUNDBERG, W. O. (1964) J. Nutr. 83, 234.

REGULATION OF THE Δ9 DESATURATION

O. Mercuri* and M. E. De Tomás*

Cátedra de Bioquímica, Instituto de Fisiología
Facultad de Ciencias Médicas
Universidad Nacional de La Plata, La Plata, Argentina

INTRODUCTION

Rat liver microsomes have the capacity to convert stearoyl CoA to oleyl CoA by an enzymatic reaction requiring both oxygen and the integrity of the NADH-linked microsomal electron transport chain of which Δ9 desaturase is the terminal component (Wilson et al., 1967).

Apparently, the capacity of the desaturation enzyme system to convert saturated into monounsaturated fatty acids depends on the amount of the terminal protein component and its control is mediated by protein synthesis and degradation (Oshino and Sato, 1972). This fact could account for the liver's adaptability to different physiological conditions in which a definite microsomal desaturation activity is required. Probably one of the best examples of this is the response of the Δ9 desaturase enzyme to fasting and refeeding.

Apparently, liver desaturation activity results in a fairly constant ratio of saturated/monounsaturated fatty acids incorporated into lipids, thus maintaining the biological functionality of the cellular lipoproteic structures.

De Tomás et al. (1975) demonstrated that feeding or refeeding a high carbohydrate diet to rats was associated with an increase in liver lipogenesis, suggesting that these metabolic activities

---

*Members of the Carrera del Investigador del Consejo Nacional de Investigaciones Científicas y Técnicas.

are interconnected. Since the saturated/monounsaturated fatty acid ratio was maintained by the liver cell and saturated fatty acids are the end product of *de novo* fatty acid synthesis, it is logical to suppose that the increase in the saturated/monounsaturated fatty acid ratio stimulates Δ9 desaturase activity, probably by *de novo* enzyme synthesis by a still unknown mechanism.

Gellhorn and Benjamin (1966) have demonstrated that oxidative desaturation of saturated fatty acids becomes depressed in the diabetic state and that this enzymatic impairment is reversed by insulin. Since insulin can also reverse the diabetic depressed lipogenic activity, it is fair to assume that the insulin-induced rise in hepatic desaturase activity could be mediated by the increase in *de novo* fatty acid synthesis. Mercuri et al. (1974) demonstrated that dietary fructose or glycerol was able to partially restore the Δ9 desaturase activity depressed by the diabetic state. Since utilization of these carbohydrates by the liver is not insulin dependent (Takeda et al., 1967 and Howard and Lowenstein, 1967) and the fatty acid synthetase activity is increased by a fructose or glycerol supplemented diet (Volpe and Vagelos, 1974 and Bruckdorfer et al., 1972), further research in this area may increase our understanding of these metabolic relationships.

MATERIAL AND METHODS

|1-^{14}C| stearic acid (54.0 Ci/mole, 99% radiochemically pure) was purchased from the Radiochemical Centre, Amersham, England. Stearic acid (99% pure) was obtained from the Lipid Preparation Laboratory, Hormel Institute, U. S. A. Streptozotocin was kindly provided by Upjohn Laboratories, U. S. A. Collagenase Type I was purchased from Sigma, U. S. A. and hyaluronidase was kindly provided by Roux-Ocefa, Argentina.

Male Wistar rats weighing 150-170 g were used throughout the experiment. The animals were maintained on a Purina chow diet and water *ad libitum* before use. Diabetes was produced by the intravenous injection of streptozotocin at a dose of 50 mg/kg body wt.

Stearoyl CoA desaturase was induced by refeeding starved rats (Strittmatter et al., 1974).

Enzyme induction by dietary glycerol and *in vivo* assay for desaturase activity have previously been described (Mercuri et al., 1974).

The incubation procedure for |1-^{14}C| stearic acid desaturation was performed as previously described (De Tomás et al., 1973) except that 25 mg of soluble protein (100,000 xg supernatant) was added to the incubation solution.

Isolation of parenchymal cells from rat liver was performed according to Berry (1974) except that the collagenase-hyaluronidase solution was not recirculated. Cell count was determined using a hemocytometer, and viability was estimated by 0.2% trypan blue staining.

Hepatocytes (15 to 20 x 10^6 cells) were incubated in 50 ml of Hanks buffer containing 20 mM glucose or glycerol and 10 μC $|1-^{14}C|$ stearic acid-albumin complex (molar ratio 2.0) at 25°C.

Aliquots of incubation medium were taken each time and centrifuged at 500 xg for 2 min. The supernatants were recovered and saved. The cells were resuspended in Hanks buffer and centrifuged at 3,000 xg for 5 min. Lipids from the cellular pellets were extracted by the method of Folch-Pi et al. (1957), saponified, esterified, and assayed for Δ9 desaturase activity with TLC-$AgNO_3$ (De Tomás and Brenner, 1964). Radioactivity assays was determined in a Packard scintillation spectrophotometer.

RESULTS AND DISCUSSION

The liver Δ9 desaturase activity measured *in vivo* in control, refed, and diabetic rats is given in Table 1. Clearly, the fatty acid desaturase enzyme is significantly depressed in the diabetic state. These results agree with previous data (Mercuri et al., 1974). On the other hand a very significant increase in the liver's capacity to desaturate stearic acid *in vivo* was elicited by feeding animals a fat-free diet.

The initial desaturation and incorporation of the injected labeled stearic acid takes place very rapidly after the assimilation of the precursor acid from the blood into the liver cells, indicating that enzymatic fatty acid activation to acyl CoA proceeds at a very high rate. On the other hand, because of the very rapid turnover of plasma-free fatty acids, the supply of labeled precursor acid to the ester pools is rapidly exhausted.

Apparently, in less than five minutes the injected labeled stearic acid is already desaturated and incorporated into lipids (Elovson, 1965).

Since the specific radioactivity of $|1-^{14}C|$ stearoyl CoA would be similar to the specific radioactivity of $|1-^{14}C|$ oleyl CoA (Holloway and Holloway, 1974), the relative distribution of the $|1-^{14}C|$ radioactivity in the stearic and oleic acid incorporated into lipids could account for the liver capacity to desaturated *in vivo* stearic acid. Nevertheless, the loss of oleic acid and its precursor, stearic acid, from the liver, excreted as lipoproteins

Table 1. *In Vivo* Desaturation of $|1-^{14}C|$ Stearic Acid in Liver from Control, Refed, and Diabetic Rats

Numbers in parentheses indicate the numbers of animals in each group. Probability (P) values are related to control group. Data are the means ± S.D.

Liver	Percentage of conversion C18:0 → C18:1
Control (5)	14.2 ± 0.5
Refed (3)	42.0 ± 5.8 P < 0.001
Diabetic (4)	3.6 ± 0.8 P < 0.001

into the plasma, could modify these values. On the other hand, the values of the *in vitro* Δ9 desaturase activity measured in the microsomal fraction of control, refed, and diabetic rats shown in Table 2 indicate that under our experimental conditions these values resemble those obtained in experiments *in vivo* (Table 1).

Table 2. *In Vitro* Oxidative Desaturation of Stearic Acid to Oleic Acid by Liver Microsomes of Control, Refed, and Diabetic Rats

Numbers in parentheses indicate the number of animals in each group. Probability (P) values are related to each control group. Data are the means ± S.D.

Liver	Percentage of conversion C18:0 → C18:1
Control (5)	15.8 ± 1.7
Refed (3)	35.6 ± 4.8 P < 0.001
Diabetic (4)	9.3 ± 2.7 P < 0.01

The activity of the Δ9 desaturase enzyme could be modified by dietary conditions. The enzyme activity which had been lowered by starvation is rapidly induced to a very high level by refeeding. The response of the enzyme can be blocked by cycloheximide administration, indicating that protein synthesis is involved in the control of the enzyme activity, probably at the level of the terminal component of the desaturation enzyme system (Oshino and Sato, 1972). To elucidate the intimate mechanism that promotes the enzyme induction we used the ability of isolated liver cells to desaturate stearic acid.

The incubation of the isolated hepatocytes was performed at 25°C to minimize fatty acid oxidation without greatly affecting fatty acid desaturation and incorporation.

Figure 1 shows the total $|1\text{-}^{14}C|$ radioactivity and $|1\text{-}^{14}C|$ oleic acid radioactivity incorporated into cellular lipids of control and refed rat livers at different times.

Evidently, the $|1\text{-}^{14}C|$ stearic acid was progressively removed from its albumin complex, desaturated, and incorporated by isolated liver cells from control and refed rats. The results clearly show not only an increase in the total radioactivity incorporated in the refed rats but also an increase in the amount of oleic acid radioactivity present at different times.

In early experiments (Mercuri et al., 1974) we demonstrated that dietary glycerol was able to restore the Δ9 desaturase activity depressed by the diabetic state.

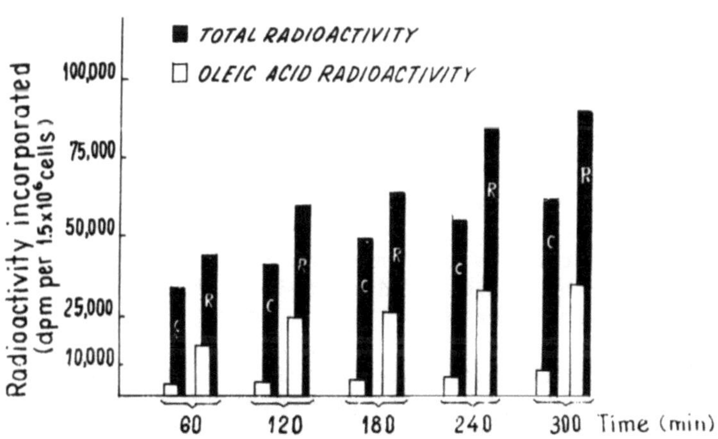

Fig. 1. Total radioactivity and oleic acid radioactivity incorporated into lipids from isolated hepatocytes of control and refed rats.

Each bar is the mean of duplicate samples.

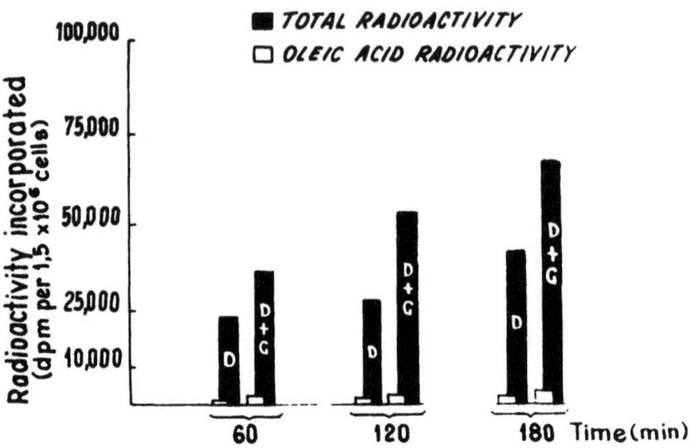

Fig. 2. *In vitro* effect of glycerol on the total radioactivity and oleic acid radioactivity incorporated into lipids from diabetic rat isolated hepatocytes.

Each bar is the mean of duplicate samples.

In the present experiment we tried to induce enzyme activity by adding glycerol to the incubation medium containing isolated diabetic rat liver cells.

Figure 2 shows the increase of the total $|1\text{-}^{14}C|$ radioactivity incorporated into the cell of the lipids at different times. Evidently, the presence of glycerol in the incubation medium stimulates not only fatty acid incorporation but also stearic acid desaturation. The results of several experiments lead us to the conclusion that the significant increase in the amount of stearic acid converted to oleic acid is promoted by glycerol in the isolated diabetic rat liver cells.

The *in vitro* and *in vivo* effects of glycerol upon the enzymatic capacity of the isolated diabetic hepatocyte to convert stearic acid to oleic acid is demonstrated in Fig. 3.

Particularly noteworthy is the increase in the amount of oleic acid incorporated into the lipids of isolated cells, evoked by dietary glycerol on diabetic rats.

The increase in the Δ9 desaturase capacity of isolated diabetic hepatocytes promoted by glycerol could be explained by the increase in the rate of fatty acid desaturation promoted by the enhancement of stearoyl CoA formation and/or by the removal of oleyl CoA by its esterification into lipids, since oleyl CoA is a competitive inhibitor of the desaturation reaction (Oshino et al., 1966).

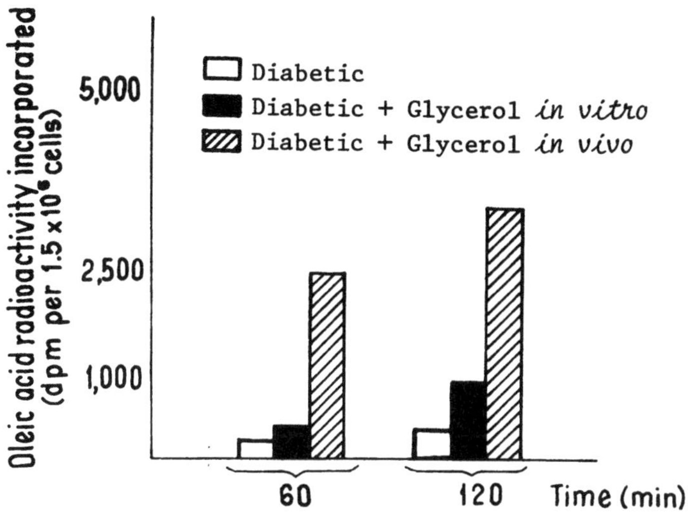

Fig. 3. *In vitro* and *in vivo* effect of glycerol on the oleic acid radioactivity incorporated into lipids from diabetic rat isolated hepatocytes.

Each bar is the mean of duplicate samples.

CONCLUSIONS

The decrease of Δ9 desaturase activity by the diabetic state and the positive response of normal rat liver microsomal desaturation activity by feeding after fasting were corroborated by *in vivo* and *in vitro* experiments.

The *in vivo* assay for Δ9 desaturase activity permits study, under physiological conditions, of the enzyme's adaptability to different nutritional and hormonal conditions.

The *in vivo* relative distribution of the $|1\text{-}^{14}C|$ radioactivity in the stearic and oleic acid incorporated into lipids would reflect liver Δ9 desaturase capacity if we assume that the specific $|1\text{-}^{14}C|$ radioactivity of the stearoyl CoA is similar to the specific radioactivity of the oleyl CoA esterified into lipids.

The microsomal-recombined fraction was useful to study liver Δ9 desaturase enzyme activity. The added soluble protein from the 100,000 xg supernatant fraction assures an adequate interaction between the substrate and the stearoyl CoA generating system.

The measurement of Δ9 desaturase activity in isolated rat hepatocytes is a useful method of studying in a simplified assay system the changes in the enzyme activity promoted by dietary conditions.

The increase in the amount of $|1-^{14}C|$ oleic acid incorporated into lipids of diabetic isolated rat liver cells promoted by glycerol in the incubation medium would reflect the increase in the ability of Δ9 desaturase to desaturate stearic acid by the enhancement of substrate formation and/or by the removal of oleyl CoA through esterification into lipids.

Apparently glycerol stimulates Δ9 desaturase capacity by enhancing cell lipogenesis through the increase of the saturated fatty acid synthesis and esterification promoted by α-glycerophosphate formation.

The simultaneous increase in the ability of the isolated cells to desaturate and esterify fatty acids suggests that both metabolic activities respond to the same regulatory factors or, at least, that Δ9 desaturase activity is promoted by the increase in the rate of fatty acid esterification.

REFERENCES

Berry, M. N. (1974) Methods in Enzymology. Vol. XXXII. Part B. (Fleischer S. & Packer L., ed.) pp. 625-632. Academic Press, London-New York.

Bruckdorfer, K. R., Khan, H. I., and Yudkin, J. (1972) Biochem. J. 129, 439-446.

De Tomás, M. E. and Brenner, R. R. (1964) Anal. Assoc. Quím. Arg. 52, 253-260.

De Tomás, M. E., Peluffo, R. O., and Mercuri, O. (1973) Biochim. Biophys. Acta 306, 149-155.

De Tomás, M. E., Mercuri, O., and Peluffo, R. O. (1975) Lipids 10, 360-362.

Elovson, J. (1965) Biochim. Biophys. Acta 106, 291-303.

Folch-Pi, J. I., Lees, M., and Sloan-Stanley, G. H. (1957) J. Biol. Chem. 226, 479-509.

Gellhorn, A. and Benjamin, W. (1966) Biochim. Biophys. Acta 116, 460-466.

Holloway, C. T. and Holloway, P. W. (1974) Lipids 9, 196-200.

Howard, C. F. Jr. and Lowenstein, J. M. (1967) Biochim. Biophys. Acta 84, 226-228.

Mercuri, O., Peluffo, R. O., and De Tomás, M. E. (1974) Biochim. Biophys. Acta 369, 264-268.

Oshino, N., Imai, Y., and Sato R. (1966) Biochim. Biophys. Acta 128, 13-28.

Oshino, N. and Sato, R. (1972) Arch. Biochem. Biophys. 149, 369-377.

Strittmatter, P., Spatz, L., Corcoran, D., Rogers, M. J., Setlow, B., and Redline, R. (1974) Proc. Nat. Acad. Sci. U. S. A. 71, 4565-4569.

Takeda, Y., Inoue, H., Honjo, H., Tanioka, H., and Daikuhara, Y. (1967) Biochim. Biophys. Acta 136, 214-222.

Uchiyama, M., Nakagawa, M., and Okui, S. (1967) J. Biochem. 62, 1-6.

Volpe, J. J. and Vagelos, R. P. (1974) Proc. Nat. Acad. Sci. U. S. A. 71, 889-893.

Waddell, M. and Gallon, H. J. (1973) J. Clin. Invest. 52, 2725-2731.

Wilson, A. C., Wakil, S. J., and Joshi, V. C. (1976) Arch. Biochem. Biophys. 173, 154-161.

ACKNOWLEDGMENTS

Technical assistance was provided by C. T. Serres and A. J. Ubici. This investigation was supported in part by grants from the Consejo Nacional de Investigaciones Científicas y Técnicas, Argentina.

REGULATORY FUNCTION OF Δ6 DESATURASE -- KEY ENZYME OF POLYUNSATURATED FATTY ACID SYNTHESIS

R. R. Brenner *

Cátedra de Bioquímica, Instituto de Fisiología, Facultad de Ciencias Médicas, Universidad Nacional de La Plata, La Plata, Argentina

INTRODUCTION

Polyunsaturated fatty acids are synthesized in animals by a sequence of desaturation and elongation reactions which, as we know now, follow special rules. The starting substrates are CoA thioesters of palmitoleic, oleic, linoleic and α linolenic acids. The acyl-CoA synthesis is faster than elongation or desaturation reactions (Marcel and Suzue, 1972). The enzymes involved in the elongation are found either in the endoplasmic reticulum or in mitochondria but the specificity of each group is different. The microsomal system prefers Δ6 unsaturated acyl-CoA and hexanoyl CoA whereas the mitochondrial prefers medium chain acyl-CoA (Podack et al, 1974; Hinsch and Seubert, 1975).

The fatty acid desaturases are exclusively microsomal. Three types of desaturation enzymes involved in polyunsaturated acid biosynthesis have been characterized: the Δ6, the Δ5 and the Δ4. They form the new double bond between carbons 6-7, 5-6 and 4-5 respectively. Δ9 desaturation mainly converts saturated acids to monoethylenic acids. The Δ6, Δ5 and Δ4 desaturations are produced by different enzymes (Brenner, 1974; Ninno et al, 1974; Ayala et al, 1973; Alaniz et al, 1975; Dunbar and Bailey, 1975).

Theoretically the biosynthesis of polyunsaturated fatty acids may begin either with a Δ6 desaturation followed by an elongation, a Δ5 desaturation, again an elongation and a final Δ4 desaturation, or start with an elongation followed by a Δ8 desaturation, a Δ5

* Member of the Carrera del Investigador Científico del Consejo Nacional de Investigaciones Científicas y Técnicas.

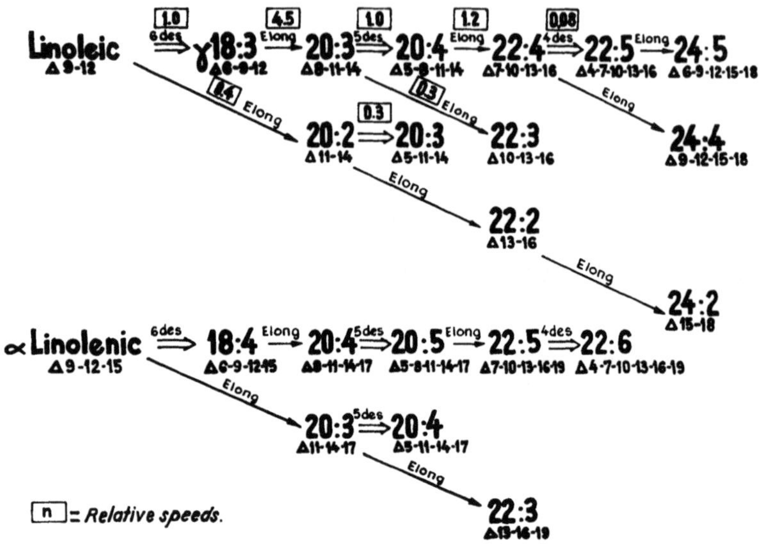

Fig. 1. Recognized pathways in the biosynthesis of polyunsaturated acids in the rat.

Relative activities were calculated from data of Bernert and Sprecher, 1975.

desaturation, etc. (Fig. 1). The existence of the first route is well proved and is the principal if not the unique way that arachidonic acid is synthesized from linoleic acid in rat liver (Marcel et al, 1968). This last conclusion is admissible since Sprecher's group (Ullman and Sprecher, 1971b, Sprecher and Lee, 1975) has shown that the elongation products of linoleic or oleic acid 20:2 (11, 14) and 20:1 (11) respectively are not converted to 20:3 (8, 11, 14) and 20:2 (8, 11) (Fig. 1). Instead they are desaturated by a Δ5 desaturase to 20:3 (5, 11, 14) and 20:2 (5, 11). These last acids are not converted to arachidonic acid and eicosa-5, 8, 11-treinoic acid respectively. Therefore the existence of a Δ8 desaturase is doubtful. Similar results were obtained by Alaniz et al (1976) in Hepatoma Tissue cultured 7288 HTC cells incubated with α-linolenate.

Therefore the biosynthesis of the important fatty acids of the animals: arachidonic, 22:4 (7, 10, 13, 16) and 22:5 (4, 7, 10, 13, 16) acids of linoleate series, 20:5 (5, 8, 11, 14, 17) and 22:6 (4, 7, 10, 13, 16, 19) acids of α-linolenate family, and 20:3 (5, 8, 11) acid of oleate family is started by the Δ6 desaturase (Fig. 1)

The comparative speeds of the elongation and desaturation

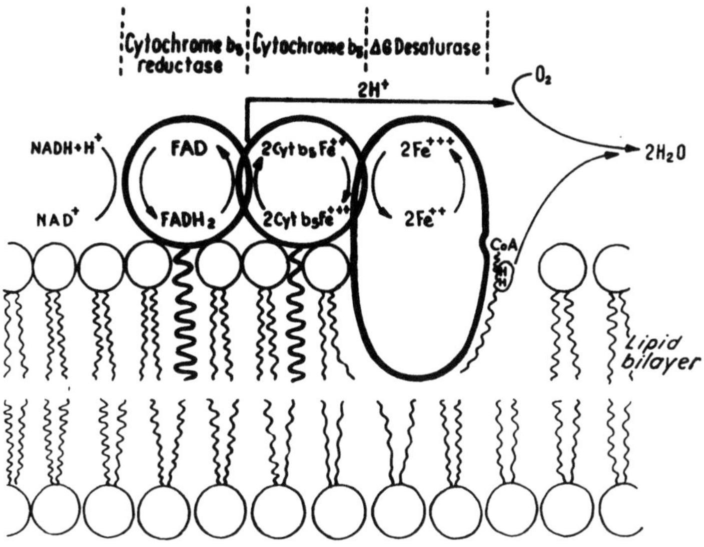

Fig. 2. Hypothetical structure of the components of Δ6 desaturation system bound to the microsomal lipid belayer.

reactions involved in linoleate conversion to higher homologs were determined by Bernert and Sprecher (1975). The values calculated from Bernert and Sprecher's work (1975) show (Fig. 1) that Δ6 desaturation of linoleate to γ-linolenate is faster than linoleate elongation to 20:2 (11, 14) and therefore is the preferred pathway. Besides Δ6 desaturation is slower than γ-linolenate elongation to 20:3 (8, 11, 14) and similar to its Δ5 desaturation to arachidonate. Therefore Δ6 desaturation is the pass-marker of arachidonic acid synthesis.

Elongation of arachidonic acid to 22:4 (7, 10, 13, 16) and the Δ4 desaturation to 22:5 (4, 7, 10, 13, 16) occur in liver in a very limited extent but may be important in adrenals and testis (Ayala et al, 1973). Similar conversion to 20:5 (5, 8, 11, 14, 17) acid that belongs to α-linolenic family to acids of 22 carbons is important in brain and it also requires a Δ4 desaturation. The Δ4 desaturase is very little active (Bernert and Sprecher, 1975; Ayala et al, 1973) and might be in consequence a second pass-marker.

Therefore polyunsaturated fatty acids would be synthesized in two big steps (Ayala et al, 1973): the first one starts with a Δ6 desaturation and forms arachidonate (linoleate family), 20:5 (5, 8, 11, 14, 17) (α-linolenate family) and 20:3 (5, 8, 11) (oleate family).

These acids are incorporated preferentially in phospholipids. The second step would be important in some tissues and it forms the polyunsaturated acids of 22 and 24 carbons that are incorporated in phospholipids and triglycerides.

Without discussing the importance of retro-conversion reactions it is obvious that the regulation of Δ6 desaturase activity would control all polyunsaturated fatty acid biosynthesis due to the key position in the pattern. (Fig. 1).

The Δ6 desaturase is a membrane bound enzyme (Catalá et al, 1975) that is inhibited by cyanide (Brenner and Peluffo, 1969). It is considered a mixed function oxidase. Both NADH and NADPH are electron donors and cytochrome b_5 reductase and cytochrome b_5 transport the electrons to the enzyme. Two hydrogens of the reduced cofactors and two hydrogens of the fatty acids are combined to oxygen to form 2 molecules of water. Acyl-CoAs are substrates for the enzyme. The full activity of the enzyme b ound to the microsome is reached when a cytosolic factor of proteical structure is present (Brenner, 1974; Catalá et al, 1975; Nervi et al, 1975). A scheme of the possible structure of the membrane bound Δ6 desaturation system is shown in Fig. 2.

REGULATORY MECHANISMS

Competition

In previous works we have already shown that a rapid modification of Δ6 desaturation of oleic, linoleic, and α-linolenic acid is evoked by mutual competition as well as by inhibition with products of reaction and higher homologs of the same or different family (Brenner and Peluffo, 1966; Brenner, 1974). These competitions and the increase of the speeds of the Δ6 desaturation from oleic to linoleic and linoleic to α-linolenic acid plays an important role in determining the fatty acid composition of animal tissues. It was already proved in experiments in vivo that these mechanisms are used by the animal tissue to retroconvert the typical fatty acid composition of essential fatty acid deficient animals to the normal composition when linoleate is again provided in the diet (Brenner and Nervi, 1965). However this mechanism is not specific of the Δ6 desaturase since the Δ5 desaturation of 20 carbon fatty acids in a further step of the biosynthetic chain is also modified by similar competitive reactions (Castuma et al, 1974; Ullman and Sprecher, 1971a). Competitive effects are also shown with the elongation reaction.

Other factors that have already been shown to modify in vitro the Δ6 desaturation of fatty acids are the concentration of ATP, (Brenner and Catalá, 1971) and acceptors of acyl CoA as lysophospholipids and glycerolphosphate (Nervi et al, 1968; Brenner and Peluffo, 1966).

However in order to recognize the existence of a key enzyme in the regulation of metabolic reactions, the enzyme must not only modify the activity with changes of substrate and product concentration but also by dietary factors and hormones. These two conditions are fulfilled for the Δ6 desaturase.

Dietary factors

Fasting decreases the Δ6 desaturation activity of rat liver microsomes and refeeding recovers the activity. This process is considered to be related with protein synthesis since it is abolished by actinomycin D injection (Brenner et al, 1968). However it is not specific of this enzyme since Δ9 desaturation and de novo synthesis of fatty acids are also modified.

Rats fed during 24 h on a hyperproteic diet (>35% protein) increase the Δ6 desaturase activity of liver microsomes enhancing the Vmax of the enzyme but not the Km (Peluffo and Brenner, 1974). This effect is less important with Δ5 desaturase and is not shown with the microsomal elongation of fatty acids and with the Δ9 desaturase (Gómez Dumm et al, 1972). Dietary glucose or fructose decreases the Δ6 desaturase activity of liver microsomes but increases the Δ9 desaturation (Brenner, 1974). However, fasted rats that are refed on a glucose diet reactivate the Δ6 desaturation for a short time. After this period, the activity falls again (Gómez Dumm et al, 1970). This transcient increase has been interpreted as the effect of insulin secretion followed by an active glycolisis (Brenner, 1974) since insulin injection reactivates the diminished Δ6 desaturase activity of diabetic rats (Mercuri et al, 1966). However insulin injection to normal rats does not modify the activity of the Δ6 desaturase (Peluffo et al, 1971).

Since the direct effect of the dietary variables: protein, and glucose may be masked in the whole animal by hormonal secretions, similar experiments were performed in cultured cells (Alaniz and Brenner, 1976). HTC 7288 C cells were cultured during 12 h in swim's 77 medium. After this period $|1-^{14}C|$ α-linolenic acid was added together with glucose or lactalbumin hydrolizate in a concentration of 3.6 g ‰ and cultured for 12 h. Other cells were cultured during the same period without the addition of carbon sources. These cells are called "fasting cells". The analysis of the distribution of the radioactivity among the different unsaturated fatty acids is shown in Fig. 3. As it was already discussed and shown in Fig. 1 HTC cells convert α-linolenic acid to higher homologs by two pathways. In "fasted" cells 96.1% of the radioactivity

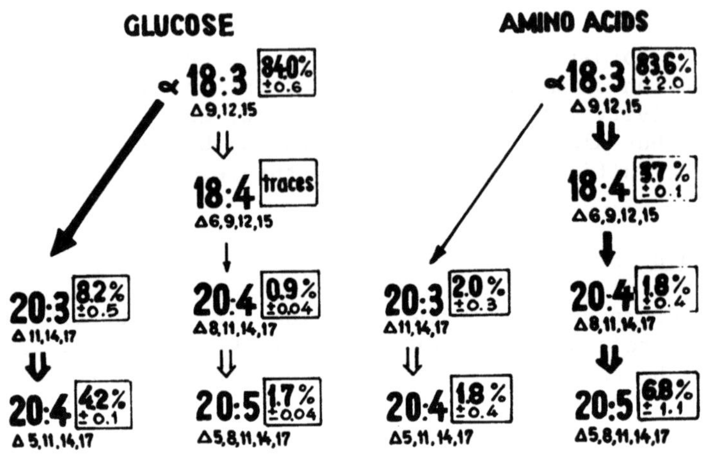

Fig. 3. Comparative effect of glucose and lactalbumin aminoacids on α-linolenate conversion to higher homologs by HTC cells.

5×10^6 cells were incubated in monolayer during 12 h with 0.5 nmoles of $|1\text{-}^{14}C|$ α-linolenate per million cells in Swim's 77 medium supplemented with 3.6 g ‰ of tested carbon sources.

incorporated in the cells remained in α-linolenic acid, only traces were detected in 18:4 (6, 9, 12, 15), 0.5 % in 20:4 (8, 11, 14, 17) and 1.5 % in 20:5 (5, 8, 11, 14, 17). All these acids belong to the normal route of synthesis. In the other route 1.0 % was detected in 20:3 (11, 14, 17) and 0.6 % in 20:4 (5, 11, 14, 17). Addition of glucose to the medium changes the patterns of distribution increasing the elongation route but not the desaturation pathway. Lactalbumin aminoacids evoke a quite different effect: they increase the Δ6 desaturation and the normal route of 20:5 acid synthesis.

Therefore these results confirm definitively that proteins activate Δ6 desaturase and carbohydrates evoke an antagonic effect without any hormonal mediation.

Diets deprived of essential fatty acids (EFA) enhance the Δ6 desaturation activity of rat liver or testis microsomes. Changes of fatty acid composition with decrease of linoleic and arachidonic acids and increase of oleic and 20:3 (ω9) acids are shown as early as 3 days after EFA deprivation (Table 1). The changes evoke a decrease of the double bond index: saturated acid ratio from 2.2 to

Table 1. Effect of EFA deficiency in lipid composition and linoleate Δ6 desaturation activity of rat liver microsomes.

Sufficient and deficient diets contain 3 g % sunflower seedoil and 3 % hydrogenated coconut oil respectively. [1] Other fatty acids make for 100 % .

Total fatty acids	Diet. Sufficient	Deficient 3 days	Deficient 15 days
18:1 ω9 (%)[1]	11.1	18.0	24.0
18:2 ω6 (%)	11.4	4.1	4.9
20:3 ω9 (%)	---	2.7	11.0
20:4 ω6 (%)	21.9	9.9	9.3
Double bond index / saturated	2.2	1.2	2.2
Triacylglycerol / Phosphatidyl choline	0.4	0.4	0.6
Vm (nmol/min per mg prot.)	0.1	0.1	0.4
ap. Km (10^{-5}M)	2.8	2.6	10.7

1.2. However the activation of the Δ6 desaturation is shown later on. At 15 days it is very important. The apparent Km and Vm of the enzyme are modified. This activation recovers the double bond index: saturated acid ratio to 2.2 and is correlative to an increase of the triacylglycerol: phosphatidyl choline ratio of the microsomes. Undoubtedly it is not correlative to the modification of the fatty acid composition of the membrane. Besides Ayala and Brenner[1] have shown that the effect is not due to substrate deprivation (linoleate or α-linolenate) since rats fed on diets containing fish oil during 4 or 6 weeks have even lower Δ6 desaturation activity in liver microsomes than animals fed on sunflower seed oil compared to rats fed on EFA free diets (Table 2). Therefore the increase of the Δ6 desaturase activity in EFA deficiency is a physiological response of the cell to maintain the unsaturated/saturated acid ratio and fluidity of the membrane.

The exact mechanism of Δ6 desaturase activation by EFA deficiency is not established yet but since it evokes an increase

---

[1] Ayala S. & Brenner R. R. (private comunication)

Table 2. Comparative effect of EFA deficient diets supplemented with fish oil, sunflower seed oil and methyl palmitate in the Δ6 desaturation (%) of linoleic and α-linolenic acid by rat liver microsomes.

80 nmoles |1-^{14}C| acids incubated 20 min at 35°C with 5 mg microsomal protein in 1.5 ml of a solution containing 1.25 μmoles of NADH, 0.1 μmoles of CoA, 4 μmoles of ATP, 5 μmoles of $MgCl_2$ and 62.5 μmoles phosphate buffer pH 7.0.
Fish oil contained: 20.2 % of 18:1 ω9, 2.0% of 18:2 ω6, 1.1% of 20:4 ω6, 7.5% of 20:5 ω3, 2.7% of 22:5 ω3 and 16.9% of 22.6 ω3.
Results expressed as % conversion ± SEM. [1] Period of diet administration after weaning.

	EFA deficient diet					
Substrates	+Palmitate		+Sunflower oil		+Fish oil	
	4 weeks[1]	6 weeks	4 weeks	6 weeks	4 weeks	6 weeks
18:2 ω6	20.0 ±0.7	25.8 ±1.0	16.3 ±0.3	15.7 ±0.5	11.6 ±0.3	10.8 ±0.7
18:3 ω3	33.3 ±0.9	45.0 ±0.7	24.8 ±0.2	28.6 ±0.2	18.9 ±0.5	19.6 ±0.5

of the V-max it may indicate enzymatic induction. EFA deficiency also enhances the Δ9 desaturation thereby the possibility a common mechanism exists for both enzymes. Considering (Fig. 2) that either the Δ6 or the Δ9 desaturation systems are constituted by components attached to the lipid belayer of the endoplasmic reticulum it is possible to suppose some connection between lipid composition of the membrane and desaturation activity. However Holloway and Holloway (1975) showed that the fractionation and reassemblage of the stearyl-CoA desaturating system by addition of lipids with different fatty acid composition do not alter the activity of the reaction. These results agree with data in Table 1.

Hormones

The effect of hormones on the Δ6 desaturation of linoleic, α-linolenic and oleic acids was shown as early as 1966 (Mercuri et al, 1966, 1967). It was found that alloxan diabetes depressed the Δ 6 desaturase activity of rat liver microsomes and insulin injection corrected the defect in less than 38 h. The simultaneous injection of actinomycin D or Puromycin that abolished protein synthesis imparred the reparative effect of insulin and suggested that insulin effect might involve Δ6 desaturase induction. However

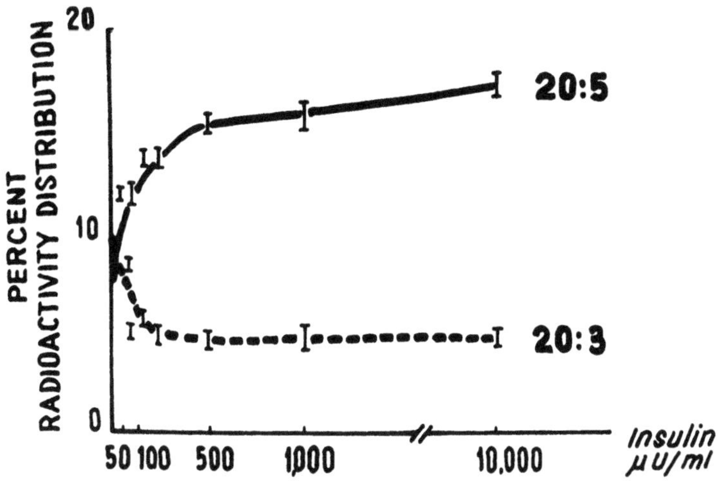

Fig. 4. Effect of insulin concentration on linolenic acid conversion to 20:5 (5, 8, 11, 14, 17) acid and 20:3 (11, 14, 17) acid by HTC cells.

Same concentration of cells and $|1\text{-}^{14}C|$ α-linolenate as in Fig. 3 incubated in monolayer in Swim's 77 medium during 24 h

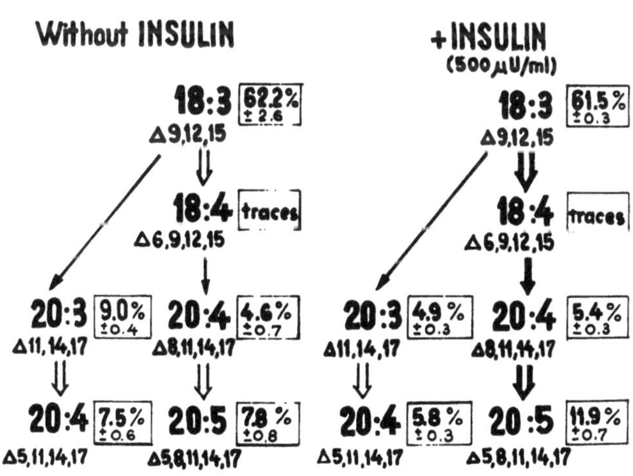

Fig. 5. Effect of insulin on α-linolenate conversion to higher homologs by HTC cells.

Same concentration of cells and $|1\text{-}^{14}C|$α-linolenic acid as in Fig. 3 incubated for 24 h in monolayer in Swim's 77 medium. Insulin concentration 500 μU/ml the effect of insulin on Δ6 desaturation is not so dramatic as with Δ9 desaturation. The problem arises in the dual effect of insulin since insulin enhances glycolisis as well, and we have shown that glucose metabolism depresses Δ6 desaturation and activates Δ9 desaturation (Brenner, 1974). Since many hormonal and regulatory mechanisms present in animals may mask the specific effect of insulin in polyunsaturated acid biosynthesis we decided to investigate the effect in cultured HTC cells (Alaniz et al, 1976).

HTC cells incubated in the presence of $|1\text{-}^{14}C|$ linolenic acid in Swim's 77 medium during 24 h were sensitive to insulin addition. A concentration effect was found (Fig. 4). However, insulin evoked different effects on the two routes of α-linolenic transformation. It increased the Δ6 desaturation route that leads to 20:4 (8, 11, 14, 17) and 20:5 (5, 8, 11, 14, 17) acids and depressed the elongation pathway that forms 20:3 (11, 14, 17) and 20:4 (5, 11, 14, 17) acids (Fig. 5).

Therefore these results compared to the others already mentioned confirm that the Δ6 desaturation is sensitive to insulin concentration enhancing polyunsaturated fatty acid biosynthesis.

Since insulin enhances Δ6 desaturation of fatty acids we suspected that hyperglucemic hormones should depress the activity through a kind of yin-yang mechanism.

The inhibitory effect of glucagon on the Δ6 desaturation activity of rat microsomes was demonstrated by Gómez Dumm et al (1975). As was already shown rats fasted for 48 h have low Δ6 desaturation of linoleic to γ-linolenic acid and α-linolenic to 18:4 (6, 9, 12, 15) acid. Refeeding a fat-free diet enhances the microsomal desaturation activity. However the simultaneous administration of glucagon (200 μg/8h/per 100 g of body weight) abolishes the reactivation of Δ6 desaturase (Table 3). This effect is shown with either linoleate or γ-linolenate but not with Δ9 desaturation of stearate or Δ5 desaturation of eicosatrienoate (8, 11, 14) (Gómez Dumm et al, 1975).

Since glucagon activates adenyl-cyclase, which converts ATP to cyclic-AMP it is possible to speculate that the inhibitory effect of glucagon on the Δ6 desaturase is produced through a rise of the intracellular level of cyclic-AMP. The data of Table 3 confirm that dibutyryl c AMP administration to refeeding rats evokes the same effect as glucagon.

Moreover the primary action of catecholamine on the liver is also the stimulation of adenyl-cyclase and cyclic-AMP formation. Therefore it is expected that epinephrine evokes a decrease of Δ6 desaturase activity. Gómez Dumm et al, (1976) have shown that an injection of 1 mg/kg body weight of epinephrine or even less

Table 3. Effect of fasting refeeding and administration of glucagon and dibutyryl-cyclic AMP on the Δ6 desaturation of |1-^{14}C| linolenic and |1-^{14}C| α-linolenic acids by rat liver microsomes.

Rats were fasted for 48 h and then refed a fat-free diet for 48 h. Glucagon was administered at a dose of 200 μg/8 h per 100 g body weight. Dibutyryl-c AMP at 5 mg/8h per 100 g of body weight and theophylline at 2 mg/8h per 100 g of body weight. Results are the means of three animals ± SEM.

Treatment	Conversion (%)	
	Linoleic	α-linolenic
Fasted	8.6 ± 0.8	26.6 ± 2.2
Refed	20.4 ± 1.3	42.1 ± 2.6
Refed+Glucagon	11.7 ± 2.2	27.6 ± 2.5
Refed+DBcAMP+theophylline	8.5 ± 0.9	28.2 ± 0.9

inhibits the microsomal activity of liver to desaturate linoleic and α-linolenic acid to γ-linolenic and 18:4 ω3 acids respectively. The effect is dose dependent and produces a decrease of the Vmax but does not modify the Km of the Δ6 desaturase (Gómez Dumm et al, 1976). A decrease of the Vmax without Km modification indicates that the amount of active Δ6 desaturase in liver microsomes has been depressed by epinephrine injection.

Curves of Fig. 6 show that an injection of epinephrine (1 mg/kg of body weight) evokes a rapid increase of c AMP (less than 30 min) followed by a depression of liver glycogen and correlative increase of plasma glucose concentration. The depression of microsomal Δ6 desaturase begins approximately 1 h later than glycogen movilization. Therefore it seems likely that epinephrine activation of adenyl-cyclase followed by an enhacement of c AMP triggers a sequence of reactions that evoke glycogen breakdown and later on Δ6 desaturase reactivation. Then comparing these results with previously discussed action of insulin and dietary glucose it is possible to suggest that the effect of c AMP may be produced by means of some glucose metabolite.

In vitro incubation of rat liver microsomes or microsomes and 100,000 xg supernatant with c AMP or c AMP and c AMP receptor protein were unable to modify the Δ6 desaturation activity of rat liver microsomes to convert linoleic to γ-linolenic acid (Table 4)

Although these results are not definitive they support the hypothesis that glucose metabolites may be involved in the depression of Δ6 desaturase activity evoked by glucagon and epinephrine through c AMP synthesis.

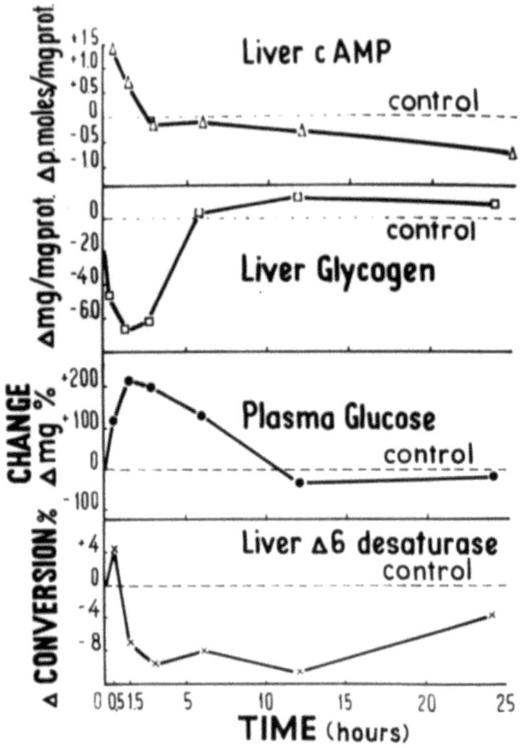

Fig. 6. Effect of epinephrine on plasma glucose concentration and concentration of glycogen, c AMP and Δ6 desaturation activity of rat liver.

Animals injected with 1 mg/kg of body wt of epinephrine at 0 time.

Table 4. Desaturation in vitro of linoleic acid to γ-linolenic by rat liver microsomes in the presence of c AMP.

5 mg microsomal protein incubated 20 min at 35°C with 100 nmoles of |1-^{14}C| linoleic acid .4 μmoles of ATP, 0.1 μmoles of CoA, 1,25 μmoles NADH and 5 μmoles of $MgCl_2$ at pH 7.0. Total volume 1.5 ml.
Supernatant = supernatant solution of 100,000 xg.
CRP = c AMP receptor protein

Additions	Conversion (%)
Microsomes	23.6
+ c AMPSupernatant	23.3
+ c AMP + CRP	21.5
+ Supernatant	27.9
+ Supernatant	21.5
+ CRP	28.2

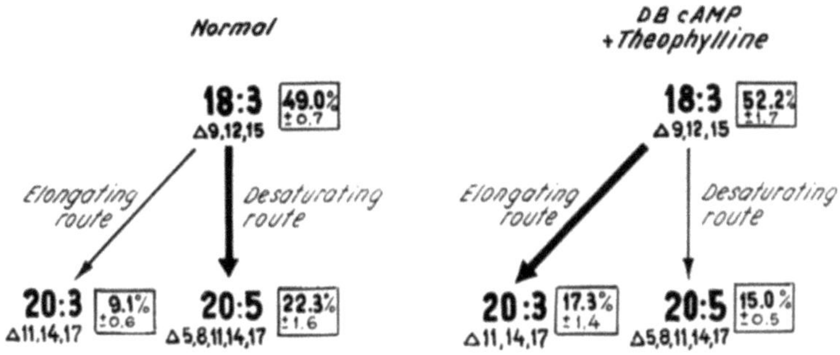

Fig. 7. Effect or dibutyryl-c AMP on the conversion of $|1-^{14}C|$ α-linolenic acid to higher homologs by HTC cells.

2.3% of 18:4 and 17.2% of both 20:4 isomers were also present in cells incubated in control medium. 1.4% 18:4 and 14.1% of both 20:4 were present in cells treated with c AMP. 5×10^6 cells were incubated during 24 h with 0.5 nmoles of $|1-^{14}C|$ α-linolenate per million cells in Swim's 77 medium. The concentrations of dibutyryl-c AMP and theophylline is 10^{-3}M for both.

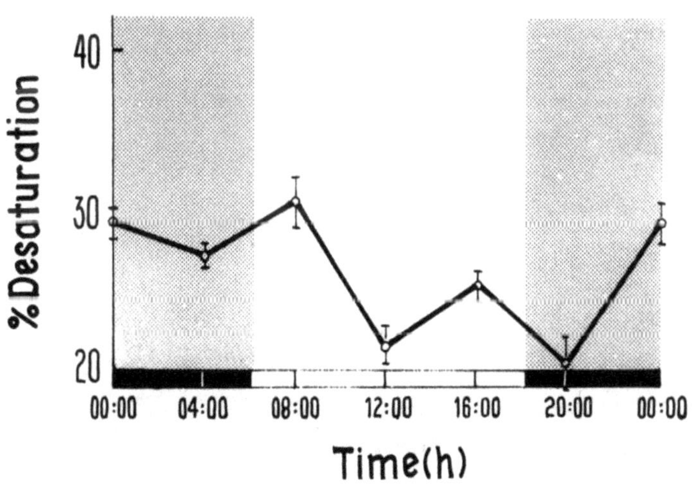

Fig. 8. Circadian changes of the Δ6 desaturation of linoleic to γ-linolenic acid in mouse liver.

Female C 3H mice 6 weeks of age maintained in Purina chow were exposed to periodical illumination (40 ω) from 6:00 to 18:00 h Δ6 desaturation was measured by aerobical incubation of $|1-^{14}C|$ linoleate with 5 mg microsomal protein of the liver and cofactors at 37°C (Actis Dato et al, 1973). Results are the means of 4 mice ± SEM.

The biosynthesis of polyunsaturated acids of α-linolenic acid family in HTC cells is also sensitive to dibutyryl-cAMP. Alaniz et al (1976) have shown that the incubation of the cells with $|1\text{-}^{14}C|$ α-linolenic acid in the presence of 10^{-3}M dibutyryl-cyclic AMP plus 10^{-3}M theophylline during 24 h depressed the Δ6 desaturation pathway whereas the elongation of 18:3 (6, 9, 12) to 20:3 (8, 11, 14) was enhanced (Fig. 7). These results complement quite well the data obtained with glucose and insulin (Fig. 3) (Fig. 5).

However we found that neither glucagon nor epinephrine added to the medium where HTC cells are incubated, modified the conversion of α-linolenic acid to higher homologs or increased glycogenolysis in measurable amount. These results suggest that HTC cells have lost very probably the receptors to these hormones but preserve intact the enzymatic mechanism from cAMP onwards.

In the present Symposium Dr. Gómez Dumm shows that thyroxine injection also depresses the Δ6 desaturase activity of rat liver microsomes. In consequence a full array of hormones modify the activity of the enzyme and prove the existence of a yin-yang mechanism in which insulin activates the Δ6 desaturase and glucagon, epinephrine and thyroxine depress the reaction.

As a corolary it is important to add that Actis Dato et al (1973) have found circadian oscillations in Δ6 desaturation activity

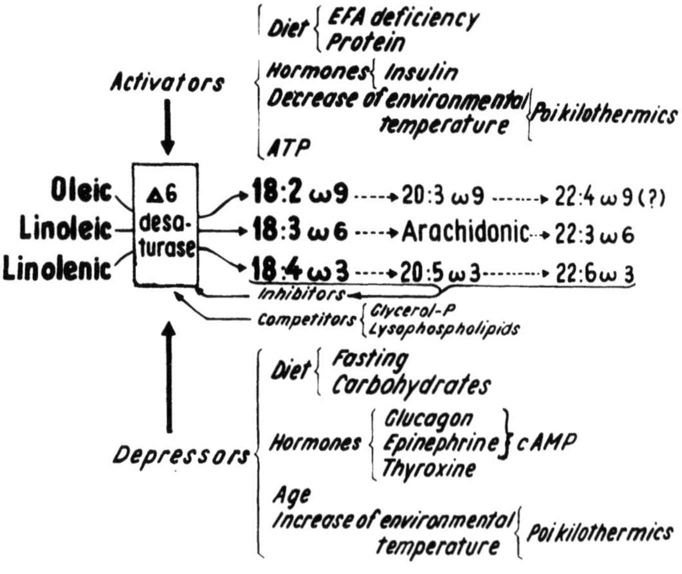

Fig. 9. Homeostatic factors.

of mouse liver microsomes (Fig. 8). These changes are similar for both substrates linoleic and α-linolenic and are undoubtedly related to an interplay of the food intake and hormone secretion.

If diet and hormones are important in the regulation of Δ6 desaturation activity as a means to increase or decrease poly-unsaturated acid biosynthesis along the day cycle not less important is the modification of the activity of the enzyme along the life span (Brenner, 1974) or in different environmental condition to adapt the cell to the new situations. The decrease of Δ9 desaturase activity of liver or testis with age has been proved (Brenner, 1974). In poikilothermic animals like fish the specific activity of the Δ6 desaturase increases with a decrease of the environmental temperature. This increase compensates the diminished velocity of the reaction due to the low temperature (Torrengo and Brenner, 1976).

In conclusion, all available data demonstrate that the Δ6 desaturase is sensitive to a full series of physiological variables (Fig. 9).

CONCLUSIONS

The Δ6 desaturation of unsaturated acyl-CoA is the first reaction involved in the normal biosynthesis of all polyunsaturated fatty acids families in animal microsomes. Due to this key position it can regulate the biosynthesis of the fatty acids of the series. The reaction is modified by competition with substrates and products, ATP, and acyl-CoA acceptors.

Dietary glucose and fructose inhibit the enzyme whereas protein diets and essential fatty acid deficient diets enhance the reaction independently of hormonal effects.

The enzyme is sensitive to hormones concentration. Insulin enhance the reaction but the effect is eliminated by protein synthesis inhibition. Hyperglucemic hormones as glucagon, and epinephrine depress the activity of the Δ6 desaturase by reactions triggers by an increase of cAMP concentration.

The lateral reaction of linoleic or α-linolenic microsomal elongation is insensitive to insulin, glucagon, epinephrine and protein.

All these effects have been proved by either in vivo experiments or cell culture using linoleic or α-linolenic acids as substrates.

REFERENCES

ACTIS DATO S. M., CATALA A. & BRENNER R. R. (1973) Lipids 8, 1-6

ALANIZ M. J. T. de & BRENNER R. R. (1976) Mol. and Cell Biochem. in press

ALANIZ M. J. T. de, GOMEZ DUMM, I. N. T. de & BRENNER R. R. (1976) Mol. and Cell Biochem. 12, 3-8

ALANIZ M. J. T. de, PONZ G. & BRENNER R. R. (1975) Acta Physiol. Latinoam. 25, 3-13

AYALA S., GASPAR G., BRENNER R. R., PELUFFO R. O. & KUNAU W. (1973) J. Lipid Res. 14, 296-305

BERNERT J. T. Jr & SPRECHER H. (1975) Biochim. Biophys. Acta 398, 354-363

BRENNER R. R. (1974) Mol. and Cell Biochem. 3, 41-52

BRENNER R. R. & CATALA A. (1971) Lipids 6, 873-881

BRENNER R. R. & NERVI A. M. (1965) J. Lipid Res. 6, 363-368

BRENNER R. R. & PELUFFO R. O. (1966) J. Biol. Chem. 241, 5213-5219

BRENNER R. R. & PELUFFO R. O. (1969) Biochim. Biophys. Acta 176, 471-479

BRENNER R. R., PELUFFO R. O., MERCURI O. and RESTELLI M. A. (1968) Amer. J. Physiol. 215, 63-70

CASTUMA J. C., CATALA A., BRENNER R. R. & CHRISTIE W. W. (1974) Acta Physiol. Latinoam. 24, 31-39

CATALA A., NERVI A. M. & BRENNER R. R. (1975) J. Biol. Chem. 250, 7481-7484

DUNBAR L. M. & BAILEY J. M. (1975) J. Biol. Chem. 250, 1152-1153

GOMEZ DUMM I. N. T. de, ALANIZ M. J. T. de & BRENNER R. R. (1970) J. Lipid Res. 11, 96-101

GOMEZ DUMM I. N. T. de, ALANIZ M. J. T. de & BRENNER R. R. (1975) J. Lipid Res. 16, 264-268

GOMEZ DUMM I. N. T. de, ALANIZ M. J. T. de & BRENNER R. R. (1976) J. Lipid Res. 17, 616-621

GOMEZ DUMM I. N. T. de, ALANIZ M. J. T. de & BRENNER R. R. Lipids in press

GOMEZ DUMM I. N. T. de, PELUFFO R. O. & BRENNER R. R. (1972) Lipids 7, 590-592

HINSCH W. & SEUBERT W. (1975) Eur. J. Biochem. 53, 437-447

MARCEL Y., CHRISTIANSEN K. & HOLMAN R. (1968) Biochim. Biophys. Acta 164, 25-34

MARCEL Y. & SUZUE G. (1972) J. Biol. Chem., 247, 4433-4436

MERCURI O., PELUFFO R. O. & BRENNER R. R. (1966) Biochim. Biophys. Acta 116, 409-411.

MERCURI O., PELUFFO R. O. & BRENNER R. R. (1967) Lipids 2, 284-285

NERVI A. M., BRENNER R. R. & PELUFFO R. O. (1968) Biochim. Biophys. Acta 152, 539-551

NERVI A. M., CATALA A., BRENNER R. R. & PELUFFO R. O. (1975) Lipids 10, 348-352

NINNO R. E., TORRENGO M. P. de, CASTUMA C. & BRENNER R. R. (1974) Biochim. Biophys. Acta 360, 124-133

PELUFFO R. O. & BRENNER R. R. (1974) J. Nutr. 104, 894-900

PELUFFO R. O., GOMEZ DUMM I. N. T. de, ALANIZ M. J. T. de & BRENNER R. R. (1971) J. Nutr. 101, 1075-1084

PELUFFO R. O., NERVI A. M. & BRENNER R. R. (1976) 441, 25-31

PODACK E. R., LAKOMEK M., SAATHOFF G. A. & SEUBERT W. (1974) Eur. J. Biochim. 45, 13-23

SPRECHER H. & LEE Ch. (1975) Biochim. Biophys. Acta 388, 113-125

TORRENGO M. P. de & BRENNER R. R. (1976) Biochim. Biophys. Acta 424, 36-44

ULLMAN D. & SPRECHER H. (1971a) Biochim. Biophys. Acta 248, 61-70

ULLMAN D. & SPRECHER H. (1971b) Biochim. Biophys. Acta 248, 186-197

ACKNOWLEDGEMENTS

These investigations were supported by the Consejo Nacional de Investigaciones Científicas y Técnicas, Comisión de Investigaciones Científicas de la Provincia de Buenos Aires, Secretaría de Ciencia y Técnica, Secretaría de Salud Pública and Comisión de Investigaciones Científicas of La Plata University.

SATURATED AND MONO-UNSATURATED FATTY ACID BIOSYNTHESIS IN BRAIN : RELATION TO DEVELOPMENT IN NORMAL AND DYSMYELINATING MUTANT MICE

J.M. Bourre, S. Pollet, M. Paturneau-Jouas, and N. Baumann

Laboratoire de Neurochimie, Inserm U. 134

Hôpital de la Salpêtrière, 75634 Paris Cédex 13, France

INTRODUCTION

Fatty acids, constituents of most lipids, play a key role in membrane structure and function. Very long chain fatty acids, either saturated or mono-unsaturated are necessary to myelin ; therefore, their biosynthesis is fundamental for normal brain maturation. Identification of some of the biosynthetic pathways has been facilitated by the use of mutant mice (Quaking and Jimpy) characterized by a defective myelination (1) and deficient in very long chain fatty acids (2, 3).

EXPERIMENTAL PROCEDURES

Investigations have been carried on 3 subcellular compartments : microsomes (4), mitochondria (5) and cytosol (6) ; their purity has been checked by electron microscopy and marker enzymes. After incubation, the radioactive fatty acids biosynthesized are extracted, methylated and analysed by a combination of thin-layer and gas-liquid chromatography (7).

RESULTS AND DISCUSSION

De novo synthesis. The de novo system is found classically in the cytosol of the cell, including brain (8). Enzymatic studies with purified enzymes have shown that this system produces free fatty acids, mainly palmitic acid. A de novo system has been also identified in microsomes, precipitating in ammonium sulphate under conditions which differ from those of elongating systems (10, 11) and soluble de novo system. Although both de novo systems synthesize the same products and use the same cofactors and substrates, they differ in their properties (optimal conditions of activity (11).

Mitochondria contain also a de novo system. Rupture of mitochondria is necessary (Triton X-100 gives the best results ; other detergents or sonication are less efficient) thus showing that this system is located inside the mitochondria and is not a contamination of cytosol or microsomal systems (5).

Microsomal elongating systems. These subcellular particles are able to synthesize long chain fatty acids (9) ; the very long chain fatty acids found in myelin are synthesized in these organelles (4). Ammonium sulphate precipitation leads to the separation of elongating systems from microsomal de novo synthesis (10, 11).

Study of the Quaking mutant, deficient only in very long chain fatty acids (over 18 carbon atoms) has lead to the hypothesis of two different elongating systems, one leading to the synthesis of stearic acid from palmityl-CoA, the other building up fatty acids with longer chain length (especially with 24 carbon atoms) from stearyl-CoA. This hypothesis has been substantiated (10, 11, 12, 13, 14). Microsomal elongases differ by optimum pH (6.6 for C16 elongase, 7.3 for C18 elongase), temperature, ionic strength and substrate concentrations. Both systems need NADPH as hydrogen donor and malonyl-CoA as C2 donor. Acetyl-CoA is inactive.

Reaction products are shown in Table I. Only the C16 elongase is normal in the Quaking mutant, the C18 elongase being reduced by approximately 60%. Moreover, enzymatic elongation of behenyl-CoA is 30% of the normal value in the Quaking mutant (15).

Nervonic acid, the homologous mono-unsaturated fatty acid of lignoceric acid, is synthesized by elongation of erucyl-CoA ; stearyl-CoA and oleyl-CoA are elongated with the same kinetics. This finding explains why myelin mono-unsaturated fatty acids are n-9 and elucidates the reduction of both mono-unsaturated and saturated very long chain fatty acids in the Quaking mutant (7).

Elongating systems in mitochondria. Brain mitochondria are able to elongate acyl-CoA in the presence of acetyl-CoA (16). Palmityl-CoA is elongated by acetyl-CoA in the presence of NADPH and NADH providing mainly stearic acid. Similar patterns for enzymatic activity, cofactors, optimal pH and substrate requirements are found for stearyl-CoA elongation presuming that the same enzyme systems elongate both acyl-CoA (contrary to microsomes) (17).

Striking are the differences between the components necessary for acyl-CoA elongation according to the organelle : in mitochondria acetyl-CoA is the immediate precursor of the two carbon elongation unit and both NADH and NADPH are necessary for synthesis of saturated fatty acids. On the contrary, microsomal enzymes require

CHAIN LENGTH			18:0	20:0	22:0	24:0
Substrates						
$1-^{14}C$ Palmityl-CoA + Malonyl-CoA	Brain	Control	96 ± 1	2.5 ± 0.4	0.5 ± 0.2	1 ± 0.2
		Quaking	98 ± 1	1.7 ± 0.3	0.2 ± 0.1	0.3 ± 0.2
	Kidney	Control	97 ± 1	3 ± 1	-	-
		Quaking	97 ± 1	3 ± 1	-	-
$1-^{14}C$ Stearyl-CoA + Malonyl-CoA	Brain	Control		75 ± 1	9 ± 1	15 ± 2
		Quaking		85 ± 1	8 ± 1	6 ± 2
	Kidney	Control		92 ± 1	6 ± 1	2 ± 1
		Quaking		93 ± 1	6 ± 1	2 ± 1

Table I. MICROSOMAL ELONGATION OF RADIOACTIVE ACYL-CoA IN BRAIN AND KIDNEY

NADPH alone and malonyl-CoA.

Synthesis during brain development. The highest level of synthetase activity in the cytosol occurs at about 6 days after birth and specific activity actually falls as the animal matures (13, 7).

In microsomes, the highest specific activity is found at 12 days for the de novo system, 15 days for the C16 elongase and 18 days for the C18 elongase (Figure I). The increase in enzyme activity parallels myelin deposition and the timing of the three microsomal systems is coherent in connection to myelin maturation (ml). In young animals, myelin contains more medium chain fatty acids (18). Relation between microsomal systems and myelin is confirmed by the studies of mutant mice : specific activity at 15-20 days of age of C18 elongase (using behenyl-CoA as acyl donor) is drastically reduced in the Quaking mutant and 5 % of its normal value in Jimpy (14). The activity of the mitochondrial systems increases regularly during development (Figure I) ; it is not correlated to myelination and not affected in both mutants (14).

Regulation : comparison between brain and kidney. Although fatty acid biosynthesis is deficient in Quaking brain microsomes, it is normal in kidney (Table I). Thus the genetic control of microsomal fatty acid synthesis varies according to the organ (19). It could involve different isoenzymes. Also brain enzymes may be activated through specific regulatory mechanisms.

CONCLUSIONS

Fatty acid biosynthesis in brain is operative in three compartments : microsomes, mitochondria and cytosol. Brain microsomes contain three different systems : de novo and two elongating systems, the C16-elongase and C18-elongase ; only this latter is disturbed in the Quaking mutant. These elongating systems need acyl-CoA, malonyl-CoA and NADPH. Moreover, saturated and mono-unsaturated very long chain fatty acids share a common elongating complex. Thus in microsomes, the following pathway is proposed :

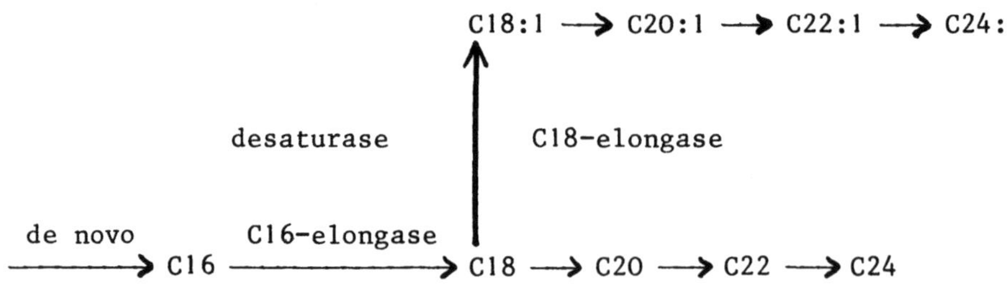

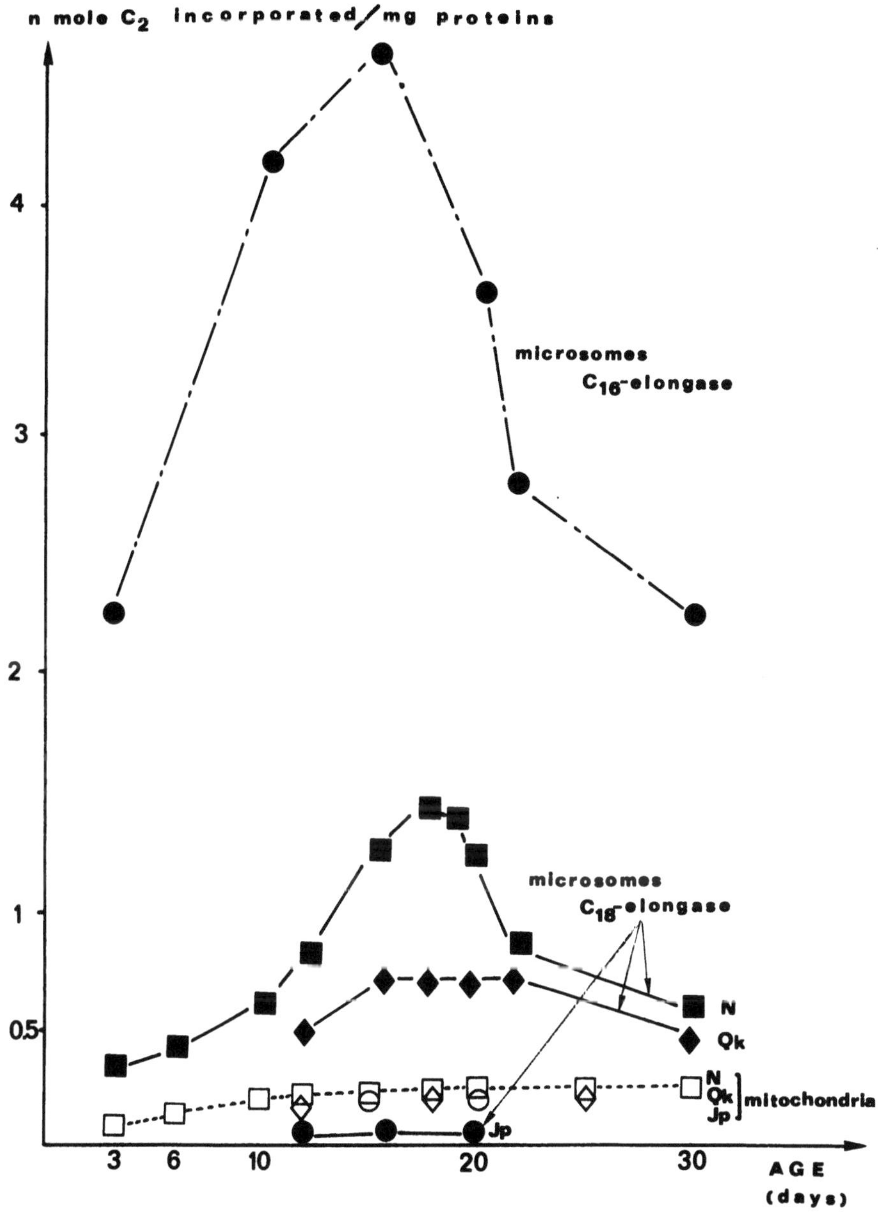

Figure I. FATTY ACID ELONGATION DURING DEVELOPMENT.

The curve for microsomal C18 elongase is identical when using C18, C20, C22 acyl-CoA. For mitochondria, behenyl-CoA was used as acyl donor.

Mitochondria contain a de novo system and only one elongating system providing C24 ; the substrates are acetyl-CoA and NADH + NADPH.

In microsomes, the specific activity for malonyl-CoA incorporation reaches a maximum at 15-20 days of age ; this peak was not obtained in the Quaking and Jimpy mutants ; the increase in enzymatic activity paralleled myelin deposition. The activity of the mitochondrial system increases regularly during development : it is not correlated to myelination and it is not affected in the Quaking and Jimpy mutants.

REFERENCES

1. Sidman R.L., Appel S.H., Fuller S.F., Science 150 (1965) 513-516.

2. Baumann N.A., Jacque C.M., Pollet S.A., Harpin M.L., Eur. J. Biochem. 4 (1968) 340-344.

3. Baumann N.A., Bourre J.M., Jacque C., Pollet S., Ciba Found. Symp. "Lipids, malnutrition and the developing brain" (1971) 91-100.

4. Bourre J.M., Pollet S.A., Daudu O.L., Baumann N.A., Brain Res. 51 (1973) 225-239.

5. Paturneau-Jouas M., Baumann N., Bourre J.M., Biochimie 58 (1976) 341-349.

6. Pollet S.A., Bourre J.M., Baumann N., C.R. Acad. Sci. 268 (1969) 2146-2149.

7. Bourre J.M., Daudu O., Baumann N.A., Biochim. Biophys. Acta 424 (1976) 1-7.

8. Brady R.O., J. Biol. Chem. 235 (1960) 3099-4005.

9. Aeberhard E., Menkes J.M., J. Biol. Chem. 243 (1968) 3834-3840.

10. Bourre J.M., Pollet S., Chaix G., Daudu O., Baumann N., Biochimie 55 (1973) 1473-1479.

11. Pollet S., Bourre J.M., Chaix G., Daudu O., Baumann N., Biochimie 55 (1973) 333-341.

12. Bourre J.M., Pollet S.A., Dubois G., Baumann N.A., C.R. Acad. Sci. 271 (1970) 1221-1223.

13. Bourre J.M., Daudu O., Baumann N., Biochimie, in the press.

14. Bourre J.M., Paturneau-Jouas M., Daudu O., Baumann N., Eur. J. Biochem., in the press.

15. Bourre J.M., Daudu O., Baumann N., Biochem. Biophys. Res. Commun. 63 (1975) 1027-1034.

16. Boone S.C., Wakil S.J., Biochemistry 9 (1970) 1470-1479.

17. Paturneau-Jouas M., Baumann N., Bourre J.M., Biochem. Biophys. Res. Commun. 71 (1976) 1326-1334.

18. Baumann N.A., Bourre J.M., Jacque C., Harpin M.L., J. Neurochem. 20 (1973) 753-759.

19. Bourre J.M., Daudu O.L., Baumann N.A., J. Neurochem. 24 (1975) 1095-1097.

PROTEIN FACTOR INVOLVED IN FATTY ACID DESATURATION OF LINOLEIC ACID

A. Catalá,* A. Leikin, A.M. Nervi,* and R.R. Brenner

Cátedra de Bioquímica, Instituto de Fisiología, Facultad de Ciencias Médicas, Universidad Nacional de La Plata, La Plata, Argentina

INTRODUCTION

It has been established that the enzyme system involved in stearyl-CoA desaturation reaction has three integral components of the microsomal membrane: the NADH-cytochrome b_5 reductase, cytochrome b_5 and the desaturase (Gaylor et al, 1970; Holloway et al, 1970; Holloway, 1971; Oshino et al, 1971, Shimakata et al, 1972). All these components have been separated and purified (Strittmatter et al, 1974; Enoch et al, 1976).

Linoleic acid and other fatty acids are considered to be desaturated in liver microsomes by a similar system, requiring ATP, CoA, Mg^{++}, NADH and oxygen (Brenner et al, 1966; Brenner et al, 1969; Castuma et al, 1972).

In our laboratory we were able to separate an additional "soluble factor" loosely bound to the microsomes that was necessary for the full linoleic acid desaturation activity of isolated microsomes "in vitro" (Catalá et al, 1972; Catalá et al, 1975; Nervi et al, 1975). Further information is reported here.

EXPERIMENTAL PROCEDURE

1-|^{14}C|-linoleic acid, 58 mCi/mmole, was provided by the Radiochemical Center, Amersham, England. Cofactors for desaturation reaction were provided by Boehringer Argentina, Buenos Aires, Argen-

* Members of the Carrera del Investigador CONICET.

tina. Sephadex and DEAE cellulose were provided by Sigma Chem. Co. St. Louis, Mo.

Microsomes were prepared as mentioned before (Catalá et al, 1975). The microsomes were extracted with a solution of 0.25 M sucrose, 0.04 M NaF, 0.15 M KCl, 1.5 mM glutathione, 0.33 mM nicotinamide and 0.04 M phosphate buffer (pH 7.0). The extraction was performed in the cold using the following proportions: 5 mg of microsomal protein suspended in the homogenizing solution (1:1 v/v) was dropped on 3 ml of the extraction solution and shaken during 15 min. They were centrifuged at 110,000 xg in the cold. Two fractions were obtained: the pellet, containing the extracted microsomes (Me), and the supernatant (Sp), which contains the "soluble factor".

Protein was determined either by a microburet method (Munkres et al, 1965) or by the method of Lowry et al (1951). Catalase activity was determined according to Beers et al (1952).

Fatty acid desaturation was assayed incubating 60 nmoles of labeled linoleic acid and 2.5 mg of microsomal protein for 25 min at 35°C in 1.5 ml of a solution containing 0.25 M sucrose, 0.15 M KCl, 1.5 mM glutathione, 0.04 M NaF, 1.3 mM ATP, 0.06 mM CoA, 0.87 mM NADH, 5 mM $MgCl_2$, 0.33 mM nicotinamide and 0.04 M phosphate buffer (pH 7.0). Reaction was stopped with KOH solution and fatty acids sterified, analyzed by gas-liquid radiochromatography and γ-linolenate formed calculated according to Catalá et al (1975).

The potency of the different fractions obtained in the purification steps to reactivate the capacity of extracted microsomes (Me) to desaturate linoleic acid was measured using the same procedure, except that the amount of Me used was the remainder after the extraction of 2.5 mg of complete microsomes (M).

The partial purification of the factor was controlled by measuring the specific reactivation capacity of the different fractions added to Me and was expressed by the increase of nmoles of γ-linolenate formed by Me after the addition of the investigated fraction. It was calculated per min per mg of protein added to Me.

Partial concentration of the material was performed with XM-100 Amicon membrane, Sephadex G-100 and DEAE-cellulose chromatography. This partially purified protein was obtained from two sources: the material loosely bound to the microsomes (Sp) and the cytosol (C).

400 ml of Sp, containing 120 mg of protein was concentrated to 20 ml through the XM-100 Amicon membrane. The final concentration was 2.2 mg/ml. 5 ml of this fraction, called Sp FI, was applied to a 1.5 x 75 cm Sephadex G-100 column equilibrated with potassium

phosphate buffer 0.02 M, 0.1 mM EDTA, (pH 7.4). The front fractions corresponding to volumes 35 to 45 ml were pooled and called Sp FII. 10 ml of this solution was directly poured on a 1 x 4 cm DEAE cellulose column, previously equilibrated with Tris-HCl buffer 0.02 M (pH 8.5). The unadsorbed material was collected washing the column with 15 ml of the same buffer. This fraction was called Sp FIII and was able to reactivate the Me in the linoleic acid desaturation reaction.

The other source of soluble factor was the cytosol. It was called C in Table 2. 3 ml of cytosol containing 90 mg of protein was directly applied to a 1.5 x 75 cm Sephadex G-100 column, equilibrated as mentioned before. The front peak, corresponding to volumes of 29-42 ml, was called C FI. 13 ml of this fraction, containing 17 mg protein, was applied to a 2.5 x 2 cm DEAE-cellulose column and the unadsorbed material was collected in the same way as mentioned before. This fraction was called C FII. Each fraction was assayed as mentioned before.

RESULTS AND DISCUSSION

Complete microsomes (M), extracted with a low ionic strength solution lost most of their linoleic acid desaturation activity. The extracted microsomes (Me) recovered the desaturation activity after the readdition of supernatant (Sp). The activity of Me was also recovered when the cytosol (C) was added to Me. Sp had no activity "per se" (Table 1).

Partial concentration of the material was obtained by XM-100 Amicon membrane, Sephadex G-100, and DEAE-cellulose, as is described in the experimental part. This partially purified protein fraction was obtained from two sources, the material loosely bound to the microsomes and the cytosol. Table 2 shows the reactivation capacity of different purified fractions when added to Me. The reactivation capacity is expressed as the increase in nmoles of γ-linolenate synthesized per min and per mg of protein fraction added to Me.

After the G-100 step, the activity of Sp was increased three times and the cytosol activity four times. The DEAE-cellulose step concentrated similarly the activity of both fractions. The active fractions were separated in both cases at the lowest concentration of Tris-HCl buffer, showing a low superficial charge. The concentration achieved was 38 times for the Sp material and 74.6 times for the cytosolic material. Quantities as low as 10 μg of protein factor of both fractions were enough to recover almost 100% of the desaturation capacity of the extracted microsomes. Therefore, it could be possible that the cytosolic supernatant factor is the same as the one found stuck to the unwashed microsomes and separated from the Sp.

Table 1. Extraction of a microsomal factor involved in linoleic acid desaturation.

Assay conditions and extraction of microsomes (see methods).

FRACTIONS	RELATIVE DESATURATION ACTIVITY
M (complete microsomes)	100.0
Me (extracted microsomes)	40.0
Sp (supernatant)	not detectable
Me + Sp	96.2
Me + C (cytosol/110.000 xg)	98.1

The capacity of Sp to reactivate Me is inhibited by trypsin digestion showing that a protein structure is necessary in Sp to reactivate Me. Besides, heated or boiled Sp are also inactives. Different diets modify Me activity but have no effect on Sp. (Nervi et al, 1975).

Table 2. Partial purification of a protein factor involved in linoleic acid desaturation.

Assay conditions and extraction of microsomes (see methods)

FRACTIONS	μg OF PROTEIN FRACTION ADDED TO Me	SPECIFIC REACTIVATIONS CAPACITY OF DIFFERENT FRACTIONS ON Me $\frac{\Delta \text{nmoles } \gamma\text{-}18{:}3}{\text{min x mg protein added to Me}}$
Me	--	--
Me + Sp FI	500	0.30
Me + Sp FII	100	0.88
Me + Sp FIII	10	11.4
Me + C	1000	0.15
Me + C FI	200	0.60
Me + C FII	10	11.2

The presence of lipids has been detected in Sp. The resolution of lipid class was achieved in silica gel G thin layer chromatography showing the presence of phosphatidyl choline, triacylglycerols, cholesterol and cholesterol esters. However the extraction of lipids with ether did not deactivate Sp. (Catala et al, 1975).

The capacity of albumin to reactivate stearate Δ9 desaturase was investigated by Jeffcoat et al (1976). Although the ability of albumin to reactivate the Me Δ6 desaturation capacity had already been studied by Catalá et al (1975), it was reinvestigated in the present work. Fig. 1 shows that Me is reactived by Sp but very little effect is shown by albumin. This means that Sp would be more active and specific than albumin to reactivate Me in this reaction. The effect exerted by albumin is only shown at low concentrations.

Fig. 2 shows the effect of linoleic acid concentration on Δ6 desaturase activity of complete and extracted microsomes. It is apparent that the specific activity of Δ6 desaturase for different substrate concentrations is lower for Me than for complete microsomes (M). Therefore Sp enhances the desaturation of linoleic acid. Maximal velocity is higher for M than for Me in our reaction conditions.

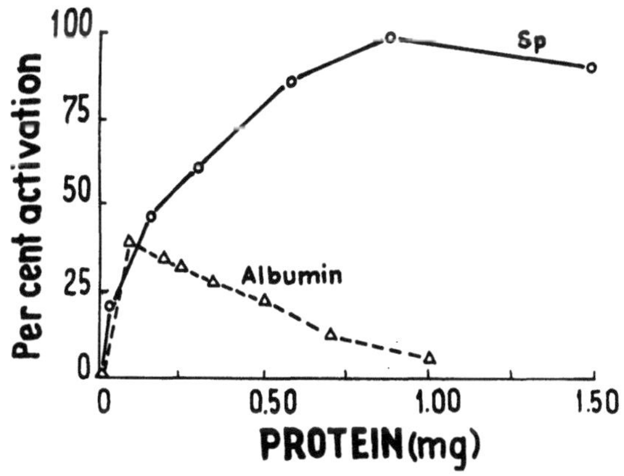

Fig. 1. Effect of different concentrations of protein factor (Sp) and bovine serum albumin on the restoration of linoleate desaturation activity.

Extraction of microsomes and assay conditions (see methods).

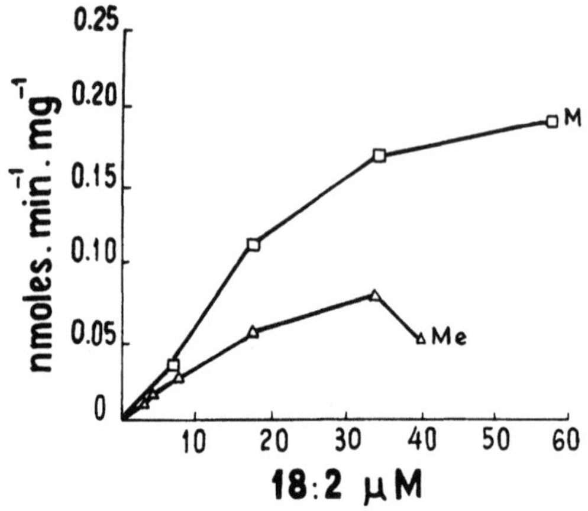

Fig. 2. Effect of linoleic acid concentration on Δ6 desaturase activity of complete microsomes (M) and extracted microsomes (Me).

Extraction of microsomes and assay conditions (see methods).

We also checked the effect of catalase on Me since a recent report of Baker et al, (1976) has shown that this enzyme enhances the stearyl-CoA desaturation. For this reason the catalase activity of Sp, Cytosol and other fractions tested in the reactivation of Me were measured. Sp and the cytosol have catalase activity but the last one was relatively more active than the first one. Fig. 3 shows the comparative reactivation capacity of Sp, cytosol, DEAE Cellulose fraction and pure catalase on Δ6 desaturation and the content of units of catalase of each fraction. Results demonstrate that there is no correlation between the Sp reactivation capacity of Δ6 desaturation reaction and their catalase activity content. Pure catalase has less reactivation capacity than Sp.

CONCLUSIONS

A protein (or proteins) loosely bound to the microsomes is necessary to obtain full activity in the linoleic acid desaturation reaction in an "in vitro" system. This protein (or proteins) is extractable from unwashed microsomes with low ionic strength

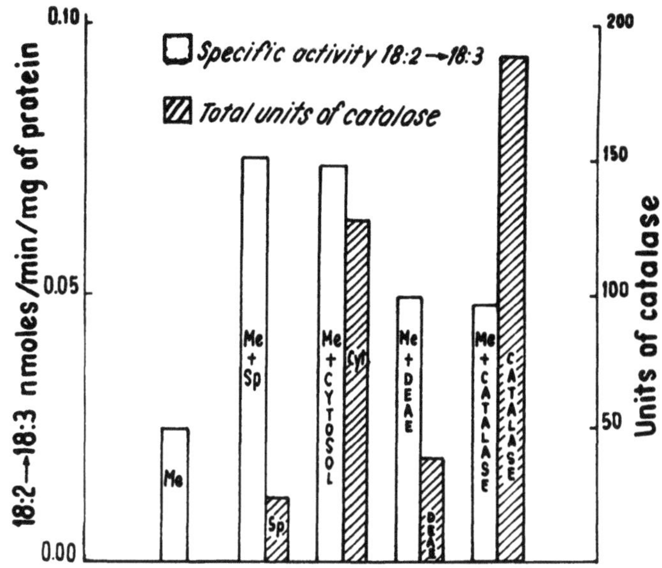

Fig. 3. Reactivation of Me by different protein fractions in the linoleic acid desaturation reaction.

Each fraction is indicated in the respective bar.

solution. Cytosol has also capacity to restore Δ6 desaturation of Me. Therefore it is considered that the protein factor present in the cytosol loosely binds to the microsomes and reactivates Δ6 desaturation. It is sensitive to temperature and trypsin treatment. It contains lipids, and sonification does not affect it. Albumin cannot substitute for the protein factor.

The protein factor contains catalase activity and pure catalase reactivates the desaturation. However there is no correlation between the protein factor reactivation capacity on Δ6 desaturation and catalase activity content. Pure catalase has less reactivation capacity than Sp although it contains many more units of catalase activity.

So the cytosol contains a protein factor able to increase the microsomal velocity of linoleic acid desaturation. This effect would be due to an unidentified protein component and to catalase.

REFERENCES

BAKER R. C., WYKLE R. L. & LOCKMILLER J. S. (1976) Fed. Proc. 35, 1625

BEERS R. F. Jr. & SIZER I. W. (1952) J. Biol. Chem. 195, 133

BRENNER R. R. & PELUFFO R. O. (1966) J. Biol. Chem. 241, 5213-5219

BRENNER R. R. & PELUFFO R. O. (1969) Biochim. Biophys. Acta 176, 471-479

CASTUMA J. C., CATALA A. & BRENNER R. R. (1972) J. Lipid Res. 13, 783-789

CATALA A. & BRENNER R. R. (1972) Ann. Assoc. Quim. Arg. 60, 149-155

CATALA A., NERVI A. M. & BRENNER R. R. (1975) J. Biol. Chem. 250, 7481-7488

ENOCH G. H., CATALA A. & STRITTMATTER P. (1976) J. Biol. Chem. 251, 5095-5103

GAYLOR J. L., MOIR N. J., SEIFRIED H. E. & JEFFCOAT C. R. (1970) J. Biol. Chem. 245, 5511-5513

HOLLOWAY P. W. & WAKIL S. J. (1970) J. Biol. Chem. 245, 1862-1865

HOLLOWAY P. W. (1971) Biochemistry. 10, 1556-1560

JEFFCOAT R., BROWN P. R. & JAMES A. T. (1976) Biochim. Biophys. Acta 431, 33.

LOWRY H. O., ROSEBROUGH N. J., FARR A. L. & RANDALL R. J. (1951) J. Biol. Chem. 193, 265-275.

MUNKRES K. O. & RICHARDS F. M. (1965) Arch. Biochem. Biophys. 109, 466-471

NERVI A. M., BRENNER R. R. & PELUFFO R. O. (1975) Lipids 10, 348-352

OSHINO N., IMAI Y. & SATO R. (1971) J. Biochem. (TOKYO) 69, 155-167

SHIMAKATA T., MIHARA K. & SATO R. (1972) J. Biochem. (TOKYO) 72, 1163-1174

STRITTMATTER P., SPATZ D., CORCORAN D., ROGERS M. J., SETLOW B. & REDLINE R. (1974) Proc. Nat. Acad. Sci. U. S. A. 71, 4565-4569

ACKNOWLEDGMENTS

This work was supported in part by the Comision de Investigaciones Científicas de la Universidad Nacional de La Plata and by the Consejo Nacional de Investigaciones Científicas y Técnicas.

The authors are indebted to Mauricio Córdoba and Susana González for their technical assistance.

BIOSYNTHESIS OF LIPIDS IN TUMORAL CELLS

M.E. De Tomás and O. Mercuri

Cátedra de Bioquímica, Instituto de Fisiología,
Facultad de Ciencias Médicas, Universidad Nacional de
La Plata, La Plata, Argentina

INTRODUCTION

Tumor cells are known to differ from normal cells in lipid metabolism. Many laboratories have focused their interest to make some insight on the relationship between malignancy and alteration in lipid metabolism, since lipids are recognized in their importance as components of membrane structures.

As early as 1966, Siperstein et al. showed negative feed-back control of cholesterol biosynthesis to be defective in transplantable hepatomas. More recently Bricker & Levey (1972) demonstrated unsupressible cholesterol and fatty acid synthesis by cyclic nucleotides in hepatomas and indicated a deletion of this regulatory mechanism of lipogenesis.

"In vitro" depressed $\Delta 6$ and $\Delta 9$ desaturases activities in Novikoff hepatoma and in hepatomas of different growth rates were reported by Chiappe et al. (1974a, 1974b). At the same time, a low activity of stearoyl-CoA desaturase in Morris hepatoma was demonstrated by Raju (1974). On the other hand, Weber & Cantero (1957) reported a marked decrease in the phospholipid content of the Novikoff hepatoma. The phospholipid composition of mitochondria and microsomes from hepatomas of different growth rates was carefully studied by Bergelson et al. (1974). The fatty acid profiles of lipid classes in plasma membranes showed a decrease in the proportion of polyunsaturated fatty acids in cancer cells (van Hoeven & Emmelot, 1973) and an abnormal distribution of saturated and unsaturated fatty acids in phosphatidylcholine (Bergelson & Dyatlovitskaya, 1973; Ruggieri & Fallani, 1973).

In an attempt to examine some properties of cancer cells we have measured the ability of Novikoff hepatoma to take up the saturated and "essential fatty acids" from the host and the capacity of this tumoral tissue to distribute these acids between the different lipid fractions. |^{3}H|-glycerol was also used as a marker for "de novo" glycerolipid biosynthesis.

MATERIALS AND METHODS

|1-^{14}C|-stearic acid (54.0 Ci/mole, 99% radiochemically pure), |1-^{14}C|-linoleic acid (52.9 Ci/mole, 98% radiochemically pure and 2% cis-trans unsaturated acid) and |2- ^{3}H|-glycerol (7.5 Ci/mole) were purchased from the Radiochemical Centre, Amersham, England. |8,9,11,12,14,15- ^{3}H|-arachidonic acid (99% radiochemically pure) (100 Ci/nmole) was purchased from New England Nuclear, USA.

The solid Novikoff hepatoma was maintained by intraperitoneal implants into male Holtzman rats and they were used 5 days after transplanting the tumor. Normal rat liver was obtained from rats of the same breed.

Two experiments were performed in which two groups of five rats each, bearing tumors and two groups of five normal rats each were intravenously injected with 0.2 ml of labelled solution containing 50 μCi|^{3}H|-glycerol and 10 μCi|1-^{14}C|stearic acid rat serum complex or 10 μCi|1-^{14}C|linoleic acid and 10 μCi|8,9,11,12, 14,15-^{3}H|-arachidonic acid rat serum complex. One hour after injection, the animals were killed by decapitation and their livers and tumors removed. The lipids were extracted by the method of Folch et al. (1957).

The lipids were recovered from the original chloroform-methanol extracts and separated into lipid classes. When |1-^{14}C|-stearic acid was injected, before making a separation in classes, a silicic acid column chromatography was used (Elovson, 1964). In both experiments, phospholipids (Wagner et al., 1961) and neutral lipids (Malins & Mangold, 1960) were resolved into their components by thin layer chromatography. The spots, identified by comparison with authentic standards were scrapped off from the plates. Choline and ethanolamine phospholipids were extracted from the silica using the two phase system of Arvidson (1968). Triacylglycerols were eluted using chloroform-methanol 2:1 (v/v).

The fatty acids from each fraction were transesterified with methanolic-HCl (Stoffel et al., 1959), extracted by light petroleum, evaporated and assayed for fatty acid radioactivity in a Packard Scintillation Spectrometer.

The |^{3}H| -glycerol radioactivity of each fraction, separated as described above was determined.

Microsomes from normal and host liver and tumoral tissues were isolated by differential centrifugation as described by Brenner & Peluffo (1966). Microsomal protein was determined by the biuret method (Gornall et al., 1949). The assay for long chain acyl CoA synthetase activity was based on the method of Suzue & Marcel (1972).

|1-^{14}C|-linoleyl CoA was prepared enzymatically as described by Kornberg & Pricer (1953). Radioactive assay for linoleyl thioesterase was based on the enzyme-catalysed release of |1-^{14}C|-linoleic acid from |1-^{14}C|-linoleyl CoA (Barnes & Wakil, 1968).

RESULTS AND DISCUSSION

The incorporation of |^{3}H|-glycerol and |1-^{14}C|-stearic acid into total liver lipids are presented in Fig. 1.

The low incorporation of |^{3}H|-glycerol into tumoral lipids would indicate a decrease of the glycerophosphate pathway for the glycerolipid biosynthesis. Nevertheless a lack in the glycerol activation to sn-glycerol-3-phosphate by the tumoral tissue cannot be ruled out.

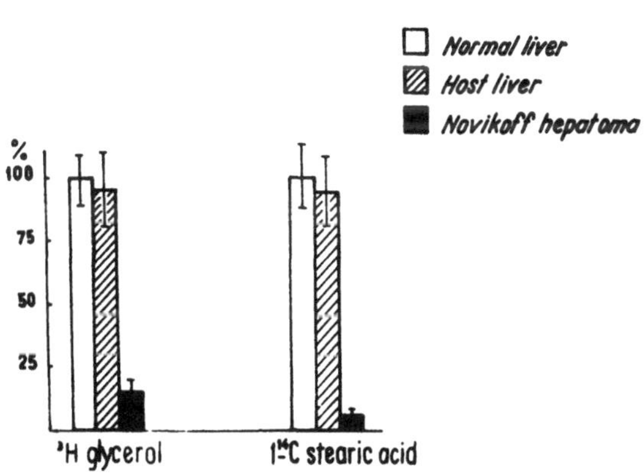

Fig. 1. |^{3}H|-glycerol and |1-^{14}C|-stearic acid incorporation into total lipids from normal liver, host liver and Novikoff hepatoma.

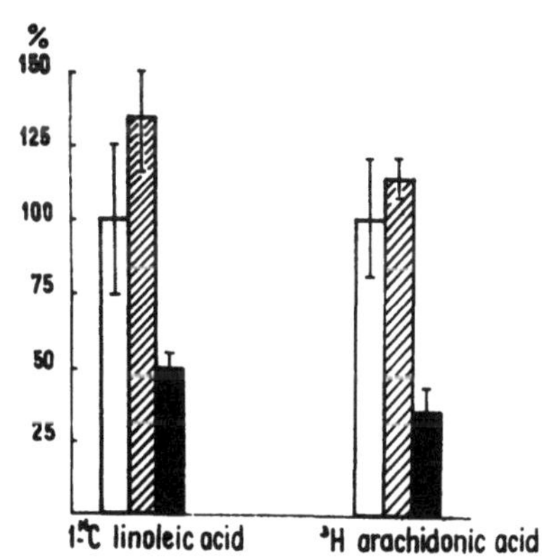

Fig. 2. |1-^{14}C|-linoleic acid and |^{3}H|-arachidonic acid incorporation into total lipids from normal liver, host liver and Novikoff hepatoma.

The values for normal liver were taken as 100%. Each bar is the mean of five determinations ± standard deviation.-

Apparently the uptake of $|^{3}H|$-glycerol into lipids from Novikoff hepatoma is independent of its high mitotic activity since regenerating liver was found to increase labelled glycerol incorporation during the period of mitotic activity (Johnson & Albert, 1960).

Incorporation of radioactivity into the tumoral lipids was considerably lower than that of the control groups. Nevertheless if values for $|1\text{-}^{14}C|$-stearic acid from Fig. 1 are compared with the data from Fig. 2, the incorporation of these "essential fatty acids" into tumoral lipids is greater than that for $|1\text{-}^{14}C|$-stearic acid.

This relatively low rate of stearic acid esterification occurring in the tumors could be explained through a relatively increased ability of acyl transferases to incorporate linoleic and arachidonic acids into the lipids. Nevertheless an increased rate of stearic acid oxidation by the tumor as an energy source, would decrease the amount of this acid available for incorporation into lipids. This assumption seems not to be the case since according to Weinhouse et al. (1973) the poorly-differentiated tumors have largely lost the capability for fatty acid oxidation.

The percent distribution of $|^{3}H|$-glycerol among the tumoral lipid fractions studied (Fig. 3) shows that it is quite different from that of normal and host liver. The $|^{3}H|$-glycerol incorporation in triacylglycerol dropped from 17.8% in the normal liver to 2.6% in the hepatoma. Nevertheless, a preferential incorporation of $|^{3}H|$-glycerol into the cholinephospholipids from the three tissues under study was observed.

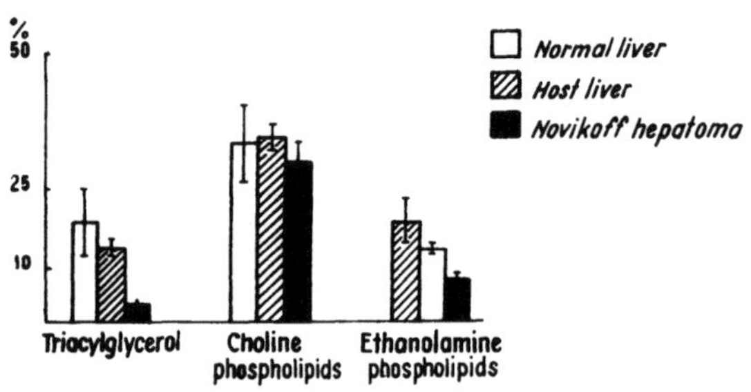

Fig. 3. Percent distribution of $|^{3}H|$-glycerol among different lipid fractions.

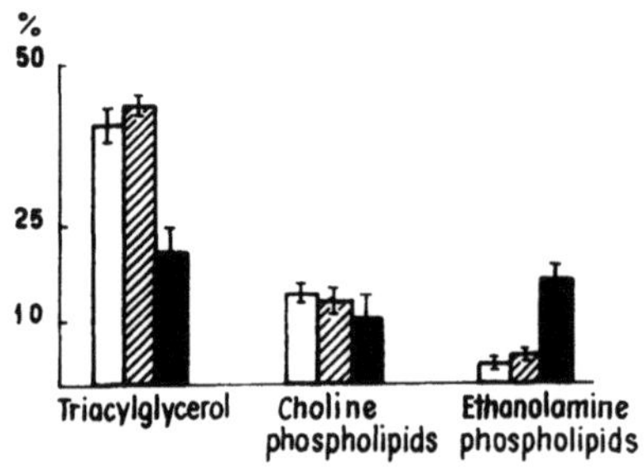

Fig. 4. Percent distribution of $|1\text{-}^{14}C|$-stearic acid among different lipid fractions.

Each bar is the mean of four determinations ± standard variation.

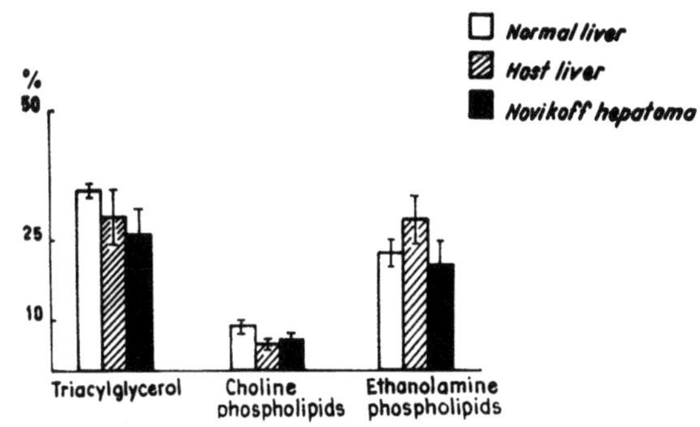

Fig. 5. Percent distribution of |1-^{14}C|-linoleic acid among different lipid fractions.

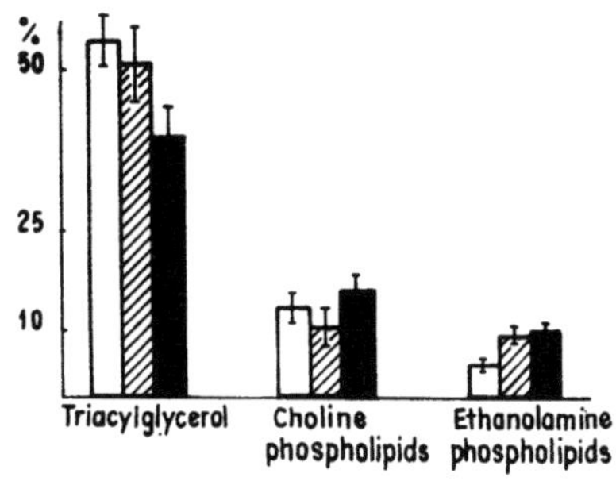

Fig. 6. Percent distribution of |^{3}H|-arachidonic acid among different lipid fractions.

Each bar is the mean of four determinations ± standard deviation.

Figs. 4,5,6 show the percent distribution of labelled fatty acids incorporated into the lipid fractions under study. While the |1-^{14}C|-stearic acid (Fig. 4) is preferentially incorporated in the triacyglycerol fraction of control tissue, a remarkable increase in the percent radioactivity in the ethanolamine phospholipid fraction from tumoral tissue was found.

As observed for |1- ^{14}C|-stearic acid, the percentage of labelled "essential fatty acids" incorporated in the triacylglycerol fraction from Novikoff hepatoma was significantly depressed when compared with the values for normal liver (Figs. 5 and 6). On the other hand, no significant differences were observed in the percent distribution of "essential fatty acids" into the phospholipid fractions studied.

As can be seen from Table 1, the specific activity of long chain acyl CoA synthetase is significantly depressed in the Novikoff hepatoma and host liver when compared with normal liver. These results agree with those of Weinhouse et al. (1973) that made it clear that the activity of the fatty acid activating enzyme is very low in poorly differentiated hepatomas. There is no doubt that the growing tumor exerts a profound effect on the enzyme system of the host liver. On the other hand, no significant differences were observed in the activity of long chain acyl CoA thioesterase from the tissues under study.

Table 1.- Specific activities of microsomal long chain acyl CoA synthetase and acyl CoA hydrolase from normal liver, host liver and Novikoff hepatoma.

a) Numbers in parentheses indicate the number of animals in each group. Data are the means ± S.D. Probability values (P) are related to normal liver. N.S. = not significant.

Tissue	Acyl CoA synthetase nmoles/h per mg protein	Acyl CoA hydrolase nmoles/h per mg protein
Normal liver (5)[a]	11,718±1,620	300±102
Host liver (4)	7,236 ± 672 P<0.01	282 ± 54 N.S
Novikoff hepatoma	2,480 ± 636 P<0.001	216 ± 78 N.S

CONCLUSIONS

According to the results of the present experiments we can conclude that the ability of the tumoral tissue to take up the labelled glycerol from the host was probably diminished by changes in the activity of enzymes leading to the biosynthesis of phospholipids and neutral lipids. Apparently the uptake of $|^3H|$-glycerol into lipids from Novikoff hepatoma was independent of its high rate of growth.

The incorporation of labelled fatty acids into tumoral lipids was significantly depressed, probably by a decrease in the ability of acyl transferases to incorporate fatty acids into lipids. Furthermore the abnormal distribution of the radioactivity incorporated among the different lipid fractions from Novikoff hepatoma indicates not only a failure of the enzymes in their capacity to incorporate the labelled substrates but also in the ability to distribute them according to a "normal" scheme. On the other hand, despite the low long chain acyl CoA synthetase activity in the tumoral tissue, no conclusion can be drawn from this fact on the esterification capacity of this tissue as a limiting step on fatty acid incorporation.

REFERENCES

ARVIDSON, G. A. E. (1968) Eur. J. Biochem. 4, 478-486.

BARNES, E. M. Jr. & WAKIL, S. J. (1968) J. Biol. Chem. 243, 2955-2962.
BERGELSON, L. D. & DYATLOVITSKAYA, E. V. (1973) In Tumor Lipids: Biochemistry and Metabolism (Wood, R. ed.) pp. 111-125, American Oil Chemists' Society Press, Illinois.
BERGELSON, L. D., DYATLOVITSKAYA, E. V., SOROKINA, I. B. & GORKOVA, N. P. (1974) Biochim. Biophys. Acta 360, 361-365.
BRENNER, R. R. & PELUFFO, R. O. (1966) J. Biol. Chem. 244, 5213-5219.
BRICKER, L. E. & LEVEY, G. S. (1972) Biochem. Biophys. Res. Commun. 48, 362-365.
CHIAPPE, L., DE TOMAS, M. E. & MERCURI, O. (1974a) Lipids 9, 360-362.
CHIAPPE, L., MERCURI, O. & DE TOMAS, M. E. (1974b) Lipids 9, 489-490.
ELOVSON, J. (1964) Biochim. Biophys. Acta 84, 275-293.
FOLCH-PI, J., LEES, M. & SLOANE-STANLEY, G. H. (1957) J. Biol. Chem. 226, 497-509.
GORNALL, A. G., BARDAWILL, C. J. & DAVID, M. M. (1949) J. Biol. Chem. 177, 751-755.
JOHNSON, R. M. & ALBERT, S. (1960) J. Biol. Chem. 235, 1299-1302.
KORNBERG, A. & PRICER, W. E.(1953) J. Biol. Chem. 204, 329-343.
MALINS, D. C. & MANGOLD, H. K. (1960) J. Am. Oil Chem. Soc. 37, 576-578.
RAJU, P. K. (1974) Lipids 9, 795-797.
RUGGIERI, S. & FALLANI, A. (1973) : In Tumor Lipids : Biochemistry and Metabolism (Wood, R. ed.) pp. 89-110, American Oil Chemists' Society Press, Illionis.
SIPERSTEIN, M. D., FAGAN, V. M. & MORRIS, H. P. (1966) Cancer Res. 26, 7-11.
STOFFEL, W., CHU, F. & AHRENS, E. H. (1959) Anal. Chem. 31, 307-308.
SUZUE, G. & MARCEL, Y. (1972) Biochemistry 11, 1704-1708.
VAN HOEVEN, R. P. & EMMELOT, P. (1973) : In Tumor Lipids: Biochemistry and Metabolism (Wood, R. ed.) pp. 127-138, American Oil Chemists' Society Press, Illinois.
WAGNER, W., HORHAMER, L. & WOLFF, P. (1961) Biochem. Z. 334, 175-184.
WEBER, G. & CANTERO, A. (1957) Exp. Cell. Res. 13, 125-131.
WEINHOUSE, S., LANGAN, J. & SHATTON, J. A. (1973) : In Tumor Lipids: Biochemistry and Metabolism (Wood, R. ed), pp. 14-20, American Oil Chemists' Society Press, Illinois

SPECIFICITY OF Δ6 DESATURASE -- EFFECT OF CHAIN LENGTH AND NUMBER OF DOUBLE BONDS

J. C. Castuma, R. R. Brenner, and W. Kunau*

Cátedra de Bioquímica, Instituto de Fisiología, Facultad de Ciencias Médicas, Universidad Nacional de La Plata, La Plata, Argentina
*Ruhr-Universität Bochum, Institute für Physiologische Chemie, Bochum-Querenburg, Germany

INTRODUCTION

Oleic, linoleic and α-linolenic acids are desaturated by the same microsomal enzyme called Δ6 desaturase since it introduces the double bond between carbons 6-7 (Brenner and Peluffo, 1966; Brenner, 1971; Inkpen et al, 1969; Brenner, 1974). This enzyme is different from the Δ5 desaturase that converts eicosa-8,11-dienoic, eicosa-9,11,14-trienoic and eicosa-8,11,14,17-tetraenoic acids to eicosa-5,8,11-trienoic, eicosa-5,8,11,14-tetraenoic and eicosa-5,8,11,14,17-pentaenoic acids respectively. (Ninno et al, 1974; Alaniz et al, 1975; Gaspar et al, 1975). It is also different from the Δ9 desaturase (Brenner, 1971). The Δ6 desaturase recognizes the number of double bonds of the substrate since α-linolenate is desaturated faster than linoleate and this one faster than oleate (Brenner and Peluffo, 1966). However, it is not known yet the effect of the chain length of the acid on the activity of the enzyme.

To solve this problem in the present experiment $|1\text{-}^{14}C|$ eicosa-9,12-dienoic and $|1\text{-}^{14}C|$ eicosa-9,12,15-trienoic acids were synthesized and their Δ6 desaturation by rat liver microsomes tested. Additional information was also gathered studying the inhibitory effect of fatty acids of different chain length and number of double bonds on the Δ6 desaturation of α-linolenic acid.

EXPERIMENTAL PROCEDURES

Substrates. All-cis |1-^{14}C| eicosa-9, 12-dienoic acid was synthesized in the laboratory by coupling of $CH_3-(CH_2)_6-C{\equiv}C-CH_2I$ with $HC{\equiv}C-(CH_2)_6-CH_2-o-\bigcirc$ by a Grignard reaction in the presence of CuCl. The starting materials for the first intermediate were $CH_3-(CH_2)_6-OH$ and $HC{\equiv}C-CH_2OH$. The starting chemicals for the second were $HO(CH_2)_7OH$. The radiochemical purity was higher than 95 % and the specific radioactivity 0.15 mCi/mmole. All-cis |1-^{14}C| eicosa-9, 12, 15-trienoic acid was synthesized similarly by coupling of $CH_3-(CH_2)_3C{\equiv}C-CH_2-C{\equiv}C-CH_2I$ with $HC{\equiv}C-(CH_2)_6-CH_2-O-\bigcirc$. The radiochemical purity was also higher than 95 % and the specific radioactivity was 0.15 mCi/mmole. The unlabeled acids were synthesized similarly.

Cis-cis-|1-^{14}C| linoleic acid (58 mCi/mmole and 99% radiochemical purity) and all cis |1-^{14}C| α-linolenic acid (58 mCi/mmole and 99 % radiochemical purity) were purchased from The Radiochemical Centre, Amersham, England. They were diluted with unlabeled acids to reach 0.15 mCi/mmole specific radioactivity. Unlabeled acids were purchased from The Hormel Institute, Austin, Min. They were 99 % pure.

Microsomal desaturation of fatty acids. Two-month-old female Wistar rats fed on Purina chow ad libitum were used. When specified, some rats were fasted for 48 hr and then refed on a diet containing 35 % protein (Peluffo and Brenner, 1974). Liver microsomes were separated by differential centrifugation at 100,000 xg in the cold in the way described by Castuma et al, 1972.

Variable amounts of labeled fatty acids were incubated 20 min at 35°C with 5 mg of microsomal protein in a total volumen of 1.5 ml. The solution contained 1 mM NADH, 0.2 mM CoA, 3.5 mM ATP, 1.5 mM $MgCl_2$ and other cofactors as has already been described by Ninno et al, 1974. When pertinent, unsaturated acids were added in different concentrations to investigate inhibitory effects. Free fatty acids were used instead of acyl-CoA since it was found that acyl-CoA formation takes place readily with either eicosa-9, 12-dienoic or eicosa-9, 12, 15-trienoic. Using Kornberg and Pricer, (1953) procedure we found that microsomes thiosterify the 20 carbon acids as easily and with similar yields as linoleic acid.

The distribution of radioactivity between substrate and product was measured by gas-liquid radiochromatography in a Packard apparatus with a proportional counter. The fatty acids were identified by comparison with standards and equivalent chain length determination. From the conversions found the speed of the reactions were calculated. Substrate concentrations that saturated the enzyme were determined and used throughout the experiment to measure the specific activities.

RESULTS

Eicosa-9, 12-dienoic acid only differs from linoleic acid by the existence of two additional carbons at the methyl end of the chain. Similarly 20:3 (9, 12, 15) has the structure of α-linolenic acid with two more carbons at the methyl end of the chain. In spite of this, both acids were desaturated by rat liver microsomes to 20:3 (6, 9, 12) and 20:4 (6, 9, 12, 15) respectively. However they were desaturated less efficiently than linoleic and α-linolenic acids showing that Δ6 desaturase recognizes the additional lengthening of the methyl end of the fatty acids (Table 1). The enzyme also recognizes the number of double bonds in the acids of 20 carbons, as was already shown in the acids of 18 carbons, since 20:3 (9, 12, 15) is a better substrate than 20:2 (9, 12) acid.

The Δ6 desaturase activity of liver microsomes increases in animals fed on hyperproteic diets (Peluffo and Brenner, 1974). Table 1 shows that the specific Δ6 desaturation of linoleic, linolenic, 20:2 (9, 12) and 20:3 (9, 12, 15) are increased by the protein whereas Δ5 desaturation of eicosa-8, 11-dienoic to eicosa-5, 8, 11-trienoic acid is changed very little, as it was demonstrated by Castuma et al (1972).

Since both synthetic acids are desaturated by the Δ6 desaturase they are expected to compete with linoleic and linolenic acid for the enzyme. Table 2 shows that linoleic acid desaturation to γ-linolenic acid is inhibited in first place by linolenic acid (18:3-9, 12, 15), in second place by 20:2 (9, 12) and in third place by 20:3 (9, 12, 15). α-linolenate that has a very strong affinity for the enzyme and is rapidly desaturated competes advantageously with labeled linoleate. The increase of chain length to 20 carbons

Table 1. Fatty acid Δ6 and Δ5 desaturation activities.

53.3 μM, 20:2 (9, 12), 20:3 (9, 12, 15) or linoleic acid or 66.6 μM, α linolenic acid or 133.3 μM, 20:2 (8, 11) acids were incubated at pH 7.0 as described under Experimental.

Substrates	New double bond	Product (nmoles/min. mg protein)	
		Purina chow	Hyperproteic diet
18:2 (9, 12)	Δ6	0.100	0.173
18:3 (9, 12, 15)	Δ6	0.185	0.340
20:2 (9, 12)	Δ6	0.025	0.064
20:3 (9, 12, 15)	Δ6	0.047	0.130
20:2 (8, 11)	Δ5	0,316	0.350

Table 2. Inhibition of linoleic acid desaturation to γ-linolenic acid.

Conditions of incubation as in Table 1. Linoleic acid concentration is 53.3 μM.

Competitors	Concentration (μM)	Fractional inhibition
18:3 (9, 12, 15)	26.6	0.73
	53.3	0.84
20:2 (9, 12)	26.6	0.36
	53.3	0.42
20:3 (9, 12, 15)	26.6	0.26
	53.3	0.38

decreases not only the desaturation capacity of the acid but also the inhibitory power. However, 20:3 (9, 12, 15) in spite of being a better substrate than 20:2 (9, 12) evokes a similar inhibitory effect than the latter.

Table 3. Inhibition of α-linolenic acid desaturation to octadeca-6, 9, 12, 15-tetraenoic acid.

Condition of incubation as in Experimental. α-linolenic acid concentration 133.3 μM.

Competitors	Concentration (μM)	Fractional inhibition
18:2 (9, 12)	133.3	0.41
	266.6	0.50
20:2 (9, 12)	133.3	0.21
	266.6	0.27
20:3 (9, 12, 15)	133.3	0.47
	266.6	0.67

Inhibition of α-linolenic acid desaturation to octadeca-6, 9, 12, 15-tetraenoic acid is shown in Table 3. Linoleate (18:2 (9, 12)) is a better substrate than 20:2 (9, 12) and 20:3 (9, 12, 15) and it is a better inhibitor than 20:2 (9, 12) but a little worse than 20:3 (9, 12). Therefore the best substrates are not necessarily the best competitors in Δ6 desaturation.

Unsaturated acids that are not substrates for the Δ6 desaturase are also inhibitors of the enzyme. In Fig. 1 we may compare the inhibitory effect of different fatty acids with variable chain length and number and position of double bonds on the Δ6 desaturation of α-linolenic acid. Similar effects were found by Castuma et al, 1974, for linoleic acid. It is shown that fatty acids of 20 carbons with the first double bond in Δ8 or Δ11 do not increase the inhibitory power by addition of double bonds in the direction of the

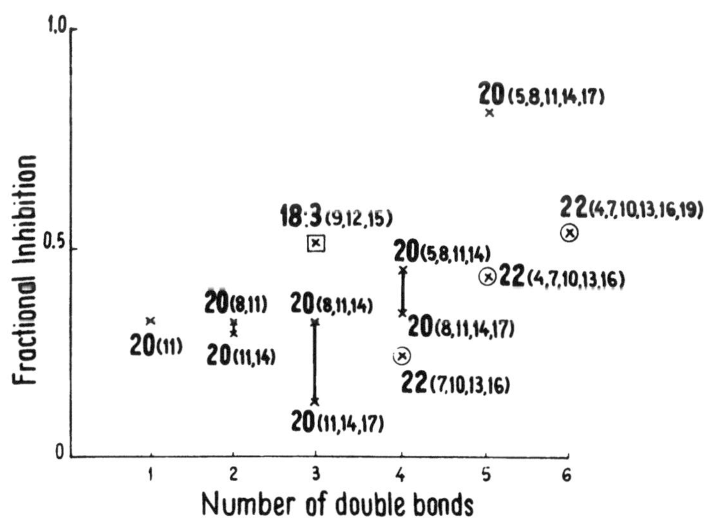

Fig. 1. Inhibitory effect of different unsaturated acids in the Δ6 desaturation of α-linolenic acid.

Equimolecular concentration of inhibitors incubated with 133.3 μM |1-^{14}C| α-linolenic acid as described in Experimental procedures.

methyl end. However, the Δ8 acids are stronger inhibitors than the Δ11 acids of same number of double bonds. The Δ5 unsaturated fatty acid of 4 double bonds is a stronger inhibitor than the equivalent Δ8 acid. In this case, the addition or a fifth double bond in Δ17 produces eicosa-5, 8, 11, 14, 17 pentaenoic acid that is most potent inhibitor of α-linolenic Δ6 desaturation found. The acids of 22 carbons of similar number of double bonds are less inhibitors than the acids of 20 carbons. The inhibitory effect increases with the number of double bonds providing that a 4 double bond is present. Therefore, the addition of double bond near both carboxylic and methylic ends of the molecule enhances the inhibitory power of polyunsaturated acids.

DISCUSSION

It may be admitted that the elements of Δ6 desaturation reactions: cytochrome b_5 reductase, cytochrome b_5 and Δ6 desaturase are partially embedded in the lipid part of the endoplasmic reticulum with the polar part outside. Therefore it may be admitted that very probably the acyl-CoAs are bound to the polar part of the enzyme by means of the CoA while the hydrocarbon tail is embedded in the lipid medium against the non polar part of the enzyme. The importance of the CoA in Δ6 desaturation was already shown (Brenner, 1974) by the inhability of the enzyme to desaturate free acids, acids bound to phospholipids, free alcohols of similar structure and pantetheine thiosters. The enzyme is absolutely specifc for a Δ6 desaturation and is different from the Δ9, Δ5 and Δ4 desaturases. Therefore the 6-7 carbons of the acid must be located in front of the active center of the enzyme, as it was already suggested (Brenner, 1971). Since as far as we know only fatty acids with preexisting cis Δ9 double bonds (not trans) are desaturated in Δ6, the cis 9-10 double bond is apparently necessary to orientate the position of the acid against the enzyme and/or to induce an activation of the hydrogens in 6-7 to evoke the synthesis of a divinyl structure (Brenner, 1971). A cis but not a trans double bond produces a kink in the molecule. Therefore the kink would be recognized by the enzyme.

An increase in the number of cis double bonds from oleic (9) to linoleic (9, 12) and α-linolenic (9, 12, 15) acid increases the desaturation (Brenner and Peluffo, 1966) and the same happens with 20:3 (9, 12) and 20:3 (9, 12, 15). This increase corresponds to an increase of the curvature of the molecule. Therefore it is possible to explain this effect if we assume that the enzyme also possesses a curve structure where the acyl-CoA will be located. An increase of curvature of the substrate would increase the fitting and bonding by cumulative London - Van der Waals interactions. An addition of 2 carbons in the tail of the acid from 18:2 (9, 12) to

20:2 (9, 12) and 18:3 (9, 12, 15) to 20:3 (9, 12, 15) would decrease the fitting and correspondingly the desaturation activity.

The results obtained studying the inhibitory effect of different fatty acids (Table 2 and 3 and Fig. 1) would coincide with this interpretation in general. Straight chain saturated acids as palmitic or stearic do not inhibit the enzyme, since in spite of the binding through the CoA, do not fit the enzyme curve (Brenner and Peluffo, 1966). Polyunsaturated acyl-CoA would fit the enzyme more or less depending on the number and position of double bonds. Acids of 20 carbons with double bonds in the middle of the chain inhibit Δ6 desaturation but the fitting increases when additional double bonds bend mainly the carboxyl end and the methyl end of the chain. Similar results are obtained with acids of 22 carbons. Therefore we may postulate that the interaction with the Δ6 desaturase occurs not only in the active centre but also through the double bonds and tail of acyl-CoA depending on the curvature of the molecule. This interpretation of the Δ6 desaturase structure is similar to the structure proposed by Brett et al (1971) for the Δ9 desaturase and Do and Sprecher (1975) for the Δ5 desaturase after testing a series of methyl branched fatty acids.

CONCLUSIONS

Rat liver microsomes contain a Δ6 desaturase that is active not only with the unsaturated acids of 18 carbons: oleic, linoleic and α-linolenic but also with similar acids of 20 carbons with the double bonds in Δ9, 12 and Δ9, 12, 15. The rate of the reaction increases with the number of double bonds of the substrate but the acids of 20 carbons are less desaturated than the corresponding homologs of 18 carbons proving that the enzyme recognizes the number and position of double bonds as well as the chain length. The investigation of the inhibitory effect of different polyunsaturated acids suggests that the acids would not only bind to the enzyme by means of the -CoA but also by other parts of the molecule different of the active center. The geometry of the molecule would be a substantial factor.

REFERENCES

ALANIZ M. J. T. de, PONZ G. & BRENNER R. R. (1975) Acta Physiol. Latinoam. 25, 3-13
BRENNER R. R. (1971) Lipids. 6, 567-575
BRENNER R. R. (1974) Mol. & Cell Biochem. 3, 41-52
BRENNER R. R. & PELUFFO R. O. (1966) J. Biol. Chem. 241 5213-5219
BRETT D., HOWLING D., MORRIS L. J. & JAMES A. T. (1971) Arch. Biochem. Biophys. 143, 535-547

CASTUMA J. C., CATALA A. & BRENNER R. R. (1972) J. Lipid Res. 13, 783-789
CASTUMA J. C., CATALA A., BRENNER R. R. & CHRISTIE W. W. (1974) Acta Physiol. Latinoam. 24, 31-39
DO U. H. & SPRECHER H. (1975) Arch. Biochem. Biophys. 171, 597-603.
GASPAR G., ALANIZ M. J. T. de & BRENNER R. R. (1975) Lipids. 10, 726-731
INKPEN C. A., HARRIS R. A. & QUACKENBUSH F. W. (1969) J. Lipid Res. 10, 277-282
KORNBERG A. & PRICER W. E. (Jr) (1953) J. Biol. Chem. 204, 329-371
NINNO R. E., TORRENGO M. A. P. de, CASTUMA J. C. & BRENNER R. R. (1974) Biochim. Biophys. Acta 360, 124-133
PELUFFO R. O. & BRENNER R. R. (1974) J. Nutrit. 104, 894-900

FETAL ACCUMULATION OF LONG-CHAIN POLYUNSATURATED FATTY ACIDS

M.A. Crawford, A.G. Hassam, G. Williams, and
W. Whitehouse

Department of Biochemistry, Nuffield Institute of
Comparative Medicine, The Zoological Society of London

INTRODUCTION

The developing brain accumulates long-chain (C20 and C22) polyenoic fatty acids, particularly during cell division (Crawford & Sinclair, 1972; Sinclair & Crawford, 1972). *De novo* synthesis of these acids does not occur in higher animals and they are derived either directly from food or by metabolism from the parent essential fatty acids, linoleate and α-linolenate.

It is difficult to alter the composition of the brain lipid fatty acids. Reduction of brain EFAs in young animals has been successfully achieved by feeding EFA-deficient diets to the mother during pregnancy. Reduction of brain EFAs using this technique was associated with retardation of brain development (Caldwell & Churchill, 1966; Paoletti & Galli, 1972; Sinclair & Crawford, 1973; Sun & Sun, 1974; Lamptey & Walker, 1976). Of special interest, Alling,Bruce, Karlsson & Svennerholm (1974) failed to alter the brain EFA in this manner, but they did observe alterations in the non-essential fatty acids involved in myelin. Their failure to change the brain EFA may have been due to the use in their diets of protein derived from fish and from which the phospholipids are not readily extracted by the neutral solvent systems in use. Fish phospholipids are particularly rich in the long-chain polyenoic acids. Dietary rehabilitation of animals deprived of EFA during brain growth has been shown to correct an early lipid deficiency but a permanent learning disability remained (Paoletti & Galli, 1972).

This evidence on lipids in particular and much other evidence on protein and on undernutrition in general (Dobbing, 1972; Winick

& Rosso, 1974) indicates that although the brain is remarkably well protected the most sensitive crucial period is during its development.

Different animal species eat different foods in which the fatty acids vary considerably. Adaptation to different food structures throughout evolution could have led either to differences in brain lipids or, if the resistance to change was sufficiently great, the same lipid profile would evolve, but associated with differnet degrees of brain development. In this paper we wish to present data from a comparative analysis of 45 different animal species together with results obtained from the study of the fetal accumulation of the brain EFA in the human and in the guinea-pig.

EXPERIMENTAL PROCEDURES

The techniques employed in the studies described in this paper have been presented in detail previously (Crawford, Casperd & Sinclair, 1976; Crawford, Hassam, Williams & Whitehouse, 1976).

Comparative Studies

We have examined the distribution of long-chain polyenoic acids in different animal species with the objective of establishing whether or not different food selection patterns were associated with differences in tissue fatty acids and in particular, brain fatty acids.

In a study on 45 different species we found substantial variation in the acyl groups of the phosphoglycerides in the liver. The wide range of polyenoic fatty acid composition in the liver ethanolamine phosphoglyceride (EPG) is shown in Fig. 1. For example, the linoleate was 47% of the total fatty acids in the liver EPG from wild zebra, but less than 1% in marine fish. By contrast, there was a remarkable constancy of the fatty acids of the brain grey matter EPG (Fig. 2)(Crawford, Casperd & Sinclair, 1976).

It was especially interesting that the large herbivores were able to accumulate substantial amounts of linoleate and α-linolenate in their liver phosphoglycerides but only relatively small amounts of the long-chain (C20 and C22) polyenoates. Yet it was only the long-chain polyenoates which were found in the brain EPG and in approximately similar amounts in all species. As it is the large mammals which have the smallest brains relative to their body mass, and have the smallest proportions of long-chain polyenoates in their tissue pools, these findings suggested to us that the long-chain polyenoates might have been important determinants of brain growth.

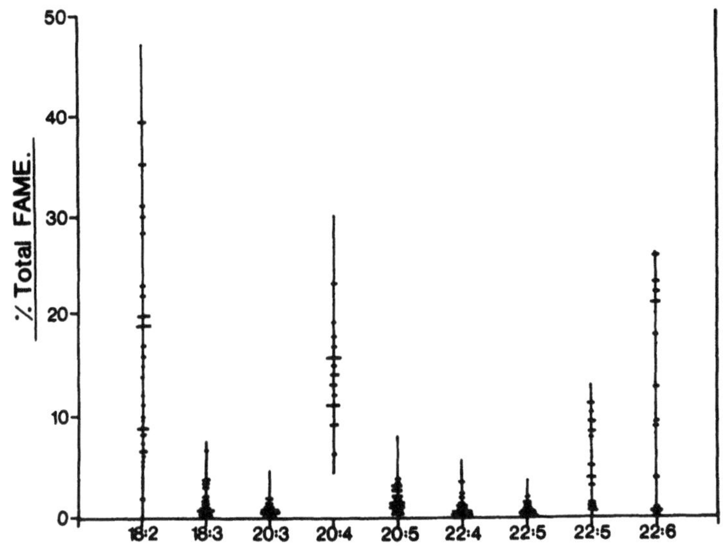

Fig. 1. Liver EPG polyenoic fatty acids in 45 species.

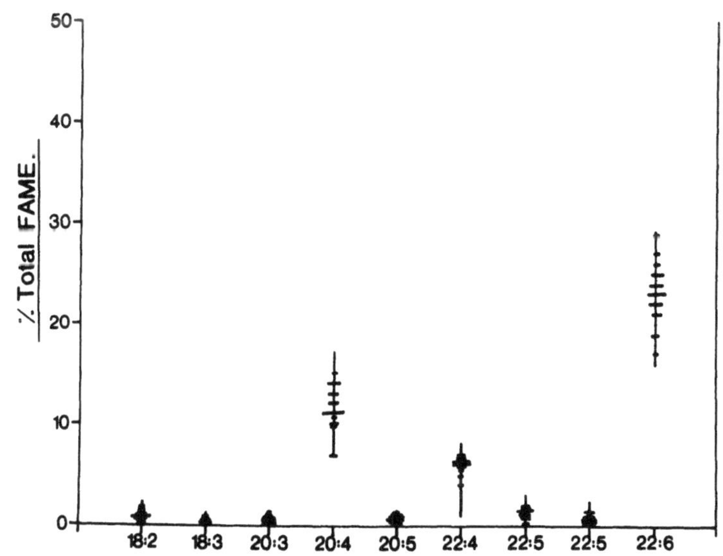

Fig. 2. Brain EPG polyenoic fatty acids in 45 species.

It seemed that the degree of brain and nervous system development was somehow related to the availability of long-chain polyenoates to the developing brain.

Human Studies

We have also examined the accumulation of long-chain polyenoates in the human fetus. Analysis of maternal cord blood, fetal liver and fetal brain, showed that the phosphoglycerides were richer in long-chain polyenoates on the fetal side than on the maternal side of the placenta and were richer still in the fetal liver and fetal brain. That is, from the maternal diet to maternal liver, placenta, fetal liver and fetal brain, there was stepwise progression in the degree of polyunsaturation and chain length. This "biomagnification" process for the long-chain polyenoates is shown in Table 1.

A computer projection curve of post-mortem material from 27 samples from human fetuses and infants of various ages is presented in Fig. 3 and gives an impression of what appears to be a reasonable generalisation, that possibly 60 - 70% of the adult complement of long-chain polyenoic acids has accumulated in the human brain by birth. Thereafter the rate at which the long-chain polyenoic acids accumulate, falls off. By contrast, the rate of accumulation of the long-chain saturated and mono-unsaturated acids (lignoceric and nervonic) involved in myelination increases during the first six months of post-natal life.

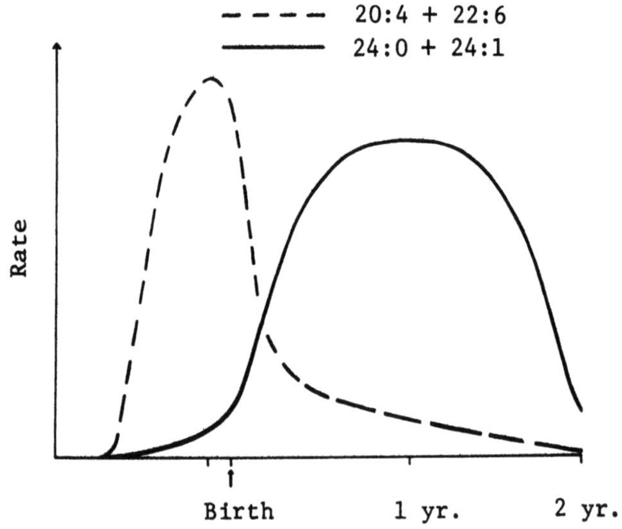

Fig. 3. Brain accumulation of long-chain polyenoic, saturated and mono-unsaturated fatty acids.

Table 1. Fatty Acids as Mean Weight % (± S.E.M.) in Phospholipid Fractions of each Tissue from the Human Fetus

	Percentage distribution of fatty acids			
	ω6 series		ω3 series	
	18:2	20:4	18:3	22:6
Choline phosphoglycerides:				
Maternal plasma (n=5)	23.0±1.4	8.2±0.71	0.93±0.14	4.4±0.31
Fetal cord blood (n=4)	8.6±0.6	15.6±0.9	0.22±0.01	6.6±0.33
Fetal liver (n=5)	5.1±0.3	17.8±1.2	0.10±0.01	6.1±0.5
Fetal brain (n=5)	1.1±0.2	16.2±1.3	0.10±0.01	7.2±0.72
Ethanolamine phosphoglycerides:				
Maternal R.B.C.s (n=5)	7.2±0.31	23.0±0.56	0.61±0.07	8.7±0.67
Fetal cord blood R.B.C.s (n=5)	2.5±0.23	27.1±1.6	0.40±0.02	11.0±0.39
Fetal liver (n=9)	0.96±0.1	30.4±0.89	0.23±0.01	14.8±0.78
Fetal brain (n=9)	0.47±0.05	22.6±1.3	0.12±0.02	18.4±1.2

Polar phospholipid of plasma is mainly choline phosphoglyceride, hence we used red blood cells (R.B.C.s) for ethanolamine phosphoglyceride analysis. Fetal material was obtained at second trimester.

Placental Transfer of Polyenoic Acids

To examine the fetal accumulation of polyenoates we have used guinea-pigs at about 25-30 days pregnant. They were dosed orally with 25-50 µCi of ^{14}C labelled α-linolenic acid. The animals were killed at time intervals up to 96 hours after administration.

The results showed that most of the isotope recovered from the maternal liver was as α-linolenic and even after 24 hours only 2% of isotope was recovered from docosapentaenoic (22:5ω3) and docosahexaenoic (22:6ω3) acids (Fig. 4).

Analysis of placental lipids showed consistently a far higher recovery of isotope from the long-chain polyenoic acids. More than 10% of the placental isotope was in the 22:5ω3 and 22:6ω3 fatty acids (Fig. 5). In the fetus the proportion of isotope in the long-chain polyenoic acids had further increased to over 25 % (Fig. 6).

The percentage of isotope recovered from the maternal liver was at its highest in the first 14 hours and by 24 hours had fallen to relatively low levels; probably much of the α-linolenic acid was oxidised. However, the isotope recovered from the placenta and fetus, although less than 1% of the administered dose, remained relatively constant over the 96-hour study period (Fig. 7). Distribution of the isotope within the ω3 fatty acids showed there was a continued shift in activity towards the long-chain fully desaturated acids, i.e. towards docosahexaenoic acid.

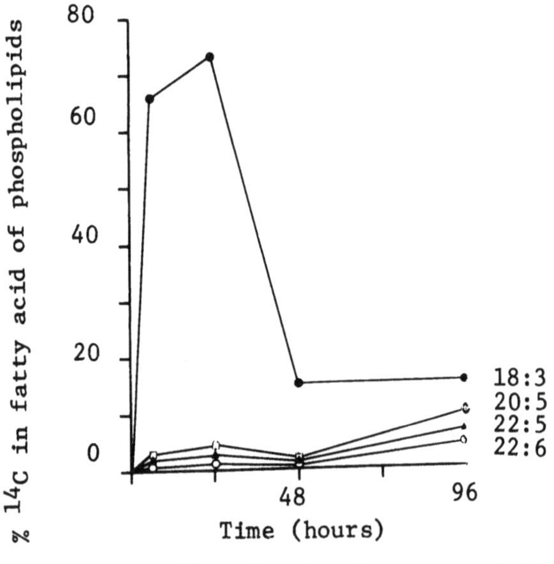

Fig. 4. Isotope distribution in maternal liver lipids.

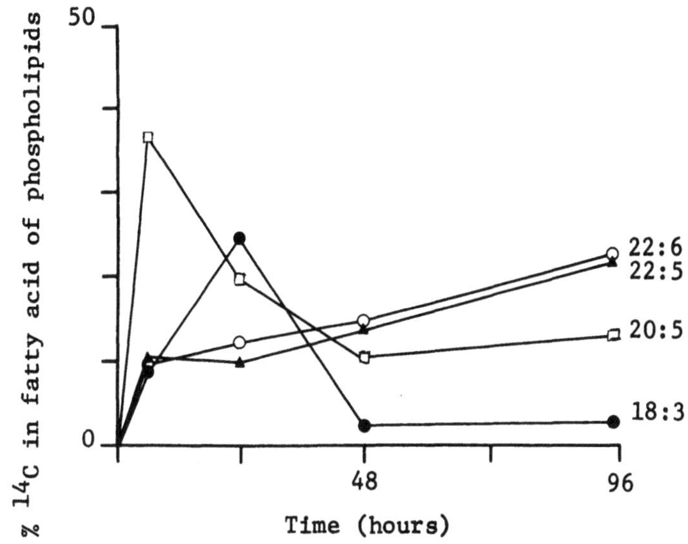

Fig. 5. Isotope distribution in placental lipids

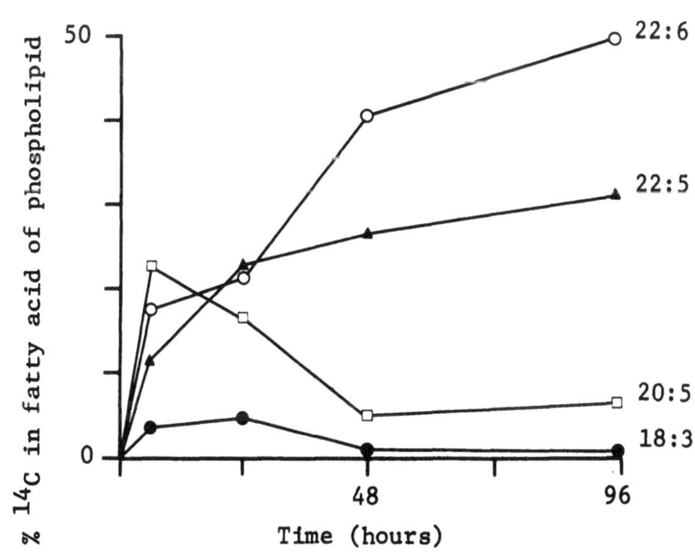

Fig. 6. Isotope distribution in fetal lipids

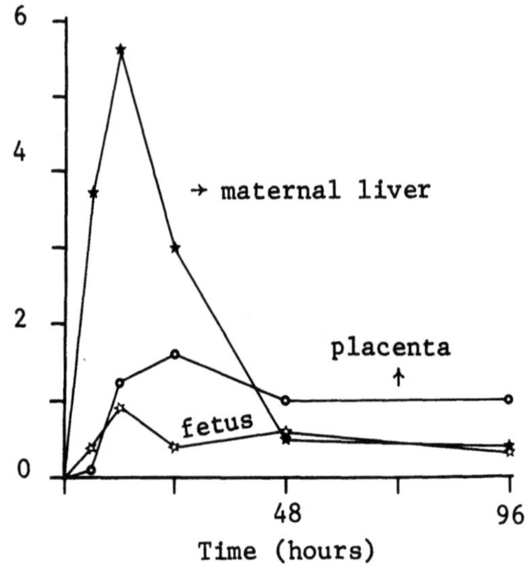

Fig. 7. Percentage recovery of isotope administered

CONCLUSIONS

The results demonstrate that the fetus is not simply dependent on maternal food intake, but that both placenta and fetus actively reprocess the essential fatty acids with the result that long-chain polyenoic acids are incorporated into the developing brain lipids. Although this process is clearly of great importance to brain development, the conversion of α-linolenic acid to docosahexaenoic acid is slow. A low desaturation rate of parent essential fatty acids is consistent with the *in vitro* studies reported by Professor Brenner and Dr. Sprecher earlier in this conference and with our *in vivo* observations in the rat (Hassam, Sinclair & Crawford, 1975; Hassam & Crawford, 1976; Hassam, Rivers & Crawford, 1976). In the cat the desaturase was not detected (Rivers, Sinclair & Crawford, 1975): *in vivo*.

Our results suggest that the placental barrier, together with the remetabolism of the EFA, provides a powerful control mechanism to ensure the supply of predominantly long-chain polyenoic acids to the fetal brain.

ACKNOWLEDGMENT

This work was supported by Medical Research Council grant no. G974/616/S.

REFERENCES

ALLING C., BRUCE A., KARLSSON I. & SVENNERHOLM L. (1974) J. Neurochem. 23, 1263

CALDWELL D. F. & CHURCHILL J. A. (1966) Psychol. Rep. 19, 99

CRAWFORD M. A., CASPERD N. M. & SINCLAIR A. J. (1976) Comp. Biochem. Physiol. 54B, 395

CRAWFORD M. A., HASSAM A. G., WILLIAMS G. & WHITEHOUSE W. (1976) Lancet, i, 452

CRAWFORD M. A. & SINCLAIR A. J. (1972) In: Lipids, Malnutrition and the Developing Brain. pp. 267. Eds. Elliot, K. & Knight, J. Associated Scientific Publishers, Amsterdam.

DOBBING J. (1972) In : Lipids, Malnutrition and the Developing Brain. pp. 9 Eds. Elliot, K. & Knight, J. Associated Scientific Publishers, Amsterdam.

HASSAM A. G. & CRAWFORD M. A. (1976) Nutr. Metabol. 20, 112

HASSAM A. G., RIVERS J. P. W. & CRAWFORD M. A. (1976) Nutr. Metabol. 20, suppl. 2. (in press)

HASSAM A. G., SINCLAIR A. J. & CRAWFORD M. A. (1975) Lipids 10, 417

MAPTEY M. G. & WALKER B. L. (1976) J. Nutr. 106, 86

PAOLETTI R. & GALLI C. (1972) In : Lipids, Malnutrition and the Developing Brain. pp. 121. Eds. Elliot, K. & Knight J. Associated Scientific Publishers, Amsterdam

RIVERS J. P. W., SINCLAIR A. J. & CRAWFORD M.A. (1975) Nature 258, 171

SINCLAIR A. J. & CRAWFORD M. A. (1972) J.Neurochem. 19, 1753

SINCLAIR A. J. & CRAWFORD M. A. (1973) Br. J. Nutr. 29, 127

SUN G. Y. & SUN A. Y. (1974) J. Neurochem. 22, 15

WINICK M. & ROSSO P. (1974) In : Early Malnutrition and Mental Development pp. 66. Eds. Cravioto, J., Hambraeus, L. & Vahlquist, B. Almquist & Wiksell, Stockholm

MATERNO-FETAL TRANSFER OF ^{14}C LINOLEIC AND ARACHIDONIC ACIDS

M. Pascaud, A. Rougier, and N. Delhaye

Laboratoire de Physiologie Métabolique et Nutrition
Université Paris 6 - 9 , quai Saint Bernard
Paris 5e, France

INTRODUCTION

Since the requirements of the developing fetus for linoleic acid are very high, it is interesting to investigate the adaptation of the pregnant mother and the fetus to this need. The present work reports a short-term investigation on the rat at the 20th day of pregnancy, i.e. one day before birth.

EXPERIMENTAL METHODS

15 µCi/100 g body weight of $|1\text{-}^{14}C|$-linoleic acid (Amersham) as the methyl ester in corn oil was given by stomach tube to pregnant Wistar rats. At 1, 3 and 6 hours after ingestion, the mothers were anaesthetized and the placentas and fetuses removed. These were decapitated and the blood collected in heparin. Lipid extraction, purification and analysis were achieved by the procedures of Lea & Rhodes (1954), Di Constanzo & Leclerq (1976) and Macheboeuf & Delsal (1943). Fatty acid analysis was carried out by gas-liquid chromatography on a 25% DEGS column at 180°C after transmethylation with 5% H_2SO_4 in methanol. The methyl esters were collected by condensation at the column outlet and specific activities were calculated and corrected according to Pascaud (1963).

RESULTS AND CONCLUSIONS

Comparison of the fatty acid compositions (Table 1) shows

Table 1. Lipid content and fatty acid composition of the maternal and fetal plasmas and of the placenta.

	Maternal Plasma	Placenta	Fetal plasma
Lipid content mg/ml or g	14	22.5	17
Fatty acid, %			
14:0	1.2	0.7	5
16:0	26	25.3	43
16:1	2.7	1.8	4
18:0	9.8	20.5	16.3
18:1	28	18.4	17.5
18:2ω6	8.3	12.2	3.5
20:4ω6	5.1	12.5	9.5
22:5ω6	3	0.8	-
22:6	9	5.3	-

that the ratio arachidonic/linoleic acids is four times higher in the fetal than in the maternal plasma, and that the placenta is the richest in these two acids.

Following the ^{14}C-retention and distribution after absorption of ^{14}C-linoleic acid, it was seen that the peak of activity in the fetal plasma (1% of the ingested dose at 3 hours) preceded the peaks of activity in the maternal plasma and placenta (8.6 and 1.2% at 6 hours respectively). Then the transfer of ^{14}C-linoleic and arachidonic acids is very rapid.

Fig. 1 shows the ^{14}C specific activities of the different fatty acids. In fact, linoleic acid is not only converted to arachidonic acid and higher ω6 polyenoic acids but is also degraded to acetate which is incorporated into the common fatty acids. The specific activities of linoleic and arachidonic acids are maximal for the maternal and fetal plasmas at about 3 hours but not in the placenta. Thus in accordance with Satomi & Matsuda (1973) this organ is not important in the synthesis of arachidonic acid. The very high peak of activity of arachidonic acid reflects a very rapid synthesis by the mother, by the fetus, or by both.

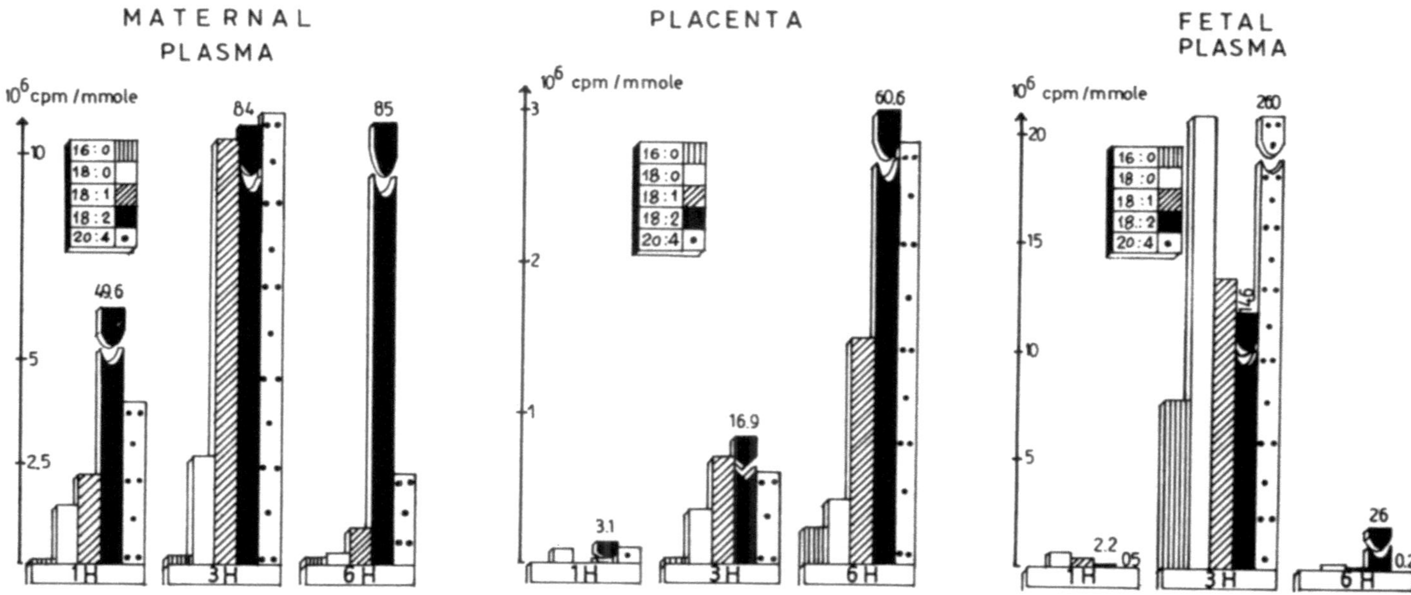

Fig. 1. ^{14}C incorporation in the fatty acids of the maternal and fetal plasmas and of the placenta after ingestion of $|1\text{-}^{14}C|$-linoleic acid.

Concerning the nature of the lipid molecules transporting linoleic and arachidonic acids, we studied two fractions, namely the phospholipid and non-phospholipid, the last one containing the free fatty acids. We should remember that with the excepcion of

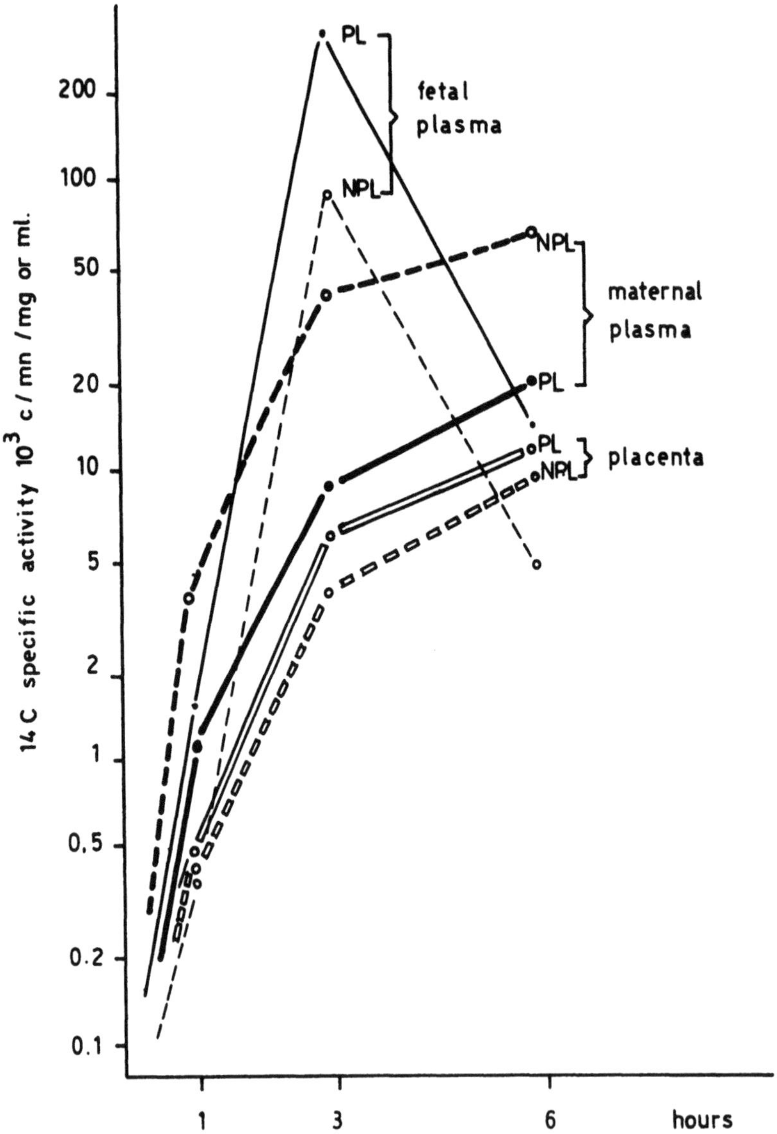

Fig. 2. ^{14}C specific activities of the phospholipid (PL) and non-phospholipid (NPL) fractions of the maternal and fetal plasmas and of the placenta after ingestion of $|1-^{14}C|$-linoleic acid.

fetus plasma arachidonic acid at 3 hs (Fig. 1), the specific activities of linoleic acid are about ten times higher than that of arachidonic acid, so the first are decisive for the total activity of the fraction. Whereas in the maternal plasma the non-phospholipid fraction is always more radioactive than the phospholipid fraction (Fig. 2) the opposite is true in the fetal plasma. In the case of the placenta, the difference is not large. These facts suggest that the placenta could function, between the mother and fetus blood, by selecting lipoproteins carrying linoleic and arachidonic acids and transferring these fatty acids from non-phospholipid molecules (probably triglycerides and cholesterol esters) to phospholipid molecules (probably lecithins) and later to new lipoproteins.

Concerning the tissular origin of fetal arachidonic acid, the question whether the mother or fetus liver or an extra-hepatic tissue is involved, is under investigation.

REFERENCES

DI CONSTANZO, G. & LECLERQ, C. (1976) in press.

LEA, CH. & RHODES, D. N. (1954) Biochem. J. 54, 467-472.

MACHEBOEUF, H. & DELSAL, J. L. (1943) Bull. Soc. Chimie Biol. 25, 116-121.

PASCAUD, M. (1963) J. Chromatog. 10, 125-130.

SATOMI, S. & MATSUDA, I. (1973) Biol. Neonate 22, 1-8.

I. LIPID INVOLVEMENT IN THE BIOGENESIS OF MEMBRANES

(B) BIOGENESIS AND ORGANIZATION OF CELLULAR MEMBRANES

ON THE ASYMMETRIC COMPOSITION OF PLASMA MEMBRANES

Eberhard G. Trams

National Institute of Neurological and Communicative Disorders and Stroke, Bethesda, Maryland 20014, USA

By intuition, we always have known that there was asymmetry in the function and composition of plasma membranes. After all, we had to accommodate into that structure specific ion channels, pumps, transporters, receptors and many other functions which clearly demand highly differentiated molecular assemblies. Until recently, we have visualized the plasma membrane primarily as a smooth contiguum which, depending somewhat on cell type, stretched over a round, polygonal or spindle-shaped cell body. This original concept of the structure of the plasma membrane had been strongly influenced by the appearance of the erythrocyte under the microscope. Reinforcement of this image occurred when the first laboratory preparations of isolated plasma membranes, cell ghosts, synaptosomes, etc. were inspected at higher resolutions with the transmission electron microscope; the prevailing image was one of a fairly smooth cytoplasmic envelope. The concurrent availability of cell ghosts and of purified plasma membrane preparations instigated numerous studies on the molecular composition of the membrane. One of the first chemical analyses of the erythrocyte membrane led to the formulation of the familiar lipid bi-layer model and such bi-layers were eventually reconstituted between two aqueous compartments (GOERTER & GRENDEL, 1925; DANIELLI & DAVSON, 1935; MUELLER _et al._, 1962). Addition of peptides, proteins, antibiotics and ionophores changed the functional characteristics of those bi-layer membranes, but on visual inspection these systems revealed remarkably smooth surfaces.

It was during this period that several investigators first raised the question of "sidedness" of certain membrane constituents. Some of our colleagues interrogated the lipid portion of the membrane, some the glycoproteins, while we primarily focused on the distribution of catalytic macromolecules, the so-called plasma membrane marker enzymes. Conceptually, the problem seemed rather simple because, 1. Different functions on the surface should be expressed by particular molecular assemblies which are spread over the surface as a mosaic (a planar asymmetry) and they should be recognizable, or separable, by those functions. 2. It was probable that the ion channels, transporters, pumps, etc. were not only distributed in a planar mosaic pattern but also subjected to a transverse asymmetry in respect to the inner and outer surface of the membrane. Thus, molecular fragments of the sodium pump might be recognizable by their capacity to bind sodium or ouabain, or by an exchange of ^{32}P on the respective surfaces of the membrane.

Improvements in methods applicable to membrane biochemistry followed. Purified plasma membrane preparations from a variety of cells became available for study and methods were developed whereby plasma membranes could be inspected from the inside or the outside. Progress was made in the in vitro reconstitution of membrane-like assemblies from subparts. The successful separation of membrane halves has now been reported and biochemical studies are underway on such preparations. In concept, the experimental approaches to the problem are simple providing the proper controls are used.

What currently complicates the molecular definition of the cell surface is the definition of "surface" itself. The scanning electron microscope has revealed a picture of the cell surface which is more complex and organized (or disorganized) than was anticipated. Evidently the question: what is external or internal in respect to the cell membrane, is not answered as easily as we thought it might be. It is apparent that numerous pseudopodia, filaments, invaginations or channels spread from a 'statistical mean surface' in either direction. In addition, there is a continuous evolution and involution of membrane matter during exocytotic or pinocytotic events. Moreover, membranes apparently undergo structural and chemical change during mitosis, migration, in viral transformation and with routine cyclical events in cellular metabolism (e.g., GRAHAM et al., 1973; KNUTTON et al., 1975; WETZEL, 1976; GOTTLIEB et al., 1974). Because such periodic changes must have a basis on the molecular level, we can look forward to a bewildering array of membrane chemistries. This means that the biochemical definition of membrane organization will require a substantial increase in the resolution of our separation and observation techniques. My review of the asymmetrical composition of the plasma membrane categorizes membrane elements essentially as if they were part of the simple envelope which we expected from our early views of the red cell.

SOME NOTES ON METHODOLOGY

Preparations for Investigating Membrane Asymmetry

One of the most useful attempts to identify specific membrane components is the isolation and characterization of so-called receptors. Affinity labeling with agonists or antagonists has led to the preparative isolation and partial characterization of several plasma membrane constituents, or what *a priori* are considered membrane components. Elsewhere in this symposium such experiments will be described in detail. Much thoughtful and elegant work has revealed in considerable detail the composition of some receptors as characterized by their binding affinities, but the mode of the signal propagation remains obscure at this time. The reconstitution of function so far has eluded many investigators and disputes about the molecular identity of receptors are commonplace.

Another approach to the study of membrane asymmetry has been based on the morphological and functional asymmetry of cells from organized tissues. Cellular polarity is expressed in the diverse biochemistry of the various membrane portions and refinements in plasma membrane isolation procedures have met with considerable success in ascribing membrane elements to topographical aspects of the cell. For instance, hepatocyte bile front has been separated from the sinus front and the contiguous membrane (EVANS & GURD, 1972; EVANS *et al.*, 1973). Differences in densities or marker molecule composition revealed topographical asymmetry in membrane composition. In some instances nature provides us with a selected portion of a cell membrane and we can bypass a tedious separation in the laboratory. During lactation, the apical membrane of the mammary gland cell forms the membrane which surrounds the milk fat globule, and this membrane can be harvested from milk with ease (PATTON & KEENAN, 1975). Our original observations on the sidedness of plasma membrane enzymes were made by comparing intact milk fat globules with membranes which had been isolated from the milk (PATTON & TRAMS, 1971).

Several other techniques have been exploited to answer queries about asymmetrical compositions of plasma membranes. To make an observation of the transverse asymmetry in the distribution of certain membrane molecules requires that one side of the membrane be excluded from a reaction medium while a constituent is chemically altered, labeled or identified by its properties. Many such experiments have been conducted with intact erythrocytes which were reacted and then lysed and reacted again. The reactant "saw" initially only the outer surface of the membrane and eventually both sides. Qualitative or quantitative differences observed under these conditions allowed conclusions about the composition or arrangement of molecules in the respective membrane halves. The method of looking alternately at one or the other side of the plasma membrane has been refined in several ways. Membrane fragments have a natural tendency to reseal into vesicles (MASON & LEE, 1973) and by choosing the proper conditions, membrane vesicles can be prepared with the cytoplasmic surface outside (inside-out) and be compared with the 'right side out' configuration. For model experiments, it is possible to reconstitute asymmetrical bi-layer vesicles in vitro (TRAUBLE & GRELL, 1971).

Another clever trick for studying plasma membrane asymmetry was by the use of virions (or Latex beads) which were encapsulated with host-specific bilayer (STOFFEL et al., 1975; LENARD & ROTHMAN, 1976; IWATA et al., 1976). Isolation of viral particles, which are thought to be enveloped in a facsimile of the plasma membrane bilayer of the host cell, allows for the harvesting of membranes which presumably have retained their native bi-layer organization. Recently, a more direct approach to the study of the membrane halves was attempted by producing batches of inner and outer membranes separately. Such separations for quantitative chemistry of inner and outer erythrocyte membrane halves has been achieved by an application of the freeze-fracture technique (FISHER, 1976).

Methods for Localizing Membrane Components

A variety of chemical probes have been inserted into membranes. Plant lectins (phytohemagglutinins) interact with specific N-acetylhexosamine-like sites on the cell surface. Concanavalin A (or its isotopic or fluorescent derivatives) has been a commonly used probe of this kind; from it we first learned about the changes in structural membrane organization in chemical or viral cell transformation (BEN-BASSAT et al., 1971). Although phytohemagglutinin binding to the cell surface is readily reversible, significant perturbation of membrane function can occur during the interaction period. For example, Concanavalin A stimulated phosphatidic acid synthesis in lymphocyte membranes (FISHER & MUELLER, 1971) but inhibited ecto-5'-nucleotidase in granulocytes (STEFANOVIC et al., 1975; SMOLEN & KARNOVSKY, 1976). Isotopic or fluorescent phytohemagglutinins might also be useful if we had to make an assessment of total surface area in a cell culture or suspension. This information could serve as a control in the determination of yields or recovery for preparative methods, but it requires that we have some information about the density of lectin binding sites for particular cells (NICOLSON et al., 1972).

The tagging of membrane receptors with traceable agonists or inhibitors is not only employed in procedures for the isolation of receptor complexes from the membrane but can be used for the identification of membrane topography which has an enriched or exclusive content of receptors. Thereby, specific membrane regions can be identified and preparative procedures adjusted to obtain fractions with selected membrane fragments.

The insertion of electron spin resonance probes into membranes is another technique; at first it provided primarily information about the physical-chemical state of the membrane (KORNBERG & McCONNELL, 1971). Several variations of this technique have been applied to the study of phospholipids in the intact erythrocyte and other membranes (WISNIESKI *et al.*, 1974; SHIMSHIK & McCONNELL, 1973). The observations suggest an asymmetric fluidity in the bi-layer phase of the membrane and a differential distribution of phospholipids, supporting data obtained by other methods.

One promising approach to the study of plasma membrane sub-assemblies has been in the use of covalently bonded reporter groups. Chemical attachment of such tracers has met with varying success. Several excellent discussions and reviews describe the methodology, reactants, results and pitfalls of the chemical labeling of membranes (KNAUF & ROTHSTEIN, 1971; GORDESKY *et al.*, 1975; CARRAWAY, 1975). The method has been most widely applied to the covalent labeling of free amino groups of plasma membrane constituents and therefore seems somewhat restrictive and non-specific. Similarly, the iodination of proteins with ^{131}I is not target specific but certainly it has yielded much insight about transverse asymmetry of membrane macromolecules.

The exchange of isotopic lipids with their unlabeled counterparts in the membrane provides us with an approach which keeps perturbations of the natural state of the membrane to a minimum, provided only trace amounts are used. The disadvantage of this technique is that given sufficient time, exchanges with other than membrane compartments will take place, a limitation which is also evident when lipophilic reagents are used for the chemical labeling of membranes. Such difficulties can be circumvented by the labeling of membrane constituents with group specific antibodies because they do not penetrate beyond the outer layers of the membrane. This technology provides us with a kind of class reaction for haptenic groups. The method has been applied mostly to membrane proteins, but it is also feasible to make antibodies against membrane lipids. A recent application in neurochemistry (RAPPORT & KARPIAK, 1976) demonstrates that we can look forward to some interesting results.

Some very elegant membrane chemistry has been based on enzymatic alteration of membrane constituents. This is a less invasive technique than the alterations produced by organic chemistry. It can be assumed that in some cases the changes which are produced by enzymatic attack are similar and perhaps identical to physiological reactions which occur on the membrane naturally. The removal of neuraminic acids from glycoprotein (or glycolipid) of the membrane by membrane affixed neuraminidases may occur in situ to produce alterations in adhesiveness, cell-cell recognition or in immunochemical phenomena. Many examples of enzymatic alterations are described in the literature and I cite only a couple of representative studies. The substitution of indigenous neuraminic acids with isotopic equivalents is a typical example (PRICER & ASHWELL, 1971). Removal of neuraminic acid from cell surface constituents has been reported to effect significant changes in the activity of ecto-enzymes (STEFANOVIC et al., 1976a; STEFANOVIC et al., 1975a), but some of our observations indicated that contaminant enzymes in the neuraminidase preparations may have been responsible for these results (TRAMS et al., 1976). Most of our knowledge about the asymmetrical composition of the lipid phases of the membrane has been deduced from experimentation with phospholipases (VERKLEIJ et al., 1973; RENOOIJ et al., 1976) and more detailed reports will be found elsewhere in this symposium. Finally, use has been made of the catalytic properties of some membrane marker enzymes to interrogate the topography of the membrane and to assign sidedness for these macromolecules. The information obtained by this method will be reviewed in more detail below.

The Asymmetric Composition of the Lipid Phase

We should evaluate our observations on the asymmetrical composition of the plasma membrane by assuming that the "statistical mean surface" of the cell represents a datum. Microelectrode work indicates that we should retain our concept of a contiguous lipid bi-layer over the entire cell. However, a part of the cell surface, or even the entire cell, may for brief periods expose the naked cytoplasm to the environment (CHAMBERS & CHAMBERS, 1961). Regeneration or biosynthesis of the membrane from pre-formed sub-assemblies in the cytoplasm can be achieved in a very brief period, as if in quantum increases (BARASSI & BAZAN, 1974; TRAMS et al., 1974). Nature has invented a very flexible apparatus to maintain or build whatever membrane is required at specific times. The liquid crystalline state of the bi-layer and the presumed fluidity of its constituent molecules are seemingly well adapted to such requirements.

Data obtained by a variety of methods show that in eukaryotic cells the choline containing phospholipids are preferentially localized in the outer leaflet of the bi-layer (70% and 80% for phosphatidylcholine and sphingomyelin, respectively) while phosphatidylinositol, phosphatidylserine and phosphatidic acid are concentrated in the inner leaflet (VERKLEIJ et al., 1973; RENOOJ et al., 1976; EMMELOT & VAN HOEVEN, 1975; KAHLENBERG et al., 1974). The distribution of phosphatidylethanolamine also is asymmetric with about 70 to 80% in the inner leaflet (BRETSCHER, 1972; GORDESKY & MARINETTI, 1973; HAEST & DEUTICKE, 1976). It should be noted that the major phospholipids of the endoplasmic reticulum are distributed in the opposite way (DePIERRE & DALLNER, 1975). This finding might lend support to the theory of plasma membrane biogenesis from the endoplasmic reticulum. There is some question about the distribution of one major membrane lipid, cholesterol. Some data indicate that the sterol is distributed about evenly between the two membrane halves (LENARD & ROTHMAN, 1976), but a partial enrichment (by about 2:1) in the outer leaflet has been suggested (FISHER, 1976). Gangliosides and related glycolipids appear to be concentrated in the outer half of the membrane (STOFFEL et al., 1975; GAHMBERG & HAKOMORI, 1975). Much of this information is inferential because the polar ganglioside moieties appear to be cryptic, i.e. not directly accessible to interrogation when the outer surface is explored for their presence by various means (TOMICH et al., 1976). This asymmetric composition of the lipid phase of the membrane has some other consequences which might be of importance to our understanding of membrane related phenomena. For instance, the PL behave in many experiments as if two different pools existed for each species (LENARD & ROTHMAN, 1976). Conceivably, these pools are created by the transverse separation of the PL, but it is probable that the matrix within each leaflet contains separate 'apparent' pools (WISNIESKI et al., 1974). The microviscosity (or fluidity) for each membrane leaflet is different (FEINSTEIN et al., 1975; KORNBERG & McCONNELL, 1971), and it appears as if the outer leaflet is more rigid (TANAKA & OHNISHI, 1976), partially as a consequence of comparative enrichment with cholesterol. Although more rigid, the exchange of lipid species by various means from the environment with the outer membrane leaflet proceeds with relative ease. By contrast, little or no exchange with the inner leaflet can be demonstrated in intact cells or sealed vesicles (TANAKA & OHNISHI, 1976; ROTHMAN & DAWIDOWICZ, 1975). When membrane biogenesis was studied with isotopic lipids, labeled fatty acids from the cytoplasmic pool predominantly were incorporated into the inner leaflet (RENOOJ et al., 1976) while lipids exchanged from lipoproteins were mostly incorporated into the outer leaflet. In addition, there is now good evidence that leaflet/leaflet exchange of lipid species, a so-called flip-flop of individual molecules from one side to the other, occurs only at very slow rates, i.e. in a matter of hours or days (ROTHMAN & DAWIDOWITZ, 1975). The plasma membrane seems quite amenable to compositional changes, be it that they are required by the physiological state of the cell or that they are induced experimentally. Thus, it is possible to increase the cholesterol content of the bi-layer (presumably in the outer leaflet) to the point where the killing power of a malignant cell is markedly inhibited (INBAR & SHINITZKY, 1974). Polar head groups of the phospholipids can be removed enzymatically from the membrane of a lobster nerve and yet the membrane lipids retain an ordered structure (SIMPKINS et al., 1971).

What we can infer from all these observations is that the two asymmetric membrane halves are maintained to a high degree to a pre-determined composition. Moreover, it is probable that the lipids in the bimolecular leaflet are assembled in an organized matrix for which lateral and transverse molecular assemblies are defined according to function. This remarkable degree of organization differs substantially from the reconstituted bi-layers and vesicles which have served so well for many experimental models of the membrane. Where is such organization derived from? Organization requires information, and I suggest that one of our impending tasks will be to identify the information bearers who direct the molecular assembly of the lipid phase in the membrane. It is evident that the membrane proteins and glycoproteins are the prime candidates for the messengers in this particular information system. I suggest that the membrane proteins have collocative properties (from the word collocation = an ordering of objects with reference to each other). The message is probably interred in the primary and secondary structure of the macromolecule (hydrophobic regions, etc.) which permits or requires the envelopment of these regions with a defined lipid phase. Systematically, the organization of the membrane lipid phase should be quite like the other information transfer mechanisms which operate in the cell, i.e.

DNA —— transcription → RNA
RNA —— translation → PROTEIN
PROTEIN — collocation → MEMBRANE BI-LAYER

Because we are not dealing with the assembly of sequences of building blocks which are covalently linked, we might expect somewhat different rules for the collocative process. Lipids, confined by natural or experimental means into an organization, may exhibit collocative properties of their own, and there is some evidence that this is so (MICHAELSON et al., 1973; ISRAELACHVILI, 1973). I think that one experimental approach will consist of some sort of "fishing expedition" whereby highly purified membrane proteins are trolled through lipid solutions with the expectation that a specific bait (= membrane protein) will attract a specific catch (= lipid). One such, apparently successful, experiment has been reported (STRUVE et al., 1975). This is an appropriate juncture to turn to a discussion of the protein composition of the plasma membrane.

Asymmetry in Membrane Protein Composition

By tradition, we have divided proteins into two classes: 1. functional proteins and 2. structural proteins, but I suspect that many so-called structural proteins are mislabeled because we have yet to discover their function. Evolution probably has selected multipurpose molecules as the most economical means to operate a cell. The reason that membrane protein structure and function have not been matched extensively is to be found in laboratory methodology. With fine tools such as disc gel electrophoresis, membrane proteins can be separated from each other with a high degree of resolution. This method, for instance, is chosen when we wish to assign a molecular weight to a particular protein which we had studied and identified by other means, or a membrane protein pattern is compared with that obtained from other membranes. This method can also be used to make assignments about the sidedness of membrane proteins because diverse functional groups can be labeled in situ and subsequently identified on a gel. The methodology is in principle the same as applied to the labeling of functional groups of lipids on the respective membrane halves. We now must standardize methods and develop an overall picture of the membrane proteins in respect to weight, composition and function.

Before we discuss the asymmetric distribution of plasma membrane proteins we might discuss briefly some generalizations about them.

Our knowledge about the composition and structure of membrane proteins comes from isolation and purification attempts as well as from experiments with model compounds (e.g., ionophores), all of which indicate that we should expect substantial diversity. In terms of size, we can anticipate a range from perhaps 1000 to several 100,000 daltons. We may also assume that most membrane proteins are amphipathic. The hydrophobic region could be provided by a ten-unit sequence of non-polar (CAPALDI & VANDERKOOI, 1972) amino acids or by a large hydrocarbon chain as in the case of the proteolipids. It is often assumed that the hydrophobic regions of the membrane proteins are immersed in the lipid bi-layer while the more polar moieties face the cytoplasmic compartments. This assumption, however, is backed insufficiently by experimentation, because proteins may form hydrophilic channels (pores) across the membrane which are lined inside with hydrophilic charged groups.

Much of our knowledge of the asymmetric composition of the membrane has come from carbohydrate chemistry. Glycosylated proteins are relatively abundant in the membrane and the carbohydrate moieties have provided convenient markers for analysis. The role of the glycoproteins in the membrane, their composition and orientation does require a separate chapter. The reader may wish to consult a number of excellent reviews which emphasize the biochemistry of membrane glycoproteins (STECK, 1974; WINTERBURN & PHELPS, 1972; QUARLES, 1975; SHUR & ROTH, 1975).

Our particular method for studying the asymmetric composition of plasma membranes employs the biocatalytic properties of some membrane proteins. Plasma membrane marker enzymes are enzymes which are thought to be characteristic solely or mainly for the plasma membrane (SOLYOM & TRAMS, 1972). They can be assayed under conditions which reveal their association with either one of the membrane halves. The same principles apply as in the localization of lipid elements: exclusion of one membrane side from the reactant medium. The constraints in this type of experiment are 1. that the cell surface is accessible to the substrate and 2. that the substrate does not penetrate into compartments other than the one which is being interrogated. The first constraint presents no problem if intact, isolated cells can be used, such as erythrocytes, lymphocytes, ascites tumor cells, etc. Cells in monolayer tissue cultures are very useful, but if such experiments were attempted with cells in an organized tissue, proof of the accessibility of the substrate to the cell surface might be difficult to obtain. Therefore, if we are working with intact suspended cells or with a tissue culture monolayer, and we observe metabolic alteration of an exogenous substrate, we presumably can infer that the reaction was catalyzed by an enzyme which functioned on the cell surface. The term ecto-enzyme is descriptive and in general use. If, on the other hand, no catalytic activity is observed with intact cells, but it is subsequently unmasked when the cells are lysed, it can be assumed that the reaction was catalyzed by an enzyme which was localized on the cytoplasmic aspect of the membrane, provided that the participation of other cytoplasmic elements has been excluded. With modifications, this rationale can be applied also to inside-out ghost cells or to sealed vesicle preparations, etc. Conditions for these experiments should be chosen carefully. Vesicles which are thought to be sealed, or cells which are believed to be intact might become permeable during the observation period. Conversely, membrane fragments which are thought to have both sides of the membrane exposed to the medium might reseal (MASON & LEE, 1973). Some of the theoretical and practical aspects of such experimentation have been discussed elsewhere (DePIERRE & KARNOVSKY, 1973; WALLACH & LIN, 1973; TRAMS & LAUTER, 1974). It is more difficult to demonstrate that the requirements of the second constraint have been met, because it is not always easy to design control experiments. These should ascertain that a substrate to be used will interact with one side of a membrane only. It is not sufficient to assume that polar substrates will not be transferred into the cell interior when they are added to a medium containing intact cells. Active or facilitated transport of substrates may occur, even though we assume that the membrane was impermeable to the substrate. Metabolic transformations may occur with such rapidity that extracellular catalysis is mimicked by the system. We had observed such a rapid transformation of medium adenosine to inosine in CNS monolayer cultures (TRAMS & LAUTER, 1975), and this observation fitted very well into a scheme of extracellular nucleotide metabolism which we had previously formulated. We assumed, therefore, that an ecto-adenosine deaminase was responsible for the metabolism of extracellular adenosine. By the use of adequate controls, we eventually demonstrated that adenosine was deaminated in the cytoplasm and that the cell excreted the newly formed inosine immediately into the surrounding medium. In such experiments, pulse labeling techniques and use of different isotopes (e.g., ^{32}P, ^{33}P) may be brought to bear on the problem of proper controls. Certainly it is not sufficient to show that the cells were viable (by dye penetration, staining of the nucleus, etc.), and then to conclude that a substrate had been excluded from the cytoplasm.

On the Asymmetric Distribution of Plasma Membrane Enzymes

To make generalizations about this topic might constitute an impropriety, because we do not yet have a good overview of the subject. The study of the "sidedness" of plasma membrane enzymes has been restricted to a few cell types only, although what appears to be a more general pattern has emerged. I shall try to project what we might find as a probable distribution pattern in eukaryotic cells. We anticipate that the plasma membrane enzymes are key building blocks in the molecular assembly of the membrane. An understanding of their distribution and functions should reveal a great deal about membrane organization. Therefore, it is disappointing that, with few exceptions, we know so little about the role of these enzymes. Cell topography should be taken into consideration in the asymmetric distribution of the membrane enzymes. Some of the membrane enzymes are more active (or abundant) on a particular cell surface as was first demonstrated by histochemical techniques (e.g., NOVIKOFF, 1957). An ATPase, for instance, was visualized on the basal surface of a cell but not on the others. On that one surface, ATPase presumably accommodated a functional demand of the cell. Membrane fractionation techniques also revealed considerable asymmetry in topical distribution of membrane enzymes (WISHER & EVANS, 1975). Hopefully, if we study a sufficient number and diversity of cell types, an element of predictability about membrane enzyme topography will emerge.

My discussion is concerned mainly with the transverse asymmetry in the distribution of plasma membrane enzymes. The scheme shown below indicates an inside-outside distribution pattern which is based on observations from several different laboratories.

Localization Pattern of Plasma Membrane Enzymes

in Eukaryotic Cells

Catalytic site at the external surface (Ecto-Enzymes)	Catalytic site at the cytoplasmic surface
5'-Nucleotidase	Adenylate cyclase
Ca^{2+}-ATPase	Na^{+}, K^{+}-ATPase
ADPase	phosphoinositide kinase
Phosphodiesterase I	Glyceraldehyde-3-Phosphate dehydrogenase
Glycosyltransferases	
Cholinesterase	

Assignments Indefinite or Ambiguous

AMP deaminase; neuraminidase;
leucyl-β-naphthylamidase;
p-nitrophenylphosphatase;
inorganic pyrophosphatase;
nucleotide pyrophosphatase
sialyltransferase; NADH oxidase

There is very good evidence that several enzymes are ecto-enzymes while others function at the cytoplasmic (internal) side of the membrane. I have defined assignments as indefinite or ambiguous where there were conflicting observations, when insufficient evidence was available or when there was a suggestion that the enzyme in question can operate either at the inside or at the outside. Furthermore, individual enzymes may not be found in the plasma membrane of all cells, or conversely, their occurrence in a few cell types constitutes a special case.

The strongest cases for ecto-enzymes as ubiquitous membrane constituents can be made for the 5'-nucleotidase and for a divalent cation-stimulated nucleoside polyphosphatase (Ca^{2+}-ATPase). If these two enzymes are not obligatory on the external surface of all eukaryotic cells, then they certainly occur rather commonly as ecto-enzymes. Moreover, they probably occur in several forms, as iso-enzymes, and their properties appear to undergo changes with the physiological or developmental state of the cell (BLOMBERG & RAFTELL, 1974; HEWLETT *et al.*, 1976; ROSENBLATT *et al.*, 1976). What follows is a brief discussion of the distribution and properties of the best characterized plasma membrane enzymes.

Ecto-5'-nucleotidase (adenosinemonophosphate-5'-phosphoester hydrolase, EC 3.1.3.5) appears to be a glycosylated lipoprotein

(WIDNELL & UNKELESS, 1968). Its lipid moiety may be constituted of a single covalently bonded sphingomyelin residue. The enzyme catalyzes the hydrolysis of the 5'-phosphoester bond of mononucleotides with fair specificity but does not discriminate well between various purine and pyrimidine riboside analogs. A variety of cations stimulate its activity but Mg^{2+} and Co^{3+} have been used as activators most often. Zn^{2+} ions are inhibitory. Recent studies have shown that ecto-5'-nucleotidase is inhibited by Concanavalin A in C-6 glioma cells (STEFANOVIC et al., 1975b) and in guinea pig granulocytes (SMOLEN & KARNOVSKY, 1976) but not in mouse spleen lymphocytes (POMMIER et al., 1975). The apparent velocity of the enzyme in different cells varies over 3 orders of magnitude. It must be recalled, though, that specific enzymatic activity is usually cited as velocity per unit time per mass unit of protein (e.g., $mol \cdot min^{-1} \cdot mg^{-1}$ of tissue protein). For ecto-enzymes, this quotation is patently unsuitable because the specific activity would be more meaningful if quoted per unit surface. This unit, however, is difficult to determine for all but the most elementary cell shapes. The specific enzymatic activity can vary drastically when cellular development or metabolism (and perhaps environment) is altered. Rate of growth in a WI-38 culture was inversely correlated with the specific activity of 5'-nucleotidase. A 6-10 fold increase in enzyme activity was noted when the cell population reached a certain age (SUN & AGGARWAL, 1974). In contrast, the activity of this enzyme was reduced about 12 fold when murine myeloma cells in culture came into contact (LeLIEVRE et al., 1971). In our laboratory it was found that the specific activity of ecto-5'-nucleotidase decreased in neonatal mouse astroblast cultures when they were propagated through 10-20 passages (TRAMS & LAUTER, unpublished data).

Antibodies have been raised against 5'-nucleotidase of mouse liver plasma membranes by using purified membrane preparations from the same source (GURD & EVANS, 1974) and the antiserum inhibited enzyme activity. Antisera raised against rat liver as well as rat fat cell plasma membranes inhibited ecto-5'-nucleotidase in isolated rat fat cells (NEWBY et al., 1975). However, it is not known whether the immune-specific and the catalytic site of the enzyme are identical.

Ecto-nucleotide polyphosphatase. This enzyme was first observed in nucleated erythrocytes (WENKSTERN & ENGELHARDT, 1957); generally, it is described as a Ca^{2+}-ATPase or Mg^{2+}-ATPase (adenosinetriphosphate-5'-phosphoester hydrolase, EC 3.6.1.3). The enzyme has a high specificity for the γ-phosphate of nucleoside-triphosphates but a broad specificity for the riboside moiety. It is not quite clear if extracellular ADP is hydrolyzed by this enzyme, but the current evidence favors the existence of a separate ecto-ADPase.

Nucleosidemonophosphates do not serve as substrates. Ecto-ATPase activity requires the presence of a divalent cation in most systems. For operational reasons, the application of the term Ca^{2+}-ATPase or Mg^{2+}-ATPase is satisfactory, because in many cells the enzyme is activated by various concentrations of either cation. We have found that the ecto-ATPase is optimally stimulated by Mn^{2+} at concentrations of 3 x 10^{-4}M or below, but we have also observed some exceptions (for instance, in mouse neuroblastoma where 10^{-3}M Ca^{2+} was most effective and Mn^{2+} did not stimulate). Mn^{2+} concentrations in excess of 3 x 10^{-4}M are inhibitory. The ecto-ATPase seems widely distributed in eukaryotic cells. There has been an earlier report that the enzyme was not present on the surface of intact neurons (CUMMINS & HYDEN, 1962) but it certainly is present on the surface of intact neuroblastoma cells (TRAMS & LAUTER, 1974; STEFANOVIC et al., 1976b). Conventional sulfhydryl compounds or inhibitors have little effect on the enzyme and ouabain is not inhibitory. We have found that ecto-ATPase of cultured CNS cells was inhibited by certain phenothiazine derivatives. Micromolar concentrations of thiazines and tricyclic antidepressants were inhibitory in rat leukocytes (MEDZIHRADSKY et al., 1975). There has been one report that adipocyte membrane Mg^{2+}-ATPase was markedly stimulated by insulin and by Concanavalin A (JARRETT & SMITH, 1974). A moderate increase in specific activity of ecto-ATPases with an increase in cell density per culture was recently reported in N-18, N-115 and NN cell lines (STEFANOVIC et al., 1976b). NN cell cultures in our laboratory exhibited a marked decline of specific enzymatic activity as cell density increased in continuous culture (TRAMS & LAUTER, unpublished data). The addition of digitonin (3 μM to 1 mM) to intact CNS cells produced a selective stimulation or inhibition of ecto-ATPase. There was, however, no apparent pattern of inhibition or stimulation by which neuronal or glial cells could be differentiated.

Ecto-ADPase. This enzyme was first described in chick embryo fibroblasts (PERDUE, 1970) and in human erythrocytes (PARKER, 1970). It apparently is distinct from the ecto-ATPase. We have observed the activity of an ecto-adenosinediphosphate phosphoesterhydrolase (EC 3.6.1.6) in various tissue culture monolayers but have not made an extensive study of the enzyme. Similarly to the ecto-ATPase, ecto-ADPase is stimulated by either Ca^{2+}, Mg^{2+} or Mn^{2+} (PERDUE, 1970) and it has a broad specificity in respect to the nucleoside residue.

Ecto-Phosphodiesterases (EC 3.1.4.1). We have described earlier the presence of an enzyme on the surface of several cell types which hydrolyzed p-nitrophenylphosphothymidine (TRAMS & LAUTER, 1974). Recently, we have incubated intact CNS derived cells with bis-p-nitrophenylphosphate and observed the apparent hydrolysis of this substrate in the medium. During the course of this investigation we found, however, that bis-p-nitrophenylphosphate enters intact cells at an appreciable rate and is hydrolyzed in the cytoplasm. Because adequate control experiments (with corrections for substrate penetration) have not been done yet, we feel that the assignment of phosphodiesterase as an ecto-enzyme requires substantiation. We have also observed the presence of a 3', 5'-cAMP ecto-phosphodiesterase in rat glioblastoma (C-6) which was not inhibited by theophylline. Elsewhere PDE was found in liver plasma membranes (LLOYD-DAVIES et al., 1972) and milk fat globule membranes (cf. PATTON & KEENAN, 1975) but an assignment of sidedness was not made.

There is a substantial amount of information on ecto-glycosyl-transferases and ecto-cholinesterases and their presumptive roles on the cell surface. We have not dealt much with these enzymes in our laboratory and the reader is referred to expert reviews.

Enzymes on the Cytoplasmic Surface of the Membrane

Adenylate cyclase (EC 4.6.1.1). Abundant information is available on the properties and distribution of this enzyme and it need not be recited here. Characteristically, this enzyme is associated with the cytoplasmic aspect of the plasma membrane (TRAMS & LAUTER, 1974; CUTLER, 1974). It, therefore, should serve as an excellent marker for preparations in which the integrity of right-side-out or inside-out vesicles has to be assayed. Adenylate cyclase, also, is one of the membrane enzymes for which a physiological role is clearly established. The localization of this enzyme on the inside of the plasma membrane makes good biological sense. We advocated the use of adenylate cyclase as a marker for the inside of the plasma membrane but must emphasize that the enzyme is not necessarily confined to that aspect of the membrane in all species. In *Bordetella pertussis* an extracyto-plasmic adenylate cyclase enables intact organisms to form cAMP from exogenous ATP (HEWLETT et al., 1976).

The Na^+, K^+-ATPase (EC 3.6.1.4). This key element of the sodium pump is localized on the cytoplasmic surface of the plasma membrane, and that is where we should expect to find it for teleologic reasons. Much structural and functional information about this enzyme is available and there is no need to elaborate here. It may suffice to recall that the lipid environment of purified (and presumably native) Na^+, K^+-ATPase is crucial for its proper functioning and that much evidence points to phosphatidyl-serine as a necessary complement for catalytic activity. There is also some evidence that the phospholipid bi-layer vicinal to the ATPase is more ordered *in situ* than the overall phospholipid bi-layer of the membrane (GRISHAM & BARNETT, 1972).

Several other enzyme proteins seem to be associated with the cytoplasmic aspect of the plasma membrane but not enough information is available to allow for a statement of preponderant sidedness.

Some characteristic plasma membrane marker enzymes for which inside-outside distribution is uncertain or ambiguous. Nucleotide pyrophosphatase (EC 3.6.1.9) is an excellent plasma membrane marker enzyme, particularly in hepatocytes (SKIDMORE & TRAMS, 1970; LAUTER et al., 1972; TOUSTER et al., 1970). It was shown that the enzyme is a sialoglycoprotein of low substrate specificity with a molecular weight of 130,000-137,000 in liver plasma membranes and that it was localized on the hepatocyte surface (EVANS, 1974; BISCHOFF et al., 1975). In the milk fat globule membrane nucleotide pyrophosphatase seemed largely confined to the inside of the membrane (PATTON & TRAMS, 1971) as it was in Chang hepatocytes and in KB cells (TRAMS & LAUTER, 1974). On the other hand, in isolated guinea pig hepatocytes this enzyme had all the characteristics of an ecto-enzyme (TRAMS & LAUTER, 1974).

Considering these discrepancies in the inside/outside distribution of nucleotide pyrophosphatase it might be prudent to defer a sidedness assignment until more data are available. A similar ambiguity exists in the case of p-nitrophenylphosphatase (p-NPPase) (EC 3.1.3.1) which we found to be preferentially localized on the outer surface of HeLa cells, KB cells and isolated guinea pig hepatocytes (TRAMS & LAUTER, 1974). In guinea pig polymorphnuclear leukocytes p-NPPase also occurred predominantly as an ecto-enzyme (DePIERRE & KARNOVSKI, 1974). Ecto-enzyme activity was ascribed to this enzyme in neonatal hamster astrocytes and mouse neuroblastoma (STEFANOVIC et al., 1976a) but our own work with these cells did not indicate that the enzyme was a definite ecto-enzyme. Nitrophenylphosphatase activity was restricted to the cytoplasmic aspect of rat kidney cortex basolateral plasma membranes (ERNST, 1975) and predominantly on the cytoplasmic side of transformed 3T6 and 3T3 cells (ROZENGURT & HEPPEL, 1975). An ambiguity of the distribution of p-NPPase can also be inferred from observations on Ehrlich ascites cells (LOEFFLER & SCHNEIDER, 1975).

There is similar ambiguity in the case of inorganic pyrophosphatase (pyrophosphate phosphohydrolase, EC 3.6.1.1). We had reported that this enzyme appeared to function as an ecto-enzyme in some cultured cells (REICHERT et al., 1974) and similar observations were made elsewhere (STEFANOVIC et al., 1975a), but the findings were more indicative than conclusive.

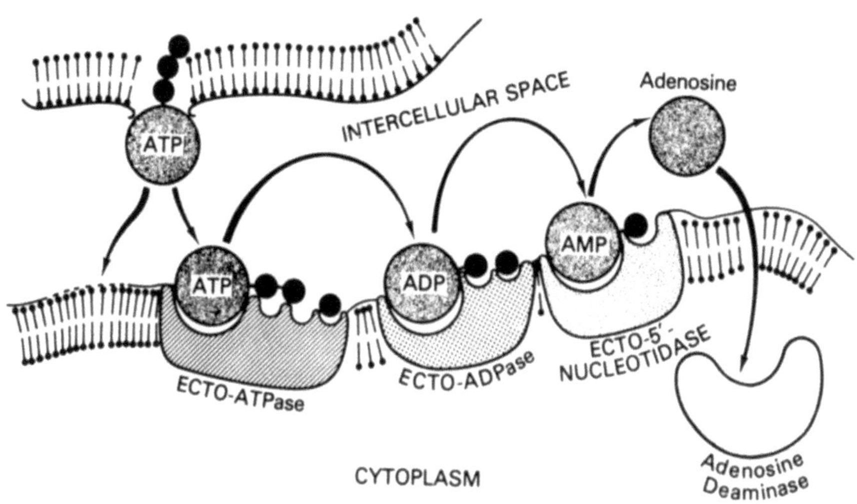

Fig. 1. When translocated cytoplasmic ATP impinges on the cell surface it is metabolized by a series of ecto-enzymes to adenosine.

SUMMARY AND CONCLUSIONS

I have emphasized plasma membrane proteins which are biocatalysts, because they are the focus of our own investigations. Many of them function as ecto-enzymes, when seemingly the proper substrates are not present. I suggest that several of the ecto-phosphoesterhydrolases (ATPase-ADPase-5'-nucleotidase) constitute part of a control system which modulates excitability threshold in the membrane. Translocated cytoplasmic ATP (ABOOD et al., 1962; TRAMS, 1974; ROSENBLATT et al., 1976) induces a transient membrane instability. Inactivation of ATP, commencing with the ecto-ATPase, may terminate with the conversion of adenosine, which also affects membrane permeability, to inosine. The analogy to the acetylcholine: cholinesterase system is evident. This may be an ancient cellular mode of communication which developed early in evolution when syncytia first formed and which may now be superseded by more finely tuned controls (adrenergic, cholinergic, etc.). See Fig. 1.

I have also reviewed briefly the asymmetric composition of the lipid bi-layer portion of the membrane. It seems that most of the plasma membrane enzymes either contain lipid moieties or require the presence of lipids for activity. Conversely, the organization of the lipid bi-layer in the membrane may depend on these proteins. Organization requires information and I suggest that we adopt the term collocation for the information transfer which directs the organization of the lipid phase in the membrane. In this review of methods and current concepts, as they relate to the study of membrane asymmetry, much information that would have been instructive could not be included.

REFERENCES

ABOOD L. G., KOKETSU K. & MIYAMOTO S. (1962) Am. J. Physiol. 202, 469-474
BARASSI C. A. & BAZAN N. G. (1974) J. Cell. Physiol. 84, 101-114
BEN-BASSAT H., INBAR M. & SACHS L. (1971) J. Membr. Biol. 6, 183-194
BISCHOFF E., TRAN-THI T-A. & DECKER K. F. A. (1975) Eur. J. Biochem. 51, 353-361
BLOMBERG F. & RAFTELL M. (1974) Eur. J. Biochem. 49, 21-29
BRETSCHER M. S. (1972) J. Mol. Biol. 71, 523-528
CAPALDI R. A. & VANDERKOOI G. (1972) Proc. Nat. Acad. Sci. USA 69, 930-932
CARRAWAY K. L. (1975) Biochim. Biophys Acta 415, 379-410
CHAMBERS R. & CHAMBERS E. L. (1961) Explorations into the nature of the living cell, Harvard Univ. Press, Cambridge, MA.
CUMMINS J. & HYDEN H. (1962) Biochim. Biophys. Acta 60, 271-283
CUTLER L. S. (1974) J. Cell. Biol. 63, 74a.
DANIELLI J. F. & DAVSON H. (1935) J. Cell. Physiol. 5, 495-508
DE PIERRE J. W. & DALLNER G. (1975) Biochim. Biophys. Acta 415, 411-472
DE PIERRE J. W. & KARNOVSKI M. L. (1973) J. Cell. Biol. 56, 275-303
EMMELOT P. & VAN HOEVEN R. P. (1975) Chem. & Phys. of Lipids 14, 236-246
ERNST S. A. (1975) J. Cell. Biol. 66, 586-608
EVANS W. H. (1974) Nature 250, 391-394
EVANS W. H., BERGERON J. J. M. & GESCHWIND I. I. (1973) FEBS Letters 34, 259-262
EVANS W. H. & GURD J. W. (1972) Biochem. J. 128, 691-700
FEINSTEIN M. B., FERNANDEZ S. M. & SHA'AFI R. I. (1975) Biochim. Biophys. Acta 413, 354-370
FISHER D. B. & MUELLER G. C. (1971) Biochim. Biophys. Acta 248, 434-448
FISHER K. A. (1976) Proc. Nat. Acad. Sci. USA 73, 173-177
GAHMBERG C. G. & HAKOMORI S-I. (1975) J. Biol. Chem. 250, 2447-2451
GORDESKY S. E. & MARINETTI G. V. (1973) Biochem. Biophys. Res. Comm. 50, 1027-1031
GORDESKY S. E., MARINETTI G. V. & LOVE R. (1975) J. Membr. Biol. 20, 111-132
GORTER E. & GRENDEL F. (1925) J. Exptl. Med. 41, 439-443
GOTTLIEB D. I., MERRELL R. & GLASER L. (1974) Proc. Nat. Acad. Sci. USA 71, 1800-1802

GRAHAM J. M., SUMNER M. C. B., CURTIS D. H. & PASTERNAK C. A. (1973) Nature 246, 291-295
GRISHAM C. M. & BARNETT R. E. (1972) Biochim. Biophys. Acta 266, 613-624
GURD J. W. & EVANS W. H. (1974) Arch. Biochem. Biophys. 164, 305-311
HAEST C. W. M. & DEUTICKE B. (1976) Biochim. Biophys. Acta 436, 353-365
HEWLETT E. L., URBAN M. A., MANCLARK C. R. & WOLFF J. (1976) Proc. Nat. Acad. Sci. USA 73, 1926-1930
INBAR M. & SHINITZKY M. (1974) Proc. Nat. Acad. Sci. USA 71, 2128-2130
ISRAELACHVILI J. N. (1973) Biochim. Biophys. Acta 323, 659-663
IWATA K. K., WISNIESKI B. J. & HUANG Y. O. (1976) Fed. Proc. 35, 1532
JARRETT L. & SMITH R. M. (1974) Fed. Proc. 33, 1361
KAHLENBERG A., WALKER C. & ROHRLICK R. (1974) Canad. J. Biochem. 52, 803-806
KNAUF P. A. & ROTHSTEIN A. (1971) J. Gen. Physiol. 58, 190-223
KNUTTON S., SUMMER M. C. B. & PASTERNAK C. A. (1975) J. Cell. Biol. 66, 568-576
KORNBERG R. D. & McCONNELL H. M. (1971) Proc. Nat. Acad. Sci. USA 68, 2564-2568
LAUTER C. J., SOLYOM A. & TRAMS E. G. (1972) Biochim. Biophys. Acta 266, 511-523
LE LIEVRE L., PRIGENT B. & PARAF A. (1971) Biochem. Biophys. Res. Comm. 45, 637-643
LENARD J. & ROTHMAN J. E. (1976) Proc. Nat. Acad. Sci. USA 73, 391-395
LLOYD-DAVIES K. A., MICHELL R. H. & COLEMAN R. (1972) Biochem. J. 127, 357-368.
LOFFLER M. & SCHNEIDER F. (1975) FEBS Letters 56, 66-69
MASON W. T. & LEE Y. F. (1973) Nature New Biol. 244, 143-145
MEDZIHRADSKY F., LIN H-L. & MARKS M. J. (1975) Life Sci. 16, 1417-1428
MICHAELSON D. M., HORWITZ A. F. & KLEIN M. P. (1973) Biochemistry 14, 2637-2645
MUELLER P. & RUDIN D. O. (1967) Biochem. Biophys. Res. Comm. 26, 398-404
MUELLER P., RUDIN D. O., TI TIEN H. & WESCOTT W. C. (1962) Nature 194, 979-980
NEWBY A. C., LUZIO J. P. & HALES C. N. (1975) Biochem. J. 146, 625-633
NICOLSON G., LACORBIERE M. & YANAGIMACHI R. (1972) Proc. Soc. Exp. Biol. & Med. 141, 661-663
NOVIKOFF A. (1957) Cancer Res. 17, 1010-1027
PARKER J. C. (1970) Am. J. Physiol. 218, 1568-1574
PATTON S. & TRAMS E. G. (1971) FEBS Letters 14, 230-232
PATTON S. & KEENAN T. W. (1975) Biochim. Biophys. Acta 415, 273-309

PERDUE J. F. (1970) Biochim. Biophys. Acta 211, 184-193.
POMMIER G., RIPERT G., AZOULAY E. & DE PIEDS R. (1975) Biochim. Biophys. Acta 389, 483-494
PRICER W. E. & ASHWELL G. (1971) J. Biol. Chem. 246, 4825-4833
QUARLES R. H. (1975) in The Nervous System (Tower, D. B., ed.), Vol. I, pp. 493-501, Raven Press, New York.
RAPPORT M. M. & KARPIAK S. E. (1976) Trans. Am. Soc. Neurochem. 7, 216.
REICHERT W. H., LAUTER C. J. & TRAMS E. G. (1974) Biochim. Biophys. Acta 370, 556-563
RENOOIJ W., VAN GOLDE L. M. G., ZWAAL R. F. A. & VAN DEENEN L. L. M. (1976) Eur. J. Biochem. 61, 53-58
ROSENBLATT D. E., LAUTER C. J. & TRAMS E. G. (1976) J. Neurochem. (in press)
ROTHMAN J. E. & DAWODOWICS E. A. (1975) Biochemistry 14, 2809-2816
ROZENGURT E. & HEPPEL L. A. (1975) Biochem. Biophys. Res. Comm. 67, 1581-1588
SHIMSHIK E. J. & McCONNEL H. M. (1973) Biochemistry 12, 2351-2360
SHUR B. D. & ROTH S. (1975) Biochim. Biophys. Acta 415, 473-512
SIMPKINS H., PANKO E. & TAY S. (1971) Biochemistry 10, 3851-3855
SKIDMORE J. R. & TRAMS E. G. (1970) Biochim. Biophys. Acta 219, 93-103
SMOLEN J. E. & KARNOVSKY M. L. (1976) Fed. Proc. 35, 1452
SOLYOM A. & TRAMS E. G. (1972) Enzyme 13, 329-372
STECK T. L. (1974) J. Cell. Biol. 62, 1-19
STEFANOVIC V., LEDIG M. & MANDEL P. (1976b) J. Neurochem. 27, 799-805
STEFANOVIC V., MANDEL P. & ROSENBERG A. (1975a) Biochemistry 14, 5257-5260
STEFANOVIC V., MANDEL P. & ROSENBERG A. (1976a) J. Biol. Chem. 251, 493-497
STEFANOVIC V., MANDEL P. & ROSENBERG A. (1975b) J. Biol. Chem. 250, 7081-7083
STOFFEL W., ANDERSON R. & STAHL J. (1975) Hoppe-Zeyler Z. Physiol. Chem. 356, 1123-1129
STRUVE W. G., WATKINS M. S., GOLDSTEIN D. J. & BULGER J. E. (1975) Fed. Prof. 34, 326.
SUN A. & AGGARWAL B. B. (1974) J. Cell. Biol. 63, 339a.
TANAKA K-I. & OHNISHI S-I. (1976) Biochim. Biophys. Acta 426, 218-231
TOMICH J. M., MATHER I. H. & KEENAN T. W. (1976) Biochim. Biophys. Acta 433, 357-364
TOUSTER O., ARONSON N. N., DULANEY J. T. & HENDRICKSON H. (1970) J. Cell. Biol. 47, 604-618
TRAMS E. G. & LAUTER C. J. (1974) Biochim. Biophys. Acta 345, 180-197
TRAMS E. G., LAUTER C. J., KOVAL G. J., RUZDIJIC S. & GLISIN V. (1974) Proc. Soc. Exp. Biol. Med. 147, 171-176.

TRAMS E. G., LAUTER C. J. & BANFIELD W. (1976) J. Neurochem. (in press)

TRAMS E. G. & LAUTER C. J. (1975) Biochem. J. 12, 681-687

TRAMS E. G. (1974) Nature 252, 480-482

TRAUBLE H. & GRELL E. (1971) Neurosci. Res. Program Bull. 9, 373-380

VERKLEIJ A. J., ZWAAL R. F. A., ROELOFSEN B., COMFURIUS P., KASTELEJEIJN D. & VAN DEENEN L. L. M. (1973) Biochim. Biophys. Acta 323, 178-193

WALLACH D. F. H. & LIN P. S. (1973) Biochim. Biophys. Acta 300, 211-254

WENKSTERN, T. V. & ENGELHARDT W. A. (1957) Biokhimia 22, 911-916

WETZEL B. (1976) Proc. Workshop Advances Biomed. Appl. SEM IIT Res. Inst., Chicago, IL, pp. 135-139

WIDNELL C. C. & UNKELESS J. C. (1968) Proc. Nat. Acad. Sci. USA 61, 1050-1057

WINTERBURN P. J. & PHELPS C. F. (1972) Nature 236, 147-151

WISHER M. H. & EVANS W. H. (1975) Biochem. J. 146, 375-378

WISNIESKI B. J., PARKES J. G., HUANG Y. O. & FOX C. F. (1974) Proc. Nat. Acad. Sci. USA 71, 4381-4385

INFLUENCE OF THE LIPID ENVIRONMENT ON THE PROPERTIES OF RHODOPSIN IN THE PHOTORECEPTOR MEMBRANE

S.L. Bonting, P.J.G.M. van Breugel, and F.J.M. Daemen

Department of Biochemistry, University of Nijmegen

Nijmegen, The Netherlands

INTRODUCTION

The vertebrate rod photoreceptor cell (Fig. 1) contains in its outer segment some 1000-2000 regularly stacked flat sacs. The membrane of these sacs, the photoreceptor membrane, consists - like all biomembranes - of a lipid bilayer and membrane protein. Its major

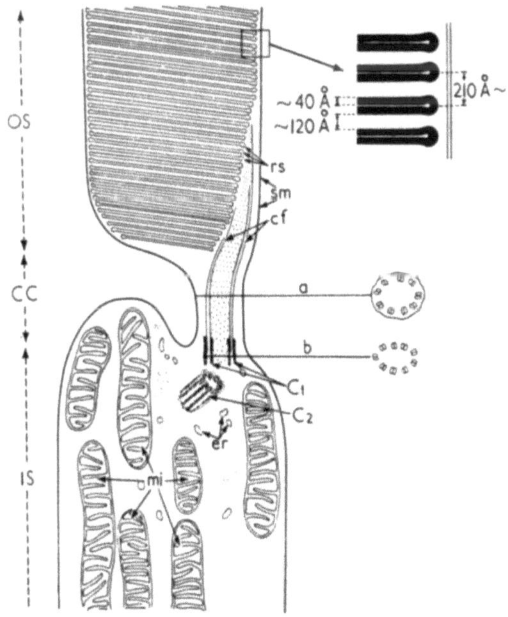

Fig. 1. Structure of vertebrate rod photoreceptor cell. Top is outer segment with rod sacs

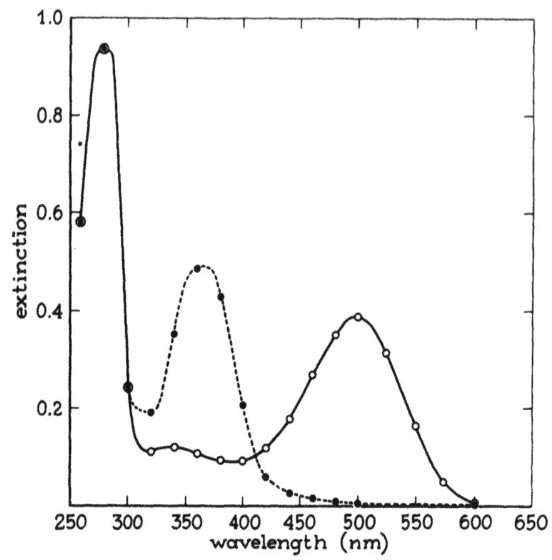

Fig. 2. Absorption spectrum of solubilized rhodopsin. Solid line: before illumination, dotted line: after illumination in the presence of hydroxylamine.

(~ 87%) membrane protein is the visual pigment rhodopsin, consisting of a protein (opsin) and a chromophoric group (11-cis retinaldehyde). The presence of rhodopsin, with its characteristic absorption spectrum (Fig. 2) and photosensitivity, as the major membrane protein and the relative ease of isolation of the photoreceptor membrane (de Grip et al., 1972; Bonting, 1973) make this membrane a suitable object for the study of lipid-protein interactions (Daemen, 1973).

All rod sacs are separated from each other, and all but the bottom few are closed and separated from the outer membrane. Autoradiographic studies have shown that the sacs are formed continuously at the bottom through invagination of the outer membrane and then move up to the top in about 10 days in warmblooded vertebrates (review: Young, 1976). There they are shed in groups of 8-30 and removed through phagocytic action of the overlying pigment epithelium. From the different labeling patterns for rhodopsin and the phospholipids it can be concluded that the pigment molecules are inserted at the time of invagination and then stay in the same membrane until removal of the sacs, whereas the phospholipids are subject to turnover during the migration of the sacs from bottom to top.

Composition of the Photoreceptor Membrane

Mild homogenization of the bovine retina and centrifugation of the homogenate in a sucrose density gradient permits isolation of an aqueous suspension of photoreceptor membranes (de Grip et al., 1972). The chemical composition of the membrane is presented in Table 1.

Table 1. Composition of bovine rod outer segment membrane

% of dry weight				
Protein	40%	of which rhodopsin is > 85%		
Phospholipids	50%	phosphatidylcholine (PC)		44%
		phosphatidylethanolamine (PE)		39%
		phosphatidylserine (PS)		14%
		fatty acids	$C_{16:0}$	12%
			$C_{18:0}$	18%
			$C_{22:0}$	50%
Cholesterol	3%			
% of wet weight				
Rhodopsin	10%	(2.5-3.5 mM)		
Phospholipids	15%	(100-200 mM)		

Lipid extraction with chloroform-methanol, followed by quantitative thin layer chromatography yields the phospholipid pattern, while fatty acids are determined by gas chromatography (Borggreven et al., 1970). Protein and cholesterol are determined colorimetrically.

Striking features are the low cholesterol content and the high content of unsaturated fatty acids (esp. C 22:6). This indicates a highly fluid lipid bilayer. In agreement with this is the finding that the rhodopsin molecules rotate freely and rapidly (rotation time 20 μsec) in the plane of the membrane (Brown, 1972; Cone, 1972), and that they diffuse in this plane at a rate of 0.3 μm/sec (Liebman & Entine, 1974). Calculations from the latter two observations indicate a viscosity equal to that of olive oil.

Some Properties of Rhodopsin

Rhodopsin is an insoluble glycoprotein with a molecular weight of 39,000 (Daemen et al., 1972). Its absorption spectrum has three absorption peaks, the α-peak at 500 nm, the β-peak at 340 nm and the γ-peak at 278 nm (Fig. 2, solid line). The first two are due to the chromophoric group, the last one to the protein part. Upon illumination the α-peak disapppears and a new one at 380 nm (or 360 nm in the presence of NH_2OH) appears (Fig. 2, dotted line). The α-peak has a molar absorbance of 40,300 (Daemen et al., 1972).

Intermediate	λ_{max}
rhodopsin	500 nm
↓ $h\nu$, $-197^{\circ}C$	
bathorhodopsin	543 nm
↓ $> -140^{\circ}C$	
lumirhodopsin	497 nm
↓ $> -40^{\circ}C$	
metarhodopsin I	478 nm
↓ H^+, $> -15^{\circ}C$	
metarhodopsin II	380 nm
↓ $> -5^{\circ}C$	
metarhodopsin III	455 nm
↓ H_2O, $+15^{\circ}C$	
retinaldehyde + opsin	383 nm

Fig. 3. Photolytic sequence of rhodopsin. Absorption maxima of the intermediates and the temperature at which they are stable are indicated.

Upon illumination rhodopsin undergoes a series of chemical changes, beginning with the isomerization of 11-cis retinaldehyde to its all-trans isomer and ending with the release of the chromophore. The intermediates have been spectrally identified by freezing the reaction sequence at various low temperatures (Fig. 3). At physiological temperature the conversion of rhodopsin to metarhodopsin I takes place in microseconds, that of metarhodopsin I ⟶ II in milliseconds, and the decay of the latter substance requires several minutes. Further details about rhodopsin and its photolytic sequence may be found in reviews (see e.g. Abrahamson & Wiesenfeld, 1972; Bonting, 1969; Bonting & Daemen, 1974).

Parameters, which are used in our study of the influence of the lipid environment on rhodopsin, are:

1. the 500 nm absorbance, which is proportional to the amount of intact rhodopsin.
2. the thermal stability, defined as the temperature at which the 500 nm absorbance is reduced to 50% in 10 min.

3. the regeneration capacity, which is the amount of rhodopsin (in percent of original amount) that can be obtained after photolysis and reaction with excess 11-cis retinaldehyde.
4. the photolytic sequence metarhodopsin I → II → III, which may be altered and which is studied by means of recording absorption spectrophotometry.

Delipidation and Reconstitution of Photoreceptor Membranes

In order to study the lipid-protein interaction in the photoreceptor membrane we have changed the micro-environment of the rhodopsin molecules by removal and replacement of lipids. Two different delipidation procedures have been applied: 1. treatment of photoreceptor membranes with phospholipase-C (Borggreven et al., 1971), and 2. affinity chromatography over a concanavalin A-sepharose-4B column (van Breugel et al., 1976).

The first procedure has the advantage of not requiring the use of detergents, but yields only partial hydrolysis of phospholipids. Incubation with phospholipase C from B. cereus can achieve hydrolysis of 80-90% of all phospholipids to diglycerides and phosphate esters without any loss of 500 nm absorbance. The resulting aqueous membrane suspension, after washing, still contains diglycerides and part of the original phosphatidylserine content. Hexane extraction can remove the diglycerides. Subsequent treatment with phospholipase A_2, followed by extraction of the hydrolysis products, removes all phospholipids to below a stoichiometric ratio of 1 mol per mol rhodopsin without loss af 500 nm molar absorbance (Borggreven et al., 1972). This result, illustrated in Fig. 4, refuted the earlier claim

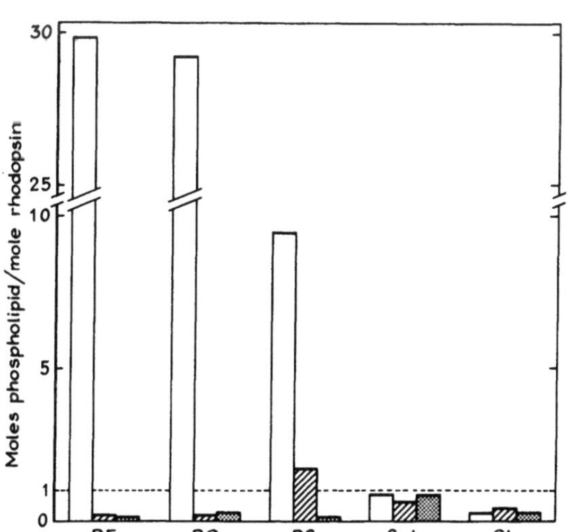

Fig. 4. Removal of phospholipids from photoreceptor membrane by treatment with phospholipase-C, hexane and phospholipase A_2, respectively (form Borggreven et al., 1972).

that the chromophoric group would be attached to an amino group-bearing phospholipid (Poincelot et al., 1969). In the present experiments treatment is merely with phospholipase-C, the hexane and phospholipase A_2 treatments being omitted.

The second procedure has the advantage of removing virtually all phospholipids, but it requires solubilization in the detergent dodecyltrimethylammoniumbromide (DTAB), which can later be removed by dialysis (Hong & Hubbell, 1972). Detergents are known to penetrate membrane proteins (Helenius & Simon, 1975) and to decrease rhodopsin stability to some extent depending on the detergent used (Daemen, 1974). However, since the properties of the preparations obtained by affinity chromatography closely resemble those of phospholipase-C treated membranes (Table II), we may rule out detergent artifacts in the former preparations. Affinity chromatography over a concanavalin A-sepharose-4B column, followed by elution with α-methylmannose (Fig. 5), yields a very pure rhodopsin preparation with an A_{278}/A_{500} ratio of 0.18-0.23, which is pure upon SDS-polyacrylamide gel electrophoresis and contains less than 0.3 mole phospholipid per mole rhodopsin (< 0.5% of original phospholipid content). The detergent can be removed by dialysis against 5 mM HEPES buffer (pH 6.6), containing dithioerythritol and EDTA (1 mM each) for 3-5 days at 4°C with at least 4 changes of dialysis buffer, yielding an aqueous suspension of pure rhodosin with a recovery of about 80%.

Reconstitution of the delipidated photoreceptor membrane or rhodopsin preparations with selected phospholipids is achieved by shaking them together with additional DTAB for 1 h (Hong & Hubbell, 1973). The detergent is then removed by the dialysis procedure described in the previous paragraph. Recovery of rhodopsin ranges from 80-100%. Freeze-fracture electronmicroscopy shows that rhodopsin is incorporated into lipid bilayer structures (Chen & Hubbell, 1973).

Table II

Rhodopsin preparation	Rhodopsin parameter: thermal stability °C	regeneration %	MI	⇒MII	⇒MIII
rod outer segment membranes	70	92	+	+	+
lipid-and detergent-free rhodopsin	53	48	+	-	-
phospholipase-C treated membranes	59	47	+	-	-
reconstituted lipid-rhodopsin vesicles	70	85	+	+	+

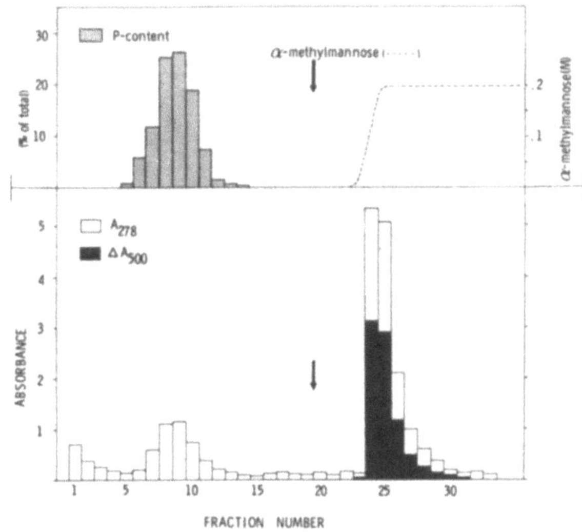

Fig. 5. Affinity chromatography of rhodopsin over a concanavalin A-sepharose 4B column in the presence of 100 mM dodecyltrimethylammoniumbromide (from van Breugel et al., 1976).

Effects of Phospholipase Treatment

Tratment with phospholipase-C leads to a reduction in amphipathic lipids by conversion of a majority of the phospholipids to neutral diglycerides. Electronmicroscopy shows that a membrane structure remains, but that the neutral lipids are to a large extent extruded in the form of lipid droplets (Fig. 6). The residual mem-

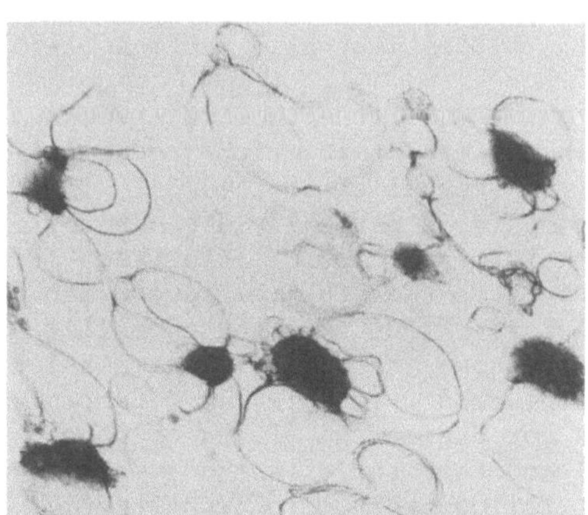

Fig. 6. Transmission electron micrograph of phospholipase-C treated photoreceptor membranes in which over 80% of the phospholipids are hydrolyzed. Dark areas represent lipid droplets (courtesy Drs. E.L. Benedetti and J. Olive, Paris).

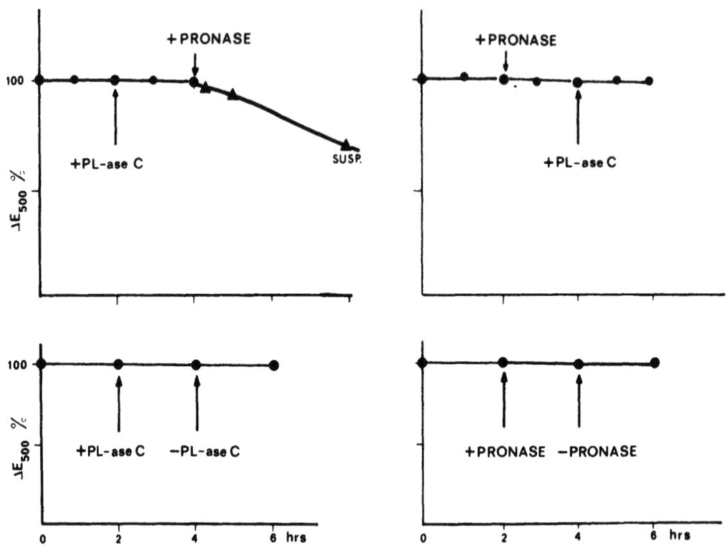

Fig. 7. Effect of treatment by pronase and phospholipase-C on the 500 nm absorption of rhodopsin in photoreceptor membranes.

brane must be composed of remaining phospholipids (mainly phosphatidylserine) and rhodopsin. The rhodopsin molecules in the residual membrane structure must be much closer packed than in the native photoreceptor membrane. This is confirmed by our finding that in freeze-fracture studies these membranes are invariably cross-fractures, and that at lower levels of lipolysis aggregation of protein molecules is observed. These observations are in agreement with those in studies of phospholipase-C treated erythrocyte membranes (Verkley et al., 1973; Limbrick & Knutton, 1975).

Phospholipase-C treatment of the photoreceptor membrane also has a prounUnced affect on the extent to which rhodopsin is attacked by proteolytic enzymes (van Breugel et al., 1975). In the case of untreated membranes rhodopsin is partially degraded by incubation with pronase without a marked loss of 500 nm absorbance (Fig. 7 , bottom right). After prior treatment of the membranes with phospholipase-C there is a subtantial loss of 500 nm absorbance (Fig. 7, top left), and gel electrophoresis shows that this is accompanied by more extensive digestion of the rhodopsin molecule. Control experiments indicate that this is not due to the effect of phospholipase per se (Fig. 7, bottom left) and that the effect does not occur when phospholipase-C treatment follows that with pronase (Fig. 7, top right). These findings indicate that removal of phospholipids makes the rhodopsin molecule more vulnerable to attack by proteolytic enzymes. Thus rhodopsin must normally be deeply embedded in the hydrophobic core of the phospholipid bilayer with only a small part of the molecule exposed to the aqueous phase.

Effects of Changes in Lipid Environment on 500 nm Absorbance

Phospholipid removal through either phospholipase-C treatment (even if followed by hexane extraction and phospholipase A_2 treatment) or affinity chromatography does not lower the 500 nm absorbance, as mentioned before. Only rather extensive proteolytic degradation of rhodopsin causes lowering of this parameter. This indicates that the chromophoric center, which is responsible for the 500 nm absorbance, is buried inside the opsin molecule. Hence, this parameter is probably too insensitive to indicate more superficial changes in configuration of the opsin molecule, which could occur upon changes in the lipid environment of rhodopsin. It is, therefore, important to see how the other three parameters react to such changes.

Effects of Changes in Lipid Environment on Thermal Stability

The thermal stability of rhodopsin, defined as the temperature at which in 10 min 50% of the 500 nm absorbance is lost, is indeed sensitive to changes in lipid environment (Table II, 2nd column). Whereas native rhodopsin has a high thermal stability (70°C), affinity chromatography and phospholipase-C treatment lower this parameter to 53°C and 59°C, respectively. Reconstitution of such delipidated preparations with egg phosphatidylcholine (66 mol per mol rhodopsin) to rhodopsin-lipid vesicles leads to full recovery of the thermal stability.

Effects of Changes in Lipid Environment on Regeneration Capacity

The regeneration capacity of rhodopsin, defined as the amount of rhodopsin (in precent of the original amount) obtained after photolysis and reaction with excess 11-cis retinaldehyde, is 92% for native photoreceptor membranes (Table II, 3rd column). This parameter is very sensitive to detergent action. It is reduced to zero in preparations solubilized in DTAB (100 mM), but can recover nearly fully upon reconstitution with phospholipids and removal of detergent by dialysis.

Removal of phospholipids by affinity chromatography, followed by dialysis to remove DTAB, lowers the regeneration capacity to 48%. Phospholipase-C treatment of detergent-free membrane suspension makes this parameter decrease to 47%.

Mere addition of phospholipids, dioleoyl-phosphatidylcholine, egg-phosphatidylcholine, phosphatidylethanolamine, phosphatidylserine, mixtures thereof or a photoreceptor membrane lipid extract, does not restore the regeneration capacity. This is the case, regardless whether the mixture is sonicated in the presence or absence of 20 mM $MgCl_2$ or 33% glycerol. Dicaproyl-phosphatidylcholine occasionally increases the regeneration capacity to about 70%.

However, true reconstitution by addition of amphipathic lipids (ca. 100 mol per mol rhodopsin) and detergent, followed by removal of detergent by dialysis, leads to nearly complete restoration of the regeneration capacity (85%). This finding confirms the earlier observations of Hong & Hubbell (1973), and is true for all amphipathic lipids tried: dioeoyl - and egg-phosphatidylcholine, phosphatidylethanolamine, phosphatidylserine, their mixtures, phosphatidic acid (obtained by treating egg-phosphatidylcholine with phospholipase-D), photoreceptor membrane lipid extract, and also the glycolipid monogalactosyl diglyceride. Dicaproyl-phosphatidylcholine, under these conditions, increases the regeneration capacity to only 70%.

Effects of Changes in Lipid Environment on Photolytic Sequence

The third parameter studied concerns the conversions of the last three intermediates in the photolytic sequence: metarhodopsin I (λ_{max} = 478 nm), metarhodopsin II (λ_{max} = 380 nm) and metarhodopsin III (λ_{max} = 455 nm). The transition metarhodopsin I ⟶ II in native photoreceptor membrane suspensions takes place within 1 sec at 25°C (pH 6.5), whereas metarhodopsin II decays to metarhodopsin III and opsin + free retinaldehyde (λ_{max} = 383 nm) with a $t_{\frac{1}{2}}$ = 5.5 min at 25°C (pH 6.5). Hence, curve 2 in Fig. 8A, which is recorded 1 min after illumination of a native membrane suspension, is dominated by an intermediate with λ_{max} = 380 nm : metarhodopsin II.

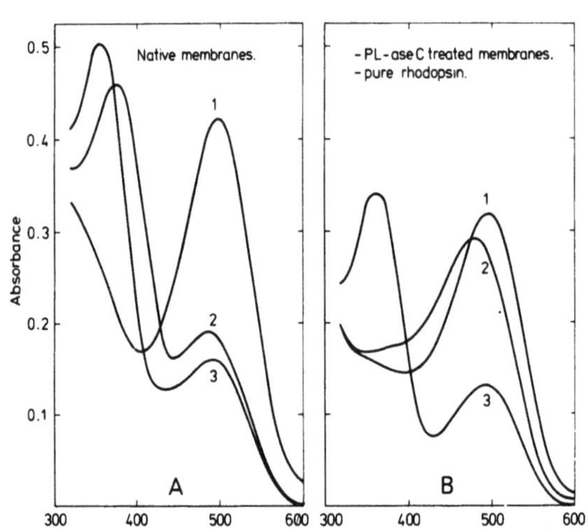

Fig. 8. Effect of delipidation on the photolytic sequence of rhodopsin. A: native photoreceptor membrane suspension; B: suspension of delipidated rhodopsin. Curves 1 show in both cases the rhodopsin spectrum before illumination. Curves 2 are taken 1 min after a 30-sec illumination. Curves 3 are taken after addition of 50 mM hydroxylamine to the illuminated preparations (to convert all free retinaldehyde to retinylidene oxim).

However, curve 2 in Fig. 8B, recorded 1 min after illumination of a delipidated preparation, is dominated by an intermediate with λ_{max} = 480 nm, most likely metarhodopsin I. Addition of hydroxylamine gives the same curves with λ_{max} = 360 nm for both the native and the delipidated preparation (Fig. 8 A and B, curves 3). Apparently, the transition of metarhodopsin I ⟶ II is blocked, and consequently also the metarhodopsin III production, when phospholipids are removed. This is true for both delipidated preparations, and agrees with a previous report for purified rhodopsin by Applebury et al. (1974). Our preliminary experiments indicate that in the delipidated preparations metarhodopsin I is slowly hydrolyzed to opsin and free chromophore, which would also explain why hydroxylamine yields the same curve as with the native preparation.

Mere addition of various phospholipids, like in the case of the regeneration capacity, does not restore the normal photolytic sequence. Addition of dicaproyl-phosphatidylcholine again occasionally restores the parameter. True reconstitution with the entire series of amphipathic lipids mentioned before, but now also including dicaproyl-phosphatidylcholine, completely restores the normal photolytic sequence (Table II, last column). Mono-, di- or triglycerides do not have this effect.

DISCUSSION

The three parameters, thermal stability, regeneration capacity and photolytic sequence, seem to be affected in parallel fashion: similar changes by both procedures of phospholipid removal, complete or nearly complete reversal after reconstitution. In no case does the 500 nm absorbance go down. This indicates that removal of most or all of the lipid bilayer causes changes in the rhodopsin molecule, but does not affect the chromophoric center as long as the pigment is not exposed to light. These effects are most easily explained as the result of aggregation and concomitant conformational changes in the rhodopsin molecule. Electronmicroscopic observations and the increased proteolytic vulnerability after phospholipase tretment, both mentioned before, support the occurrence of a process of aggregation and increased exposure of rhodopsin to the aqueous phase upon phospholipase-C treatment. This will be true a fortiori for the rhodopsin suspensions obtained by affinity chromatography.

The effects on the regeneration capacity can be explained by aggregation alone. Regeneration is an all or none process: a molecule of rhodopsin is either regenerated or not. Hence, a lowering of the regeneration capacity suggests that 11-cis retinaldehyde is unable to reach its binding site in a fraction of the opsin molecules. This is more likely due to their aggregation than to the occurrence of two populations of rhodopsin of different configuration.

The transition of metarhodopsin I to metarhodopsin II is generally believed to be accompanied by rather large conformational changes of the protein molecule. Apparently in delipidated preparations such changes are blocked. We tend to ascribe this to conformational changes of the rhodopsin molecule, when it is exposed to what is for the greater part of its surface an unfamiliar polar environment during delipidation. Apolar regions of the protein molecule may have to retract with simultaneous exposure of polar groups normally buried inside. This may restrict the motion of those parts of the protein which are involved in the conformational transitions during the metarhodopsin I to II transition. This could also explain the observation that mere addition of lipids does not restore the affected parameters, since the new contacts between lipids and proteins would not be the original ones. Preliminary freeze-fracture studies indicate that rhodopsin under these circumstances is not even incorporated into the lipid bilayer.

Obviously a remobilization of the protein structure is demanded, requiring treatment with a detergent like DTAB. This can break up aggregates and penetrate proteins, giving greater freedom of motion to the peptide chain. Evidence for this is that solubilization of rhodopsin in detergents like Triton X-100 or DTAB leads to marked reduction of thermal stability, loss of regeneration capacity, and acceleration of the photolytic sequence. ESR measurements of detergent solubilized rhodosin, labeled with a paramagnetic probe, point in the same direction (Pontus & Delmelle, 1975).

When amphipathic lipids are added to detergent solubilized rhodopsin and the detergent is removed by dialysis, the lipids can reoccupy their proper positions around the rhodopsin molecule. The latter molecule is then enabled to return to its original conformation, in which it is maintained by the surrounding lipids when the detergent is gone. Amphipathic lipids may be needed, because the polarity of the lipids, regardless of the nature of the polar headgroup or the fatty acids, is important for the interaction with the rhodopsin molecule, which certainly has polar regions: its sugar groups and the hydrophylic parts which can be attacked by proteolytic enzymes.

The occasional, partial recovery of the parameters by the mere addition of dicaproyl-phosphatidylcholine may be explained by the fact that this lipid has the greatest detergent effect of all lipids used. The limited extent of the recovery with this lipid may be due to the fatty acid chain being too short to span the entire polar surface of rhodopsin (Chen & Hubbell, 1973).

CONCLUSIONS

Our results are interpreted and summarized in Fig. 9, where A represents rhodopsin embedded in either the native or a reconstituted

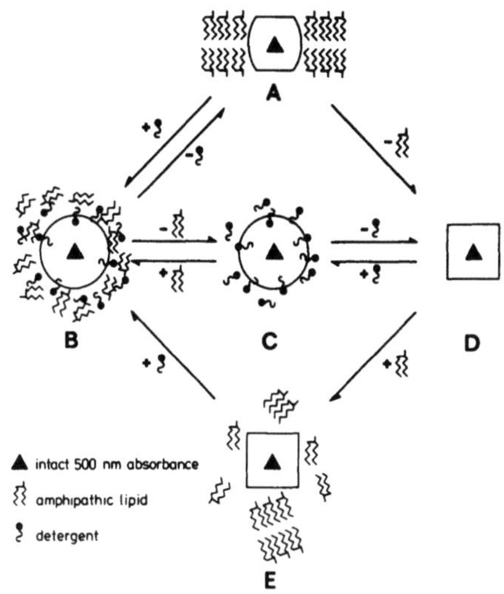

Fig. 9. Interrelations between various rhodopsin preparations. A. Functionally intact rhodopsin embedded in native or reconstituted lipid-bilayer. B. Rhodopsin in detergent solution in the presence of phospholipids. C. Detergent solubilized rhodopsin obtained after affinity chromatography. D. Lipid- and detergent-free rhodopsin, obtained either from C by detergent dialysis or by phospholipase-C treatment of native membranes (A). E. Phospholipids merely added to preparation D without detergent. Circle indicates conformational mobility of rhodopsin in presence of detergent: square indicates rigid structure of detergent-free rhodopsin not incorporated in a lipid bilayer.

lipid bilayer. When the membranes are solubilized in detergent, rhodopsin is made labile (B). The phospholipids are to some extent competitive with the detergent molecules, since the thermal stability of rodopsin solubilized by DTAB is much higher in the presence of phospholipids (B) than in their absence (C) (van Breugel et al., 1976). Lipid removal by affinity chromatography (B → C) and removal of detergent by dialysis (C → D) yields a rigid rhodopsin molecule (D) of altered conformation. This is also achieved by phospholipase-C treatment (A → D), but less completely - hence the slightly smaller loss in thermal stability in the latter case. Similar changes in protein conformation upon phospholipase-C treatment can probably explain the decreases in various membrane-bound enzyme activities (Mavis et al., 1972; Martonosi et al., 1968; Duttera et al., 1968; Zwaal et al., 1973).

Mere addition of phospholipids to delipidated preparations (D → E) does not influence the rigid, altered structure, hence does not normalize the parameters. The only way back to normally functional rhodopsin is thorough temporary exposure to detergent (D → E → B → A or D → C → B → A).

The deleterious effect of phospholipid removal appears to be the exposure of apolar regions of the membrane protein to a polar environment. This may lead to retractions of these apolar regions with simultaneous exposure of polar groups normally buried inside. The protein molecule may thus be locked in an abnormal, inactive conformation. Reconstitution through addition of phospholipids and temporary exposure to detergent seems to allow the peptide chain to refold to its natural conformation in association with the amphipathic lipids, leading to normalization of the functional parameters.

ACKNOWLEDGEMENTS

This work is supported in part by the Netherlands Organization for Basic Research (ZWO) through the Netherlands Foundation for Chemical Research (SON).

REFERENCES

ABRAHAMSON, E. W. & WIESENFELD, J. R. (1972) in Handbook of Sensory Physiology (Dartnall, H.J.A., ed.) , Vol. VII/1, pp. 69-121, Springer-Verlag, Berlin-Heidelberg, New York.
APPLEBURY, M. L., ZUCKERMAN, D. M., LAMOLA, A. A. & JOVIN, T. M. (1974) Biochemistry 13, 3448-3458.
BONTING, S. L. (1969) Curr. Topics Bioenergetics 3, 351-415.
BONTING, S. L. (1973) in Protides of the Biological Fluids, 21st Colloquium (Peeters, H., ed.) , pp. 97-108, Pergamon Press, Oxford.
BONTING, S. L. & DAEMEN, F. J. M. (1974) in Biochemistry of Sensory Functions (Jaenicke, L., ed.), pp. 1-21, Springer-Verlag, Berlin.
BORGGREVEN, J. M. P. M., DAEMEN, F. J. M. & BONTING, S. L. (1970) Biochim. Biophys. Acta 202, 374-381.
BORGGREVEN, J. M. P. M., ROTMANS, J. P., BONTING, S. L. & DAEMEN, F. J. M. (1971) Arch. Biochem. Biophys. 145, 290-299.
BORGGREVEN, J. M. P. M., DAEMEN, F. J. M. & BONTING, S. L. (1972) Arch. Biochem. Biophys. 151, 1-7.
BROWN, P. K. (1972) Nature 236, 35-38.
CHEN, Y. S. & HUBBELL, W. L. (1973) Exp. Eye Res. 17, 517-532.
CONE, R. A. (1972) Nature 236, 39-43.
DAEMEN, F. J. M. (1973) Biochim. Biophys. Acta 300, 255-288.
DAEMEN, F. J. M. (1974) in Biomembranes - Lipids, Proteins and Receptors (Burton, K.M. & Packer, L., eds.), pp. 319-410, Bioscience Publ. Div. Webster Groves.
DAEMEN, F. J. M., De GRIP, W. J. & JANSEN, P. A. A. (1972) Biochim. Biophys. Acta 271, 419-428.
DE GRIP, W. J., DAEMEN, F. J. M. & BONTING, S. L. (1972) Vision

Res. 12, 1697-1707.
DUTTERA, S. M., BYRNE, W. L. & GANOZA, M. C. (1968) J. Biol. Chem. 243, 2216-2228.
HELENIUS, A. & SIMON, K. (1975) Biochim. Biophys. Acta 415, 29-79.
HONG, K. & HUBBELL, W. L. (1972) Proc. Nat. Acad. Sci. US 69, 2617-2621.
HONG, K. & HUBBELL, W. L. (1973) Biochemistry 12, 4517-4523.
LIEBMAN, P. A. & ENTINE, G. (1974) Science 185, 457-459.
LIMBRICK, A. R. & KNUTTON, S. (1975) J. Cell Sci. 19, 341-355.
MARTONOSI, A., DONLEY, J. & HALPIN, R. A. (1968) J. Biol. Chem. 243, 61-70.
MAVIS, R. M., BELL, R. M. & VAGELOS, P. R. (1972) J. Biol. Chem. 247, 2835-2841.
POINCELOT, R., MILLAR, P. G., KIMBEL, R. L. & ABRAHAMSON, E. W. (1969) Nature 221, 256-257.
PONTUS, M. & DELMELLE, M. (1975) Exp. Eye Res. 20, 599-603.
Van BREUGEL, P. J. G. M., DAEMEN, F. J. M. & BONTING, S. L. (1975) Exp. Eye Res. 21, 315-324.
Van BREUGEL, P. J. G. M., DAEMEN, F. J. M. & BONTING, S. L. (1976) Exp. Eye Res., (in press).
VERKLEY, A. J., ZWAAL, R. F. A., ROELOFSEN, B., COMCURIUS, P. KASTELEYN, D. & Van DEENEN, L. L. M. (1973) Biochim. Biophys. Acta 323, 178-193.
YOUNG, R. W. (1976) Invest. Ophthal. 15, 700-725.
ZWAAL, R. F. A., ROELOFSEN, B. & COLLEY, C. M. (1973) Biochim. Biophys. Acta 300, 159-182.

APPROACHES TO THE BIOCHEMISTRY OF REGENERATION IN THE CENTRAL NERVOUS SYSTEM

Bernard W. Agranoff

Department of Biological Chemistry and
Neuroscience Laboratory
University of Michigan, Ann Arbor, Michigan, USA

Our laboratory has a long-standing interest in the biochemical basis of central nervous system plasticity. We have come to recognize that many of the ultimate answers we seek are imbedded in the more general question of the biochemical basis of neural development. That is, until we learn how cells recognize and mark one another during development, we will probably not understand fully how neurons connect with one another. Present experimental approaches to the general problem of cell recognition in the nervous system, as well as in a number of other eukaryotic preparations, range from the whole animal model--a biochemical extension of classical embryology--to reductionist models. An example of the latter is the reaggregation of dissociated cells or their selective adhesion to other cells or to cell membranes. Lipid biochemists have been drawn to the area of cell recognition since it has become increasingly apparent that some cell surface determinants, at least at the immunological level, may be lipid in nature.

The nervous system has proven useful for the study of another timely problem: that of intracellular distribution of enzymes and substrates of structural lipid formation and degradation in eukaryotic cells and the mechanisms of their assembly into membranes. The neuron constitutes a particularly useful model, by virtue of its peculiar shape. Because its axon may be larger than the perikaryon by several magnitudes, we can take advantage of the process of axonal flow to study the rate of appearance of labeled lipid at some distance from sites of synthesis. We should then be able to distinguish those lipids in the nerve which are synthesized within the confines of the cytoplasm and are transported, from those that are synthesized locally or are chemically modified after they have reached their final destination.

The foregoing remarks are by way of introduction and explanation as to why a lipid biochemist is studying neural development, and more cogently, why he is presenting his findings to other lipid chemists.

REGENERATION OF THE GOLDFISH OPTIC NERVE

The goldfish visual system is completely crossed, each retina connecting with the contralateral tectum by means of a myelinated optic nerve. It is well-known that following crush or cut of an optic nerve of an adult goldfish, an orderly retinotectal map will be reestablished, as is inferred from the observation that the fish will eventually regain vision (Sperry, 1963). The specificity of reconnection has been documented in elegant experiments in which light is flashed onto a particular region of the visual field and electrophysiological measurements are recorded from the optic tectum (Jacobson and Gaze, 1965). It should be noted that the recorded electrophysiological responses are generally presynaptic. That is, the retinal ganglion cells are discharged via stimulation of photoreceptors; and their axons, having coursed along the optic nerve, are electrophysiologically demonstrated to be present at their proper tectal sites. Such experiments have not indicated, however, whether or not functional synapses have been formed.

In a series of recent experiments, we have examined regeneration of the goldfish visual system at three levels. (1) In vivo experiments: goldfish in which an optic nerve has previously been crushed are studied behaviorally, anatomically and biochemically. (2) "Short-term" in vitro experiments: the post-crush retina is incubated for an hour with radioisotopic precursor and is compared with a control retinal incubation. This is usually the retina on the opposite side of the same animal, but in which the optic nerve has not been crushed. (3) "Long-term" in vitro experiments: retinal explants from regenerating and control sides of the goldfish are compared in culture.

IN VIVO EXPERIMENTS

In these experiments, one eye was removed and the other optic nerve was crushed in a single operation, rendering the fish temporarily blind. Recovery of vision in these one-eyed fish was tested periodically. In a series of behavioral and radioautographic studies, Springer et al. (in press, a, b, c) reported the reappearance of vision as evidenced by four measures of visual function over a period of several weeks. They returned in the following order: optokinetic nystagmus and startle response to a light flash (respiratory suppression--an autonomic measure), food-catching, and finally, the optomotor response (swimming to track a rotating, striped drum). By means of such experiments we were able to demonstrate, for example, that the rate of recovery of vision is a function of the temperature in which the fish have been stored following nerve crush. In fish stored at 30° C, first indications of behavioral recovery are seen about 10 days following crush. At the same time, we have been able to trace the course of regenerating fibers by means of [^{3}H]proline radioautography (Neale et al., 1972). When [^{3}H]proline is injected intraocularly (IO), it is converted into protein in the retina and is then transported axonally with at least two major velocities, so-called "fast" and "slow" transport. Proline is particularly suitable because it does not give rise to significant amounts of systemic labeling in the brain from proline which has escaped the eye and enters the brain via the blood. Due to proline's high "signal-to-noise" ratio (Elam & Agranoff, 1971), 8-24 h after its IO injection, radioactivity in the brain is virtually all in the primary projections of the retinal ganglion cells.

The [^{3}H]proline axonal flow studies on the regenerating, one-eyed fish yielded several surprising results. By radioautographic criteria the nerve appeared to regrow more rapidly than the behavioral response might have suggested. The delayed return of function indicates either that synapse formation is rate-limiting in functional recovery, or that additional anatomical reorganization must take place after reconnection occurs. It is also of interest that while the goldfish optic nerve is ordinarily completely myelinated, the functioning regenerated nerve is unmyelinated; and remyelination takes place only after several months, long after vision has been restored by all of the criteria we have used (Murray, 1976).

Another interesting outcome of these experiments was the observation that upon regrowth, some optic nerve fibers were diverted to the now vacant ipsilateral tract. One eye then anomalously projects fibers to both sides of the brain, in contrast with the normal goldfish, in which the tectal innervation appears to be completely contralateral (Springer et al., in press c).

SHORT-TERM IN VITRO EXPERIMENTS

Studies on the retina itself indicate that following optic nerve crush or section, increased amounts of RNA are seen in the ganglion cell, and in radioautographic studies there is increased protein labeling in the ganglion cell layer (Murray & Grafstein, 1969). Histological changes in the goldfish ganglion cell are somewhat like the chromatolytic response noted in the homeothermic central neurons following nerve section.

We therefore examined possible differences in protein metabolism in control and post-crush retinas, by means of brief incubations with labeled amino acids. Both retinas were removed from fish 2-45 days after unilateral retro-orbital crush (Heacock & Agranoff, 1976). The two retinas were incubated separately for one hour: one with ^{3}H-methionine and the other with ^{35}S-methionine, after which the retinas were combined and homogenized. The 100,000 x g supernatants were combined and analyzed by SDS polyacrylamide gel electrophoresis (PAGE). By this means, it is possible to detect whether a protein or a group of proteins is labeled uniquely as a result of the regeneration process. In such experiments Heacock & Agranoff (1976) were able to demonstrate a selective labeling in the 55,000 MW region of the gel. (Fig. 1). By means of several procedures including vinblastine precipitation and reassembly (Fig. 2), it was shown that a major difference between the two conditions was that following nerve crush there was an increase in tubulin synthesis in the post-crush retina. Assuming that the ganglion cells accounted for all of the observed change and that the latter represent only about 5 percent of the retina, the inferred increase in tubulin synthesis in ganglion cells is much greater than is apparent from the 3-fold increase in isotopic ratios observed in whole retina (Fig. 2).

The result is of interest since it indicates that the increased tubulin synthesis may be considered a marker in the study of factors which signal the cell response to axotomy. In addition to its well-known role in neurotubule formation, tubulin may have a direct role in membrane function since tubulin-like substance appears to be present in membranes (Bhattacharyya and Wolff, 1975).

Possible differences in lipid composition of regenerating and normal retinas are under investigation, but as noted above are rendered difficult because the fraction of retina known to be undergoing the greatest change (the ganglion cell layer) constitutes a small fraction of the total retinal mass. Some of these problems may ultimately be resolved by the use of neurites harvested from explant cultures of retina, a possibility proposed below.

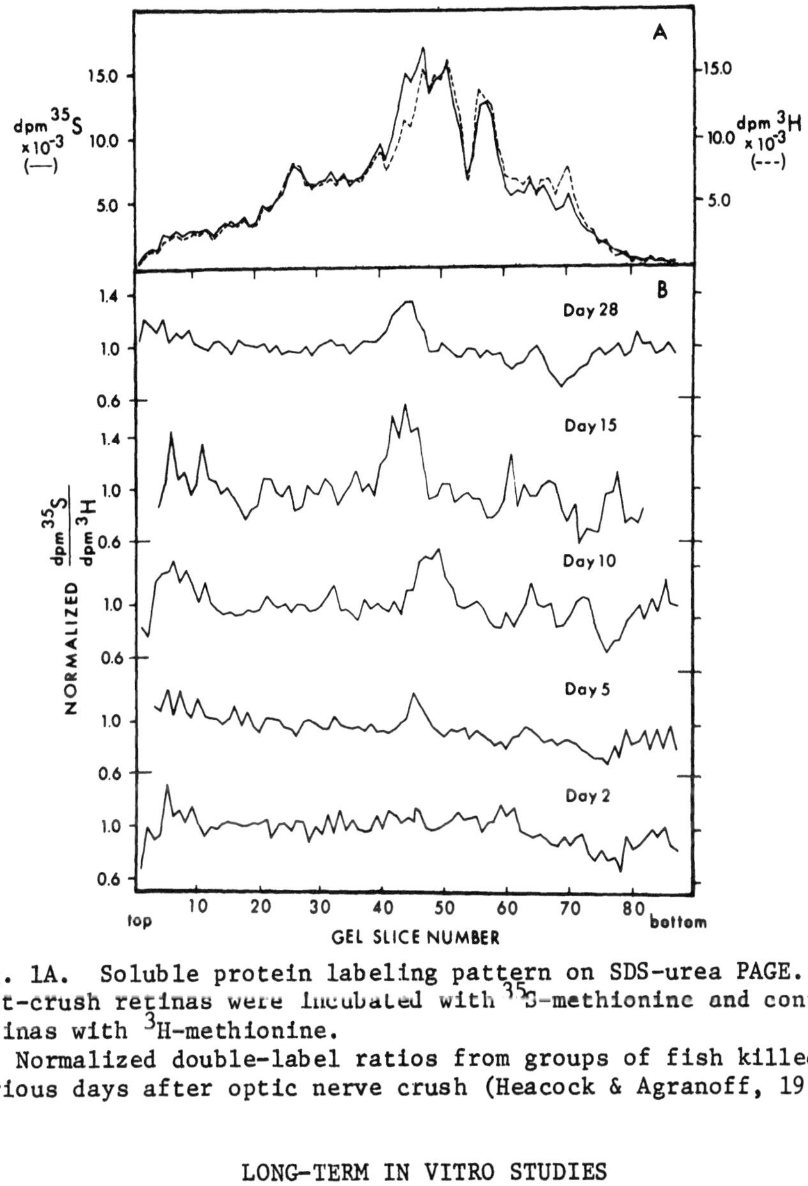

Fig. 1A. Soluble protein labeling pattern on SDS-urea PAGE. Post-crush retinas were incubated with ^{35}S-methionine and control retinas with ^{3}H-methionine.
B. Normalized double-label ratios from groups of fish killed various days after optic nerve crush (Heacock & Agranoff, 1976).

LONG-TERM IN VITRO STUDIES

We have previously observed that prior section of the optic nerve greatly enhanced outgrowth of neurites from the explanted retina of <u>Xenopus laevis</u>, the African clawed toad (Agranoff et al., 1976). A similar finding was obtained with goldfish (Landreth & Agranoff, in press). In this case, retinas are removed under sterile conditions and are cut into 600 µm wide ribbons. The retina is turned 90°, and the process is repeated. The resultant full-thickness squares of retina are now set out in petri dishes on a suitable substratum, such as a collagen gel, or

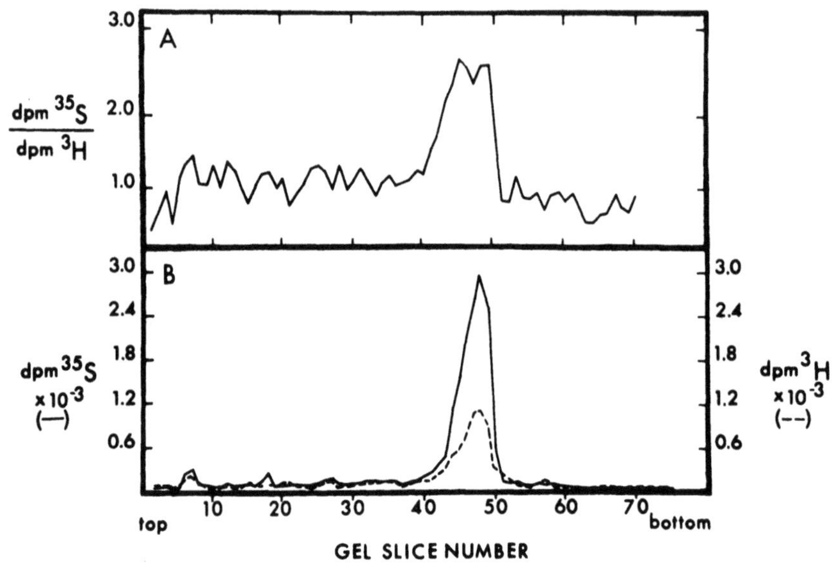

Fig. 2A. Radioactivity in SDS gels from 28 days post-crush and control retinas labeled as in Fig. 1, but after reassembly with carrier tubulin.
B. Double label ratios (Heacock & Agranoff, 1976).

a film of collagen or poly-lysine on glass or plastic. The explants, about 15 per 35 mm dish, are supplied with a suitable antibiotic-containing nutrient medium. A marked difference was noted between post-crush and control retinas. After 1-3 days in culture, the post-crush retinal explants sent out profuse neuritic outgrowth, presumed to be extensions of the retinal ganglion cells. They are relatively free of non-neuritic elements and tend to fasciculate as well as to have large numbers of prominent growth cones (Fig. 3, Fig. 4). After several days in culture the body of the retinal explants still retains its *in vivo* morphology, with recognizable cell layers. Unlike conventional homeothermic cultures, the fish explants thrive at room temperature (about 20° C) providing they are stored in a moist chamber. No gassing is required.

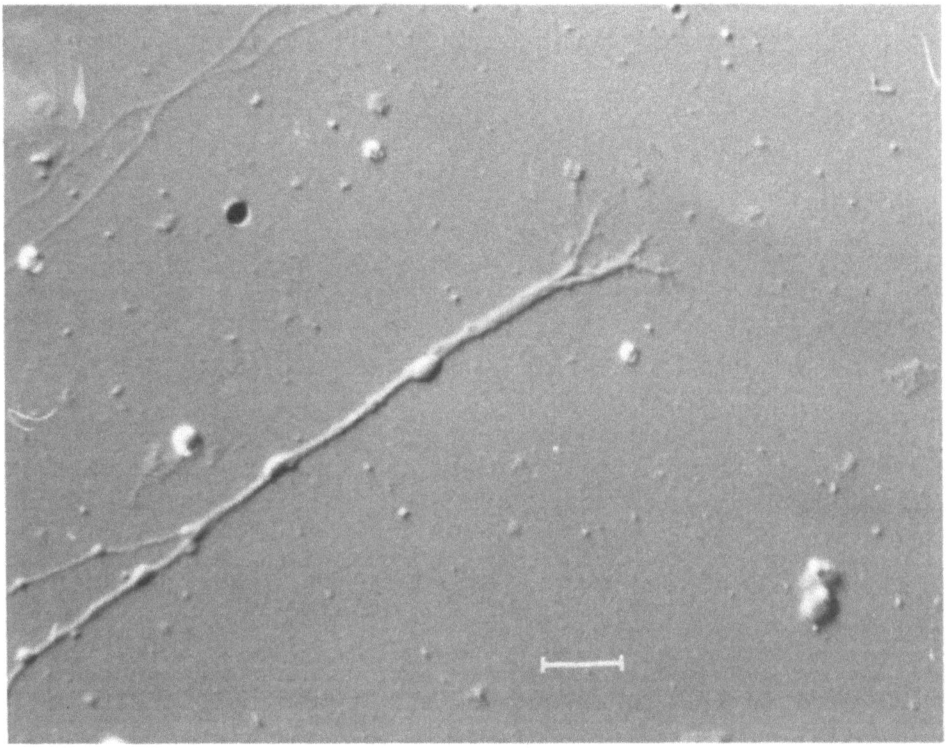

Fig. 3. Interference contrast photomicrograph of neurites from a post-crush (regenerating for 12 days in vivo prior to explantation) retinal explant demonstrating fasciculation and prominent growth cones. Three days in culture. Bar = 10 μm. (Landreth & Agranoff, 1976)

AXONAL FLOW OF LIPIDS AND THE BIOCHEMISTRY OF REGENERATION

Among the first reports of axonal transport were those of Miani (1962) who observed a proximodistal gradient of phospholipid labeling along the vagus and hypoglossal nerves following injection of $^{32}P_i$ in the region of the fourth ventricle. A complication of such experiments is the possibility of simultaneous transport of $^{32}P_i$ itself and of other labeled molecules which can enter into exchange reactions that lead to in situ lipid labeling at the same time that lipids are being transported. Other marker molecules that have been used to study lipid labeling include glycerol, choline, serine, as well as a number of carbohydrate precursors for the study of axonal transport of glycolipids (for reviews, see Heslop, 1975; Grafstein et al., 1975). A rather complicated picture emerges. In the normal goldfish visual system

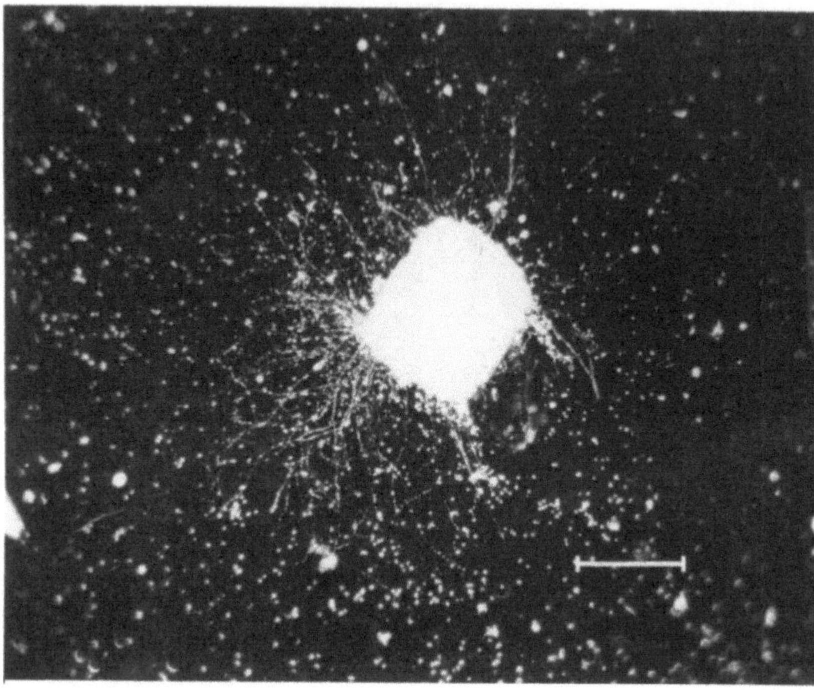

Fig. 4. Dark-field photomicrograph of regenerating retinal explant after several days in culture. Bar = 500 μm. (Landreth & Agranoff, 1976)

it appears that glycerolipids migrate via fast and, perhaps to a lesser extent, slow flow and that the former is at least partially dependent on protein synthesis (Grafstein et al., 1975).

The use of labeled cholesterol or its precursor, mevalonate, has the appeal that a limited number of products are presumably formed and that the lipid is believed to turn over very slowly within the nervous system. It should be noted, however, that the observed slow cholesterol turnover reflects primarily the major brain pool of this lipid, myelin. Axonal flow studies are however directed at neurons, not at the glial cells that synthesize myelin. MacGregor et al., (1973) noted that following injection of ^{14}C cholesterol into the lumbar region of the chick, a proximodistal gradient of cholesterol was found in the sciatic nerve. The rate was thought to be about that observed for protein. Both cholesterol and cholesterol ester were detected, but the relative proportions were variable. A slow and fast rate of axonal flow were reported. Studies from the same laboratory (Rostas et al., 1975) observed the transport of cholesterol following injection into the vitreous humor of the young chick to be only of the slow variety. Labeled cholesterol and cholesterol ester were found in relatively equal amounts in the ipsilateral tectum, but no labeled cholesterol ester was found in the contralateral tectum. In an abstract published in 1972, Griffith and Larramendi reported axonal transport of cholesterol following intraocular (IO) injection of ^{3}H mevalonic acid in the goldfish. A peak of tectal labeling was observed 12 hours after injection, falling to background levels by 36 hours. Much later, radioactivity began to accumulate again in the tectum, reaching a plateau at 4 weeks and remaining unchanged for an additional 16 weeks.

We have repeated and extended this report to the regenerating visual system of the goldfish. The right optic nerve was crushed, and 7 days later the right eye was injected with 5 μC ^{3}H mevalonate. Unoperated fish served as controls. The fish were killed 4-28 days after the injection. The left tectum, right tectum and right retina were extracted, and pools of three of each tissue were counted. We failed to find a transient peak in the tectal lipid in either the control or in the experimental fish, but did observe arrival of labeled lipid via a slow transport process. There was an enhanced transport of labeled cholesterol as a result of prior optic nerve crush (Fig. 5). At all times examined, the retina contained much more lipid-soluble radioactivity than did the tectum. Even 28 days after injection of precursor there was five times more radioactivity in the retina than in the tectum of "post-crush" goldfish. The nature of the radioactivity was examined by thin layer chromatography. At a time point studied (14 days after injection of ^{3}H mevalonate) there appeared to be no difference in distribution of radioactivity between post-crush and control retina. Approximately equal amounts of material cochromatographing with cholesterol and with cholesterol ester are observed, as identified by thin layer chromatography. There is also a third, minor component, having a migration rate intermediate between that of cholesterol and cholesterol ester. There are no significant differences in amounts of the three substances between the control and experimental fish retinas. Tectal radioactivity appears to contain little or no "cholesterol ester," suggesting that the latter is not transported axonally; or alternately, the ester is broken down during or after transport to liberate cholesterol. In the tectum of fish undergoing regeneration of the visual system, we find indications of small amounts of labeled material cochromatographing with cholesterol ester.

To examine more directly the possible role of cholesterol in neurite outgrowth, an inhibitor, diazacholesterol, was applied to explants of regenerating retina (Heacock & Agranoff, unpublished).

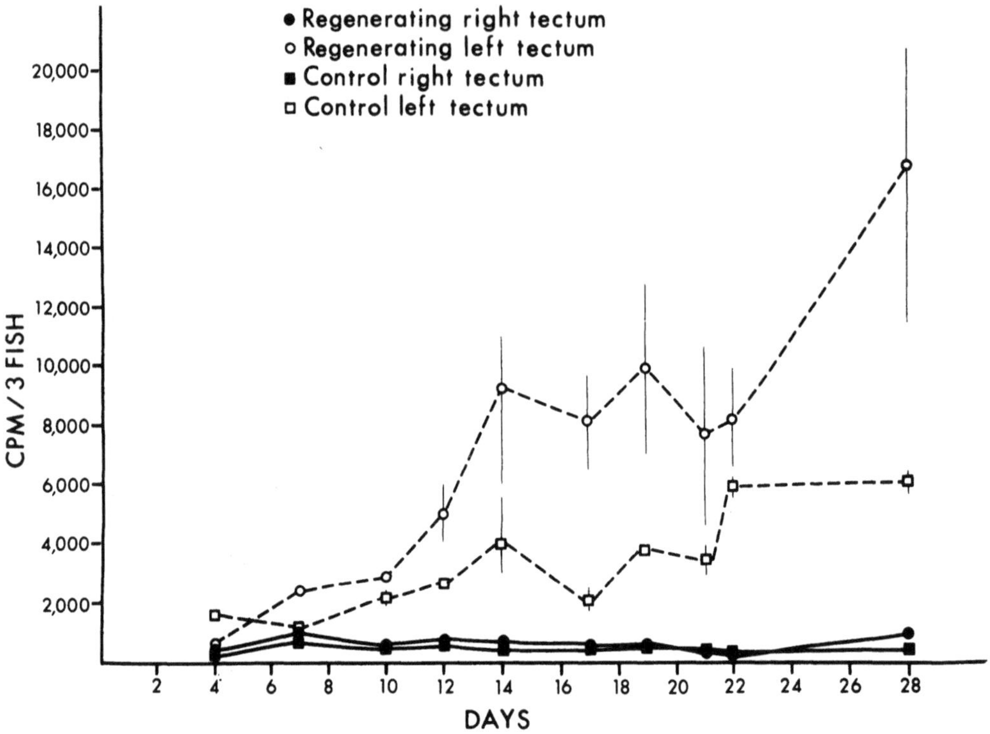

Fig. 5. Comparison of radioactive lipid in tectum following injection of 5μc of ^{3}H-mevalonate into right eye in control goldfish and into goldfish whose optic nerve had been crushed 7 days prior to injection. Two groups of 3 fish each were averaged for each time point. Four groups were used for the 14-day point.

In preliminary experiments 3 x 10^{-5}M concentrations were sufficient to block neurite outgrowth. Further experiments are directed at testing whether neurites, removed from the explant, are capable of cholesterol synthesis, in order to distinguish the possibility that some cholesterol synthesis occurs within the growing neurite itself.

CONCLUSION

A number of model systems are presented for the study of lipid synthesis which permit the use of conventional biochemical techniques for the study of the complex functions of the nerve cell.

REFERENCES

Agranoff, B.W., Field, P. and Gaze, R.M. (1976) Brain Res. 113 225-234.

Bhattacharyya, B. and Wolff, J. (1975) J. Biol. Chem. 250, 7639-7646.

Elam, J.S. and Agranoff, B.W. (1971) J. Neurochem. 18, 375-387.

Grafstein, B., Miller, J.A., Ledeen, R.W., Haley, J. and Specht, S.C. (1975) Exp. Neurol. 46, 261-281.

Griffith, A. and Larramendi, L.M.H. (1972) Anat. Rec. 173, 320.

Heacock, A.M. and Agranoff, B.W. (1976) Proc. Nat. Acad. Sci. U.S. 73, 828-832.

Heacock. A.M. and Agranoff, B.W. (unpublished).

Heslop, J.P. (1975) in Advances in Comp. Physiol. and Biochem. (Lowenstein, O., ed.), vol. 6, pp. 75-163, Acad. Press, New York.

Jacobson, M. and Gaze, R.M. (1965) Exp. Neurol. 13, 418-430.

Landreth, G.E. and Agranoff, B.W. (1976) Neuroscience Abstracts, Vol. II, Part 1, p. 1025.

Landreth, G.E. and Agranoff, B.W. (in press).

McGregor, A., Jeffrey, P.L., Klingman, J.D. and Austin, L. (1973) Brain Res. 63, 644-649.

Miani, N. (1962) Nature 193, 887-888.

Murray, M. (1976) J. Comp. Neurol. 168, 175-196.

Murray, M. and Grafstein, B. (1969) Exp. Neurol. 23, 544-560.

Neale, J.H., Neale, E.A. and Agranoff, B.W. (1972) Science 176, 407-410.

Rostas, J.A.P., McGregor, A., Jeffrey, P.L. and Austin, L. (1975) J. Neurochem. 24, 295-302.

Sperry, R.W. (1963) Proc. Nat. Acad. Sci. U.S. 50, 703-710.

Springer, A.D. and Agranoff, B.W. (in press, a).

Springer, A.D., Easter, Jr., S.S. and Agranoff, B.W. (in press, b).

Springer, A.D., Heacock, A.M., Schmidt, J.T. and Agranoff, B.W. (in press, c).

3-D-(-) HYDROXYBUTYRATE DEHYDROGENASE FROM RAT LIVER MITOCHONDRIA -- PURIFICATION AND INTERACTION WITH PHOSPHOLIPIDS

J. C. Vidal, E. A. Guglielmucci, and A. O. M. Stoppani

Instituto de Química Biológica, Facultad de Medicina

Paraguay 2155, 1121-C.F. Buenos Aires, Argentina

The existence of membrane-bound enzymes has been known for many years and it has been proposed that most particulate enzymes are associated with the lipid matrix of biological membranes. Fleischer & Klouwen (1961) were the first to show the lipid requirement in enzyme reactions of the mitochondrial electron transport system. It is now recognized that there are a number of membrane-bound enzymes which require phospholipid for function. Any lipid-requiring enzyme must fulfill two minimal experimental criteria for proof of lipid involvement, namely (a) removal of lipid by mild procedures result in loss of enzymic activity, and (b) the activity is restored when lipids are added in a suitable physical state to the delipidized enzymes. Point (b) entails the demonstration that activity is restored after rebinding of phospholipid to the enzyme.

ABBREVIATIONS

In addition to standard abbreviations, the following are employed: HBD, 3-D-(-) hydroxybutyrate dehydrogenase; apoHBD, 3-D-(-) hydroxybutyrate dehydrogenase apoenzyme; SMP, submitochondrial particles; MPL, mitochondrial phospholipids. The 1,2-diacyl-sn-glycero-3-phosphorylcholines will be referred to according to the number of carbons of the fatty acyl chains, namely L-diC_7 (diheptanoyl-), L-diC_{10}(didecanoyl-), L-diC_{14}(dimirystoyl-) and L-diC_{16}(dipalmitoyl-). The 2,3-diacyl-sn-glycero-3-phosphorylcholines will be referred to as D-diC_7(diheptanoyl-), D-diC_{10} (didecanoyl-). LysoC_{14}, 1-myristoyl-sn-glycero-3-phosphorylcholine.

3-Hydroxybutyrate dehydrogenase (3-D-(-) hydroxybutyrate: NAD oxido reductase EC 1.1.1.30) is a good example of this group of phospholipid-requiring enzymes. It is tightly bound to the inner mitochondrial membrane and shows a specific requirement for phosphatidyl choline (Green et al., 1937; Sekuzu et al., 1963). The apodehydrogenase cannot be released by osmotic shock (Bendall & de Duve, 1960) or sonication (Ziegler & Linnane, 1958), although it can be solubilized either by detergents such as cholate (Sekuzu et al., 1963; Menzel & Hammes, 1973) or by digestion of membrane phospholipids with phospholipase A (Fleischer et al., 1966; Nielsen & Fleischer, 1973).

ApoHBD from beef heart mitochondria has been solubilized using phospholipase A and purified to homogeneity as a soluble, phospholipid-free protein, with a subunit molecular weight of 31500 dalton (Bock & Fleischer, 1975). Its lipid requirement has been extensively studied (Grover et al., 1975; Gazzotti et al., 1975; Houslay et al., 1975).

Rat liver mitochondria HBD has received less attention (Gotterer, 1967; Hexter & Goldman, 1973) and only recently Levy et al. (1976) obtained a highly purified enzyme-lecithin complex and determined some physical properties of the enzyme.

Since definition of the precise role(s) of phospholipids in enzyme activity is basic to an understanding of the structure and function of biological membranes, we shall describe our studies on the rat liver mitochondria HBD in order to determine (a) the specificity of phospholipid requirement at the molecular level and (b) the influence of phospholipid fatty acyl composition on HBD activity. Point (a) involves the purification to homogeneity of the lipid-free apoHBD and the examination of reactivation of the soluble apoenzyme by different phospholipids or analogs. Point (b) was studied either using specific phospholipids and purified apoHBD or by exchanging the endogenous MPL by lecithins with a defined fatty acid composition.

Studies on Specificity of Phospholipid Requirement

Apo HBD was solubilized from rat liver mitochondria or SMP by digestion of the membrane phospholipids with porcine pancreas phospholipase A and centrifugation of membrane residues. The supernatant was catalytically inactive and HBD activity was regained after addition of lecithin or phospholipid mixtures containing lecithin as shown in Fig. 1. Reactivation occurs after rebinding of the apodehydrogenase to phospholipid. As shown in Fig. 2, the HBD activity is eluted from Sephadex G-100 together with the liposomes as a symmetric peak, the position of which depended on the elution volume of the phospholipid aggregates.

The apoHBD can be specifically displaced from the enzyme-lecithin complex with *Bothrops atrox* venom phospholipase A in the absence of Ca^{2+} ions, because the higher affinity of the latter for

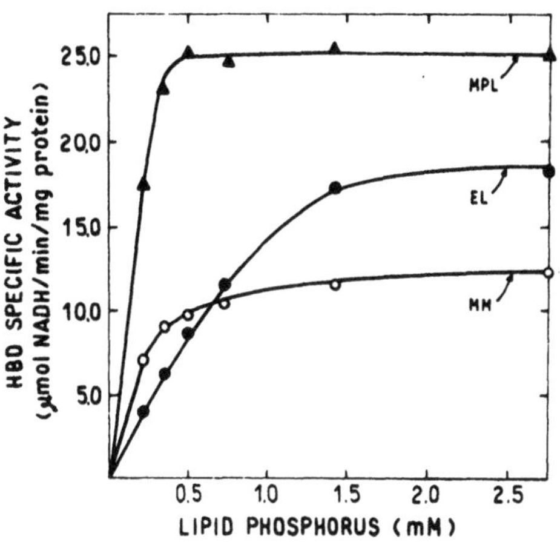

Fig. 1. Reactivation of apoHBD with MPL, egg lecithin (EL) and mixed micelles (diC_{10} plus $lysoC_{14}$, molar ratio 1:1) (MM)

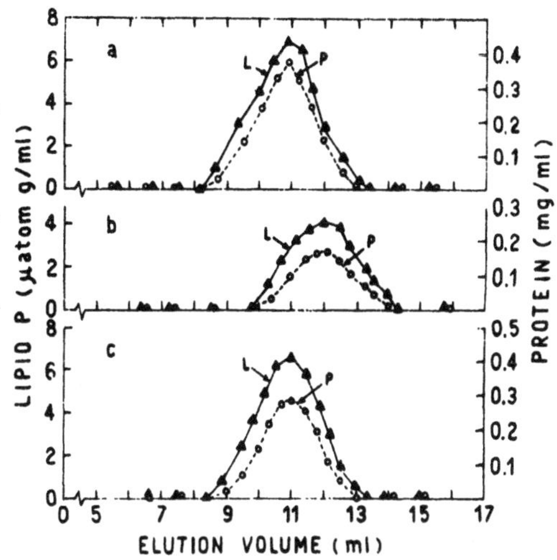

Fig. 2. Chromatographic behavior of apoHBD-phospholipid complexes on Sephadex G-100. (a) MPL; (b) mixed micelles; (c) egg lecithin. L, phospholipid; P, protein

the lecithin-water interfaces. Therefore, the displaced apoenzyme can be separated by gel filtration on Sephadex G-100 (Vidal et al., 1976).

The apoHBD purified by about 170-fold and shows a single sharp band on SDS-polyacrylamide gel-electrophoresis, with a mobility corresponding to a molecular weight 38000 dalton. Gel-filtration on Sephadex G-100 shows two peaks with HBD activity, corresponding to molecular weights of 76000 (dimer) and 38500 dalton (monomer) (Vidal et al., 1976).

In agreement with previous reports (Grover et al., 1975; Gazzotti et al., 1975) phosphatidyl cholines were specifically required for activity; so the specificity of the apoenzyme would be primarily for the chemical parameters (net charge and/or precise molecular fit) of the polar moiety of the phospholipid. The distance between the two ionic groups seems to be critical since mixed micelles of dicetylphosphate plus cetyltrimethylammonium failed to reactivate the apoHBD.

ApoHBD can be reactivated by all the phosphatidyl cholines tested. From titration curves such as those shown in Fig. 1, two reactivating parameters were obtained for each individual phosphatidyl choline, namely (a) the relative maximal activity (compared to the activity obtained with egg lecithin, which was set equal to one) and (b) the efficiency for reactivation, defined as the number of phospholipid molecules per apoHBD monomer required to obtain half-maximal reactivation.

The activation energy (E_a) for each apoHBD-phospholipid complex was calculated from plots of the logarithm of initial velocity as a function of the reciprocal of absolute temperature in the range between 18 and 35°C. The apoHBD-phospholipid complexes were prepared at 30° using 10-15 μg of apoHBD and the phospholipid concentration at which maximal activity was obtained. The results are presented in Table 1

Table 1. Properties of different apoHBD-lecithin complexes

Phospholipid	Relative maximal activity	Reactivating efficiency (PC/apoHBD)	Activation energy (kcal/mol)
L-diC_7	0.42	5200.0	15.4
D-diC_7	0.40	5300.0	15.0
L-diC_{10}	2.40	13.2	14.8
D-diC_{10}	2.30	13.0	14.6
L-diC_{14}	1.60	51.8	15.6
DL-diC_{16}	0.58	195.0	15.4
Egg lecithin[a]	1.00	46.2	14.1
Soybean lecithin[a]	1.20	29.8	--
Mitochondrial lecithin[a]	1.40	26.4	13.6
MPL[a]	1.48	2.2	14.1
Mixed micelles (diC_{10} plus lysoC_{14} (1:1))	0.6	2.6	--
LysoC_{14}	0.08	11.3	--

[a]The corresponding fatty acid compositions are described elsewhere (see Grover et al., 1975; Vidal et al., 1972 and Guglielmucci et al., unpublished).

From these data, some conclusions can be drawn:

(a) Inasmuch as lysoC_{14} is an activator, only one fatty acyl chain is necessary for reactivation. The presence in the molecule of the glycerol moiety is not critical, since the synthetic analog cetylphosphorylcholine is able to reactivate the apoenzyme. On the other hand, either glycerylphosphoryl choline or phosphoryl choline were no activators. Then, the minimal molecular requirement for reactivation are the presence of phosphoryl choline bound to, at least, one hydrophobic chain. The fact that lysoC_7 did not reactivate the apoenzyme also indicate a lower limit of hydrophobicity required for this process (see also Grover et al., 1975).

(b) The stereochemistry of phosphatidyl cholines is not critical since either the 1,2-diacyl-sn-glycero-3-phosphorylcholine (L-) or the 1-sn-phosphorylcholine (D-) isomers reactivate apoHBD with identical reactivating parameters.

(c) Short-chain lecithins (i.e., diC_7) reactive the apoHBD in concentrations lower than the critical micellar concentration (1.3-1.5 mM, see Tansk et al., 1974a); the existence of a lecithin-water interface is not a prerequisite for reactivation.

(d) The similarity of the activation energies for different apoHBD-lecithin complexes (the values do not differ significantly, relative to the experimental uncertainties) (Grover et al., 1975; Gazzotti et al., 1975) suggests that the nature of the reactivation process is essentially the same in all cases. However, the value for the membrane-bound enzyme is lower (about 10 kcal/mol, see also Latruffe & Gaudemer, 1974).

(e) The extremely low reactivating efficiency of DiC_7 can be adscribed to its weak binding to the apoHBD as demonstrated by the almost complete dissociation of the complex upon gel-filtration on Sephadex G-100 (Fig. 3)

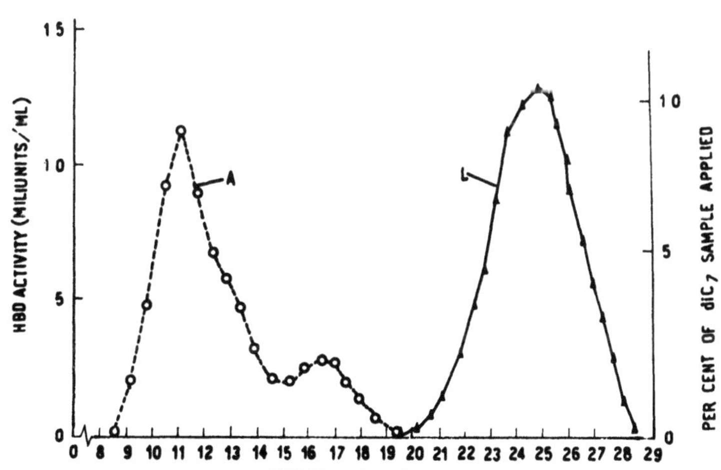

Fig. 3. Chromatographic behavior of apoHBD-diC_7 complex. L, lipid; A, HBD-activity after addition of egg lecithin.

Conversely, apoHBD form tightly-bound complexes with MPL, liposomal lecithins or mixed diC_{10} plus $lysoC_{14}$ micelles as demonstrated by gel-filtration experiments (see Fig. 2). However, binding and reactivation did not relate in a strictly proportional manner.

(f) It is worthnoting the very high reactivating efficiency of MPL, where lecithin constitutes only about 45 %. It has been indicated (Gazzotti et al., 1975) that anionic mitochondrial phospholipids although unable to reactivate apoHBD, increases lecithin reactivating efficiency, probably through changes in the properties of the lecithin-water interface. The small amount of lecithin required is at the level of a cofactor. On the other hand, the reactivating efficiency of L- or $D\text{-}diC_{10}$ increases from 1.3 to 2.6 moles per apoHBD monomer when in the form of mixed micelles with $lysoC_{14}$ (Vidal et al., 1976; Guglielmucci et al., unpublished). This enhancement of the reactivating efficiency produced by $lysoC_{14}$ may be related to a spacing effect which would increase the accessibility of lecithin in the mixed micelle. This interpretation could explain the lower efficiency of reactivation of the tightly packed diC_{16} liposomes at temperatures which are below the gel-to-liquid-crystalline transition temperature, as well as the higher reactivating efficiency of the more spaced liposomes of the unsaturated lecithins (their transition temperatures are below 0°C) (Grant et al., 1974; Williams & Chapman, 1970; Phillips et al., 1969).

<u>Effect of Phospholipid Fatty Acyl Composition</u>

Although no strict structural requirement for the hydrophobic chains becomes apparent from the data presented, the nature of the hydrocarbon moieties is also important. The lecithins employed shows different physical state in aqueous media (monomers, pure or mixed micelles (Tansk et al., 1974a; Tansk et al., 1974b) and liposomes for diC_{10} or higher homologs (Sheetz & Chan, 1972)). At the same time, these lecithins covers a broad range of gel-to-liquid-crystalline phase transition temperatures (Latruffe & Gaudemer, 1974; Grant et al., 1974; Williams & Chapman, 1969).

(a) Pure micelles ($lysoC_{14}$, cetylphosphoryl choline) are poor reactivators and commonly inhibit HBD activity at higher concentrations (Grover et al., 1975; Gazzotti et al., 1975). The extreme case is diC_7 micelles which inhibits the HBD activity. This result suggests that apoHBD has more affinity for micelles than for monomers, notwithstanding the fact that monomers reactivate the apoHBD and micelles inactivated it.

(b) Liposomes forming lecithins shows no inhibitory effect at high concentrations, being diC_{10} the only exception. There are evidences (de Gier et al., 1968; Hauser & Barratt, 1973) indicating that diC_{10} form very fluid and relatively unstable liposomes. This is due to the low interaction energy between lipid molecules (Phillips et al., 1969; Phillips, 1972) favoured by the low gel to liquid-crystalline transition temperature (Grant et al., 1974; Williams & Chapman, 1970; Phillips et al., 1969) as well as to the strong curvature of these liposomes (Phillips, 1972). These effects leads to a collision induced rupture and resealing of the lipid bilayer resulting in formation of multilamellar particles (Hauser & Barratt, 1973) which are inhibitory.

(c) Additional information is given by pre steady-state kinetic studies. In fact, when the lipid-free apoenzyme is incubated with NAD and 3-hydroxybutyrate no HBD activity can be detected. After addition of lecithin the reaction starts, the rate increasing with time approaching to a limit, as shown in Fig. 4, curve A. The limit value is the reaction rate obtained when apoHBD has been preincubated with the same lipid (Fig. 4, curve B)

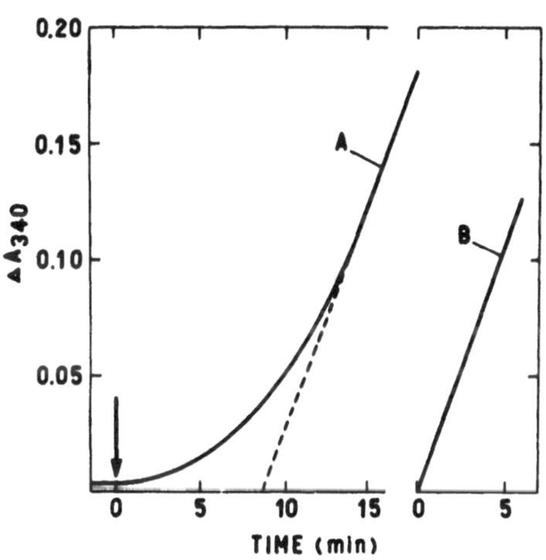

Fig. 4. HBD activity in the presence of egg lecithin.

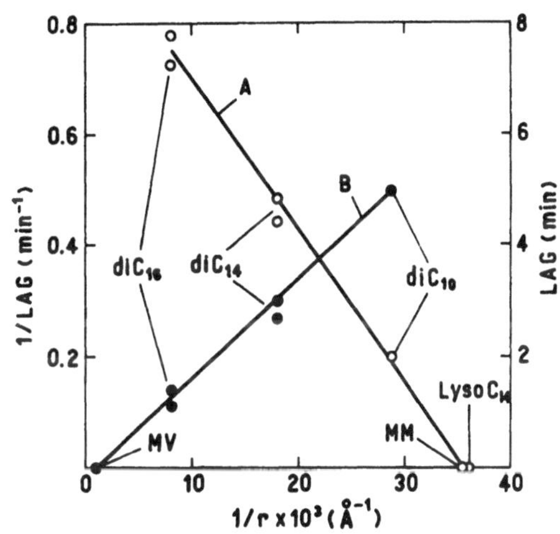

Fig. 5. Dependence of lag period and reciprocal of lag period as a function of the reciprocal radius of the aggregates.

We define lag period as the intercept on the time axis of the steady velocity. This lag period is thought to reflect the slow specific interaction of apoHBD with the lipid and its value depends on the aggregate radius of the lecithin employed. As shown in Fig. 5 when the lag period (right ordinate) is plotted as a function of the reciprocal radius, a straight line with a negative slope is obtained (curve A) the intercept with the abscissa being the reciprocal radius of the aggregates to which lag period is zero. In fact, mixed diC_{10} plus $lysoC_{14}$ micelles or pure $lysoC_{14}$ micelles (radius about 30 Å) produced instantaneous activation. On the other hand, when the reciprocal of lag period (left ordinate) is plotted against the reciprocal radius (curve B) a straight line with a positive slope is obtained; the intercept with the abscissa being the reciprocal radius of the aggregates to which lag period is infinite. This particle size is in the range of multilayered lecithin vesicles and, in fact, when apoHBD is incubated with these vesicles no reactivation is found, irrespective of the fatty acid composition of the lecithin used. Further addition of a three-fold excess of sonicated lecithin restores the HBD activity only after several hours incubation. This means that apoHBD is bound reversibly to the outer lamella of the multilayer vesicles, presumably only to the polar head of these tightly packed structures and in an inactive conformation. In the presence of added liposomes, the binding of the apoHBD is displaced toward the liposomes probably because the interaction with the apolar chains stabilizes the apoenzyme-liposome complex and HBD activity is regained (Guglielmucci et al., unpublished).

Lipid Substitution of Mitochondrial Membrane

Different results are obtained when mitochondrial membrane phospholipids are substituted by diC_{14} (Chapman, 1973) using the fusion technique as described by Houslay et al. (1975). The substituted mitochondrial membranes (95 % diC_{14} and 5 % of a mixture of endogenous phospholipid) showed 60 % of the initial HBD activity and more than 65 % of succinatecytochrome c reductase activity. The apoHBD could be released from these substituted membranes by phospholipase A treatment in an inactive form which could be reactivated by addition of lecithin liposomes. This indicate that no coarse alterations of membrane proteins occurred as a result of lipid substitution (Guglielmucci et al., unpublished).

In contrast with the results obtained with the apoHBD-diC_{14} liposomes complexes (cf Table 1), with the diC_{14} substituted membranes, no HBD activity is detected at temperatures below 24°C (Fig. 6) which is near the gel-to-liquid crystalline transition temperature of diC_{14} (the difference could be due to the effect to the protein). HBD activity appeared at 26°C and increased sharply with temperature; the activation energy being about 7.4 kcal/mol.

We can rationalise these differences assuming that when in liposomal complex, the apoHBD interacts with the polar head and with the proximal part of the hydrophobic chains (may be up to 4-6 carbon atoms) which is the region less modified by the gel-to-liquid-crystalline phase transition (Hubbell & Mc Connell, 1970; Levine et al., 1972; Barton & Gunstone, 1975); therefore the enzyme-liposome interaction and, consequently, the HBD activity of these complexes is not sensitive to phase transitions. Conversely, in the diC_{14} substituted membranes (and may be in the original mitochondria) the apoHBD could be deeply inserted in the lipid matrix and would

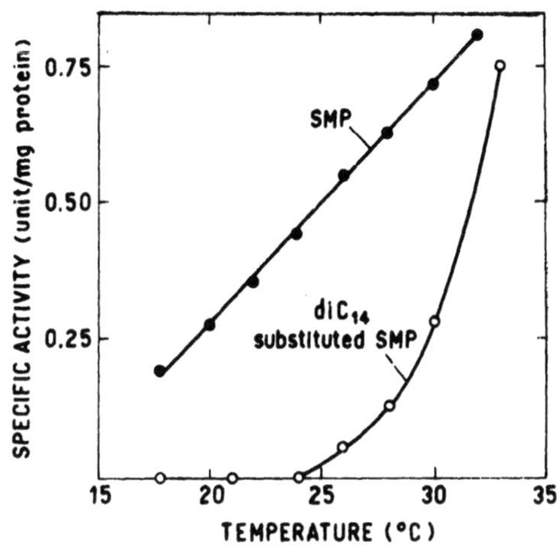

Fig. 6. Temperature dependence of HBD activity in original SMP and diC14 substituted mitochondrial membrane.

interact with the whole fatty acyl chain, so becoming sensitive to phase transitions (see also Houslay et al., 1975).

This interpretation is supported by the different mechanism of apoHBD release by phospholipase A in the successive steps of purification of the apoHBD (Vidal et al., 1976). In fact, solubilization of apoHBD from the mitochondrial membrane require phospholipid hydrolysis, while apoHBD release from the enzyme-lecithin complex is achieved by competitive binding of phospholipase to the lecithin-water interface without phospholipid hydrolysis.

Another line of evidences comes from studies on experimental diabetes. Liver mitochondria from rats made diabetic with the cytotoxic antibiotic streptozotocin show after four weeks (when the blood glucose is over 300 mg/100 ml) only a small fraction (20-30%) of the HBD activity, the magnitude of this decrease seems to be related with the severity of the diabetic condition (Boveris et al., 1969; Roldán et al., 1971). This decrease in activity is not modified by addition of insulin to mitochondria but can be prevented or reversed by insulin administration to diabetic animals.

In spite of this decrease in HBD activity, treatment of SMP with phospholipase A results in solubilization of a similar amount of apoHBD per mg of mitochondrial protein either from control or diabetic mitochondria. As shown in Fig. 7, when HBD activity in phospholipase supernatants from control or diabetic mitochondria is titrated with the same lecithin, similar hyperbolic curves are obtained (Fig. 7,1). When these curves are transformed into double reciprocal plots (Fig. 7,2), straight lines are obtained with a common intercept on the ordinate axis, indicating that at infinite lecithin concentration HBD activity of the control or diabetic

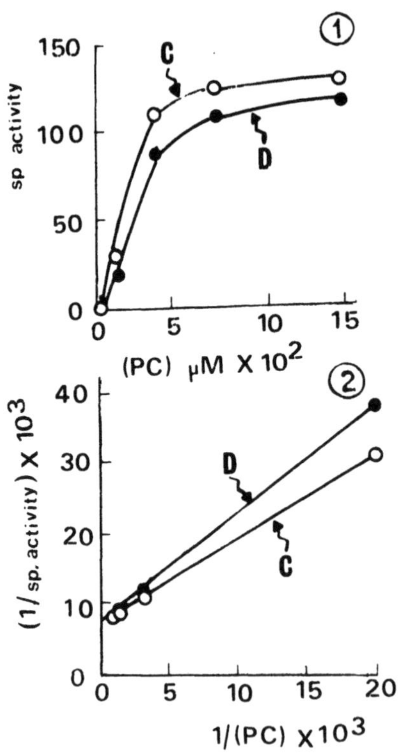

Fig. 7. (1) HBD activity in phospholipase A supernatants of control (C) and diabetic (D) rat liver mitochondria as a function of egg lecithin (PC) concentration. (2) Double reciprocal plot of the data presented in (1).

supernatants are the same. Moreover, lipid bustitution of control as well as diabetic SMP resulted in identical HBD activities (Guglielmucci et al., unpublished[b])

Similar results are obtained when the supernatants are titrated with other lecithins, indicating that the apoenzyme released from diabetic mitochondria is not defective with respect to lipid binding capacity. This results also is a proof that the apoenzymes obtained from normal or diabetic rat liver mitochondria are functionally (and, consequently, structurally) identical.

Fig. 8 shows the titration of apoHBD with MPL extracted from either control or diabetic rat liver mitochondria. It can be seen that "diabetic MPL" reactivate apoHBD with the same efficiency but to a lower extent compared to the control MPL (Fig. 9). However, this difference does not account for the four- to five-fold decrease of HBD activity in mitochondria from diabetic animals.

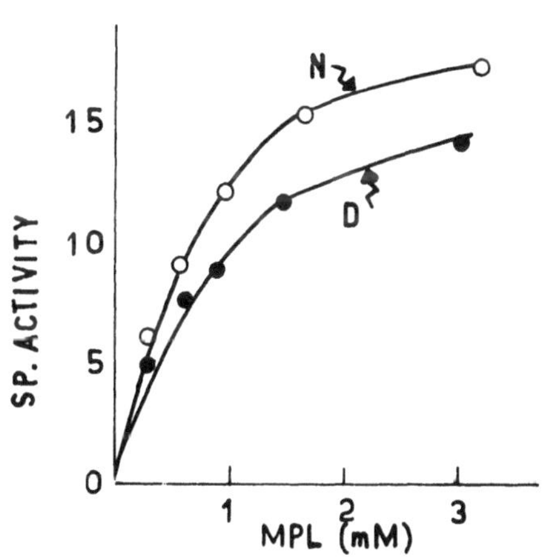

Fig. 8. Reactivation of 50 µg of apoHBD with MPL extracted from either normal (N) or diabetic (D) rat liver mitochondria.

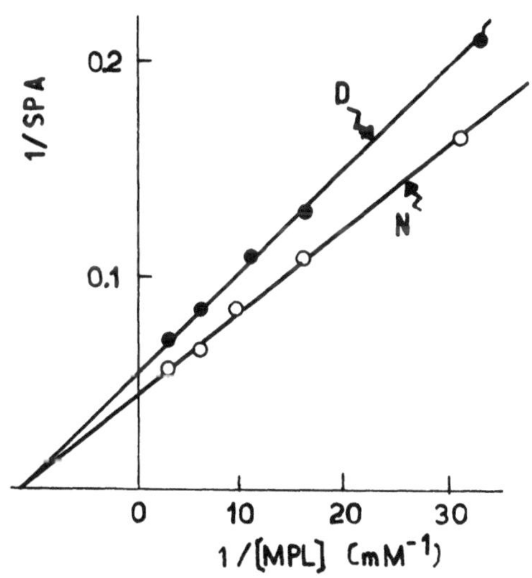

Fig. 9. Double reciprocal plots of data presented in Fig. 8.

This difference is clearly shown in Table 2, where the kinetic parameters for the NAD-linked oxidation of 3-hydroxybutyrate to acetoacetate are presented for the membrane-bound enzymes as well as for the apoHBD-MPL complexes.

Table 2. Kinetic parameters of NAD-linked oxidation of 3-hydroxybutyrate

Reaction mixtures contained 40 μg of mitochondrial protein in the form of SMP or 50 μg of purified apoHBD with saturating amounts of MPL in a medium of constant ionic strength and pH 8.1

Kinetic parameter		Membrane - bound		ApoHBD (control or diabetic)	
		Control	Diabetic	Control MPL	Diabetic MPL
Vm ($mM.min^{-1}.mg\ protein^{-1}$)		0.22	0.067	4.40	3.08
NAD^+	K_{iNAD} (mM)	0.31	0.80	0.40	0.36
	K_{NAD} (mM)	0.08	0.33	0.15	0.21
3-hydroxy-butyrate	K_{HOB} (mM)	0.44	0.45	0.67	0.80

Comparing firstly the membrane-bound enzymes, in SMP from diabetic rat livers, there is a 3-fold decrease in Vm accompanied with a nearly 3-fold increase in K_{iNAD} (the dissociation constant of the enzyme-NAD complex) and a 4-fold increase in K_{NAD} (the Km for NAD at infinite concentration of 3-hydroxybutyrate). No changes are found in K_{HOB} (the Km for 3-hydroxybutyrate at infinite concentration of NAD). On the other hand, the complex of apoHBD with MPL from diabetic rat liver mitochondria shows only a 30% decrease in Vm compared to the complex of apoHBD with MPL from control rat liver mitochondria. This means that the kinetic parameters of the reaction catalyzed by the complexes of apoHBD with MPL from either control or diabetic are more similar between them that to those of the corresponding membrane-bound enzymes.

This discrepancy could arise from the fact that in the enzyme-lyposome complexes HBD activity is poorly affected by the fatty acyl composition of the phospholipids, while it is quite sensitive to changes in the fatty acyl composition of the phospholipids within the membrane.

Table 3 shows the fatty acid composition of MPL extracted from control and diabetic rat liver mitochondrial membranes.

Table 3. Fatty acid composition and arachidonic:linoleic acid ratio in MPL extracted from control and diabetic rat liver mitochondria

Fatty acids	Control (%)	Diabetic (%)
14:0	0.4	0.6
16:0	17.5	15.2
16:1	1.6	1.2
18:0	23.2	22.8
18:1	10.8	9.6
18:2	18.0	26.6
20:3 (ω 9)	--	0.6
20:3 (ω 6)	1.0	1.3
20:4	26.9	20.5
Ratio $\frac{20:4}{18:2}$	1.49	0.77

Results are expressed as percent of total fatty acids. Minor components make for 100 %. Fatty acids of 14 to 20 carbons only measured.

Although no significant alteration is found either in the total amount or in the relative proportions of the different phospholipid classes in MPL from diabetic rat liver mitochondria compared to that of control rat liver mitochondria (Guglielmucci et al., unpublished[b]) Table 3 shows that significant differences are observed in fatty acid composition of MPL from control and diabetic rat liver mitochondria.

The observed changes are due, at least in part, to a decrease in the activity of fatty acid desaturases (Mercuri et al., 1966; Mercuri et al., 1967; Brenner, 1974). The most apparent results (see Table 3) are the decrease in the arachidonic:linoleic acid ratio, which is reduced to about one half the value found in the MPL from control rats and that a small amount of 20:3 (ω 9) is present in the MPL extracted from liver mitochondria of diabetic rats, showing a typical manifestation of an essential fatty acid defficiency. The conclusion is that the proportion of molecular specimens of phospholipids has been changed by the diabetic condition, with a reduction of higly unsaturated forms containing arachidonic acid.

Then, it can be concluded that the decrease in HBD activity in liver mitochondria from diabetic rats could be adscribed to alterations in the interaction of apoHBD with the structural components of the mitochondrial membrane as a consequence of an abnormal metabolism of the phospholipid fatty acids.

REFERENCES

BARTON P. G. & GUNSTONE F. D. (1975) J. Biol. Chem. 250, 4470-4475
BENDALL D. S. & de DUVE C. (1960) Biochem. J. 74, 444-450
BOCK H. G. & FLEISCHER S. (1975) J. Biol. Chem. 250, 5774-5781
BOVERIS A., CATTANEO de PERALTA RAMOS M., STOPPANI A. O. M. & FOGLIA V. (1969) Proc. Soc. Exptl. Biol. Med. 132, 170-175
BRENNER R. R. (1974) Mol. Cell. Biochem. 3, 41-52
CHAPMAN D. in Form and Function of Phospholipids (ANSELL G. B., HAWTHORNE J. N. & DAWSON R. M. C., editors), pp 117-142, Elsevier, Amsterdam. 1973
de GIER J., MANDERSLOOT J. G. & van DEENEN L. L. M. (1968) Biochim. Biophys. Acta 150, 666-675
FLEISCHER B., CASU A. & FLEISCHER S. (1966) Biochem. Biophys. Res. Commun. 24, 189-194
FLEISCHER S. & KLOUWEN H. (1961) Biochem. Biophys. Res. Commun. 5, 378-383
GAZZOTTI P., BOCK H. G. & FLEISCHER S. (1975) J. Biol. Chem. 250, 5782-5790
GOTTERER G. S. (1967) Biochemistry 6, 2139-2146
GRANT C. W. M., WU S. H-W. & Mc CONNELL H. M. (1974) Biochim. Biophys. Acta 363, 151-158
GREEN D. E., DEWAN J. G. & LELOIR L. F. (1937) Biochem. J. 31, 934-939
GROVER A. K., SLOTBOOM A. J., DE HAAS G. H. & HAMMES G. G. (1975) J. Biol. Chem. 250, 31-38
GUGLIELMUCCI E. A., VIDAL J. C., BRENNER R. R. & STOPPANI A. O. M. (Unpublished)
GUGLIELMUCCI E. A., VIDAL J. C. & STOPPANI A. O. M. (Unpublished)
HAUSER H. & BARRATT M. D. (1973) Biochem. Biophys. Res. Commun. 53, 399-405
HEXTER C. S. & GOLDMAN R. (1973) Biochim. Biophys. Acta 307, 421-427
HOUSLAY M. D., WARREN G. B., BIRDSALL N. J. M. & METCALFE J. C. (1975) FEBS Letters 51, 146-151
HUBBELL W. L. & Mc CONNELL H. M. (1970) J. Am. Chem. Soc. 43, 314-326

LATRUFFE N. & GAUDEMER Y. (1974) Biochemie 56, 435-439
LEVINE Y. K., BIRDSALL N. J. M., LEE A. G. & METCALFE J. C. (1972) Biochemistry 11, 1416-1421
LEVY M., JONCOURT M. & THIESSARD J. (1976) Biochim. Biophys. Acta 424, 57-65
MENZEL H. M. & HAMMES G. G. (1973) J. Biol. Chem. 248, 4885-4889
MERCURI O., PELUFFO R. O. & BRENNER R. R. (1966) Biochim. Biophys. Acta 116, 284-290
MERCURI O., PELUFFO R. O. & BRENNER R. R. (1967) Lipids 2, 284-290
NIELSEN N. C. & FLEISCHER S. (1973) J. Biol. Chem. 248, 2549-2555
PHILLIPS M. C. in Progress in Surface and Membrane Science (DANIELLI F. J., ROSENBERG M. D. & CADENHEAD D. A., editors) Vol. 5, pp 139-221, Academic Press, New York, 1972
PHILLIPS M. C., WILLIAMS R. M. & CHAPMAN D. (1969) Chem. Phys. Lipids 3, 234-244
ROLDAN A. G., DEL CASTILLO E. J., BOVERIS A., GARAZA PEREIRA A. M. & STOPPANI A. O. M. (1971) Proc. Soc. Exptl. Biol. Med. 137, 791-805
SEKUZU I., JURTSHUK P. & GREEN D. E. (1963) J. Biol. Chem. 238, 975-982
SHEETZ M. P. & CHAN S. I. (1972) Biochemistry 11, 4573-4579
TAUSK R. J. M., KARMIGGELT J., OUDSHOORN C. & OVERBEEK J. Th. G. (1974) Biophys. Chem. 1, 175-183
TAUSK R. J. M., van ESCH J., KARMIGGELT J., VOORDOUW G. & OVERBEEK J. Th. G. (1974) Biophys. Chem. 1, 184-203
VIDAL J. C., CATTANEO P., STOPPANI A. O. M. (1972) Arch. Biochem. Biophys. 151, 168-179
VIDAL J. C., GUGLIELMUCCI E. A. & STOPPANI A. O. M., Mol. Cell. Biochem., 1976 (in press)
WARREN G. B., TOON P. A., BIRDSALL N. J. M., LEE A. G. & METCALFE J. C. (1974) Proc. Natl. Acad. Sci. U.S.A. 71, 622-626
WILLIAMS R. M. & CHAPMAN D. (1970), Prog. Chem. Fats other Lipids 11, 1-79
ZIEGLER D. M. & LINNANE A. W. (1958) Biochim. Biophys. Acta 30, 53-63

THE ROLE OF PHOSPHOLIPIDS IN Na-K ATPase

J.J.H.H.M. de Pont and S.L. Bonting

Department of Biochemistry, University of Nijmegen

Nijmegen, The Netherlands

It has been known for a long time that lipids present in the plasma membrane influence the activity of the Na-K ATPase system. Treatment of Na-K ATPase preparations with detergents, phospholipases and organic solvents leads to partial or complete inactivation of the enzyme activity (for a recent review see Kimelberg, 1976). Moreover Arrhenius plots for the Na-K ATPase activity show a marked discontinuity around 20°C, which is generally attributed to a change in the solid/fluid transition of the fatty acid chains of the phospholipids. This interpretation has received support from studies of electron paramagnetic resonance spectra with spin-labeled fatty acids (Grisham and Barnett, 1973) and of fluorescent probe studies (Charnock and Bashford, 1975).Two questions remain: what is the molecular mechanism of this phospholipid effect, and is the interaction due to specific phospholipids?

The latter question has been studied primarily by reconstitution studies. The Na-K ATPase preparation is delipidated to a certain extent by one of several approaches, and is then incubated with various phospholipids; the activity is determined before and after reconstitution. With the exception of organic solvent extraction at low temperatures, when cholesterol gives the highest degree of reactivation (Noguchi and Freed, 1971; Järnefelt, 1972), negatively charged phospholipids are necessary for obtaining good reconstitution of the Na-K ATPase activity in nearly all other reactivation studies. Phosphatidylserine has mainly been used in these experiments, but reconstitution is also possible with phosphatidylglycerol, phosphatidylinositol and phosphatidic acid.

In an elegant study with phospholipids, prepared enzymatically from one parent phospholipid, Walker and Wheeler (1975) have shown that the fatty acid chains of the phospholipids also have an effect on the kinetics of the reconstitution process. Since phosphatidylserine in the only one of the acidic phospholipids, which is present in significant amounts in all mammalian plasma membranes, several authors have concluded that this phospholipid has a specific importance for the lipid-protein interaction and the functioning of tha Na-K ATPase system (see Kimelberg, 1976). Zwaal et al. (1973) see further support for this in the fact that phosphatidylserine is in the human erythrocyte located on the inner face of the bilayer, and the substrate catalytic side of the Na-K ATPase system on the inside of the membrane. Kimelberg and Papahadjopoulos (1972) correlate the specific reactivating properties of phosphatidylserine and phosphatidylglycerol with the fact that liposomes prepared from these phospholipids have a high K^+/Na^+ permeability ratio (Papahadjopoulos, 1971).

Since we felt that these observations do not constitute absolute proof that phosphatidylserine is the essential phospholipid in native Na-K ATPase preparations, we have applied another approach. Membranes of E. coli contain an enzyme, which catalyses the conversion of phosphatidylserine into phosphatidylethanolamine (Kanfer and Kennedy, 1964; Dowhan et al., 1974). Treatment of Na-K ATPase preparations by this enzyme could remove phosphatidylserine without removing any other phospholipids. There is, however, a technical complication: the isolated phosphatidylserine decarboxylase requires a non-ionic detergent, such as Triton-X-100, for activity, but this detergent inactivates Na-K ATPase. We were able to overcome this problem, since inorganic phosphate and EDTA reduce the inactivating effect of Triton-X-100 on Na-K ATPase. In order to eliminate the effects of factors present in the phosphatidylserine decarboxylase preparation, we have always used a preheated phosphatidylserine decarboxylase preparation as a control.

In our initial experiments we have treated a partially purified Na-K ATPase preparation from cattle brain with a partially purified preparation of the decarboxylase (de Pont et al., 1973). We were indeed able to convert "all" phosphatidylserine into phosphatidylethanolamine without reduction of the Na-K ATPase activity, when comparing it with the control described above. However, since our starting Na-K ATPase preparation contained 10^3 molecules phosphatidylserine (out of 10^4 molecules of phospholipid) per molecule of Na-K ATPase, after a "complete" conversion of phosphatidylserine up to ten molecules of this phospholipid could still be present per molecule of the enzyme.

For this reason we have tried to improve the sensitivity of the approach by prior removal of an important part of less essential phospholipids. Extraction of lyophilized cattle brain Na-K ATPase with n-hexane leads to removal of all cholesterol and of 68% (SE:1.8, n=5) of the phospholipids (Fig. 1). There is relatively little difference in extractability of the individual phospholipids, varying from 77% (SE:4.0, n=5) for phosphatidylserine to 61% (SE:4.6, n=5) for sphingomyelin. The Na-K ATPase activity after hexane extraction is 68% (SE:3.7, n=16) of the control activity (Fig. 1).

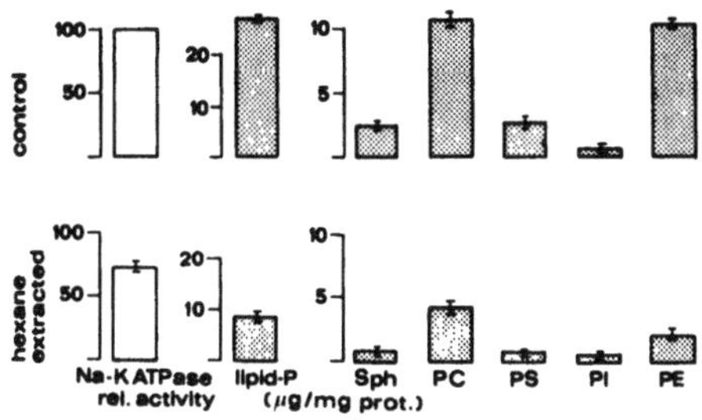

Fig. 1. Effect of hexane extraction on the lipid-P content and five individual phospholipids in a cattle brain Na-K ATPase preparation. Values, representing averages of five or more experiments, are given with the standard errors. Abbreviations: Sph = sphingomyelin; PC = phosphatidylcholine; PS = phosphatidylserine; PI = phosphatidylinositol; PE = phosphatidylethanolamine.

Most properties of the hexane extracted Na-K ATPase preparation (K_m values for Na^+ and K^+, pI_{50} for ouabain) do not differ from those of the control preparation. Only the K_m value for Mg-ATP of the hexane treated preparation (0.82 mM; SE:0.14 mM; n=4) is significantly lower than that of the control preparation (1.22 mM; SE:0.05 mM; n=7). The activation energies and the transition temperatures, calculated from Arrhenius plots, do not differ significantly (Fig. 2). This suggests that cholesterol and the ex-

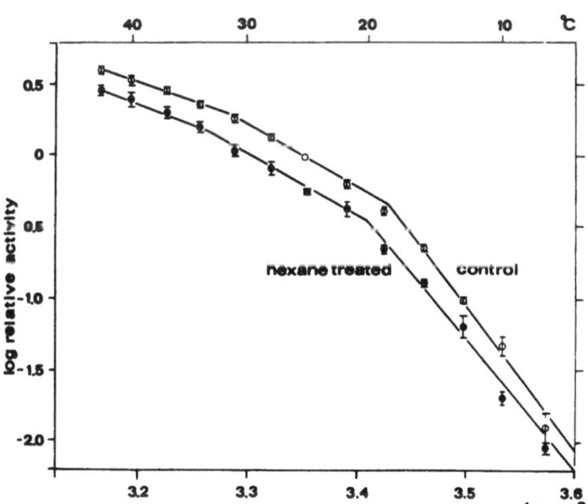

Fig. 2. Relation between the logarithm of the Na-K ATPase activity in untreated (-o-o-) and hexane extracted (-•-•-) preparations and the reciprocal value of the absolute temperature. In each experiment the activity of the control preparation at 25°C is set at 100 and all activities are expressed relative to this value. Averages from three experiments with standard errors are shown.

tracted phospholipids do not affect these characteristic of the Na-K ATPase system and thus do not appear to be present in the lipid annulus around the enzyme. Since cholesterol reduces the fluidity of the plasma membrane, this finding confirms that of Grisham and Barnett (1972), according to which the interior or the phospholipid bilayer adjacent to the ATPase is more fluid than in the rest of the membrane.

After hexane treatment of cattle brain Na-K ATPase again complete conversion of phosphatidylserine into phosphatidylethanolamine has been achived without decrease in Na-K ATPase activity relative to the control. Although the sensitivity of these experiments has increased, there is still the possibility that up to three molecules of phosphatidylserine per molecule of Na-K ATPase are left.

The only possibility for further improvement of the sensitivity is a further purification on the Na-K ATPase preparation so that the number of phospholipids per molecule of Na-K ATPase in the starting preparation is further decreased. The Na-K ATPase from rabbit kidney outer medulla has been purified by the method of Jørgensen (1974). This preparation is free of other ATPase activity and with SDS electrophoresis it shows only the two characteristic protein bands. This preparation contains per molecule of Na-K ATPase some 300 phospholipid molecules, the composition of which is given in Table 1.

Incubation of this Na-K ATPase preparation with phosphatidylserine decarboxylase leads to complete conversion of the 31 phosphatidyl serine molecules present per molecule of Na-K ATPase, again without significant reduction in Na-K ATPase activity. In this case the sensitivity of the phosphatidylserine assay (thin layer chromatography and amino acid analysis) is such that it can be stated that less than one phosphatidylserine molecule per molecule of enzyme is left.

These experiments indicate that phosphatidylserine per se is not necessary for proper functioning of the Na-K ATPase system. Since phosphatidylinositol is still present in this enzyme preparation, it might be that this acidic phospholipid takes over the place of phosphatidylserine in this system. However, quite recently Hilden and Hokin (1976) have replaced all phospholipids by phosphatidylcholine with a lipid replacement method (Warren et al., 1974) and in this system the enzyme is still functional. Moreover, Racker and Fisher (1975) have reported a reconstitution of the sodium pump with liposomes of phosphatidylcholine and phosphatidylethanolamine and mixtures of these phospholipids (a 1:4 ratio being the most effective) but they do not give an analysis of the residual phospholipids in the enzyme preparation.

Table 1. Effect of phosphatidylserine decarboxylase on phospholipid composition and activity of purified Na-K ATPase from rabbit kidney outer medulla

	- PS decarb. (11 expts.)	+ PS decarb. (5 expts.)	Difference
Sphingomyelin	55±4	56±6	n.s.
Phosphatidylcholine	116±4	117±2	n.s.
Phosphatidylserine	31±3	0	- 31±3
Phosphatidylinositol	14±2	16±1	n.s.
Phosphatidylethanolamine	89±5	116±7	+ 27±9
Na-K ATPase activity	≡100%	91±6%	n.s.

Average numbers of moles phospholipid per mole Na-K ATPase (250.000 g protein) are given with standard errors.

This suggests that for maintenance of the activity in native Na-K ATPase there is no preference for specific phospholipids, but rather a requirement for a fluid bilayer. The question now arises why negatively charged phospholipids are so efficient in reconstitution experiments. Only a speculative answer can be given to this question. There is considerable evidence that in the Na-K ATPase molecule there is an interaction between at least two subunits and that a proper interaction between these two subunits is necessary for activity of the enzyme. If the function of the phospholipids is to create a fluid microenvironment, which facilitates the interaction between the subunits, it is very likely that delipidation decreases the interaction between the subunits. Acidic phospholipids might be very effective in reestablishing subunit interaction, but they need not be essential for the functioning of the *native* enzyme.

ACKNOWLEDGEMENTS

The excellent technical assistance of Mrs. Ans van Prooyen - van Eeden is gratefully acknowledged.

REFERENCES

CHARNOCK, J. S. & BASHFORD, C. L. (1975) Mol. Pharmacol. 11, 766-774.
DOWHAN, W., WICKNER, W. T. & KENNEDY, E. P. (1974) J. Biol. Chem. 249, 3079-3084.
GRISHAM, C. M. & BARNETT, R. E. (1972) Biochim. Biophys. Acta 226, 613-624.
GRISHAM, C. M. & BARNETT, R. E. (1973) Biochemistry 12, 2635-2637.
HILDEN, S. & HOKIN, L. E. (1976) Biochem. Biophys. Res. Commun. 69, 521-527.
JÄRNEFELT, J. (1972) Biochim. Biophys. Acta 266, 91-96.
JØRGENSEN, P. L. (1974) Biochim. Biophys. Acta 356, 36-52.
KANFER, J. & KENNEDY, E. P. (1964) J. Biol. Chem. 239, 1720-1726.
KIMELBERG, H. K. (1976) Mol. Cell. Biochem. 10, 171-190.
KIMELBERG, H. K. & PAPAHAJOPOULOS, D. (1972) Biochim. Biophys. Acta 282, 277-292.
NOGUCHI, T. & FREED, S. (1971) Nature New Biology 230, 148-150.
PAPAHADJOPOULOS, D. (1971) Biochim. Biophys. Acta 241, 254-259.
de PONT, J. J. H. H. M., van PROOYEN-van EEDEN, A. & BONTING, S. L. (1973) Biochim. Biophys. Acta 323, 487-494.
RACKER, E. & FISHER, L. W. (1975) Biochem. Biophys. Res. Commun. 67, 1144-1150.
WALKER, J. A. & WHEELER, K. P. (1975) Biochim. Biophys. Acta 394, 135-144.
WARREN, G. B., TOON, P. A., BIRDSALL, N. J. M., LEE, A. G. & METCALFE, J. C. (1974) Proc. Natl. Acad. Sci. U.S.A. 71, 622-626.
ZWAAL, R. F. A., ROELOFSEN, B. & COLLEY, C. M. (1973) Biochim. Biophys. Acta 300, 159-182.

INTERACTIONS WITH PHOSPHOLIPID MONOLAYERS OF LIPIDS AND WATER-SOLUBLE COMPOUNDS THAT INDUCE MEMBRANE FUSION

B. Maggio* and J. A. Lucy

Department of Biochemistry & Chemistry. Royal Free Hospital School of Medicine. University of London
8 Hunter Street. London WC1N 1BP. U.K.
* Present address: Departamento de Química Biológica, Facultad de Ciencias Químicas, Córdoba, Argentina

INTRODUCTION

Hen erythrocytes fuse into multinucleated cells on incubation under defined conditions "in vitro" with a variety of chemical agents. These include various unsaturated and low-melting fat-soluble substance (Ahkong et al., 1973; 1974), dimethyl sulphoxide, glycerol and other polyols, and poly(ethyleneglycol)(Ahkong et al., 1975a; 1975b). The present communication reports some effects of these compounds on phospholipid monolayers that may be related to the molecular mechanism of chemically-induced membrane fusion.

EXPERIMENTAL PROCEDURES

In experiments with water-soluble fusogens the phospholipid monolayers were spread on subphases containing different amounts of the organic solute under study and NaCl (145 mM) at pH 5.6. The surface pressure, molecular areas and surface potential of the monolayers were obtained in an automated equipment and compared with those obtained on subphases containing only NaCl (Maggio et al., 1976). The effects of fusogenic lipids were studied by comparing the surface properties of mixed monolayers containing different molar proportions of phospholipids and fusogenic lipids, with the properties expected for ideally-mixed films of noninteracting molecules (Maggio & Lucy, 1975; 1976).

RESULTS AND DISCUSSION

Water-soluble fusogens. Large decreases of the surface potential were observed when dipalmitoylphosphatidylcholine (dpPC) was spread on subphases containing glycerol, dimethyl sulphoxide, sucrose, sorbitol or poly(ethyleneglycol) of M.W. 1500; 6000; 3×10^5 or 5×10^6 (Fig. 1). Similar results were obtained for dipalmitoylphosphatidylethanolamine (dpPE).

These observations indicate that the dipolar properties of phospholipids are changed in presence of the organic solutes. Interactions of the polar head group of the phospholipids with the water-soluble fusogens probably occur, together with various structural changes in the bulk-phase water which may induce alterations in the orientation and hydration of the phospholipid head groups (Maggio et al., 1976).

Fusogenic lipids. Mixed monolayers of fusogenic lipids with dpPC showed ion-dipole interactions with decreases in the surface potential that occurred together with negative deviations in the area occupied per molecule, with respect to ideal films.

Other natural and synthetic choline-containing phospholipids behaved like dpPC in mixed monolayers with a variety of fusogenic

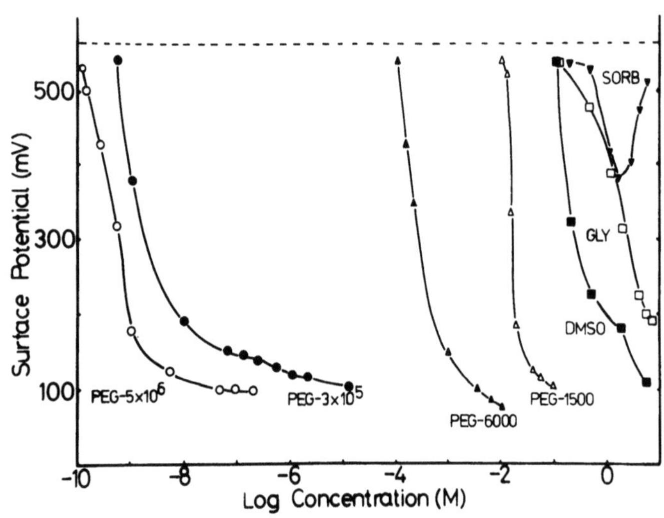

Fig. 1. Surface potential of dpPC (at 35 $mN.m^{-1}$) on subphases containing differing concentrations of water-soluble fusogens. Horizontal dashed line represent the value of surface potential on 145 mM NaCl, pH 5.6.

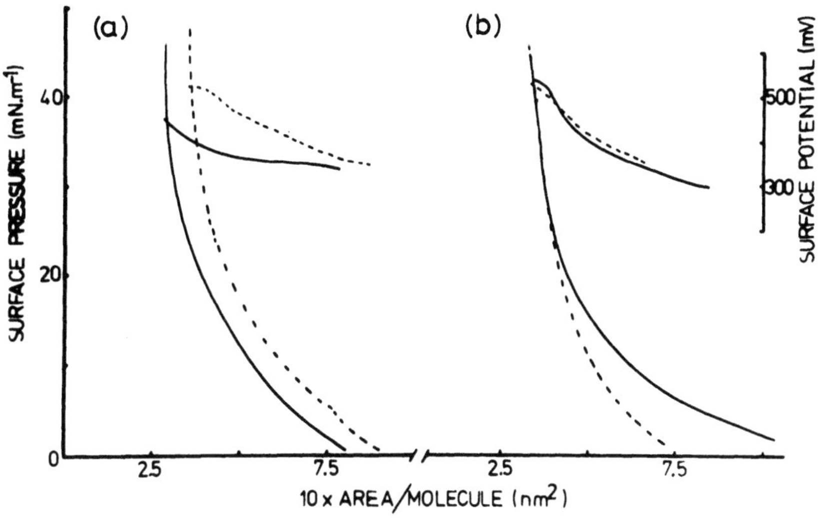

Fig. 2. Surface pressure and surface potential-area curves for mixed monolayers of GMO with dpPC (a) or dpPE (b) on 145 mM NaCl, pH 5.6. Dashed curves represent ideal films.

lipids but the interactions were different in mixed films with dpPE (Fig. 2). The characteristic interactions found in mixed monolayers of fusogenic lipids with phosphatidylcholine were independent of the type of hydrophobic chain of the phospholipid. Conversely, a single unsaturated or low-melting hydrophobic chain, independent of the type of polar head group, appears to be required for lipids to be both fusogenic and give the behaviour described with phosphatidylcholine (Maggio & Lucy, 1975; 1976).

The specific interactions of fusogenic lipids with dpPC may be attributed to an effect of the low-melting chain of the lipid which is mediated by the polar head group of dpPC. In these conditions the polar group of dpPE behaves differently. The lowering of the pH of the subphase from 5.6 to 2 virtually abolished the negative deviations in surface potential and molecular area seen for mixed films of glycerylmono-oleate (GMO) with dpPC and this may be attributed to protonation of the primary phosphate group of the phospholipid. However, the complete head group of phosphatidylcholine (with two CH_2 groups between the positive and negative charges) is involved in the behaviour described since the molecular interactions are quite different in a mixed monolayer of the fusogenic lipid with an equimolar mixture of dihexadecylphosphate-hexadecyltrimethyl ammonium, which contains essentially the same ionized components as those present in the polar moiety of dpPC (Maggio & Lucy, 1976).

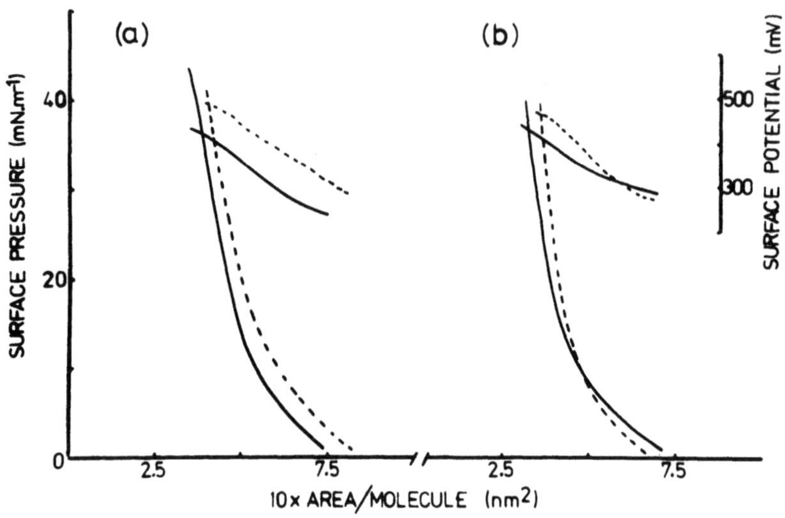

Fig. 3. Surface pressure and surface potential-area curves for mixed monolayers of GMO with dpPC (a) or dpPE (b) on 145 mM NaCl, pH 5.6, containing 10 mM $CaCl_2$. Dashed curves represent ideal films.

When Ca^{2+} (or UO_2^{2+}) was present in the subphase (Fig. 3), a mixed monolayer of GMO with dpPE also showed negative deviations in the molecular area and surface potential and behaved, at high pressures, like a mixed monolayer with dpPC. Such experiments indicated that dpPE can show interactions with fusogenic lipids similar to dpPC, depending on the surface pressure of the film and on the degree to which the dipolar properties are constrained by the influence of an external bivalent metal ion (Maggio & Lucy, 1976).

A low-melting lipid in a mixed monolayer with phosphatidylcholine may facilitate molecular movement in the polar region of the phospholipid (Traub1e & Haynes, 1971; Seelig & Seelig, 1974) so that a more stable arrangement is produced, with decreases in the surface potential and a closer packing of the molecules. In order to obtain similar interactions with phosphatidylethanolamine it seems necessary first to disturb the orientation of its polar group by adding bivalent cations to the subphase. Fig. 4 shows some of these possibilities with space-filling Corey-Pauling-Koltum molecular models (Maggio & Lucy, 1976).

The effects described here in a model membrane may have some implications for membrane fusion. When a fusogenic lipid is introduced into the asymmetric bilayer structure of an erythrocyte membrane (Zwaal et al., 1973) it may initially interact with choline-containing phospholipids in the outer half of the bilayer (Maggio & Lucy, 1975). This will alter the phospholipid head groups and

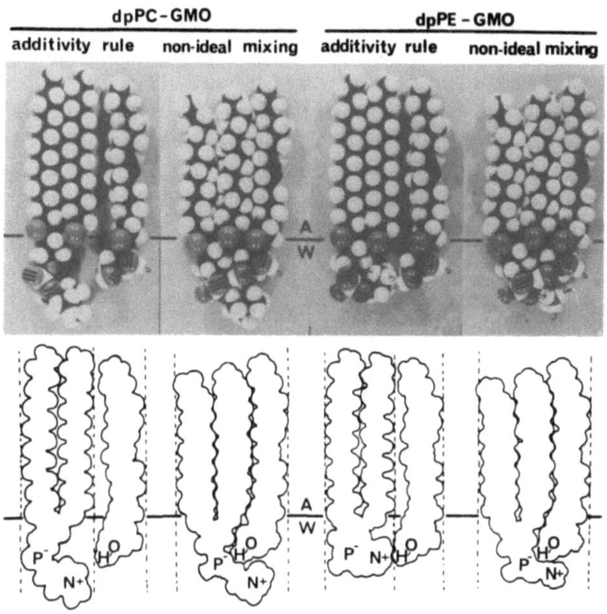

Fig. 4. Molecular models of glycerylmono-oleate and phospholipids.

produce a decrease of the surface potential of the lipid bilayer which, as in the case of the water-soluble fusogens, may be a physical perturbance of major importance for membrane fusion. In relation to the effects of water-soluble fusogens, Maroudas (1975) suggested that glycerol and hydrophillic polymers would produce lateral displacement of membrane glycoproteins by decreasing their exclusion volumes, thus facilitating cell adhesion. Cellular aggregation would also be expected from the reductions of the electrostatic field perpendicular to a naked surface of the lipid molecules of the membrane caused by the decrease of the surface potential. The decrease is smaller for the fusogenic lipids which are, comparatively, more toxic for the membrane than the water-soluble fusogens. This may result from undue perturbantions of the hydrocarbon chains in the lipid bilayer. The low-melting lipid molecules might co-operatively increase the number of rotational isomers in clusters of hydrocarbon chains in the membrane interior (Trauble & Haynes, 1971) giving a well-ordered bilayer with disordered chains (Seelig & Seelig, 1974).

Changes in surface potential and molecular packing can control the ionic permeabilities of a membrane (Gingell, 1967; Trauble & Haynes, 1971) possibly facilitating the entry of Ca^{2+} into the cell which is required for membrane fusion (Ahkong et al., 1975c; Maggio et al., 1976). In presence of a fusogenic lipid, it may be speculated on the basis of our findings that the properties of the inner half of the lipid bilayer (relatively rich in phsophatidylethanolamine) could be altered and behave like the outer half (relatively rich in phosphatidylcholine) if the concentration of cytoplasmic Ca^{2+} is raised to about 1 mM. This would increase the symmetry of the membrane and decrease the energy barrier for the intermixing of the constituents of membranes in closely adjacent cells, thus facilitating membrane fusion.

CONCLUSIONS

Fusogenic lipids interact with phosphatidylcholine at the air-water interface showing reductions in the surface potential and area per molecule. These effects may reflect an arrangement of phospholipid dipoles that allows a closer packing in the monolayer. A decrease of the surface potential of a naked lipid bilayer may be of major importance for membrane fusion since large decreases of surface potential of phospholipid monolayers are also induced by water-soluble fusogens. A reduced electrostatic field perpendicular to the surface of lipid molecules of the bilayer may allow cellular aggregation and entry of Ca^{2+} into the cell which might then facilitate the fusion of membranes. In presence of Ca^{2+}, interactions of fusogenic lipids with phosphatidylethanolamine show similar characteristics to those seen with phosphatidylcholine. Since these phospholipids are asymmetrically located in the erythrocyte membrane, the action of fusogenic lipids may be interpreted as inducing changes in the membrane permeability and, in presence of an increased concentration of intracellular Ca^{2+}, leading to the formation of similar molecular arrangements with phospholipids in the inner and outer half of the lipid bilayer, thus reducing asymmetry and facilitating membrane fusion.

This work was undertaken with the aid of a British Council Fellowship to Dr. Maggio.

REFERENCES

AHKONG Q. F., FISHER D., TAMPION W. & LUCY J. A. (1973) Biochem. J. 136, 147-155

AHKONG Q. F., HOWELL J. I., TAMPION W. & LUCY J. A. (1974) FEBS Lett. 41, 206-210

AHKONG Q. F., FISHER D., TAMPION W. & LUCY J. A. (1975a) Nature (London) 253,194-195

AHKONG Q. F., HOWELL J. I., LUCY J. A., SAFWAT F., DAVEY M. R. & COCKING E. C. (1975b) Nature (London) 255, 66-67

AHKONG Q. F., TAMPION W. & LUCY J. A. (1975c) Nature (London) 256, 208-209
GINGELL D. (1967) J. Theoret. Biol. 17, 451-482
MAGGIO B. & LUCY J. A. (1975) Biochem. J. 149, 597-608
MAGGIO B. & LUCY J. A. (1976) Biochem. J. 155, 353-364
MAGGIO B., AHKONG Q. F. & LUCY J. A. (1976) Biochem. J. 158, 647-650
SEELIG A. & SEELIG J. (1974) Biochemistry 13, 4839-4845
TRAUBLE H. & HAYNES D. H. (1971) Chem. Phys. Lipids 7, 324-335
ZWAAL R. F., ROELOFSEN B. & COLLEY C. M. (1973) Biochim. Biophys. Acta 300, 159-182

THE LIPID COMPOSITION AND Ca TRANSPORT FUNCTION OF SARCOPLASMIC RETICULUM(SR) MEMBRANES DURING DEVELOPMENT IN VIVO AND IN VITRO

R. L. Boland and A. Martonosi

Department of Biochemistry
St. Louis University School of Medicine
Saint Louis, Missouri 63104, USA

INTRODUCTION

Lipids play an essential role in the structure and function of biological membranes. The concept that the lipid enviroment influences the activity of membrane bound enzymes is now generally accepted. The sarcoplasmic reticulum(SR) of skeletal muscle is an intracellular membranous system with an important physiological role in contraction and relaxation (Hasselbach, 1964; Martonosi, 1971). The Ca^{2+} transport and (Ca^{2+} + Mg^{2+})-ATPase activities which exhibit isolated SR preparations have been shown to be dependent on membrane phospholipids (Martonosi, 1968). An involvement of the fatty acid component of phospholipids in the functioning of the Ca^{2+} pump has also been suggested (Seiler et al., 1970; Seiler & Hasselbach, 1971). With the purpose of further exploring the relationships existing between lipids and other components of sarcoplasmic reticulum and its Ca transport function, changes in biochemical composition were correlated with variations in Ca^{2+} uptake during development of skeletal muscle in vivo and in vitro.

EXPERIMENTAL PROCEDURES

Chicken skeletal muscle developed in vivo and in vitro was used to prepare whole homogenates and microsomal fractions isolated therefrom. For the in vivo studies, embryos and chicks of 10 to 50 days of age, begining with the start of incubation, were employed. The superficial pectoralis muscles were excised, cleaned of fat and connective tissue, and homogenized in 0.1M KCl, 10 mM imidazole, 0.3 M sucrose, pH 7.4. The homogenate was stored at -70°until analyzed. For the isolation of sarcoplasmic reticulum membranes the homogenate was subjected to differential centrifugation inmediately after preparation (Boland et al., 1974).

Cultures of muscle cells were prepared from superficial pectoralis muscles of 12- to 13-day-old chicken embryos as indicated in a previous report (Martonosi et al., 1976). The cultures were harvested between 2 and 13 days. The collected muscle cells were repeatedly washed with Earl's balanced salt solution (Bischoff & Holzer, 1969) and finally homogenized in 0.1 M KCl, 10 mM imidazole, 0,3 M sucrose and stored at -70°.

The techniques employed to measure the ATP dependent, azide insensitive calcium uptake at 25° by whole cell muscle homogenates and isolated SR are described elsewhere (Boland et al., 1974; Martonosi et al., 1976). The concentration of Ca-ATPase in homogenates and sarcoplasmic reticulum membranes was determined by specific labeling of the enzyme with ^{32}P-ATP (Martonosi et al., 1976).

The phospholipid and fatty acid composition of total lipid extracts of muscle homogenates and SR (Folch et al., 1959) was analyzed by two-dimentional thin layer chromatography and gas liquid chromatography respectively (Boland et al., 1974).

RESULTS AND DISCUSSION

The calcium transport function of sarcoplasmic reticulum markedly increases during in vivo and in vitro development of chicken

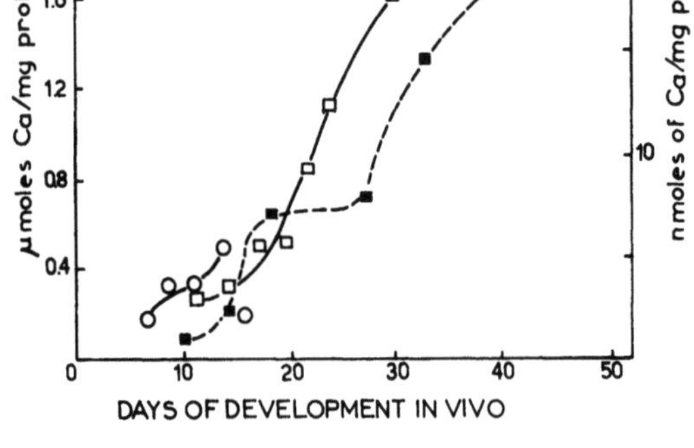

Figure 1. Calcium uptake by whole homogenates and microsome (SR) during muscle development in vitro and in vivo. o——o, homogenates of cultured muscle; □—□, homogenates of muscle developed in vivo; ■——■, microsomes from muscle developed in vivo.

pectoralis muscle (Fig. 1). In muscle developed in vivo there is a rapid increase with age until 30-50 days of development. Essentially the same trend is observed when the measurements of Ca uptake are performed in either isolated SR membranes or in whole homogenates under conditions (5 mM azide) in which Ca transport by mitochondria is inhibited. Due to the limited amounts of material available in tissue culture, the determinations of Ca transport (and compositional analysis) were carried out only in whole homogenates. The ATP mediated azide-insensitive Ca uptake increases until 9 days of culture. However, the maximum amounts of Ca taken up were less than one third of the Ca accumulated by homogenates of chicken skeletal muscle at 50 days of development.

The above described variations in Ca transport are roughly proportional to developmental changes in steady state concentration of the phosphoprotein intermediate involved in the hydrolysis of ATP by the ($Ca^{2+}+Mg^{2+}$)-ATPase, which may be considered an indicator of enzyme concentration (Figure 2). The total amounts (nmoles ^{32}P/plate) of Ca transport ATPase in cultured muscle increase between the second and the 8th day close to 20 fold (Fig. 2A).

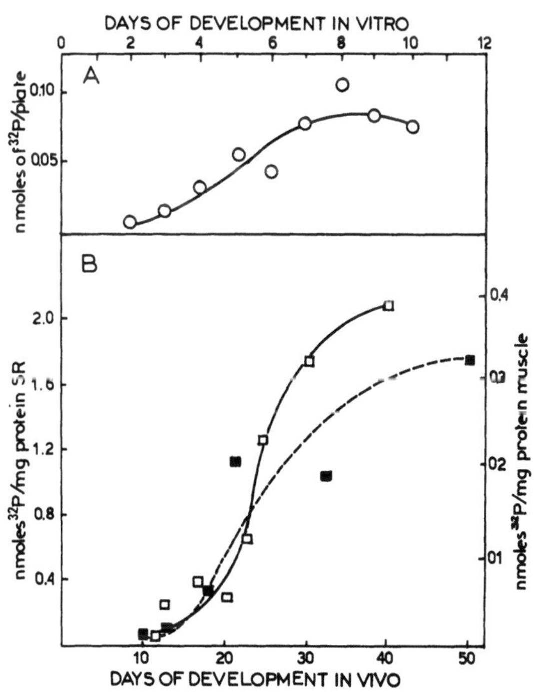

Figure 2. Phosphoprotein concentration in whole homogenates and microsomes(SR) of developing muscle. A,o——o, homogenates of cultured muscle; B,□——□, homogenates of muscle developed in vivo; ■--■, microsomes from muscle developed in vivo.

In whole homogenates and microsomal fractions from pectoralis muscle of developing embryos and chicks there is also a 30-40 fold increase in the concentration of phosphoprotein (Fig. 2B).

The parallel rise in Ca transport and the steady state concentration of phosphoprotein intermediate during development both in vivo and in vitro, suggests that the concentration of the Ca ATPase is a major factor in the regulation of the Ca transport activity of sarcoplasmic reticulum.

A phospholipid requirement in connection with Ca transport and ATPase activity of SR has been clearly demonstrated (Martonosi, 1968). In addition phospholipids contribute to the permeability characteristics of sarcoplasmic reticulum which is important for the retention of accumulated calcium (Boland et al., 1975). Therefore, developmental changes in lipid composition of skeletal muscle cells were also investigated. During development in vivo of skeletal muscle the phospholipid composition of whole homogenates and SR does not fluctuate considerably (Table 1). Similary, no significant variations in the relative amounts of phospholipids were detected in cultures muscle, from 3 to 13 days of culture. There are marked changes, however, in the fatty acid composition of membrane lipids. During development in vivo palmitate decreases and linoleate increases (Table 2). These variations are less pronounced in whole homogenates than in isolated SR membranes, reflecting probably the contribution of developmental changes in other muscle cell membrane systems.

Table 1. The phospholipid composition of chicken pectoralis muscle and sarcoplasmic reticulum during development in vivo

Days of Development	PS		SM+I		PC		PE	
	WH	SR	WH	SR	WH	SR	WH	SR
12 days	8.5	---	9.5	---	55.9	---	24.2	---
14 days	7.6	9.7	9.8	13.5	56.3	44.9	23.5	25.0
17 days	6.4	---	11.9	---	56.2	---	25.4	---
19 days	11.6	7.5	12.2	12.5	47.2	50.8	27.4	26.2
22 days	7.0	7.7	11.7	11.2	48.4	45.2	31.3	32.4
27 days	---	5.9	---	6.2	---	41.4	---	28.7
30 days	7.3	---	7.2	---	53.5	---	29.5	---
50 days	8.3	5.5	7.8	7.0	51.9	48.1	22.0	31.1

PS, phosphatidylserine; SM+I, sphingomyelin plus inositides; PC, phosphatidylcholine; PE, phosphatidylethanolamine; WH, whole homogenate; SR, sarcoplasmic reticulum. Values correspond to percent of total phospholipid.

Table 2. The fatty acid composition of chicken pectoralis and sarcoplasmic reticulum lipids during development in vivo

Days of Develop-ment	16:0		18:0		18:1		18:2		20:4	
	WH	RS	WH	RS	WH	RS	WH	RS	WH	RS
12 days	36.1	---	11.9	---	25.5	---	8.8	---	14.7	---
14 days	35.8	37.1	14.5	18.6	24.4	16.4	8.7	7.0	12.4	16.0
17 days	34.4	---	16.2	---	22.7	---	10.7	---	13.3	---
19 days	33.2	27.6	15.8	19.2	26.4	16.3	12.1	14.0	10.5	18.8
22 days	26.5	23.4	13.6	18.1	28.9	21.3	17.1	17.0	10.3	17.9
27 days	---	21.1	---	19.7	---	18.4	---	21.5	---	16.3
30 days	25.5	---	11.4	---	30.0	---	17.5	---	6.7	---
50 days	29.9	19.4	14.2	17.6	25.5	21.3	16.1	27.3	8.2	11.2

WH, whole homogenate; SR, sarcoplasmic reticulum. Short chain fatty acids (14:0, 14:1 and 16:1) did not represent more than 10% of total fatty acids measured.

Other fatty acids varied to a lesser extent. The accumulation of unsaturated fatty acids occurs with an increase in chain length which would mantain the fluidity of the membrane constant, suggesting that these variations in fatty acid composition do not contribute significantly to the observed changes in Ca transport during muscle development in vivo. This is supported by the lack of changes in passive Ca permeability observed in sarcoplasmic reticulum membranes isolated from developing embryos and chicks (Martonosi, 1975).

A different developmental trend in fatty acid composition is observed during culture of muscle cells (Table 3). Stearate steadily increases from 2 to 13 days of culture while linoleate decreases. Simultaneously, there is an increase in arachidonic acid. The standard medium used to grow muscle cells in vitro includes horse serum which is rich in linoleate (42 % of total fatty acids) and contains low amounts of arachidonate (5%). Therefore, the changes observed in fatty acids from muscle developed in vitro might be related to enzymes involved in fatty acid biosynthesis or uptake. If horse serum is replaced in the medium by fetal calf serum with a low content of linoleate (5%), this fatty acid is almost totally replaced in cell phospholipids by oleate. The linoleate content of the cells can be replenished by supplementation of the medium with free linoleate (Fig. 3). This shows a remarkable plasticity of cultured muscle cells to modify their fatty acid composition. The in vitro system could

Table 3. The fatty acid composition of total lipids from cultured muscle tissue

Days of Development	16:0	18:0	18:1	18:2	20:4
1-2 days	27.1	16.1	15.2	24.3	9.6
3 days	22.8	22.2	13.7	28.5	8.1
4 days	22.6	23.1	15.1	30.1	5.2
5 days	21.5	25.3	16.3	26.3	8.2
7 days	21.5	25.9	14.2	23.1	13.0
9 days	23.0	26.2	13.6	19.4	15.0
11 days	22.7	27.7	12.1	16.6	17.5
13 days	25.0	29.2	10.6	14.2	19.4

represent, therefore, an interesting model to study effects of lipid physical state on the calcium transport function of muscle membranes.

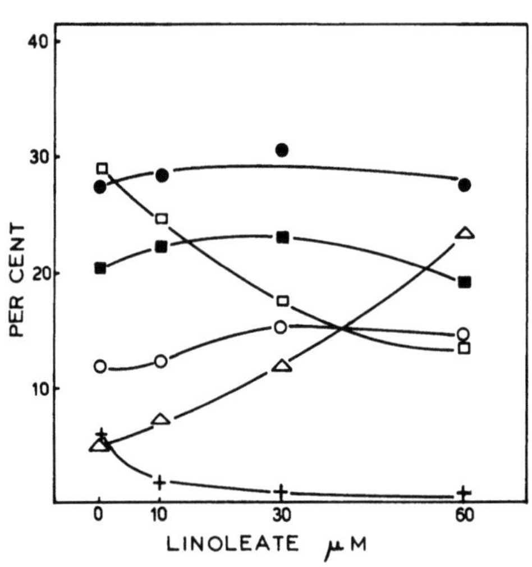

Figure 3. Fatty acid composition of cultured muscle cells suplemented with linoleate ●—●, 16:0; ■—■, 18:0; □—□, 18:1; Δ—Δ, 18:2; o—o, 20:4; +—+, 16:1. The cultures were grown 7 days in medium containing fetal calf serum with the additions of linoleate indicated in the graph.

CONCLUSIONS

During development of chicken skeletal muscle cells, the marked increase in Ca transport of sarcoplasmic reticulum, observed both in vivo and in vitro systems, can be interpreted mainly as the result of an increase in the concentration of Ca transport ATPase. Changes in the fatty acid composition of muscle membranes developed in vivo occur with a balance between chain length and unsaturation, without affecting significantly their Ca permeability. In cultured muscle cells their fatty acid composition can be manipulated to a great extent by lipid supplementation of the culture medium. The effects of these in vitro modifications in lipid composition on the calcium transport function of muscle membranes should be investigated.

REFERENCES

BISCHOFF, R. & HOLZER, H. (1969) J. Cell Biol. 41, 188-200.

BOLAND, A. R. de, JILKA, R.L. & MARTONOSI, A. (1975) J. Biol.Chem. 250, 7501-7510.

BOLAND, R., MARTONOSI, A. & TILLACK, T.W. (1974) J. Biol. Chem. 249, 612-623.

FOLCH, J., LEES, M. & SLOANE-STANLEY, G.H. (1957) J. Biol. Chem. 226, 497-509.

HASSELBACH, W. (1964) Progr. Biophys. Mol. Biol. 14, 167.

MARTONOSI, A. (1971) in The structure and function of sarcoplasmic reticulum membranes. Biomembranes vol. 1 (Mason, L. A., ed.), pp. 191-256, Plenum Press, New York.

MARTONOSI, A. (1975) Biochim. Biophys. Acta 415, 311-333.

MARTONOSI, A., DONLEY, J. & HALPIN, R. A. (1968) J. Biol. Chem. 243, 61-70.

MARTONOSI, A., ROUFA, D., BOLAND, R. & REYES, E. (1976) J. Biol. Chem. (in press).

SEILER, D. & HASSELBACH, W. (1971) Eur. J. Biochem. 21, 385-387

SEILER, D. KUHN, E., FIEHN, W. & HASSELBACH, W. (1970) Eur. J. Biochem. 12, 375-379.

LIPID METABOLISM DURING DEVELOPMENT IN INSECTS

A. M. Municio

Department of Biochemistry, Faculty of Sciences

Complutensis University, Madrid, Spain

INTRODUCTION

Lipids are very suitable energy sources in insects for processes so important as flight, pupation, diapause, etc. On the other hand, very fundamental biochemical processes have to be present behind the great morphological changes that occur during insect metamorphosis. As development is accompanied by a greater structural complexity, one can expect membrane systems to undergo extensive structural changes which may be subsidiary to alterations in their phospholipid metabolism to serve the newly acquired functions.

Some information concerning differences in lipid metabolism involved in the development of the insect *Ceratitis capitata* is given in this paper.

EXPERIMENTAL PROCEDURES

Insect Culture. *Ceratitis capitata* (Wiedeman) was used during the egg, larval, pharate adult and adult stages of development (Hinton, 1968). Diet, temperature and humidity conditions of culturing were carefully controlled (Municio et al., 1971a). Insects were reared at various development stages and periods according to the particular experiments.

Isolation and Fractionation of Lipids. Insects were homogenized and extracted according to standard procedures (Fernández-Souza et al., 1971; Municio et al., 1971b). Fractionation in lipid classes was accomplished according to the aim of the experiment (Fernández-Souza et al., 1971; Municio et al., 1971b; Municio et al., 1975a).

Positional Distribution of Fatty Acids in Phosphoglycerides. These determinations were carried out through hydrolysis by phospholipase A from *Crotalus adamanteus* venom based on the method described by Van Golde & Van Deenen (1967).

Stereospecific Distribution of Fatty Acids in Triacylglycerols. It has been carried out by an adaptation of the procedures: devised by Brockerhoff (1967) and Christie & Moore (1969).

Incorporation Experiments. Different labelled precursors were incubated with either homogenates or subcellular preparations from various stages of development of the insect according to the type of the experiment (Municio et al., 1975a; Municio et al., 1975b; Pérez-Albarsanz, 1976).

RESULTS AND DISCUSSION

The overall fatty acid composition of 1,2-diacyl-*sn*-glyceryl-3-phosphoryl-choline and ethanolamine from four development stages of the Dipterous *Ceratitis capitata* is recorded in Table 1. A great similarity is shown in the fatty acid composition of both individual phosphoglycerides from eggs whereas specific changes leading to a variety in the fatty acid patterns are exhibited in the subsequent stages of development. This has been interpreted as being caused by the absence in eggs of transacylating reactions and their participation in the other stages. Both lipid classes from eggs

Table 1. Percentages of fatty acids of diacyl-glyceryl-phosphoryl-choline (PC) and -ethanolamine (PE) in various stages of development of the insect

Fatty Acids	egg		larva		ph. adult		adult	
	PC	PE	PC	PE	PC	PE	PC	PE
12:0	tr	-	1.5	-	tr	-	0.4	-
12:1	tr	0.7	1.7	tr	tr	-	tr	-
14:0	1.3	1.3	3.3	tr	1.1	tr	0.9	tr
14:1	0.9	1.3	1.6	tr	0.9	tr	tr	-
16:0	18.3	17.2	26.3	26.7	31.2	33.8	21.7	26.1
16:1n7	42.2	42.1	16.0	12.3	21.4	8.8	22.3	10.9
16:1n9	-	-	-	tr	-	tr	-	-
18:0	2.2	2.6	2.7	4.1	2.4	5.3	2.6	5.0
18:1n7	8.6	6.6	-	tr	-	tr	-	-
18:1n9	25.0	24.8	16.9	35.1	21.2	30.0	23.7	29.8
18:2	1.3	1.9	24.8	17.3	20.2	14.2	25.5	20.0
18:3	tr	tr	3.3	2.7	1.7	4.9	3.0	5.9

have in common the very low levels or nearly complete absence of polyunsaturated fatty acids (18:2 and 18:3); however, changes in unsaturated fatty acids during development are different according to their presence in both lipid classes. The levels of 18:2 from both phosphoglycerides show a notable increase during development whereas the levels of 18:1 show qualitative variations. In terms of the molar degree of unsaturation, there is a net tendency to increase the degree of unsaturation during metamorphosis in both phosphoglycerides (Fernández-Souza et al., 1971; Municio et al., 1971b); in both cases it is during larval and adult development when the most marked shift towards the highest values of molar unsaturation occurs.

Concerning the fatty acid distribution in positions 1 and 2, the results obtained (Fernández-Souza et al., 1971; Municio et al., 1971b) agree with the general rule of distribution of fatty acids in animal phosphoglycerides; thus, in all developmental stages palmitic and stearic acids are confined to the position 1 whereas the remaining fatty acids are distributed in both positions 1 and 2. Nevertheless, some differences are revealed by the patterns (Municio et al., 1971b) of fatty acid distribution in both phosphoglycerides; the most salient feature on these differences is the more pronounced presence of unsaturated fatty acids joining the position 1 of diacyl-glyceryl-phosphorylcholine than in the same position of diacyl-glyceryl-phosphorylethanolamine.

To provide new data on the structural changes involved in the development of the insect and to obtain information on the metabolic relations between phosphoglycerides and triacylglycerols during the sequential stages of the insect development, a stereospecific analysis has been performed on triacylglycerols from various stages of development of the insect (García 1976). Table 2 shows the stereospecific analyses of total triacylglycerols from eggs, larvae, pharate adults and adults of *Ceratitis capitata*.

Position 1 is mainly occupied by palmitic acid that increases during the development. Therefore, position 1 in triacylglycerols from eggs is more unsaturated than in triacylglycerols from other stages. Palmitic acid shares position 3 with palmitoleic and oleic acids; palmitoleic acid decreases during development of the insect and it is counterbalanced by the increase of the relative proportions of palmitic acid and also oleic acid at the last stage of development. Palmitoleic acid decrease also at the position 2 being replaced by the unsaturated 18C, oleic and linoleic acids, during development.

Regarding the stereospecific distribution of fatty acids in triacylglycerols in relation to the positional distribution in phosphoglycerides, it can be stated that oleic acid is randomly distributed in the three positions of triacylglycerols from eggs

Table 2. Stereospecific distribution of fatty acids in triacyl glycerol during development of *Ceratitis capitata.*

Fatty Acids	eggs			larvae			ph. adults			adults		
	1	2	3	1	2	3	1	2	3	1	2	3
12:0	tr	tr	0.6	2.0	tr	4.3	0.8	tr	3.1	1.7	0.7	2.5
14:0	2.1	0.9	4.3	4.3	1.4	6.4	3.4	1.3	6.4	2.3	1.0	6.3
14:1	tr	1.7	0.6	tr	1.1	tr	tr	0.9	tr	tr	0.6	tr
16:0	37.2	5.8	24.5	66.6	3.1	33.1	73.6	3.6	39.1	72.2	3.9	36.4
16:1n9	0.8	0.6	1.0	-	-	-	-	-	-	-	-	-
16:1n7	27.1	68.7	40.5	16.8	48.1	28.7	12.7	46.0	25.0	9.1	30.2	22.0
18:0	4.9	tr	2.7	2.5	tr	8.4	3.0	tr	7.3	3.6	0.6	5.0
18:1n9	17.3	18.8	16.0	5.3	35.4	14.2	5.5	39.6	16.0	8.3	41.2	21.5
18:1n7	10.7	1.4	9.7	-	-	-	-	-	-	-	-	-
18:2	tr	2.1	tr	2.5	11.0	5.0	1.0	8.7	3.0	2.2	18.3	6.2
18:3	-	-	-	-	-	-	-	-	-	-	-	-

in agreement with the random oleic acid utilization for phosphoglyceride biosynthesis. The great diminution during development of the relative proportion of oleic acid at the position 1 is counterbalanced by its increase at the position 2 whereas it remains fairly constant at the position 3. Thus, unsaturated fatty acids at the position 2 of triacylglycerols exhibits during development a metabolic mobility similar to that shown by the same position of phosphoglycerides.

To obtain further information on the participation of the various positions of the glycerol moiety in the acyltransferase mechanism, the stereospecific distribution of fatty acids in triacylglycerols obtained after incubating insect homogenates with labelled fatty acids was investigated.

To test previously the different metabolic behavior of various stages of development of the insect concerning the fatty acid incorporating capacity into triacylglycerols, a general study on the incorporation of different labelled fatty acids by insect homogenates has been carried out (Municio et al., 1975b). All labelled fatty acids used in the incorporation experiments, decanoic, lauric, myristic, palmitic, stearic, oleic and linoleic acids, are efficiently incorporated by larval homogenate whereas most of the fatty acid used remains as free fatty acid in the presence of pharate adult homogenate; palmitic and stearic acids are the most scarcely incorporated into triacylglycerols by this stage of development of the insect.

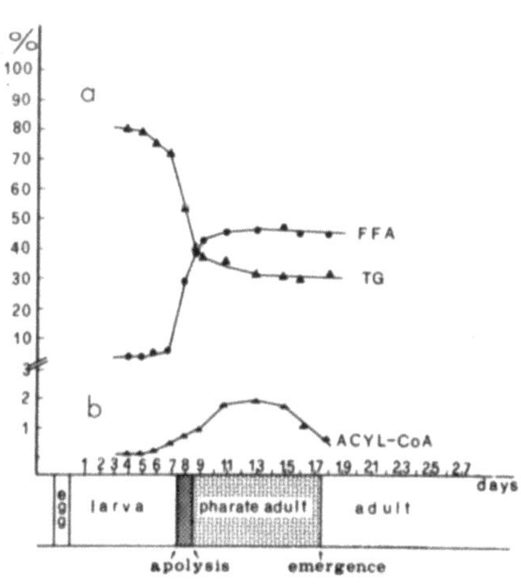

Figure 1. Incorporation of 10:0 and 12:0 into triacylglycerols (▲) Free fatty acids (●). Acyl-CoA synthetase activity (B).

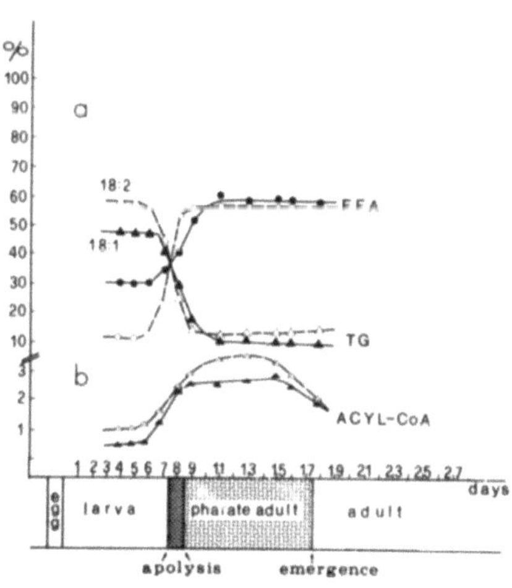

Figure 2. Incorporation of 18:1 (▲) and 18:2 (Δ) into triacylglycerols. Free fatty acids (●). Acyl-Coa synthetase activity (B).

Figures 1 and 2 show the capacity of insect homogenates from various stages of development to incorporate medium chain fatty acids (10:0 and 12:0) (fig. 1A) as well as unsaturated 18C (18:1 and 18:2) (fig. 2A). In all cases a crossing-zone between the radioactivity patterns of free fatty acids and triacylglycerols is defined in clear coincidence with the larval-pupal apolysis stage. This metabolic difference between larval and pharate adult homogenates could not be explained through differences in the acyl-CoA synthetase activity of the insect; furthermore,the enzyme activity increases notably from the larval to the adult stage of development in a manner similar for each fatty acids (figs. 1B and 2B).

Results on the stereospecific incorporation of saturated and unsaturated labelled fatty acids into the positions *sn*-1, 2 and 3 of triacylglycerols by larval homogenates are given in fig. 3 (Municio et al., 1975a). Palmitic acid is mainly incorporated into *sn*-1 and 3 positions, whereas *sn*-2 position exhibits a low incorporation; myristic acid acylates *sn*-3 position at a highter rate than it acylates he other *sn*-positions. Oleate is distributed more specifically than palmitic acid; linoleic acid is incorporated more efficiently than the monounsaturated acid. Kinetics of incorporation of labelled fatty acids into the *sn*-positions points to a non-random distribution with respect to the major saturated and unsaturated fatty acids in tiacylglycerols of larvae. However, the stereospecific distribution changes during incubation intervals; at the earliest time point, the presence of labelled fatty acids at the position *sn*-3 is the highest one to decrease, afterwards, to the mass values of triacylglycerols. This decrease was counterbalanced by an increase of both the saturated fatty acids at the position *sn*-1 and the unsaturated fatty acids at the position *sn*-2; this increase exhibits also a clear tendency towards the higher mass analytical values. Percentage of saturated fatty acids does not change prectically at the position *sn*-2 during the time-course experiment; changes in the relative proportions of unsaturated fatty acids at the position *sn*-1 are not also significant.

These data are consistend with an initial rapid labelling at the position *sn*-3 that most likely accounted for by acylation of endogenous larval diacylglycerols; the lower initial labelling at positions *sn*-1 and *sn*-2 could suggest the fatty acid incorporation by a *de novo* synthesis of diacylglycerols followed by an acyl-exchange mechanism that points to a non-random distribution. Pharate adult homogenates incorporate fatty acids very scarcely and mainly into positions 1-3 (Municio et al., 1975a).

In order to acquire new data on the biosynthetic pathways of triacylglycerols and diacylphosphoglycerides by the larval and pharate adult stages of development of the insect, a set of double-label experiments using either $|^3H|$ oleate or $|^3H|$ palmitate and

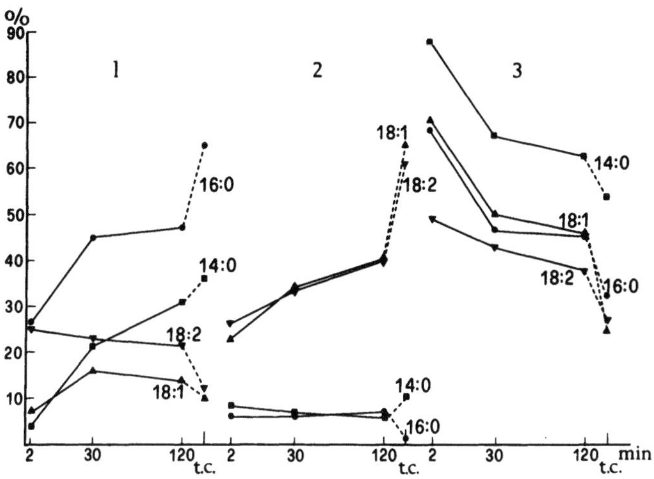

Figure 3. Stereospecific incorporation of fatty acids into triacylglycerols by larval homogenates of *Ceratitis capitata*.

|^{14}C| glycerol-3-phosphate has been carried out. The $^{14}C/^{3}H$ ratio exhibited by labelled glycerides is given in fig. 4. The results presented here for the synthesis of triacylglycerols, as well as the values for independent isotopic incorporation, show that either larval or pharate adult homogenates incorporated ^{14}C and ^{3}H precursors during the early periods of incubation according to a *de novo* pattern of synthesis; however, the progressive increase of the $^{14}C/^{3}H$ molar ratio versus incubation intervals suggests that endogenous fatty acids are freely available for exchange with *de novo* synthesized triacylglycerols. This effect provides clear differences between either the stages of development of the insect or the fatty acid used in the incorporating experiments (Municio et al., 1975b). The increase of the molar ratio $^{14}C/^{3}H$ versus time of incubation is more rapid using oleic as fatty acid precursor than using palmitic acid which indicates an easier exchange of oleic acid previously introduced through a *de novo* pathway. On the other hand, incorporation into triacyglycerols by pharate adult homogenates shows patterns with a slower tendency to carry out the fatty acid exchange than that exhibited by the larval stage of development. The fact that *de novo* triacylglycerols sinthesized by either larval or pharate adult homogenates have a $^{14}C/^{3}H$ molar ratio 1/3 smaller than that of phosphoglycerides at the early times of incubation indicates their common origin through the glycerol-3-phosphate acylation pathway.

Metabolic differences during the development of the insect are

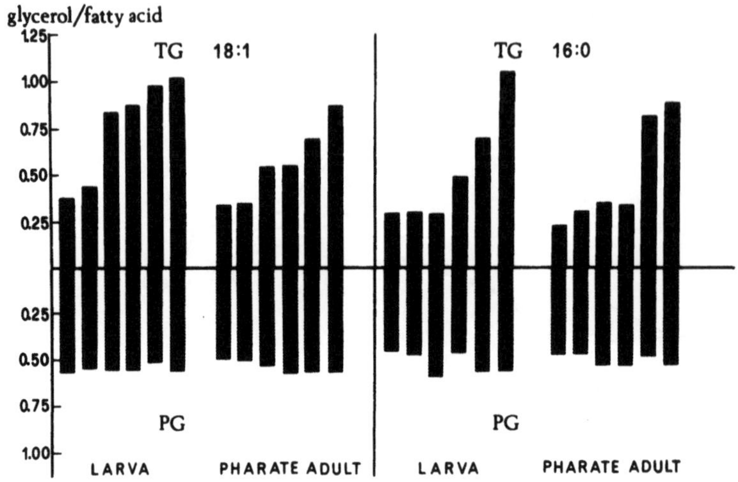

Figure 4. Double-label incorporation by triacylglycerols and phosphoglycerides with larval or pharate adult homogenates.

mainly concerning triacylglycerol metabolism. Results (Municio et al., 1975b; Pérez-Albarsanz 1976; García 1976) are consistent with the presence of acyltransferase activity in larval and pharate adult stages of development.

REFERENCES

BROCKERHOFF H. (1967) J. Lipid Res. 8, 167
CHRISTIE W. W. & MOORE J. H. (1969) Biochim. Biophys. Acta 176, 445
FERNANDEZ-SOUZA J. M., MUNICIO A. M. & RIBERA A. (1971) Biochim. Biophys. Acta 231, 527
GARCIA R. (1976) Doctoral Thesis, University Madrid
HINTON H. E. (1968) Adv. Insect.Physiol. 5, 65
MUNICIO A. M., ODRIOZOLA J. M., PIÑEIRO A. & RIBERA A. (1971a) Biochim. Biophys. Acta 248, 212
MUNICIO A. M., ODRIOZOLA J. M., PIÑEIRO A. & RIBERA A. (1971b) Biochim. Biophys. Acta 248, 226
MUNICIO A. M., GARCIA R. & PEREZ-ALBARSANZ M. A. (1975a) Eur. J. Biochem. 60, 117.
MUNICIO A. M., ODRIOZOLA J. M. & PEREZ-ALBARSANZ M. A. (1975b) Eur. J. Biochem. 60, 123
PEREZ-ALBARSANZ M. A. (1976) Doctoral Thesis, University Madrid
VAN GOLDE L. M. G. & VAN DEENEN L.L.M. (1967) Chem. Phys. Lipids 1, 157

COMPOSITION AND METABOLISM OF PHOSPHOLIPIDS DURING EARLY STAGES OF VERTEBRATE EMBRYONIC DEVELOPMENT

A.M. Pechén de D'Angelo, I.C. Bonini de Romanelli,
T.S. Alonso, and N.G. Bazán

Instituto de Investigaciones Bioquímicas, Universidad
Nacional del Sur y Consejo Nacional de Investigaciones
Científicas y Técnicas, Bahía Blanca, Argentina

Cell cleavage during the early stages of embryonic development involves deep modifications in the membranes, notably in the plasma membrane. Several of these changes are triggered by fertilization (e.g. Selman & Perry, 1970; Tupper, 1972; De Laat et al., 1974; Trams et al., 1974; Mazia et al., 1975; Marber & Mead, 1975) . However they were not yet defined in terms of membrane biogenesis and on the manner by which polar lipid requirements are met. It has been proposed that phospholipids derive from a storage site and that lipid biosynthesis is scarce or absent during early development (Pechén & Bazán, 1973; Crupkin et al., 1973; Pechén et al., 1974; Barassi & Bazán, 1974a, 1974b). A transport system may redistribute the lipids from stores for the assembly of nascent membranes (Pechén & Bazán, 1973; Pechén et al., 1974; Barassi & Bazán, 1974b).

In this paper we describe lipid labeling from ^{14}C-glycerol, ^{14}C-acetate and ^{14}C-glycerol-3-phosphate during amphibian embryogenesis. In addition the fatty acid pattern of phospholipids in microsomal and mitochondrial fractions is presented.

METHODS

Obtention of embryos

Ovulation of toads (*Bufo arenarum* Hensel) as well as artificial fertilization was carried out as previously described (Pechén & Bazán, 1974).

Incorporation of radioactive precursors in vitro

($U-{}^{14}C$)-glycerol(15µCi specific activity 7,4 mCi/mM, New England Nuclear Corp., Mass) incorporation was performed using 2500 oocyte or embryo homogenates in 30 ml of Tris buffer 50 mM, pH 7.4 containing NaCl 11.1 mM, KCl 0.13 mM, $CaCl_2$ 0.27 mM and EDTA 1 mM. The homogenate was incubated 60 minutes at 27°C, then 2.5 ml aliquots were taken and thoroughly mixed with chloroform-methanol (1:2, v/v) according to Bligh & Dyer (1959). After removing the proteins from the interphase, two additional extractions with chloroform-methanol (2:1 v/v) were performed to insure complete extraction.

Samples of 50 oocytes or late blastula embryos were incubated in phosphate buffer 50 mM pH 7.4 KCl 90 mM, $MgCl_2$ 5mM, ATP 10mM, NADH 5mM, EDTA 1mM and 50 µCi of ($1-{}^{14}C$) acetate (New England Nuclear Corp. Mass, specific activity 1-3mCi/mM). After 30 minutes of incubation with gentle shaking at 25°C, the reaction was stopped by adding chloroform-methanol (1:2 , v/v) (Bligh & Dyer, 1959). Neutral and polar lipids were separated by TLC as described above. Phospholipids were eluted several times with chloroform - methanol - formic acid (1:4:0.1) and neutral lipids with chloroform. Radioactivity was measured in a gas flow counter (Alfanuclear, Argentina) with 50% of counting efficiency.

Incorporation of ($U-{}^{14}C$) glycerol-3-phosphate was carried out using 2000 oocytes or embryos homogenated in Tris - HCl buffer 50mM pH 7.4 made up in diluted Ringer containing EDTA 10^{-3} M sucrose 0.3 M. The homogenates were centrifuged for 20 minutes at 1500 x g and the yolk platelet supernatant was separated.

In vivo experiments

They were performed as previously described (Barassi & Bazân, 1974b). 40 µCi of ${}^{14}C$-glycerol were injected to the female toad along with the pituitary extract. Oocytes and embryos at heart beat stage were dejellied by brief exposure to 2% neutralized thioglycholic acid and then lipid extracted by chloroform-methanol (2:1 , v/v) (Folch et al., 1957). Each sample was made up of 250 oocytes or embryos.

Total phospholipids and triacylglycerols were separated by thin-layer chromatography on Silica Gel G, using hexane-ethyl ether-acetic acid (60:40:2.5). Spots were visualized by iodine vapour and the radioactivity was counted in a Packard Tricarb Scintillation Spectrometer after suspending the uneluted spots in 4% Omnifluor (New England Nuclear Corp., Mass.) in toluene.

Fatty acid composition of phosphoglycerides

Phospholipids were isolated by two-dimensional thin-layer chromatography (Rouser et al., 1970), methanolyzed, and analyzed by gas-liquid chromatography, following the details previously described (Barassi & Bazán, 1974b).

RESULTS

^{14}C-glycerol labeling of lipids

Our attempts to obtain ^{14}C-glycerol incorporation in lipids during early cleavage were unsuccessful unless the precursor was administered to the female toad at the time of inducing oogenesis (Figure 1).

Oocyte or embryo homogenates failed to incorporate ^{14}C-glycerol in lipids when incubated in an unsupplemented Tris-Ringer solution; however, from gastrulation onwards labeling began to be measurable under this condition (Pechén & Bazán, 1977). In Fig. 1 is shown the incorporation of ^{14}C-glycerol in vivo and in vitro by unfertilized oocytes and embryos at heart beat stage. Phospholipids and triacylglycerols are labeled in unfertilized oocytes in vivo whereas the precursor fails to label the lipids in experiments in vitro. At heart beat stage in vivo and in vitro uptake of ^{14}C-glycerol in lipids is apparent.

The embryos at heart beat stage derived from oocytes prelabeled during oogenesis display high radioactivity in triacylglycerol (65%). On the other hand, when cell-free homogenates of the same stage were incubated in the presence of the precursor most of the radioactivity (80%) was found in phospholipids after 40 minutes of incubation.

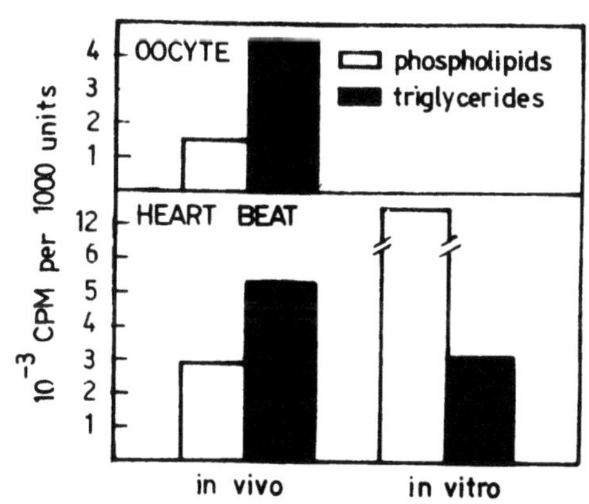

Fig. 1. ^{14}C-glycerol incorporation in lipids. Three samples made of 1000 oocytes or embryos were analyzed.

Negligible lipid labeling was found employing ^{14}C-Glycerol-3-phosphate as precursor. These experiments were carried out using the post yolk platelet supernatant in the presence and in the absence of several cofactors. The following additions failed to yield lipid labeling: 300 µg/2000 oocytes or embryos of palmitic or arachidonic acids complexed with bovine serum albumin, 10 mM ATP, 30 mM $MgCl_2$ and 8.5 mM cysteine.

^{14}C-acetate incorporation

Labeling of lipids by ^{14}C-acetate was detected in homogenates as early as the oocyte stage. Using late blastula homogenates a 70% increment was evidenced (Table 1). The percent distribution of radioactivity changes from one stage to the other. The phospholipids increase from 22 to 30%, cholesterol and monoacylglycerides remain unmodified, and the chromatographic fraction that includes diacylglycerols, free fatty acids and triacylglycerols decreases. The enhanced incorporation in phospholipids may be related to lipid requirements of new membranes.

Fatty acid composition of major polar lipid classes

The fatty acid profiles of individual phospholipids are not altered between oviposition and gastrulation. Moreover we have

Table 1. Labeling of lipids by ^{14}C-acetate

	Oocyte	Late blastula
a) Incorporation (cpm/mg of lipid)		
Total lipids	1079	1843 (170%)
b) Per cent of radioactivity		
Total phospholipids	22	30
Monoglycerides	20	20
Cholesterol	15	14
Free fatty acids, diglycerides and triglycerides	42	37

Table 2. Fatty acid composition of microsomal phospholipids

Fatty acids	Oocytes				Gastrula			
	PC	PE	PS	PI LPE	PC	PE	PS	PI LPE
	(percent)							
16:0	31.2	7.8	4.4	12.2	27.8	9.7	6.1	12.0
16:1	9.4	1.3	1.5	1.0	9.5	7.1	2.4	1.0
18:0	7.0	16.7	55.7	54.9	8.3	18.7	62.3	60.2
18:1	20.4	16.0	7.4	8.3	23.6	17.0	10.5	10.0
18:2	10.0	6.5	4.5	0.6	10.9	6.0	2.2	0.8
18:3	1.1	1.0	1.1	0.3	1.4	1.3	3.0	0.7
20:4	8.8	21.0	14.6	16.0	6.7	16.0	4.1	7.2
22:5	1.5	7.1	0.8	0.6	1.3	4.9		0.3
22:6	2.7	10.5	2.2	0.8	1.9	6.9		0.3

PC, phosphatidylcholine; PE, phoaphatidylethanolamine; PS, phosphatidylserine; PI, phosphatidylinositol and LPE, lysophosphatidyl ethanolamine. The presence of LPE along with PI is suspected due to the consistent appearance of a ninhidrin-positive spot slightly overlapping the inner edge of the PI area. In addition, pure standards of LPE closely migrate to PI. Microsomal fraction was obtained using a modified (unpublished) procedure of Kamath after preparing the homogenate as described elsewhere (Pechên et al., 1974). Data are averages of determinations in two independent samples derived from 2000 oocytes or embryos each.

surveyed also two subcellular fractions in order to avoid a possible masking effect due to the yolk platelet lipids. In Table 2 the fatty acid composition of microsomal phospholipids from oocytes and gastrula embryos is shown. An increase in the relative concentration of 16:1 from phosphatidylethanolamine was observed followed by a decrease in the polyenoic fatty acids. Phosphatidylserine showed an increase in the percent distributions of 18:0, while the phosphatidylinositol fraction has a lower relative concentration of 20:4 than the corresponding oocyte phospholipid. In Table 3 the fatty acid composition of mitochondrial phospholipids is presented. Slight changes were detected in the fatty acid moiety of the phospholipids. Phosphatidylserine contains larger proportions of 18:0 at gastrula stage. A decrease in the relative

Table 3. Fatty acid composition of mitochondrial phospholipids

Fatty acids	Oocytes					Gastrula				
	PC	PE	PS	PI LPE	DPG	PC	PE	PS	PI LPE	DPG
	(percent)									
16:0	23.7	7.9	6.9	13.5	2.1	25.3	7.3	8.0	15.2	1.7
16:1	7.1	1.4	4.6	1.0	5.9	8.4	1.5	3.4	0.7	6.8
18:0	7.3	16.4	46.1	43.5	1.5	6.4	17.4	59.2	49.0	1.1
18:1	22.6	21.6	10.5	7.3	19.9	26.3	24.5	8.1	10.8	22.3
18:2	13.2	8.6	5.9	0.9	53.5	11.6	7.2	2.6	1.0	48.1
18:3	1.4	1.2	5.4	0.3	3.9	1.5	1.3	2.3	0.5	3.5
20:4	10.9	18.8	6.2	24.9		7.7	18.2	6.4	15.0	
22:5	2.3	4.8	0.7	1.3		1.8	4.6	0.7	0.6	
22:6	4.0	8.1	1.8	1.3		2.8	7.6	1.6	0.9	

DPG, diphosphatidylglycerol. Details as in Table 2 and in Methods.

concentration of 20:4 from the phosphatidylinositol fraction occurs during this period. However, in general, the fatty acids of each of the phospholipids analyzed from gastrula are richer in monoenoic fatty acids than those of the oocyte.

DISCUSSION

Striking differences were encountered in the incorporation of various precursors of lipids in toad oocytes and embryos. The lack of lipid synthesis from ^{14}C-glycerol or ^{14}C-glycerol-3-phosphate in cell-broken preparations until late gastrula supports the view that such process is absent or very limited during early embryogenesis (Barassi & Bazán, 1974a; 1974b; Pechén et al., 1974; Mes-Hartree & Amstrong, 1976). The only way to obtain lipid labeling from ^{14}C-glycerol is by injecting the precursor along with the hormone to induce ovulation. However at neural fold an increase of only about 25% in phospholipid labeling from ^{14}C-glycerol was observed whereas a 280% increment in ^{32}P takes place (Pechén & Bazán, 1977). The enhanced radioactive choline uptake (Pasternak, 1973) may be related to turnover rather than to de novo synthesis.

The ^{14}C-acetate labeling of lipids may likely be related to fatty acid synthesis rather than to de novo synthesis of entire glycerolipid molecules. Our results show ^{14}C-acetate incorporation in lipids by an oocyte or embryo homogenate somewhat higher than that reported by Miceli & Brenner (1976). Such differences may arise from the system used: while we have worked with a cell-free homogenate to perform the incorporation they used an in vivo system. However the relative increases between the analyzed stages are very similar. Miceli & Brenner (1976) postulated that the oocyte would be able to synthesize saturated fatty acids, whereas at gastrulation a significant increment in desaturation due to the activity of $\Delta 9$ desaturase may occur. In addition we have observed an increase in the labeling of phospholipids when ^{14}C-acetate was used as precursor (Table 1).

In sea urchin embryos ^{14}C-glycerol labeling of lipid, although low, was detected (Mohri, 1964). More recently Schmell & Lennarz (1974) found, using sea urchin prior to first cleavage, that ^{3}H-choline and ^{3}H-ethanolamine were taken up by the egg; however only the latter was incorporated into the corresponding phospholipid. ^{32}P on the other hand mainly labeled the phosphoinositides. In agreement with this observation we found a higher labeling of the fraction containing these phospholipids by ^{32}P up to gastrula stage (Barassi & Bazán, 1974a).

The unchanged acyl group distribution in phospholipids of microsomal and mitochondrial fractions up to gastrula also agree with the observations previously made on fatty acids of lipids in entire embryos (Barassi & Bazán, 1973; 1974). Even in oocytes phosphatidylethanolamine was the more unsaturated lipid. Mitochondrial diphosphatidylglycerol (cardiolipin) in oocyte already displays a fatty acid profile similar to that observed in mature tissues (e.g. Comte et al., 1976).

In summary, during the early developmental stages of the toad Bufo arenarum Hensel apparently there is no de novo lipid biosynthesis. In entire embryos as well as in subcellular fractions there is a constancy in the content and composition of individual lipids. The labeling by ^{14}C-acetate suggests that some fatty acid synthesis takes place; however when the specific activities of phospholipids labeled both by ^{14}C-glycerol and ^{32}P were compared it was disclosed that while the former greatly increases, the latter augments slowly. This may indicate that during early development an active turnover of the polar moiety of lipids does occur without de novo synthesis. The working hypothesis being evaluated sequentially includes a) a storage site (yolk platelet); b) a glycosylating place (Golgi complex); c) a flow of subunits and a transport system; and d) assemblage in nascent membranes. The turnover of the phosphorylcholine moiety may be involved in the last step.

ACKNOWLEDGMENT

To PLAMIRH for grants 24.55.2.75 and 24.55.2.75 R.

REFERENCES

BARASSI, C. A. & BAZAN, N. G. (1974a) Lipids 9, 27-34
BARASSI, C. A. & BAZAN, N. G. (1974b) J. Cell. Physiol. 84, 101-114.
BARBER, M. L. & MEAD, J. F. (1975) Wilhelm Roux' Archiv. 177, 19-27.
BLIGH, E. G. & DYER, W. S. (1959) Can. J. Biochem. Physiol. 37, 911-917.
BYRD, W. E. (1975) Develop. Biol. 47, 309-315.
COMTE, J., MAISTERRENA, B. & GANTHERON, D. C. (1976) Biochim. Biophys. Acta 419, 271-284.
CRUPKIN, M., BARASSI, C. A. & BAZAN, N. G. (1973) Comp. Biochem. Physiol. 452, 523-528.
DE LAAT, S. W., BUWALDA, R. J. A. & HABETS, A. M. C. (1974) Exp. Cell. Res. 989, 1-4.
FOLCH, J., LEES, M. & SLOANE-STANLEY, G. H. (1957) J. Biol. Chem. 226, 497-509.
MAZIA, D., SCHATTEN, G. & STEINHARDT, R. (1975) Proc. Natl. Acad. Sci. USA 72, 4469-4473.
MES-HARTREE, M. & ARMSTRONG, J. B. (1976) Can. J. Biochem. 54, 578-582.
MICELI, D. & BRENNER, R. R. (1976) Lipids 11, 291-295.
MOHRI, H. (1964) Biol. Bull. 126, 440-445.
PASTERNAK, C. (1973) Develop. Biol. 30, 403-410.
PECHEN, A. M. & BAZAN, N. G. (1973) 9th Int. Congr. Biochem. Stockholm. pp. 291.
PECHEN, A. M. & BAZAN, N. G. (1974) Exp. Cell. Res. 88, 432-435.
PECHEN, A. M., BONINI, I. C. & BAZAN, N. G. (1974) Biochim. Biophys. Acta 372, 388-399.
PECHEN, A. M. & BAZAN, N. G. (1977) Lipids (in press).
ROUSER, G., FLEISCHER, S. & YAMAMOTO, A. (1970) Lipids 5, 494-496
SCHMELL, E. & LENNARZ, W. J. (1974) Biochemistry 13, 4114-4121.
SELMAN, G. G. & PERRY, M. M. (1970) J. Cell. Sci. 6, 207-227.
TRAMS, E. G., LAUTER, C. J., KOVAL, G. J., RUZDIJIC, S. & GLISIN, V. (1974) Proc. Soc. Exp. Biol. Med. 147, 171-176.
TUPPER, J. T. (1972) Develop. Biol. 29, 273-282.

II. GLYCOSPHINGOLIPIDS—GANGLIOSIDES

STRUCTURAL SPECIFICITY OF GANGLIOSIDES*

H. Wiegandt

Physiolog. Chem. Inst. University

355 Marburg/L., FRG

Like the molecules of other amphipathic lipids, such as phospholipids, which constitute the building blocks of biological membranes, the ganglioside molecule consists of a lipophilic part and a hydrophilic part, i.e., a ceramide and a sialo-oligosaccharide. As compared to phospholipids, the gangliosides show much more variation in their polar head group region than in their hydrocarbon chains. (Wiegandt, 1968; 1971; 1973). In recent times increasingly observations are reported that point towards the potential biological significances of the gangliosides in view of the large diversity of highly specific interactions possible for their carbohydrate residues. Thus gangliosides are implicated in the interactions of hormones (Wolley & Gommie, 1965; Mullin et al., 1976), lectins (Surola et al., 1975), viruses (Haywood, 1974) and interferon (Besancon et al., 1976). The longest known instance of a specific binding of ganglioside dates back to the observation made in 1898 by Wassermann & Takaki that brain tissue could fix tetanus toxin. This phenomenon was later shown by van Heyningen (1974) to be due to ganglioside, which may fix the toxin without however inhibiting its in vivo toxicity. More recently other bacterial toxins were found also to bind specifically to various gangliosides and it was postulated that the sialoglycolipids may actually constitute the natural cell membrane receptors for the toxins. Cholera toxin is fixed by ganglioside II^3NeuAc-$GgOse_4$-Cer (van Heyningen, 1974; Stärk, et al., 1974). *Escherichia coli* toxin can be prevented from entering polyacrylamide gel in electrophoresis when it is in the presence of the trisialoganglioside IV^3NeuAc-,II^3($\leftarrow$2αNeuAc8)$_2$ $GgOse_4$-Cer (Stärk, J., Sattler, J., and Wiegandt, H., unpublished observation). Recently Kato and Naiki (1976) reported that staphylococcus toxin was bound by ganglioside IV^3NeuAc-$LnOse_4$-Cer. The structure of the latter ganglioside, i.e., NeuAcα2$\rightarrow$3Galβ1$\rightarrow$4GlcNAcβ1$\rightarrow$

3Galβ1→4Glcβ1→Cer (Wiegandt, 1973), is in its terminal mono-saccharide residues identical to sialo-oligosaccharide moieties found frequently with glycoproteins, in particular in those of the human plasma (Kato & Naiki, 1976).

With all these more or less specific interactions shown by gangliosides, the question may be raised as to the involvement and the role played in those events by the two heterophilic molecular moieties of the glycolipid. We therefore chose two bacterial toxins, tetanus toxin and cholera toxin, to perform binding studies which were aimed at a further identification of the nature of the protein fixation by the sialo-glycolipid.

Tetanus Toxin

Earlier studies with tetanus toxin as described by van Heyningen & Miller (1961) showed complex formation with gangliosides by ultracentrifugal analysis at rather high component concentrations. In order to measure at very low concentrations - more similar to physiological conditions - a new binding assay was developed. It involves the use of small columns of Sephadex equilibrated with protective colloid and ^{3}H-labelled ganglioside (Helting et al., 1976). A protective colloid is added in order to secure the specificity of the reaction. It does not interfere with the assay. This binding assay makes use of the fact that small amounts of tetanus toxin added to Sephadex become highly adsorbed without losing the capacity to fix ganglioside. Ganglioside alone is not adsorbed to the gel. The protein-ganglioside complex, which adheres to the Sephadex, is very stable. However, it can be removed by methanol-water mixtures or at extreme pH-values. The binding as tested also with other proteins appears to be very specific. Only the native tetanus toxin and its heavy peptide chain show the characteristic ganglioside binding. With this assay system it was found that with a fixed load of 0.8nmole tetanus toxin a 50% saturation was achieved at $5x10^{-8}$ M ganglioside $II^{3}(\leftarrow 2\alpha NeuAc8)_{2}GgOse_{4}$-Cer concentration. For binding to the toxin a lipid moiety was required of the sialo-glyco-compound. Of all sialoglycolipids tested, natural ganglioside with an intact ceramide moiety bound with highest affinity to the toxin. Sialoglycolipids containing only a single aliphatic hydro-carbon chain could also fix tetanus toxin, but to a significantly lower degree. Interestingly, no comparable binding to tetanus toxin occurred with the free sialo-oligosaccharide in the Sephadex system.

The specific structural requirements for the sialo-oligo-saccharide portion of gangliosides for binding to tetanus toxin were found to be rather limited. Thus removal of the terminal sialic acid residue from ganglioside $II^{3}(\leftarrow 2\alpha NeuAc8)_{2}GgOse_{4}$-Cer had no effect on toxin fixation. Furthermore, gangliosides with carbohydrate chains reduced in size, namely $II^{3}NeuAc-GgOse_{3}$-Cer or $II^{3}NeuAc$-Lac-Cer, all showed similar binding capacities. One sialic acid residue however is essential for fixation to tetanus toxin. No comparable binding occurred with gangliotetraosyl-ceramide. We are therefore unable to confirm earlier reports that tetanus toxin binds preferentially to gangliosides carrying a $(\leftarrow 2\alpha NeuAc8)_2$ disialo-group (van Heyningen & Miller, 1961). Moreover, from older studies with the analytical ultracentrifuge it was calculated that on a molar basis 1 ganglioside could fix 4 tetanus toxins (mol. weight 143,000), a value which would appear unlikely. It was now estimated that, under the conditions of the new assay, only one molecule of ganglioside binds to one molecule of the toxin (Helting et al., 1976).

On the part of the tetanus toxin, it was found that only its heavy peptide chain (mol. weight 95,000) binds to ganglioside. No fixation occurs with the light chain (mol. weight 48,000). Peptide Fragment C (mol. weight 47,000), which originates from the heavy chain and is obtained from tetanus toxin by papain digestion (Helting & Zwisler, 1976), shows for itself only marginal binding of ganglioside. However antibodies directed against Fragment C uniquely interfere with the binding of ganglioside to tetanus toxin. One must therefore conclude that the binding site for ganglioside is located on the heavy chain portion of the toxin, in or near the region comprised by Fragment C (Helting & Zwisler, 1976).

Cholera Toxin

The reaction between ganglioside and cholera toxin, as was also first shown by van Heyningen (1974), differs from that with tetanus toxin in that precipitation occurs with ganglioside $II^3NeuAc\text{-}GgOse_4\text{-}Cer$ from aqueous media leading to complete detoxification. This precipitation, which is preceded by the formation of soluble high molecular weight complexes, obviously results from multivalent interactions with higher structured, possibly micellar ganglioside (Stärk et al., 1974). A multivalent interaction between cholera toxin B-subunit and ganglioside can also be deduced from the fact that erythrocytes which were coated with B-subunit could be aglutinated by addition of ganglioside $II^3NeuAc\text{-}GgOse_4\text{-}Cer$ (Stärk & Wiegandt, unpublished results).

In order to test for the role played by the lipophilic molecular moiety of ganglioside $II^3NeuAc\text{-}GgOse_4\text{-}Cer$, a number of derivatives were synthesized. They all carried the terminal $Gal\beta 1 \rightarrow 3GalNAc\beta 1 \rightarrow 4Gal|3 \leftarrow 2\alpha NeuAc|\beta 1 \rightarrow$ residue of the native ganglioside but varied in important aspects of their lipophilic moiety (Wiegandt & Ziegler, 1974). It was found that for the formation of high molecular weight complexes with the toxin and for precipitation, the sialoglycolipid had to have a lipid moiety of at least the size of an aliphatic C-14 hydrocarbon chain (Wiegandt et al., 1976).

Monosialo-gangliotetraose, the free carbohydrate moiety of ganglioside II^{3}NeuAc-GgOse$_4$-Cer, neither precipitated nor detoxified cholera toxin as measured with in vivo systems. In fact, the only indication that the sialo-sugar would bind to the toxin at all was the finding by Holmgren et al. (1974) that a much higher concentration of ganglioside was necessary to inactivate the toxin if it was in the presence of monosialo-gangliotetraose. Also, a slight increase in the electrophoretic mobility of the toxin was noted after addition of the sialo-oligosaccharide (Stärk et al., 1974). It was shown with reductaminated fluresceine-labelled monosialo-gangliotetraose that the sugar comigrated with the toxin in polyacrylamide gel eletrophoresis (Sattler et al., 1976).

In order to study the binding of sialo-oligosaccharides to cholera toxin or its B-subunit, equilibrium displacement dialysis was used (Sattler et al., 1976).

In this test system it could be shown that monosialo-ganglio-tetraose did bind specifically and strongly to the cholera toxin. The binding isotherm shows a saturable binding (Fig. 1).

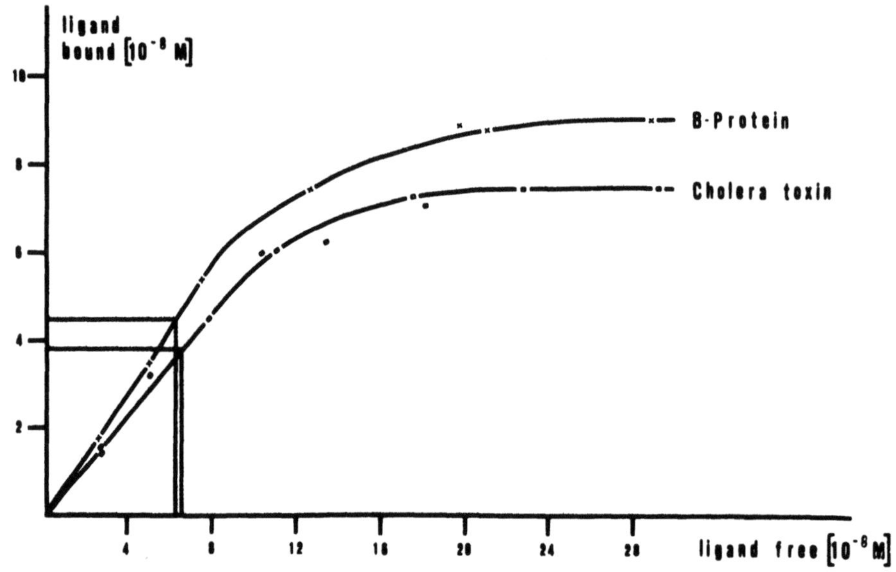

Fig. 1 . Saturation curve of cholera toxin by ganglioside II^{3}NeuAc-GgOse$_4$-Cer.

When the data are given according to the method of Scatchard (1949), a dramatically nonlinear plot is obtained indicating a high degree of cooperativity in the interaction (Fig. 2).

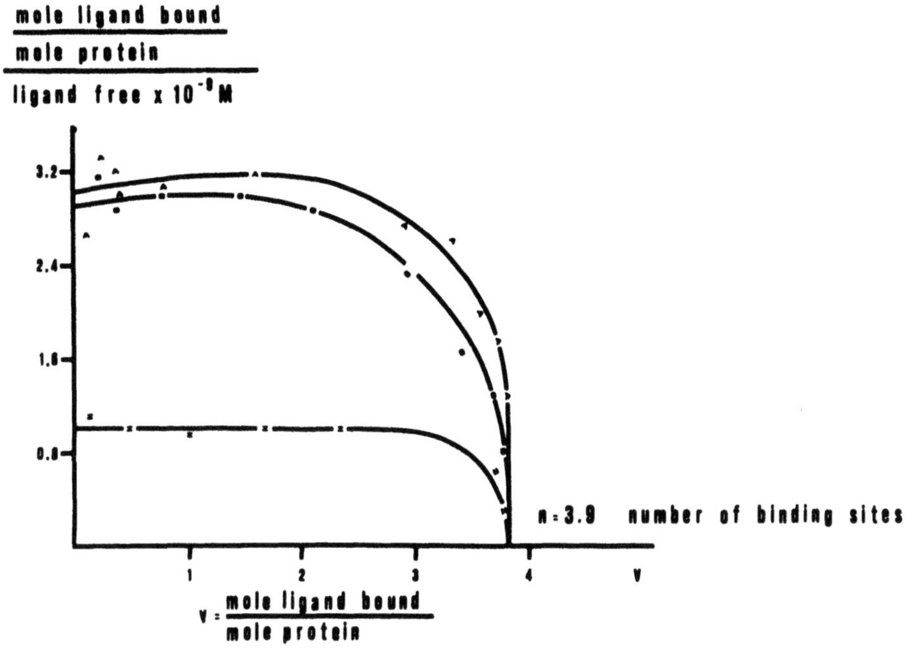

Fig. 2. Scatchard plot of the binding of cholera toxin by ganglioside II^3 NeuAc-GgOse$_4$-Cer. -A-A-=$2x10^{-8}$M, -z-z-=$8x10^{-8}$M cholera toxin; -o-o-=B-subunit, $2.5x10^{-8}$M.

This becomes particularly obvious when the data of the binding isotherm are presented according to Hill (1913). At a 50 % saturation of the protein by the ligand, a Hill-coefficient of 2 is obtained. If the number of binding sites of the toxin is taken from the Scatchard plot to be 4, the Hill equation $\bar{V} = nK_a v^h/1+K_a v^h$ ($\bar{V}$ = mole ligand bound per mole protein; n = number of cooperative binding sites; K_a = association constant) would yield a dissociation constant of approximately $5x10^{-15}mol.1^{-1}$ at 50 % protein saturation with the ligand (Sattler et al., 1976). A 20 times higher concentration of intact ganglioside (conc. < cmc) was necessary to replace 50 % of monosialo-gangliotetraose bound to the cholera toxin.

The structural requirements for sialo-oligosaccharide to bind to cholera toxin were tested with a number of relevant carbohydrates. The following sugars showed no or only marginal binding to the toxin: monosialo-gangliotriose (II^3NeuAc-GgOse$_3$=GalNAcβ1→4Gal|3←2αNeuAc| β1→4Glc), ganglio-tetraose (GgOse$_4$=Galβ1→3GalNAcβ1→4Galβ1→4Glc), and neuraminulosyl-gangliotetraitol; the last carbohydrate was obtained from the methylester of monosialo-gangliotetraose by reduction with borohydride (Sattler et al., 1976). These saccharides differed from monosialo-gangliotetraose by lacking either the terminal galac-

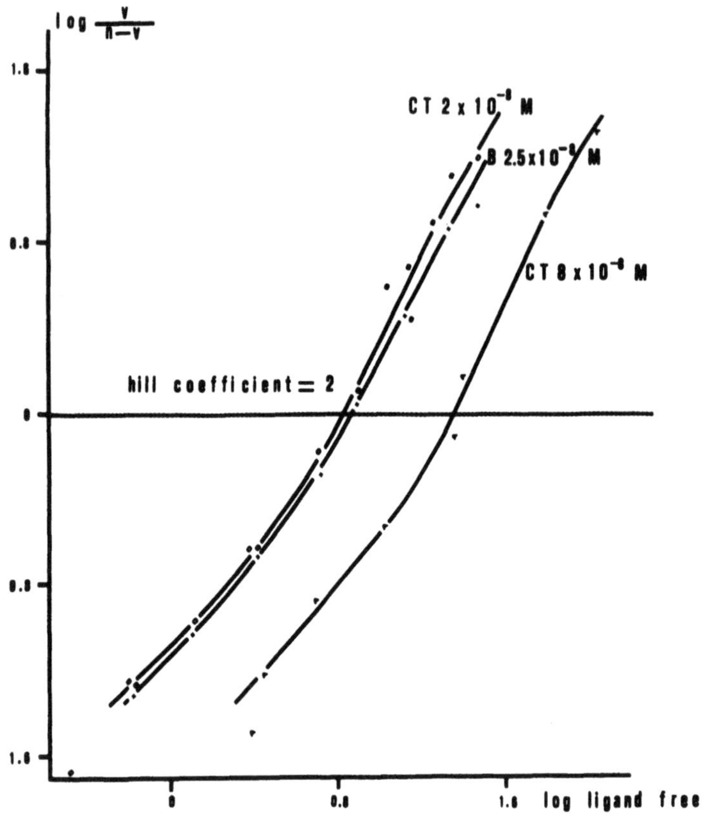

Fig. 3.

Hill graph of data of Fig. 2. CT= cholera toxin; B=B-subunit.

tose-residue, the sialic acid-residue or only by substitution of the $-CO_2H$-group by $-CH_2.OH$. All these alterations lead to a loss of the ability to bind to the toxin. In contrast to this, reduction of only the glucose moiety of monosialo-gangliotetraose does not interfere with fixation of the toxin.

From these studies with cholera toxin it becomes apparent that the lipophilic moiety of ganglioside II^3NeuAc-$GgOse_4$-Cer does not interact directly with the protein. Its functional significance therefore may rather be seen in the fact that it provides a multivalent anchorage for the toxin onto the lipid bilayer of the cellular surface membrane (Stärk et al., 1974). This step appears to be necessary for the introduction of the cholera toxin A-subunit into the cell interior, where it may activate adenylate cyclase (Kimberg et al., 1971). Similar mechanisms of fixation of a binding subunit to receptors located at the cell surface followed by insertion of a biologically active subunit into the cell interior can also be postulated for a number of other toxins. Among those are diphtheria toxin (Pappenheimer & Gill, 1973), staphylococcus toxin (Kato & Naiki, 1976), *Escherichia coli* toxin (Schenkein et al., 1976), abrin, and ricin (Olsnes et al., 1974).

REFERENCES

BESANCON, F., ANKEL, H. & BASU, S. (1976) Nature 259, 576-578.
HAYWOOD, A. M. (1974) J. Mol. Biol. 87, 625-628.
HELTING, T. B. & ZWISLER, O. (1976) J. Biol. Chem. (in press).
HELTING, T. B. ZWISLER, O. & WIEGANDT, H. (1976) J. Biol. Chem. (in press).
HILL, A. J. (1913) Biochem. J. 7, 471.
HOLMGREN, J., MANSON, J. E. & SVENNERHOLM, L. (1974) Med. Biol. 52, 229-233.
KATO, J. & NAIKI, M. (1976) Infection and Immunity 13, 289-291.
KIMBERG, D. V., FIELD, M., JOHNSON, J., HENDERSSON, A. & GERSHON, E. (1971) Clin. Invest. 50, 1218-1230.
MULLIN, B. R., ALOJ, S. M., FISHMAN, P. H., LEE, G., KOHN, L.D. & BRADY, R. O. (1976) Proc. Natl. Acad. Sci. 73, 1679-1683.
OLSNES, S., REFSNES, K. & PIHL, A. (1974) Nature 249, 627-631.
PAPPENHEIMER, Jr. A. M. & GILL, D. M. (1973) Science 182, 353-364.
SATTLER, J., SCHWARZMANN, G., ZIEGLER, W. & WIEGANDT, H. (1976) Hoppe-Seyler's Z. Physiol. Chem. submitted.
SCATCHARD, G. (1949) Ann. N. Y. Acad. Sci. 51, 660-672.
SCHENKEIN, I., GREEN, R. F., SANTOS, D. S. & MAAS, W. K. (1976) Infection and Immunity 13, 1710-1720.
SHARP, G. W. G.& HYNIE, S. (1971) Nature 229, 266-269.
STÄRK, J., RONNEBERGER,H. J., WIEGANDT, H. & ZIEGLER, W. (1974) Eur. J. Biochem. 48, 103-110.
SUROLA, A., BACHHAWAT, B. K. & PADDER, S. K. (1975) Nature 257, 802-804.
SWAMINATHAN,N. & ALADJEM, F. (1976) Biochemistry 15, 1516-1522.
VAN HEYNINGEN, W. E. (1974) Nature 249, 415-417.
VAN HEYNINGEN, W. E. & MILLER, P. (1961) J. Gen. Microbiol. 24, 107-119.
VASSERMANN, A. & TAKAKI, T.(1898) Berl. Klin. Wschr. 35, 5.
WIEGANDT, H. (1968) Ang. Chem. Int. Edt. 7, 87-96.
WIEGANDT, H. (1971) Adv. Lipid Res. 9, 249-289.
WIEGANDT, H. (1973) Hoppe-Seyler's Z. Physiol. Chem. 354, 1049-1056.
WIEGANDT, H. & ZIEGLER, W. (1974) Hoppe-Seyler's Z. Physiol. Chem. 355, 11-18.
WIEGANDT, H. ZIEGLER, W., STAERK, J., KRANS, Th., RONNEBERGER, H. J., ZILG, H., KARLSSON, K. A. & SAMUELSSON, B. E. (1976) Hoppe-Seyler's Z. Physiol. Chem. (in press)
WOLLEY, D. W. & GOMMIE, B. W. (1965) Proc. Natl. Acad. Sci. 53, 959-963.

* Abbreviations of gangliosides are given according to recommendations of IUPAC-IUB lipid nomenclature document 1976.

THE BIOSYNTHESIS OF BRAIN GANGLIOSIDES — EVIDENCE FOR A "TRANSIENT POOL" AND AN "END PRODUCT POOL" OF GANGLIOSIDES

H.J. Maccioni, C.Landa, A. Arce, and R. Caputto

Departamento de Química Biológica
Facultad de Ciencias Químicas
Universidad Nacional de Córdoba, Córdoba, Argentina

Incubation of sugar nucleotides with brain membranes in which endogenous, incomplete gangliosides are glycosylated, and addition of sugars to exogenous acceptors indicated the same pathway of synthesis of gangliosides (for a review, see Caputto et al. 1974; 1976). An exception was the pathway of completion of GD1b, for which a different route was worked out using endogenous (Arce et al., 1971) or exogenous acceptors (Cumar et al., 1971; 1972). The pathway of synthesis found for GT1b (Arce et al., 1971) was recently confirmed by using exogenous acceptors (Mestrallet et al., 1974).

Studies "in vitro" of labeling of endogenous acceptors and studies of labeling "in vivo" indicate also that, on a metabolic basis, two different ganglioside pools can be distinguished. One of these pools, which we have defined as the "transient pool" of gangliosides, is formed by gangliosides in the path to completion closely related with the multiglycosyltransferase system that built them. Once a ganglioside has been completed, it leaves its position in the "transient pool" and passes to form part of the "end product pool" of gangliosides. Gangliosides from the "end product pool" do not mix with gangliosides of the same chemical structure but belonging to the "transient pool". In a quantitative sense, the "end product pool" of gangliosides is bigger than the "transient pool" of gangliosides, which was inferred to be almost nonexistent.

This concept of two pools of gangliosides arose principally from the following observations:

a) No precursor-product relationship between simple and complex gangliosides was found in experiments of labeling endogenous gangliosides "in vitro" (Arce et al., 1971) or labeling of gangliosides "in vivo" (Maccioni et al.,1971; 1972). Since simple and complex gangliosides differ mainly in the degree of complexity of their carbohydrate chain, and this is built up by the stepwise addition of carbohydrates to a common backbone (Roseman, 1970; see also Caputto et al., 1974), this means that the different gangliosides do not interact freely with the glycosyltransferases. This was also indicated by the fact that the relative amount of each ganglioside could not be noticeably decreased by incubation of brain membranes with the sugar nucleotide that passes it to the next more complex ganglioside (Maccioni et al., 1974).

b) The labeling "in vivo" of repeated units in the major ganglioside molecule (NeuNAc, galactose) was found essentially equal at any time after suitable precursors were injected into rats. This indicates that the size of the "transient pool" of gangliosides was very small, almost nonexistent. In fact, it was found that only 2/10,000 of the total ganglioside content of the membrane preparations acted as acceptor of glycosyl groups "in vitro".

c) From the equality of labeling "in vivo" of repeated units in the carbohydrate chain of gangliosides, it was concluded that partial turnover of the carbohydrate chain, if it occurs, is a minor event. Once a ganglioside is built, it passes to another position ("end product pool") in the membrane so that if it becomes degraded it cannot be repaired by the glycosyltransferase system.

Experiments "in vitro" based on the accessibility to *Clostridium perfringens* neuraminidase confirmed the presence of two ganglioside populations. *C. perfringens* neuraminidase converted most of the membrane-bound polysialogangliosides into monosialoganglioside. However, the endogenous sialytransferase(s) were unable to resialylate the empty sites created by neuraminidase treatment, since the endogenous acceptor capacity of the neuraminidase-treated membranes was not increased in comparison with untreated membranes. These experiments confirmed the notion of a "transient pool" of gangliosides accessible to the sialyltransferase(s) and of an "end product pool" of gangliosides inaccessible to the glycosylating system. They also indicate that, as "in vivo", there is no partial turnover of the ganglioside molecule "in vitro" (Maccioni et al., 1974).

The experiments with *C. perfringens* neuraminidase showed, in addition, that the accessibility to neuraminidase of the gangliosides of the "transient pool" was lower than that of the gangliosides of the "end product pool" . This was concluded from results of neuraminidase treatment of microsomal membranes that were previously incubated with CMP-^{3}H-NeuNAc to label their endogenous gangliosides and comparing the release of radioactive and nonradioactive ganglioside-NeuNAC. Labeled and unlabeled ganglioside-NeuNAc was released at very different rates unless the labeled membranes were, prior to the neuraminidase treatment, disorganized by treatment with a detergent.

No information was obtained on whether the two pools of gangliosides were in the same membrane (Maccioni et al., 1974) and we decided to investigate this point. Different subcellular fractions were prepared and in each the activity of CMP-NeuNAc:lactosylceramide sialyltransferase (sialyltransferase activity, STA) and the amount of endogenous ganglioside acceptors of NeuNAc (endogenous acceptor capacity, EAC) was determined. From the concept of two pools of gangliosides defined above, these determinations indicate ganglioside sialylating activity, that is, the presence of membranes enriched in the "transient pool" of gangliosides and in their related glycosyltransferases. In addition to "in vitro" experiments, "in vivo" experiments were carried out in an attempt to determine the relationship between the two ganglioside pools.

STUDIES "IN VITRO"

Distribution of gangliosides and ganglioside sialylating activity among the primary subcellular fractions.

Most of the EAC and STA were found in the crude mitochondrial fraction when this was prepared at 17,000g for 60 min (Eichberg et al., 1964). The remaining was found mainly in the microsomal fraction. About 50 percent of gangliosides was associated with the crude mitochondrial fraction and approximately 30 percent with the microsomal fraction. When the crude mitochondrial fraction was prepared by centrifugation at 10,000g for 20 min, thus diminishing the microsomal contamination of crude mitochondria (Whittaker et al., 1964) the bulk of the EAC and STA was found associated with the microsomal fraction, only 25 percent being associated with crude mitochondria. The gangliosides were also found mainly in the microsomal fraction and only 32 percent was in the crude mitochondria. In terms of concentration, these results indicate a slight enrichment of EAC and STA in relation to gangliosides in the crude mitochondrial fraction collected at 17,000g for 60 min (ratios, 1.65 and 1.73) in contrast with the lower ratios observed when it was collected at 10,000g for 20 min (ratios, 0.78 and 0.70) (Table 1).

It was reported that repeated washings of the crude mitochondria obtained at 10,000g for 20 min removed additional microsomal contamination (Morgan et al., 1971; Gurd et al., 1974; Raghupathy et al., 1972). Table 1 shows that most of the ganglioside sialylating system and a part of the ganglioside-containing membranes were removed after the crude mitochondrial fraction was washed three times.

Table 1. Gangliosides, endogenous acceptor capacity and sialyltransferase activity in differently prepared crude mitochondrial fractions

Values are the mean of two experiments. Absolute values per g of brain were: ganglioside NeuNAc, 1.4 µmol; endogenous acceptor capacity (E A C), 245.616 cpm of ^{3}H-NeuNAc incorporated into endogenous gangliosides after 60 min of incubation of membranes with CMP-^{3}H-NeuNAc; CMP-NeuNAc: lactosylceramide sialyltransferase activity (STA), 3,380,951 cpm of ^{3}H-NeuNAc incorporated into lactosylceramide after 60 min of incubation with CMP-^{3}H-NeuNAc. Relative concentration (RC) means:

$$\frac{^3\text{H-NeuNAc incorporated, \% recovered}}{\text{ganglioside NeuNAc, \% recovered}}$$

For determination of ganglioside NeuNAc, ganglioside radioactivity, and incorporation of NeuNAc into endogenous gangliosides and lactosylceramide see Maccioni et al., (1974).

Crude mitochondrial fraction	Gangliosides %	EAC %	RC	STA %	RC
17,000g, 60 min	51.9	85.9	1.65	90.2	1.73
10,000g, 20 min	32.9	25.8	0.78	23.2	0.70
10,000g, 20 min, washed	13.0	1.1	0.08	2.0	0.15

That the washing steps removed membranes enriched in ganglioside sialylating system in relation to gangliosides was evident from the ratios of EAC and STA to gangliosides, which decreased from 0.78 and 0.70 to 0.08 and 0.15, respectively.

Distribution of gangliosides and ganglioside sialylating activity among the subfractions from the crude mitochondrial fraction.

The crude mitochondrial fraction obtained at 17,000g for 60 min and that obtained at 10,000g for 20 min were washed three times and resolved in myelin, synaptosome, and mitochondrial fractions in the density gradient procedure of Eichberg et al., (1964). As expected (Whittaker et al., 1964), in gradient from the mitochondrial fraction obtained at 10,000g and washed, the amount of material

Table 2. Comparison of the subfractions from crude mitochondrial fractions obtained at different gravitational fields

Submitochondrial fractions were obtained according to Eichberg et al., (1964). Fraction A floated between 0.32 and 0.8 M sucrose, fraction B between 0.8 and 1.2 M sucrose, fraction C is the pellet below 1.2 M sucrose. According to Whittaker these fractions are enriched in, respectively, small myelin fragments, synaptosomes and mitochondria (see Whittaker, 1969).

Crude mitochondrial fraction	Sub-fraction	Ganglioside-NeuNAc (nmol/mg prot)	Labeling* "in vivo" (cpm/nmol NeuNAc)
17,000g, 60 min			
	A	18.8	187.4
	B	20.9	184.9
	C	8.8	236.0
10,000g, 20 min			
	A	#	#
	B	23.0	186.6
	C	10.0	203.6

* Rats were sacrificed 48 h after intracerebral injection with N-^{3}H-acetylmannosamine.
# Insufficient material for assay purposes.

banding at the myelin and synaptosome layers was markedly lower than the corresponding bands in the gradient from the crude mitochondrial fraction obtained at 17,000g. No differences were observed under the electron microscope in the synaptosomes obtained in the gradient from the two crude mitochondrial fractions. The synaptosomes were also indistinguishable in their ganglioside content or in the specific radioactivity they reached 48 h after N-^{3}H-acetylmannosamine was injected intracerebrally into the rats (Table 2).

The synaptosome fractions coming from the two crude mitochondrial fractions were, however, clearly different in their capacity to sialylate endogenous or exogenous gangliosides "in vitro". The

Table 3. Labeling "in vitro" of gangliosides of subfractions from crude mitochondrial fractions obtained at different gravitational fields

Values are cpm/nmol ganglioside-NeuNAc. Other details, symbols, and nomenclature of subfractions as in Table 1.

Crude mitochondrial fraction	Sub-fraction	Sialylation of endogenous gangliosides		
		in total homogenate*	in isolated subfraction	STA
17,000g, 60 min				
	A	195.9	606	5711
	B	203.6	273	3017
	C	37.0	177	1687
10,000g, 20 min				
	A	#	#	#
	B	4.4	48	306
	C	10.2	18	318

* Immediately after the total homogenate was obtained it was labeled by incubation with CMP-^{3}H-NeuNAc.

synaptosomes from the washed crude mitochondrial fraction showed less than 10 percent of the sialylating activity found in those synaptosomes from the crude mitochondrial fraction obtained at 17,000g when both were related to their respective ganglioside content (Table 3).

It was possible that fractions with relatively low STA and EAC were selectively inactivated during the time elapsed prior to the enzymic assay (about 7 h). To discard this possibility we ran an experiment in which endogenous gangliosides were labeled by incubation of total homogenate with CMP-^{3}H-NeuNAc for 15 min immediately after it was obtained as usual. Subfractionation of this labeled homogenate showed the sedimentation properties of the membranes carrying the labeled gangliosides and eliminated the possibility that the low ganglioside sialylating activity of some subcellular fractions was due to the lack of some component of the brain necessary for the sialylation reaction in these subfractions. The results obtained using this experimental design were essentially the same as those obtained when the subfractions were prepared first and then tested for their EAC and STA. (Table 3).

Distribution of gangliosides and ganglioside sialylating activity in subsynaptosomal fractions.

The distribution of the ganglioside sialylating activity associated with the unwashed mitochondrial fractions was studied after they were disrupted hypoosmotically and the membranes separated by the density gradient procedure of Whittaker et al., (1964). The membranes with the highest ganglioside sialylating activity floated

Table 4. Labeling "in vitro" of gangliosides from subfractions of crude mitochondrial fraction obtained at 17,000g for 60 min and disrupted hypoosmotically

Values are cpm/nmol of ganglioside-NeuNAc. The crude mitochondrial fraction from either a labeled "in vitro" total homogenate (see Table 3) or an unlabeled total homogenate was subjected to a hypo-osmotic shock and subfractionated in a density gradient (Whittaker et al., 1964). For other details see Table 1.-

Crude mitochondrial subfraction	Sialylation of endogenous gangliosides		
	in total homogenate	in isolated subfraction	STA
0	65	0.7	—
D	20	50.0	100
E	25	50.0	60
F	80	200.0	400
G	175	300.0	800
H	60	70.0	700
I	20	40.0	300

at the interphases 0.6-0.8 and 0.8-1.0 M sucrose (subfractions F and G) (Table 4). Here again, it was noticeable that the membrane subfraction floating at the interphase 0.4-0.6 M sucrose (subfraction E) which has a ganglioside content as high as in the membranes of subfractions F and G, has a markedly lower activity for ganglioside sialylation. The result of this experiment was essentially the same whether the EAC, the STA or the distribution of membranes labeled by incubation of total homogenate with CMP-^{3}H-NeuNAc were considered.

From the results discussed above, it appears that in brain there are membranes that contain comparable amounts of gangliosides but have diverse ganglioside sialylating activities "in vitro". Synaptosomes prepared from a washed, crude mitochondrial fraction obtained at 10,000g for 20 min appears to belong to the group of gangliosides containing structures with very limited, if any, ganglioside sialylating activity. In contrast, synaptosomes obtained from the crude mitochondrial pellet obtained by centrifugation at 17,000g for 60 min were found highly active for ganglioside sialylation.

We don't know at present whether the synaptosomes prepared from a crude mitochondrial fraction obtained at low gravitational fields and washed belong to a kind of synaptosomes with very limited activity to sialylate their gangliosides "in vitro" or are representative of the total synaptosomes but it is certain that they are freed of membranes carrying the ganglioside sialylating activity. In any case, these results indicate that it is possible to obtain brain membranes containing principally gangliosides with the metabolic characteristics of the "end product pool" of gangliosides and membranes relatively enriched in the "transient pool" of gangliosides.

STUDIES "IN VIVO"

The results just described of experiments "in vitro" showed that some synaptosomal preparations have a very low capacity to sialylate their gangliosides. "In vivo", starting from N-^{3}H-acetylmannosamine, the gangliosides from these synaptosomes were equally labeled as those from synaptosomal preparations that "in vitro" showed sialylating activity. It was considered possible that they obtained their gangliosides from other structures, for instance, those that "in vitro" were found enriched in the "transient pool" of gangliosides.

The labeling of gangliosides present in subfractions from brain of rats that received an injection of N-^{3}H-acetylmannosamine was compared. "In vitro", the highest labeling was found in subfraction G after it was obtained from the hypoosmotically disrupted 17,000g pellet; for this reason, the labeling "in vivo" of this subfraction at different times after injection was compared with the labeling of synaptosomes obtained from the washed mitochondrial fraction or with the subfraction E.

Results from Table 5 showed that 1 h after N-^{3}H-acetylmannosamine was injected the gangliosides of subfraction G had higher specific radioactivities than those from the subfraction E or synaptosomes; 48 h after injection, the differences had disappeared. In another experiment the time course of the labeling was followed from 6 to 48 h (Fig. 1). It was observed that up to 12 h subfraction G gained radioactivity at a higher rate than subfraction E or the total microsomal fraction. From 12 to 24 h, subfraction G lost radioactivity faster than in the period from 24 to 48 h. On the other hand, in the period 12-24 h subfraction E and total microsomal still continued increasing their specific radioactivity. The shapes of the curves were consistent with the possibility that, floating at the same interphase as subfraction G, there is a membrane population mantaining a precursor-product relationship with subfraction E, total microsomal fraction, and synaptosomes (see Table 5).

Table 5. Ratios of the specific radioactivities of gangliosides from subfraction G and subfraction E or synaptosomal at short and long times after injection of N-^{3}H-acetylmannosamine to rats

Rats were injected with N-^{3}H-acetylmannosamine intracerebrally. At the times indicated they were killed and synaptosomes and subfractions of crude mitochondria were prepared after hypoosmotic shock (see Tables 2 and 4).

time after injection (hours)	Specific radioactivity ratio	
	G / E	G / Synaptosome
1	1.92	2.37
48	1.27	1.07

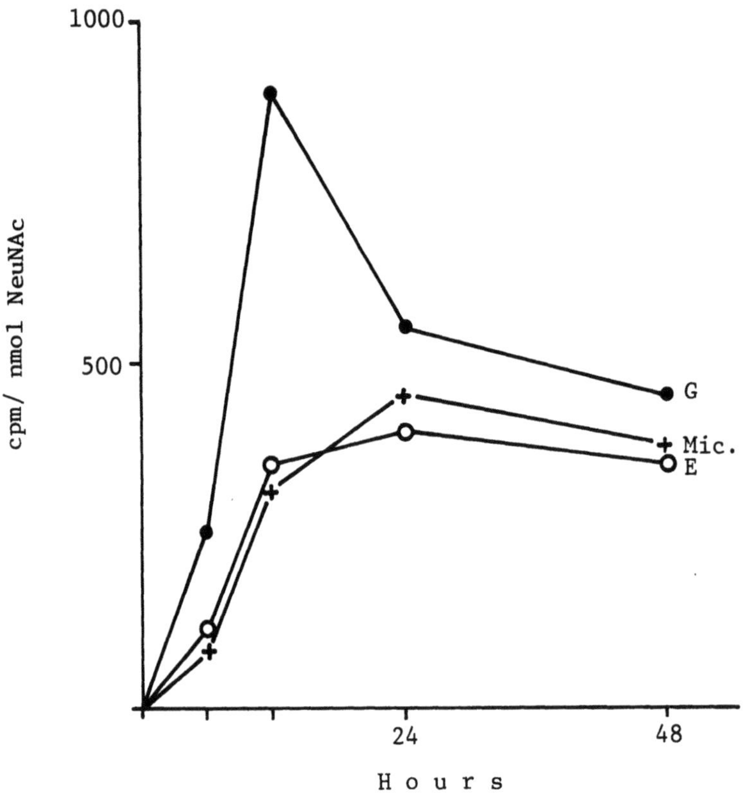

Figure 1 . Labeling of gangliosides of subfractions G and E and total microsomal fraction asa function of time after injection of N-^{3}H-acetylmannosamine to rats.
A crude mitochondrial fraction was obtained by centrifugation at 17,000g for 60 min. The supernatant of this centrifugation was centrifuged at 100,000g for 60 min to obtain the total microsomal fraction. The crude mitochondrial fraction was disrupted hypo-osmotically and processed according to Whittaker et al., (1964) to obtain the subsynaptosomal fractions.

Whatever the nature of the brain subfractions we are considering here, the experiments "in vitro" and "in vivo" discussed above are consistent with the possibility that the ganglioside containing structures that "in vitro" showed low activities to sialylate gangliosides gained their gangliosides from other structures (e.g. Golgi apparatus, endoplasmic reticulum) with high activity. This would mean that the "transient pool" and the "end product pool" of gangliosides are located in different subcellular fractions. This in turn would explain the lack of precursor-product relationship between the major gangliosides, the lack of partial turnover of the carbohydrate chain of gangliosides, and the inability of endogenous sialyltransferase(s) to sialylate the monosialogangliosides created in the membranes by the action of C. perfringens neuraminidase (Arce et al., 1971; Maccioni et al., 1971; 1972; 1974).

We have no information on the mechanism by which both pools of gangliosides are connected "in vivo". Translocation of lipids from one structure to another is known to occur for several lipids (see Rothfield et al., 1972). Axonal transport of gangliosides has been claimed to occur (Forman and Ledeen, 1972; Ledeen et al., 1976). Other investigators were not able to demostrate it (Holm, 1972) or to decide on whether the precursor only or the precursor plus the ganglioside is transported (Rosner et al., 1973; Rosner, 1975).

We carried out experiments to test the possibility that the "transient pool" and the "end product pool" of gangliosides were connected by axonal transport; the visual system of chickens was used for this purpose. N-^{3}H-acetylmannosamine was injected into the left eye, and radioactivity of both TCA-soluble substances and gangliosides was measured in the tectum contralateral and ipsilateral (Fig. 2). The incorporation into gangliosides was higher in the contralateral than in the ipsilateral tectum. The incorporation into TCA-soluble substances was very low and almost identical in both tecta throughout the entire duration of the experiment. This would indicate transport of gangliosides from the retinal cell to the tectum, but we cannot exclude the possibility of arrival of precursors that are quickly exhausted by local synthesis of gangliosides or other sialic acid containing molecules. However, it should be noted that, as in brain, the synaptosomes isolated from the optic tectum showed very low activity to sialylate gangliosides "in vitro" (see Table 6).

Table 6. Effect of colchicine on the labeling "in vivo" and "in vitro" of gangliosides from the visual system of the chicken

	Incorporation (cpm/mg prot)			
	"in vivo"		"in vitro"	
	Control	Colchicine	Control	Colchicine
Retina	5560	4268	363	278
Optic tectum	906*	80*	2840	2910
Optic tectum synaptosomes	1206*	29*	354	424

* Differences between contralateral and ipsilateral tectum (C-I).

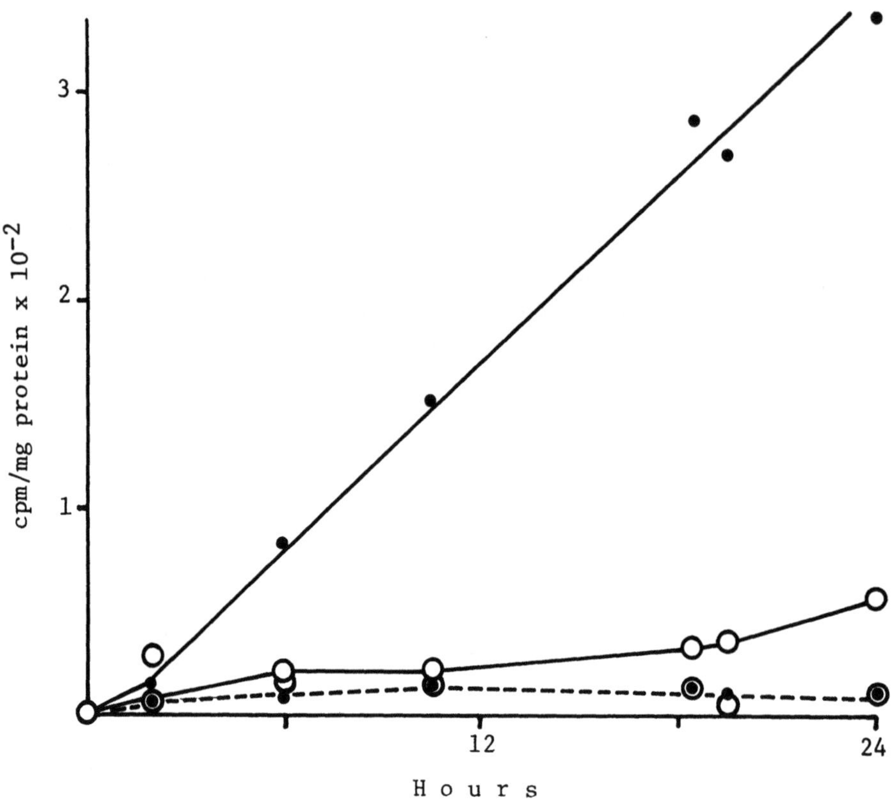

Figure 2. N-^{3}H-acetylmannosamine was injected into the left eye of chickens and the radioactivity of ganglioside and TCA soluble substances was measured in the contralateral and ipsilateral tectum.
● — ●, radioactivity in gangliosides from the contralateral tectum;
o — o, radioactivity in gangliosides from the ipsilateral tectum;
● --- ●, radioactivity in acid soluble from the contralateral tectum;
o --- o, radioactivity in acid soluble from the ipsilateral tectum.

It has been reported that colchicine inhibits the migration of labeled proteins in peripheral nerves (Sjostrand et al., 1970; James et al., 1970). The labeling of gangliosides of the contralateral optic tectum and of its synaptosomal fraction was inhibited when colchicine was injected into the eye prior to the injection of labeled N-acetylmannosamine (Table 6). Since the sialylation "in vitro" of both retina and tectum gangliosides was not inhibited by the injection of colchicine (Table 6), it seems that colchicine acted on the mechanism of transport of gangliosides or of its precursors.

Several studies indicate that the bulk of brain ganglioside glycosyltransferases are associated with the synaptosome-rich fractions (Roseman, 1970; Di Cesare and Dain, 1972) and after hypoosmotic shock of the crude mitochondrial fraction they are found in the subfractions enriched in synaptic plasma membranes (Den et al., 1975; Di Cesare and Dain, 1972). On the other hand, UDP-glucose:ceramide glucosyltransferase, the first glycosyltransferase acting in the building of the carbohydrate chain of gangliosides (Kaufman et al., 1966; Curtino and Caputto, 1972,1974), was found concentrated in the microsomal fraction of brain (Shah, 1973; Neskovic et al., 1973). Studies from this laboratory indicate that the sialyl transferase activity toward endogenous or exogenous acceptors was rather evenly distributed in all the brain subcellular particles (Arce et al., 1966; 1971).

In liver, the ganglioside glycosyltransferases are enriched in the Golgi apparatus (Keenan et al., 1974). From the results we reported here it is possible that also in brain some subcellular fractions are specialized for the addition of sugars in the synthesis of glycoconjugates. Probably because of the cellular heterogeneity of brain tissue, no method of obtaining Golgi apparatus is yet available. Seijo and Rodríguez de Lores Arnaiz (1970) described the presence of curved membrane fragments accompanying the nerve-ending-enriched fraction that may be identified as disrupted Golgi complexes. Membranes carrying the fucosyglycoproteins labeled after short term incorporation of fucose "in vivo" were considered as Golgi derived membranes and could be washed from the crude mitochondrial fraction obtained at 10,000g for 20 min (Gurd et al., 1974). It was reported by Fishman (1974) that the best specific activities of several ganglioside glycosyltransferases were in the myelin subfraction, including that from premyelinating rats.

CONCLUDING REMARKS

It seems clear that there are in brain structures relatively enriched in the "end product pool" of gangliosides and that at least some synaptosomal membranes are included in this group of structures. Evidence has also been provided that shows that the "transient pool" of gangliosides, which in our original concept was linked to the ganglioside glycosylating system, is localized in membranes with sedimentation properties different to those of membranes enriched in the "end product pool" of gangliosides. The nature of these structures has proven to be an elusive problem so far. By analogy with other organs it is tempting to assume that these structures are from the Golgi apparatus. But to prove that this assumption is correct it will be required to purify such structures, to verify their morphology under the electron microscope and to investigate whether in those structures occur other activities normally present in this apparatus. Another interesting point to investigate is whether the synthesizing apparatus is localized in certain regions or spread all over the neuron. Experiments with the retino tectal system of chickens indicate that the gangliosides can be synthesized in the neuronal soma. This is not by itself good evidence that all the synaptosomal gangliosides are synthesized in the soma, but if the poor capacity as endogenous acceptor of the nerve endings and the poor activity of these endings in ganglioside sialylating enzymes is considered, the evidence becomes stronger.

REFERENCES

ARCE, A., MACCIONI, H. J. & CAPUTTO, R. (1966) Arch. Biochem. Biophys., 116, 52.
ARCE, A., MACCIONI, H. J. & CAPUTTO, R. (1971) Biochem. J., 121, 483.
CAPUTTO, R., MACCIONI, H. J. & ARCE, A. (1974) Mol. Cell. Biochem., 4, 97.
CAPUTTO, R., MACCIONI, H. J., ARCE, A. & CUMAR, F. A. (1976) Adv. Exptl. Med. Biol., 71, 27.
CUMAR, F. A., FISHMAN, P. H. & BRADY, R. O. (1971) J. Biol. Chem., 246, 5075.
CUMAR, F. A., TALLMAN, J. F. & BRADY, R. O. (1972) J. Biol. Chem., 247, 2322.
CURTINO, J. A., & CAPUTTO, R. (1972) Lipids, 8, 525.
CURTINO, J. A., & CAPUTTO, R. (1974) Biochem. Biophys. Res. Commun., 56, 142.
DEN, H., KAUFMAN, B., Mc GUIRE, E. J. & ROSEMAN, S. (1975) J. Biol. Chem., 250, 739.
DI CESARE, J. L., & DAIN, J. A. (1972) J. Neurochem., 19, 403.
EICHBERG, J., WHITTAKER, V. P. & DAWSON, R. M. C. (1964) Biochem. J. 92, 91.
FISHMAN, P. H. (1974) Chem. Phys. Lipids, 13, 305.
FORMAN, D. S. & LEDEEN, R. W. (1972) Science, N. Y., 177, 630.
GURD, J. W. , JONES, L. R., MAHLER, H. R. & MOORE, W. J. (1974) J. Neurochem., 22, 281.
HOLM, M. (1972) J. Neurochem., 19, 623.
JAMES, A. C. K., BRAY, J. J., MORGAN, I. G. & AUSTIN, L. (1970) Biochem. J., 117, 767.
KAUFMAN, B. BASU, S. & ROSEMAN, S. (1966) Proc. 3rd. int. Sypm. Cerebral Sphingolipidosis, p. 193. Ed. by Aronson, S. M. & Volk, B. N. New York, Pergamon Press Inc.
KEENAN, T. W., MORRE, D. J. & BASU, S. (1974) J. Biol. Chem. 249, 310.
LEDEEN, R. W., SKRIVANEK, J. A., TIRRI, L.J. , MARGOLIS, R. K. & MARGOLIS, R. U. (1976) Adv. Exptl. Med. Biol., 71, 83.
MACCIONI, H. J., ARCE, A. & CAPUTTO, R. (1971) Bichem. J., 125, 1131.
MACCIONI, H. J., ARCE, A. & CAPUTTO, R. (1972) in Biochemistry of

the Glycoside linkage (R. Piras & H. Pontis, eds.) p. 413. Academic Press, New York and London.
MACCIONI, H. J., ARCE, A., LANDA, C. & CAPUTTO, R. (1974) Biochem. J., 138, 291.
MESTRALLET, M. G., CUMAR, F. A. & CAPUTTO, R. (1974) Biochem. Biophys. Res. Commun., 59, 1.
MORGAN, I. G., WOLFE, L. S., MANDEL, P. & GOMBOS, G. (1971) Biochim. Biophys. Acta, 241, 737.
NESKOVIC, N. M., SARLIVE, L.L. & MANDEL, P. (1973) J. Neurochem. 20, 1419.
RAGHUPATHY, E., KO, G. K W. & PETERSON, N. A. (1972) Biochim. Biophys. Acta, 286, 339.
ROSEMAN, S. (1970) Chem. Phys. Lipids, 5, 270.
ROSNER, H., WIEGANDT, H. & RAHMANN, H. (1973) J. Neurochem., 21, 655.
ROSNER, H., (1975) Brain Research, 97, 107.
ROTHFIELD, L., ROMEO, D & HINCLEY, A. (1972) Fed. Proc., 31, 12.
SEIJO, L. & RODRIGUEZ DE LORES ARNAIZ, G. (1970) Biochim. Biophys. Acta, 211, 595.
SHAH, S. N. (1973) Arch. Bichem. Biophys., 159, 143.
SJOSTRAND, J., FRIZELL, M. & HASSELGREN, P. O. (1970) J. Neurochem. 17, 1563.
WHITTAKER, V. P., MICHAELSON, I. A. & KIRKLAND, J. A. (1964) Biochem. J., 90, 293.
WHITTAKER, V. P., (1969) Handbook of Neurochemistry (Lajtha A., ed.) Vol. II, p. 327.

EFFECTS OF BRAIN GANGLIOSIDES ON REINNERVATION OF FAST-TWITCH RAT SKELETAL MUSCLE

B. Ceccarelli,* F. Aporti, and M. Finesso

*Department of Pharmacology C.N.R. Center of Cytopharmacology, University of Milano, Milano, Italy
Fidia Res. Laboratory, Abano Terme, Padova, Italy

INTRODUCTION

We have recently produced evidence showing that treatment with gangliosides substantially increases the rate and degree of functional recovery that follows the process of regeneration of nerve fibers (Ceccarelli et al., 1976). Our previous results were obtained in two different experimental models involving the regeneration of peripheral sympathetic cholinergic and adrenergic nerve fibers (see also Ceccarelli et al., 1971; 1972). The results of these experiments clearly indicated that both in a cholinergic type (preganglionic) and in an adrenergic type (postganglionic) of anastomosis prolonged treatment with a mixture of brain cortex gangliosides induced an increased rate and degree of functional recovery, as measured by electrophysiological means. This might be explained by a more efficient reinnervation on the effector organs(superior cervical ganglion and smooth muscle of nictitating membrane). Although it is difficult at the present to account for this observation we thought that it might be useful to investigate the pharmacological effect of gangliosides in the physiological characteristics of reinnervation also of mammalian skeletal muscle fibers. We report here an account of our preliminar experiments, using the fast-twitch extensor digitorum longus (hereafter called EDL) muscle and its nerve supply.

The left sciatic nerve branch to EDL muscle was transected in adult female rats and the ends of the two stumps of the transected nerve were tied together. Some of the operated animals were treated daily, during the entire post-operative period, with 50mg/kg i.p. of a mixture of brain cortex gangliosides (Fidia S.p.A., Abano Terme,

Italy), whose composition was previously given (Ceccarelli et al., 1976). The time-course of the recovery that follows regeneration and reformation of functioning neuromuscular junctions on EDL muscle fibers was followed and evaluated by recording "in situ" the maximal isometric twitch tension (g/mg muscle wt.) developed in response to indirect repetitive stimulation. When skeletal muscle is surgically denervated, muscle fibers develop extrajunctional sensitivity to acetylcholine (ACh) (Axelsson & Thesleff, 1956; Miledi, 1960; Albuquerque & Mc Isaac, 1970; Jones & Vrbová, 1974). Since during reinnervation of mammalian skeletal muscle extrajunctional sensitivity to neurotransmitter persists until after nerve stimulation is able to evoke action potentials in the muscle fibers (Tonge, 1974), we hoped to obtain an additional and independent estimate of the amount of functional reinnervation in the muscle by measuring the time-course of the decline in the extrajunctional sensitivity to ACh. For this purpose, in all the operated animals at the conclusion of the "in situ" mechanical recording, the left EDL muscle was removed and transferred to a bath containing oxygenated Krebs solution. Isometric tension developed by the muscle in response to ACh added to the bath at 10 min intervals was recorded and dose-response curves obtained for each muscle.

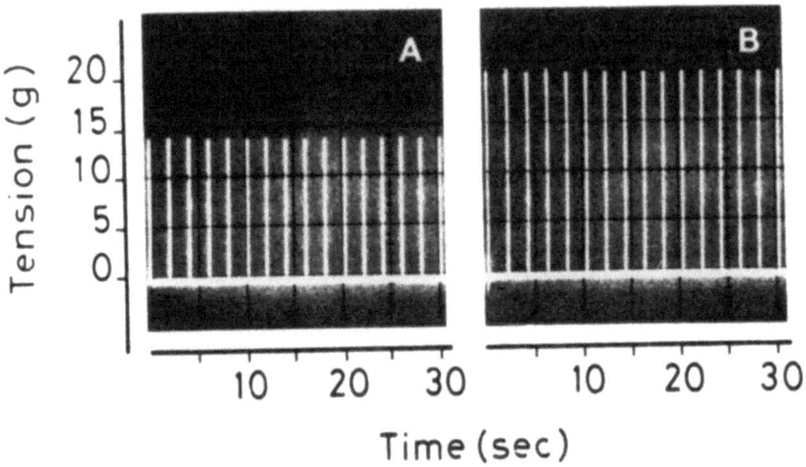

Fig. 1. Maximal isometric twitch-tension developed in response to indirect repetitive stimulation. A: record taken twenty days after the surgery in an untreated animal. B: record taken after the same time after the surgery in an animal treated for the entire post-operative period with 50 mg/kg i.p. of gangliosides. These two records were taken from EDL muscles of about the same weight(100mg).

RESULTS AND DISCUSSION

A representative example of a record, taken twenty days after the surgery, in an animal treated for the entire post-operative period with 50 mg/kg i.p. of gangliosides is given in Fig. 1.

Fig. 2 summarizes the average values of our estimate of the time-course of functional recovery of EDL muscles in animals operated and treated with gangliosides and compares this estimate with the one obtained in operated but untreated animals. The average percentages of maximun tension developed by indirect stimulation (ratio of operated and unoperated side in each animal) are plotted as a function of time after the surgery. It can be seen that the time-course of functional recovery of EDL muscle is very rapid in both group of animals and virtually complete recovery is obtained by 30 days. However in the group of animals treated with ganglio-

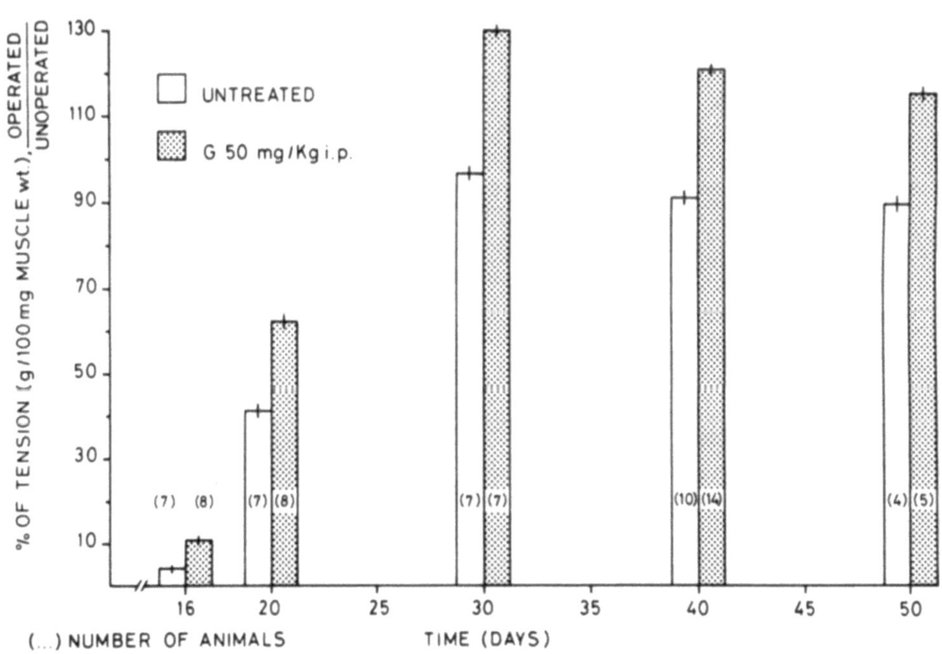

Fig. 2. Time-course of the maximal isometric twitch-tension developed in response to indirect repetitive stimulation. Percentages of tension (ratio of operated and unoperated muscles) are plotted as a function of time after the surgery. Results are expressed as mean ± s.e.

sides the percentage of tension developed by the muscles is higher than the one observed in the untreated animals, during the overall time-course of our experiments. The peak difference in the twitch-tension appears to occur 30 days after the operation and then gradually subsides but is still evident 50 days after surgery. No data are so far available beyond this period. Of particular interest are the results we obtained 20 days after the surgery, i.e. after a period in which functional recovery is still incomplete in both groups of operated animals. In fact twenty days after the surgery the amount of functional recovery in the gangliosides treated animals is about 60% of control values whereas in the untreated animals the average recovery is about 40%. Thus it appears that at this time the treatment with the mixture of gangliosides increases the amount of functional reinnervation in the EDL muscle by about 50%.

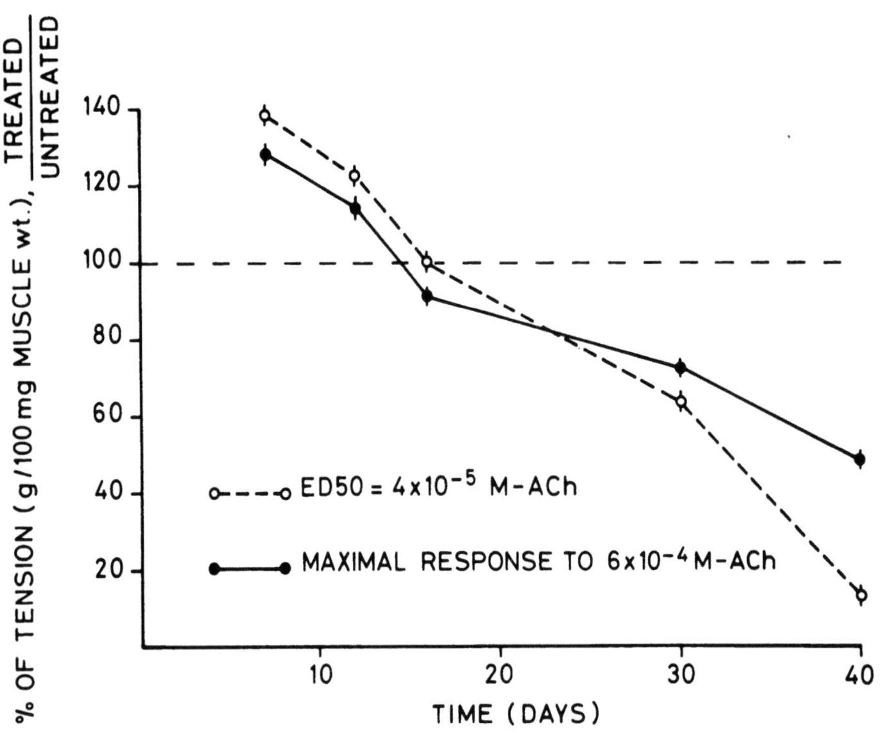

Fig. 3. Percentages of isometric tension evoked by ACh applied to the bath. The ratio of treated (50 mg/kg i.p. of gangliosides) and untreated are plotted as a function of time after the operation. Data of ED50 and maximal response to ACh are shown. Results are expressed as mean ± s.e.

Additional evidence supporting the view that the treatment with gangliosides increases the degree of functional reinnervation in the EDL muscle is given by the experiments in which extrajunctional sensitivity to ACh has been determined "in vitro". The results of all these experiments are summarized in Fig. 3. The estimated percentages of tension evoked by ACh applied to the bath (ratio of treated and untreated) are plotted, in the operated animals, as a function of time after the operation. Fig. 3 clearly shows the relative decline of the extrajunctional sensitivity to ACh in the group of treated animals with respect to the untreated ones which occurs during the time-course of the experiments. This decline in the extrajunctional sensitivity of the muscles to their normal chemical transmitter can be taken as additional indirect evidence that in the muscles of those animals treated with gangliosides the extent of functional reinnervation was higher than in the untreated animals. We therefore concluded that the differences in extrajunctional sensitivity could actually reflect the real differences in the amount of functioning neuromuscular junctions on the EDL muscle fibers in the two groups of operated animals.

Thus, taken together, the results of both sets of experiments suggest that the daily treatment with brain cortex gangliosides during the entire post-operative period consistently increases the extent of functional reinnervation in the fast-twitch EDL skeletal muscle. These findings corroborate and extend our previous pharmacological study on the effects of gangliosides on peripheral sympathetic regeneration and reinnervation.

ACKNOWLEDGMENT

We are happy to acknowledge the skillful technical assitance of Paolo Tinelli and Franco Crippa who prepared the illustrations.

REFERENCES

ALBUQUERQUE E.N. & McISSAC R.J. (1970) Exp. Neurol. 26, 183-202
AXELSSON J. & THESLEFF S. (1959) J. Physiol. 147, 178
CECCARELLI B., CLEMENTI F. & MANTEGAZZA P. (1971) J.Physiol. 216, 87-98
CECCARELLI B., CLEMENTI F. & MANTEGAZZA P. (1972) J.Physiol. 220, 211-227
CECCARELLI B., APORTI F. & FINESSO M. (1976) In ganglioside function: biochemical and pharmacological implications. pp. 275-293. Plenum Press, New York and London.
JONES R. & VRBOVA G. (1974) J.Physiol. 236, 517-538
MILEDI R. (1960) J. Physiol. 151, 1-23
TONGE D.A. (1974) J. Physiol. 241, 141-153

STUDIES ON THE FUNCTIONS OF GANGLIOSIDES IN THE CENTRAL NERVOUS SYSTEM

R. Caputto, A. H. R. de Maccioni, and B. L. Caputto

Departamento de Química Biológica, Facultad de Ciencias Químicas, Universidad Nacional de Córdoba, Córdoba, Argentina

No specific functions have been found for most of the great variety of complex lipids, including the gangliosides. These last compounds, at a concentration of 3 per thousand in gray matter are, in this part of the nervous tissue about twice as concentrated as in white matter and at least ten fold higher than in any other tissue of extra-neural origin.

STUDIES "IN VIVO"

Due to their localization in grey matter gangliosides have been investigated by several workers in relation to nervous activity. Provided that they are involved in cerebral function, it was expected to find modifications in situations in which the activity of large brain regions are affected. Irwing & Samson (1971) reported that behavioural stimulation of rats modifies the pattern of incorporation of labelled precursors into cerebral gangliosides. Injecting rats with ^{14}C-glucosamine and submitting them to swimming trials, they found that disialogangliosides obtained from the brains of those rats showed a decreased incorporation of radioactivity compared with their controls.

Also, Dunn et al. (1973) obtained a change in the labelling of mice brain gangliosides during conditioned avoidance training; an increase in the specific radioactivity of gangliosides extracted from the brains of animals subjected to the learning experience was observed, but no difference for any particular ganglioside species was detected.

We approached the study of the role of gangliosides searching for changes in their metabolism as a response to a more defined input. We used light as stimulus and looked for changes in brain subcellular particles and in retina. The experimental design consisted in the exposure of animals from different species to 1000 lux, previous injection of tritiated glucosamine, and comparing the radiochemical incorporated into the gangliosides of several subcellular organelles in the animals exposed to light with that incorporated in animals maintained in the dark. The results obtained indicated that the incorporation of radioactivity into the brain synaptosomal gangliosides in light and in dark run parallel to the effect of the light on the motor activity of the animals. In the rat, which has nocturnal habits, the labelling of synaptosomal gangliosides was lower in light exposed animals (Maccioni et al., 1971). The opposite occurred in chickens which awake under light: in the animals exposed to illumination the incorporation of radioactivity in the gangliosides from telencephalic synaptosomes was higher than that obtained from dark maintained chickens (Table 1) (Maccioni et al., 1974).

Table 1. Labelling of synaptosomal gangliosides after visual stimulation

Rats and chickens maintained in the dark for several days were injected with 6-^{3}H-glucosamine and divided in 2 groups: one remained in the dark while the other was exposed to 1000 lux. Rats were exposed to illumination for 1 h; chickens for 2 h. Then animals were killed and the nervous tissue fractionated according to Eichberg et al. (1964). Gangliosides and sialic acid determinations as described previously. (Maccioni et al., 1971; Maccioni A. H. R. et al., 1974). p was calculated by Student's test for correlated data. Values for animals in the dark were taken as 100 %.

Animal Species	Darkness	Light	Change %	p
	c.p.m. gangliosides / µmol sialic acid			
Rat (brain)	5309	4304	-15	<0.01
	c.p.m. gangliosides / c.p.m. acid soluble fraction			
Chicken (telencephalon)	0.16	0.21	33	<0.001

Since it could not be decided whether the difference in ganglioside labelling was due to a direct effect of light or to an indirect effect mediated by the general activity of the animal we carried out some studies with chicken retina which is directly stimulated by light (Caputto et al., 1975). It also offers the advantage that it is possible to perform comparative experiments in the same animal by occluding one eye and exposing the other to light. The experiments with retina showed that light influences directly the labelling of gangliosides; in this organ the incorporation of radioactivity from 6-^{3}H glucosamine was higher in dark maintained than in the light exposed retinas (Table 2). In experiments designed to investigate whether the differences were limited to one or several gangliosidic species, it was found that no individual or group of ganglisides but all the species were responsable for the observed changes.

Table 2. Labelling of retinal gangliosides after visual stimulation

Chickens maintained in the dark for 2 days were exposed to 1000 lux previous subcutaneous injection of 6-^{3}H glucosamine while others remained in the dark. For the experiments with chickens with one eye occluded, they had one eye covered with black tape while the other eye was exposed to light, previous injection of the radiochemical. Retinal gangliosides were obtained by the TCA-PTA (trichloroacetic acid - phosphotungstic acid) method (Maccioni et al., 1974). Statistical analysis as in Table 1. Values for retinas in light were considered 100 %.

Time of exposure	Chickens in Darkness	Light	Change %	p
	$\frac{\text{c.p.m. gangliosides}}{\text{c.p.m. acid soluble fraction}} \times 10^5$			
0.5 h	1384	932	51	<0.01
2 h	4310	3160	36	<0.001
	Chickens with one eye occluded			
	c.p.m. gangliosides/mg protein			
	Occluded eye	Exposed eye		
1 h	160	141	18	<0.01

The labelling of the acid soluble fraction in retina or in synaptosomes obtained from brains of animals in dark or light showed no difference between them, which suggests that light does not alter the permeability of cell membranes for labelled glucosamine. However, since the time variation curve of the acid soluble fraction was not determined it cannot be discarded that the changes obtained are a reflection of modifications in the level of the pool of precursor.

ENZYMATIC STUDIES

Indications respect to the functions of gangliosides may be found by studying their influence on enzyme activity. Igarashi & Suzuki (1976) found that the ganglioside species commonly known as GM1 and its derivative asialoganglioside strongly inhibited the cholesterol-ester-hydrolase from myelin. In considering the possible mechanism of this inhibition the authors favor the idea that the negative charge or the carbohydrate chain (which has to be longer than three sugars since lactosyl-ceramide does not inhibit) in the surface of the co-micelle formed by the glycosphingolipid and the substrate may hinder the accesibility of the enzyme to the substrate. Since they found that the amount of detergent (taurocholate) used in the assay was not enough to dissolve the enzyme they consider that the interaction of the lipid with the enzyme is only a minor factor in the inhibition. However, the lack of solubilization of an enzyme does not prove that the integrity of the membrane has not been impaired or that the enzyme is not exposed to extra-membrane effectors.

Activation of ATPases by gangliosides

It has been known for some time that Na^{+}- K^{+} -ATPases are partially or totally inactivated when they are solubilized by detergents. Preparations of this enzyme from brain tissue, obtained in 0.33 per cent sodium deoxycholate can be reactivated by phosphatides (Tanaka & Strickland, 1965) specially phosphatidylserine (Fenster & Copenhaver, 1967). This observation indicates that the lipids can be of importance in determining the activity of enzymes embedded in their membrane medium. Since in nervous tissue the ratio of the content of gangliosides to that of any single phosphatide may vary from 0.2 to 1 or higher, the gangliosides were considered as possible effectors of similar enzymes. We, consequently, proceeded to test the following gangliosides or ganglioside derivatives: a mixture of rat brain gangliosides, a mixture from the brain of a patient with Tay Sachs desease, hematoside (GM3), monosialoganglioside (GM1), disialoganglioside (GD1a) and asialo-ganglioside. All of them were found effective in activating, in the presence of Mg^{2+} salts, preparations of sodium deoxycholate solubilized ATPase obtained from rat brain (Caputto et al., 1976). Ceramide, glucosyl-ceramide, galactosyl-ceramide and lactosyl-ceramide tested as water suspensions or solubilized with detergent (Table 3) did not activate the enzyme. Sialyl-lactose also failed to activate ATPase. Since hematoside (sialyl-lactosyl-ceramide) activated the enzyme whereas lactosyl-ceramide and sialyl-lactose did not, it is concluded that the glycosphingolipid must contain a lipophilic moiety and a hydrophilic moiety containing a carbohydrate chain of more than two units to be an activator. It has not yet been determined if a neutral carbohydrate can substitute for sialic acid in a chain of three carbohydrate units but it has been proven that a sialyl group is not required when the carbohydrate chain contains four units (see above, asialo-ganglioside).

Comparing the activity of the Mg^{2+}-ATPase in the presence of water suspensions of gangliosides and phosphatidylserine it was found that only the gangliosides activated. However, when sodium-deoxycholate (0.06 mM) was added to the incubation medium phosphatidylserine also activated the enzyme.

Table 3. Effect of phosphatides, glycosphingolipids and derivatives on Mg^{2+} -dependent ATPase activity

Additions	Activity (%)	
	-Doc	+Doc
None	100	100
Gangliosides	220	220
Phosphatidylserine	100	220
Ceramide	110	100
Galactosyl-ceramide	100	90
Glucosyl-ceramide	90	90
Lactosyl-ceramide	100	100
Sialyl-lactose	100	100

Doc: sodium-deoxycholate

Nature of the ATPase(s) activated by gangliosides

It is widely known that the ATPases can be distinguished on the basis of the cations that activate them. Mg^{2+}-, Mg^{2+} plus Ca^{2+} and Mg^{2+} plus Na^{+} and K^{+}-ATPases are probably the best known. They are also distinguished by their localization in different subcellular organelles and by their functions, especially on account of their participation or failure to participate in a proton or Ca^{2+} or Na^{+}-K^{+} pump. We know of no specific function assigned to the microsomal ATPases from nervous tissue. Recently Trams & Lauter (1974) and Stefanovic et al. (1976) have shown that in cultured mouse neuroblastoma and glial cells the ATPases are present in the external surface of the cell. These ATPases, for which the normal relationship to their assumed substrates still has to be determined, are inactive in the presence of 1 mM EDTA and are reactivated by Ca^{2+}, Mg^{2+}, Mn^{2+}, Co^{2+}, Cd^{2+} and other divalent cations.

From Na-deoxycholate solubilized brain microsomes the fraction that precipitates between 45 and 60 per cent ammonium sulfate saturation was obtained. This fraction contains ATPases that can be activated with Mg^{2+} or Ca^{2+} or Mn^{2+} in the presence of EDTA and ouabaine (Table 4). In the presence of EGTA (ethanedioxy bis (ethyl amine tetra-acetic acid)) but not ouabaine, and Mg^{2+} the addition

Table 4. Activation of ATPase by cations and gangliosides*

Cation added		No gangliosides %	Gangliosides added
None		6	11
$MgCl_2$	(2.8 μM)	55	108
$CaCl_2$	(0.8 μM)	100	134
$MnCl_2$	(0.4 μM)	19	53
$MgCl_2$ $MnCl_2$	(2.5 μM) plus (0.4 μM)	53	103

* The value in the presence of 0.8 μM $CaCl_2$ without gangliosides was considered as 100 per cent.

of Na^+ plus K^+ also increased slightly the enzyme activity. In all of these conditions the addition of gangliosides usually brings about an increase of the ATPase activity that has varied in several experiments from 1.3 to 12 fold of the activity without gangliosides.

Whether the activation of ATPase by gangliosides is relevant to the functions of these last compounds still has to be determined. A substantial fraction, if not all, of the hydrophilic moiety of the microsomal gangliosides has been shown to be in the surface of the membrane and accesible to the action of neuraminidase (Maccioni et al., 1974). As a first approach to the possible relationship between the functions of ATPase and gangliosides it may be interesting to attempt to determine whether they are in the same side of the membrane surface.

REFERENCES

CAPUTTO B. L., MACCIONI A. H. R. & CAPUTTO R. (1975) Nature 257, 492-493
CAPUTTO R., MACCIONI A. H. R. & CAPUTTO B. L. (1976) Biochem. Biophys. Res. Commun. (In press)
DUNN A., BROGAN L., ENTINGH T., GISPEN W. H., MACHLUS B., PERUMAL R. & RESS H. D. In F. O. Schmidt (Ed). The Neurosciences, Third Study Program, M.I.T. Press, Cambridge, Mass. 1973, pp. 679-684
EICHBERG J., WHITTAKER V. P. & DAWSON R. M. C. (1964) Biochem. J. 92, 91-96
FENSTER L. J. & COPENHAVER J. H. (1967) Biochim. Biophys. Acta 137, 406-408
IGARASHI M. & SUZUKI K. (1976) J. Neurochem. 27, 859-866
IRWIN L. N. & SAMSON F. E. (1971) J. Neurochem. 18, 203-211
MACCIONI A. H. R., GIMENEZ M. S. & CAPUTTO R. (1971) J. Neurochem. 18, 2362-2370
MACCIONI A. H. R., GIMENEZ M. S., CAPUTTO B. L. & CAPUTTO R. (1974) Brain Research 73, 503-511
MACCIONI H. J. F., ARCE A., LANDA C. & CAPUTTO R. (1974) Biochem. J. 138, 291-298
STEFANOVIC V., LEDIG M. & MANDEL P. (1976) J. Neurochem. 27, 799-805
TANAKA R. & STRICKLAND K. P. (1965) Arch. Biochem. Biophys. 111, 583-592
TRAMS E. G. & LAUTER C. G. (1974) Biochim. Biophys. Acta 345, 180-197

THE BIOSYNTHESIS OF GANGLIOSIDES AND SIALOGLYCOPROTEINS IN BOVINE RETINA

J. A. Curtino, H. J. Maccioni, and R. Caputto

Departamento de Química Biológica
Facultad de Ciencias Químicas
Universidad Nacional de Córdoba
Córdoba, Argentina

The biosynthesis of gangliosides and sialoglycoproteins in retina has been studied "in vitro" measuring the incorporation of ^{3}H-NeuNAc from CMP-^{3}H-NeuNAc into endogenous acceptors of bovine retina.

A general picture of the fractionation of retina membranes is shown in Fig. 1. The ROS (rod outer segment) membranes were isolated by the procedure of McConnell (1965) with slight modifications. In the sucrose density gradient that we used the ROS membranes floated as two well defined red colored layers, at the junction between 0.77 and 0.94 M sucrose and between 0.94 and 1.10 M sucrose. The proteins of these two membrane layers showed similar patterns by sodium dodecyl-sulfate polyacrylamide gel electrophoresis, being rhodopsin its major component. The retina membranes that did not sediment with the crude ROS fraction, namely ROS free membranes, were subfractionated by centrifugation in sucrose density gradients into subfractions Pla,Plb,P2,P3 and P4. The ROS and Plb layers were red colored while the sediment at the bottom of the tube was brownish.

The incorporation of ^{3}H-NeuNAc into endogenous acceptors of the two primary retina fractions is shown in Table 1. The labelling of lipids and proteins in total ROS free membranes was about six fold greater than the labelling of lipids and proteins found in total ROS membranes. Subfractionation of the ROS free membranes in a sucrose density gradient resulted in the isolation of membranes with different activities for the incorporation of NeuNAc. The highest activity was in subfraction Pla (Table 2). The incorporation of radioactivity into lipids paralleled the labelling of proteins in all the subfractions, from the more active Pla to the less active

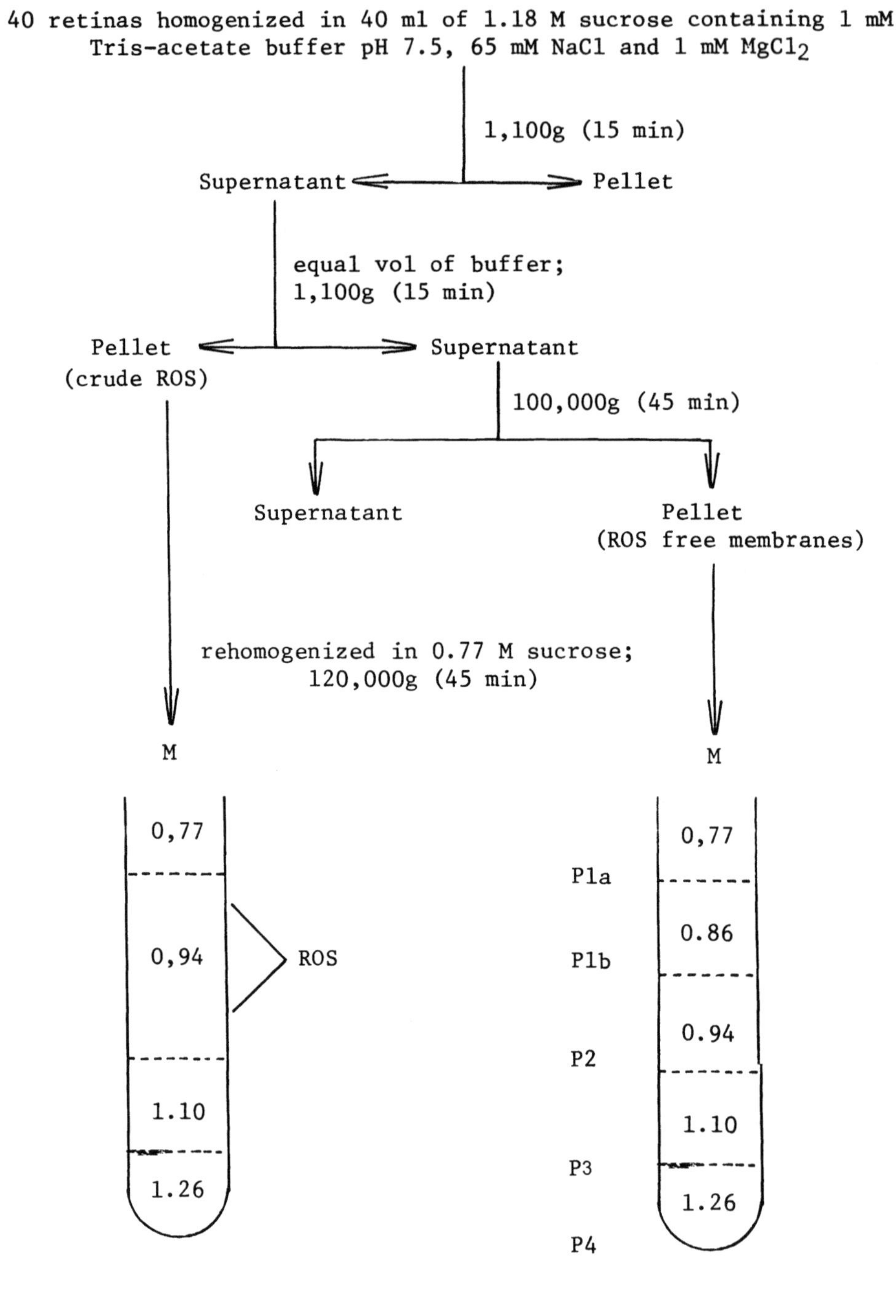

Fig. 1. Diagram of retina fractionation.

Table 1. ^{3}H-NeuNAc incorporation into the primary membrane fractions of bovine retina

About 0.4 mg. of protein of each retina fraction was incubated for 30 min at 37°in a system containing, in a final volume of 50 µl, 10 mM $MgCl_2$, 10 mM $MnCl_2$ and 1.3 µM CMP-^{3}H-NeuNAc. The precipitate obtained after addition of cold 5 % (w/v) trichloroacetic acid containing 0.5 % (w/v) phosphotungstic acid was washed with the same solution and extracted with chloroform:methanol (2:1, v/v) (C:M extract). The protein residue was washed with chloroform : methanol : water (1:1:0.3, by vol) and dispersed in dioxane-scintillation mixture with the aid of sonication. The c.p.m. values are radioactivity incorporated into lipids and proteins of the total of each fraction isolated from forty retinas.

Membrane fraction	Total c.p.m. C:M extract	Total c.p.m. Protein
ROS (40 mg of protein)	20,000	24,000
ROS free (85 mg of protein)	119,000	145,000

P4. The specific activity expressed in c.p.m. per mg of protein, was about fourteen times higher in the Pla subfraction than in the ROS fraction.

With the purpose of characterizing the lipidic material, Pla plus Plb membranes were incubated with CMP-^{3}H-NeuNAc, extracted with chloroform:methanol and passed through a column of Sephadex G 25 (Arce et al., 1971). The Sephadex effluent was partitioned with water and the radioactive material in the upper phase, which accounted for 90 per cent of total radioactivity of the effluent, was analyzed by TLC. Bovine retina GD3 and total gangliosides from bovine brain were used as standards. About half of the radioactivity was found at the level of GD3, while the other half moved behind GT1. The deffated protein residue obtained after incubation of Pla plus Plb membranes with CMP-^{3}H-NeuNAc was analyzed by sodium dodecylsulfate polyacrylamide gel electrophoresis. Four peaks of radioactivity were found; three of them were coincident with three PAS positive bands.

The transfer activity of retina membranes was preserved for at least six months upon storage at -15° in 0.24 M sucrose. The incorporation of NeuNAc into gangliosides and sialoglycoproteins was inhibited by 1 mM ATP(Table 3). The inhibition of NeuNAc transfer

Table 2. Incorporation of ^{3}H-NeuNAc into endogenous acceptors from retina membrane fractions

Incubations and isolation of reaction products were as indicated in Table 1.

Membrane fraction	(c.p.m. / mg of protein) C:M extract	Protein
ROS	500	600
Pla	7,700	7,100
Plb	3,400	2,400
P2	1,300	1,600
P3	700	800
P4	600	500

Table 3. Inhibition of the incorporation of ^{3}H-NeuNAc into gangliosides and sialoglycoproteins by ATP

Incubation for 10 min (0.4 mg of protein of Pla membranes) and isolation of reaction products were as in Table 1.

Additions	Gangliosides (c.p.m.)	Sialoglycoproteins
None	6,600	5,400
ATP (1 mM)	3,000	3,600

activity by mM concentrations of cytidine nucleotides has been described in rat liver microsomes using exogenous glycoprotein acceptor; this activity was not affected by ATP (Bernacki, 1975)

These results clearly indicate that the bulk of the "in vitro" sialic acid incorporation into endogenous gangliosides and sialoglycoproteins was in retina membranes other than ROS. Centrifugation of ROS free membranes in sucrose density gradients resulted in the isolation of the more active membranes for the transfer of ^{3}H-NeuNAc from CMP-^{3}H-NeuNAc, which were collected at the junction between the sucrose layers of densities 1.10 - 1.11. The labelled lipids and proteins found in the ROS membranes after they were incubated with CMP-^{3}H-NeuNAc could arise from Pla membranes contaminating the crude ROS fraction during fractionation of retinas.

Most of the incorporation of sialic acid into both gangliosides and sialoglycoproteins was found in the same membrane fraction and was similarly inhibited by ATP. Even if it is difficult to draw general conclusions about the organization of membrane components with this type of experiment, a possible interpretation on the basis of our results may be that the two glycoconjugates are, at least for their sialylation, in a close metabolic relationship, in the same kind of membrane.

REFERENCES

ARCE, A., MACCIONI, H. J. & CAPUTTO, R. (1971) Biochem. J., 121, 483-493

BERNACKI, R. J. (1975) Eur. J. Biochem., 58, 477-481.

McCONNELL, D. G. (1965) J. Cell Biol., 27, 459-473.

POSSIBLE PRESENCE OF PSYCHOSINE IN BRAIN

J.M. Pasquini,* F.B. Jungalwala,** and R.H. McCluer**

* Departamento Química Biológica, Facultad de Farmacia y Bioquímica, U.B.A., Buenos Aires, Argentina

** E.K. Shriver Center for Mental Retardation, Waltham, Mass., U.S.A.

INTRODUCTION

Jungalwala et al. (1975) described a very sensitive method for the separation and quantitation of phospholipids containing primary amino groups such as phosphatidylethanolamine (PE), phosphatidylserine (PS) and lysophosphatidylethanolamine by high performance liquid chromatography. Compounds which contain primary amino groups can be converted into their U.V. absorbing N-biphenylcarbonyl derivatives. Based on this property, we developed a high performance liquid chromatographic method for the analysis of brain psychosine, with the use of U.V. detection at 280 nm.

It is our interest to demonstrate the presence of a free psychosine or galactosyl-sphingosine pool in brain, because enzymic reactions for the synthesis of galactosyl sphingosine and also glucosylsphingosine have been described (Cleland & Kennedy, 1960; Kanfer, 1969; Curtino & Caputto, 1972). These reactions offer the possibility of an alternative pathway for the biosynthesis of cerebrosides. The N-acylation of galactosyl sphingosine in an incubation system was reported by Brady (1962) and Hammarstrom (1971). Curtino & Caputto (1974) have demonstrated the enzymatic acylation of glycosyl sphingosine and galactosyl sphingosine by stearoyl CoA using a chicken embryo brain microsomal fraction.

EXPERIMENTAL PROCEDURES

The psychosine standard was commercially available from Supelco,

Belefonte, Pa, U.S.A. or obtained from brain cerebrosides by hydrolysis. Adult rat brain cerebroside was obtained by extraction with chloroform:methanol 2:1. The extract was washed and passed through a silicic acid column and eluted with acetone:methanol 9:1. Hydrolysis was carried out using the method described by Radin (1974) using butanol and aqueous KOH.

The procedure for derivative preparation was that described by Jungalwala et al. (1975). Standards (10-100 nmol) were dried and 50 µl of 1% solution of 4-biphenylcarbonyl chloride in tetrahydrofurane was added, (4-biphenylcarbonyl chloride was obtained from Aldrich Chemical Co., Milwaukee, Wi. U.S.A.) followed by 100 µl of aqueous 50% sodium acetate solution. The mixture was shaken vigorously at room temperature for sixty minutes. The reaction products were purified as follows: 5 ml of lower phase (1 vol. of chloroform:methanol 2:1 was equilibrated with 0.2 vol. of water, the phases separated and the upper phase was discarded) was added to the reaction mixture and the excess of reagent was removed by washing three times with 1 ml of methanol:water:chloroform:15 M-NH_3 (96:92:2:3). The lower phase was evaporated to dryness under N_2 and the residue was then dissolved in 250 µl of chloroform:methanol 1:1.

Adult rat brains (Wistar) were homogenized in methanol and incubated at 50°C for 15 minutes. Chloroform was added and the tissue extract was filtered and washed consecutively according to Raghavan et al. (1973) (Fig. 1). The lower phases (where the psychosine should be present) were combined.

Alkaline hydrolysis was carried out in order to remove PS and PE that can be extracted together with the psychosine. After the alkaline hydrolysis, the extract was washed three times with upper phase containing ammonium hydroxide. If the extract (checked by TLC) had still some contaminants as cerebrosides or sulphatides, a preparative TLC was carried out using chloroform:methanol:2 N-NH_3 60:35:8 as solvent in order to remove them. The spots were identified by spraying the plate with fluorescamine reagent in acetone (4-phenyl-spiro furan-2 (3H) 1'-phtalan 3,3-dione) Hoffmann La Roche, Nutley, U.S.A.

The final preparation which was assumed to contain the tissue psychosine, was then dried under N_2 and used for derivative formation as described before.

High performance liquid chromatography was performed with a high pressure liquid chromatography apparatus equipped with a U.V. detector (280 nm). The U.V. monitor has an 8 µl flow cell. The chromatographic column was a 20 cm x 2.1 mm (internal diameter) pre-packed stainless-steel "Micro-Pack" column (Varian, Palo Alto, Calif., U.S.A.) with an average particle diameter of silica gel of 10 µm. The solvent mixture used was dichloromethane:methanol:NH_3 (85:8.5:1.5) by vol. The elution was performed at a flow rate of 2 ml/min. The

Tissue Homogenized in CH_3OH and Incubated at 50°C 15 Min.
Add Cl_3CH. Mix and Filter.
Wash the C:M 2:1 Extract Consecutively with:
0.5 Vol 0.1M KCl in 0.5M NaOH and
C:M:0.1M KCl in 0.5M NaOH (3:48:47)
C:M:0.5M NaOH (3:48:47)
Discard Each Upper Phase.

Lower Chloroform Phase:
Add CH_3OH + Glacial Acetic Acid.
Wash with:
0.5 Vol 0.1M KCl + 0.5 ml 1N H_2SO_4
C:M:0.1M KCl (3:48:47) + 0.5 ml 1N H_2SO_4
C:M:W (3:48:47) + 0.5 ml 1N H_2SO_4
The Upper Phases Adjusted to pH 9.0 with 10N NaOH.
Extract Twice with Equal Volumes of C:M:W (84:14:1)
Combine Lower Phases.
Alkaline Hydrolysis.

Fig. 1. Flow sheet of the procedure used for the isolation of Psychosine from adult rat brain.

attenuator on the U.V. detector was normally set at 0.04 absorbance unit full scale.

RESULTS

The psychosine derivatives were characterized by TLC on a silica gel G.F. plate developed with chloroform:methanol:2 N NH_3, 100: 25:25. When the plate was viewed under short wavelength U.V. light, the biphenylcarbonyl derivatives appeared as quenched spots on a fluorescent back-ground. By spraying the plate with ANSA (8-anilino naphtyl sulphonic acid reagent) we checked that all the psychosine was converted into the derivative. Fig. 2 shows an elution pattern of the standard psychosine derivatives. Two different peaks were eluted from the column with a retention time of 7 and 10 minutes respectively.

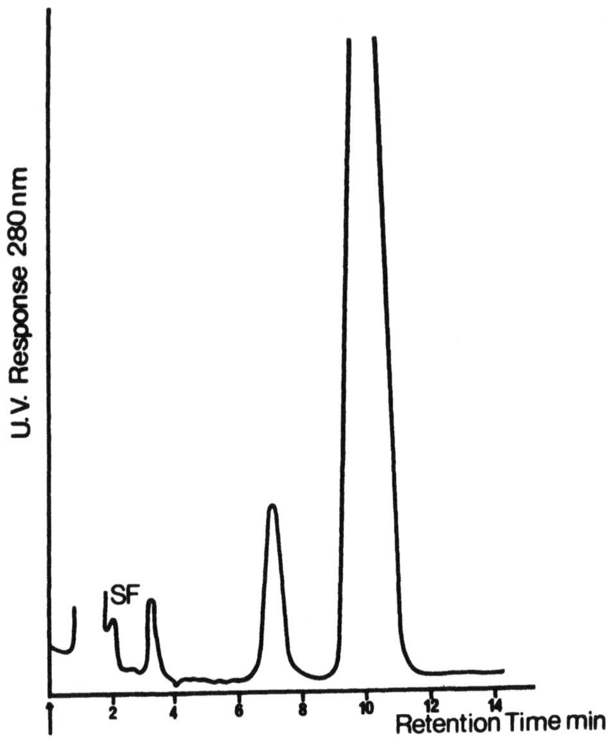

Fig. 2. High-performance liquid chromatographic analysis of biphenyl-carbonyl derivative of psychosine standards obtained by cerebroside hydrolysis. ↑ represents point of injection. Abbreviation: S.F.=solvent front.

Radioactive psychosines (erythro and threo 3-^{3}H psychosine 38.2 Ci/mole - 6-^{3}H psychosine 11.15 uCi/umol) were used in order to know the recovery of psychosine after the extraction procedure. The purity of the radioactive psychosine used was tested by TLC. Radioactivity scanning of a TLC plate spotted with labeled compounds showed only a single radioactive spot. The recovery obtained in two different experiments were as follows: 88-89% in the first lower phase, 68-72% in the second lower phase and 62-70% after alkaline hydrolysis. Radioactive psychosine spotted, scrapped off the plate and eluted, gave a recovery ranging between 88-97% in three different experiments.

Derivatives from brain tissue extracts were injected and two different peaks were eluted from the column with the same retention time as the standard. Each peak was collected separately, concentrated, and re-chromatographed. Fig. 3 show the elution pattern of each of the peaks obtained from brain.

Preliminary experiments of mass spectrometry of trimethylsilyl ether derivatives of standard psychosine show that the first peak obtained with a retention time of 7 minutes is dehydropsychosine and the second one with a retention time of 10 minutes is psychosine.

Further experiments should be done in order to know the mass spectrum of the two different peaks obtained from brain, to prove definitively the presence of a brain psychosine pool.

DISCUSSION

The pathways for the "in vivo" biosynthesis of cerebrosides that we can suggest are the following: either sphingosine in the presence of an acyl CoA, gives ceramide and this ceramide, in the presence of UDP galactose, cerebroside; or sphingosine forms first psychosine, which by acylation gives cerebroside. Due to the impossibility of isolating galactosyl sphingosine or ceramide from normal mammalian tissue, it is difficult to know the role of these postulated inter-

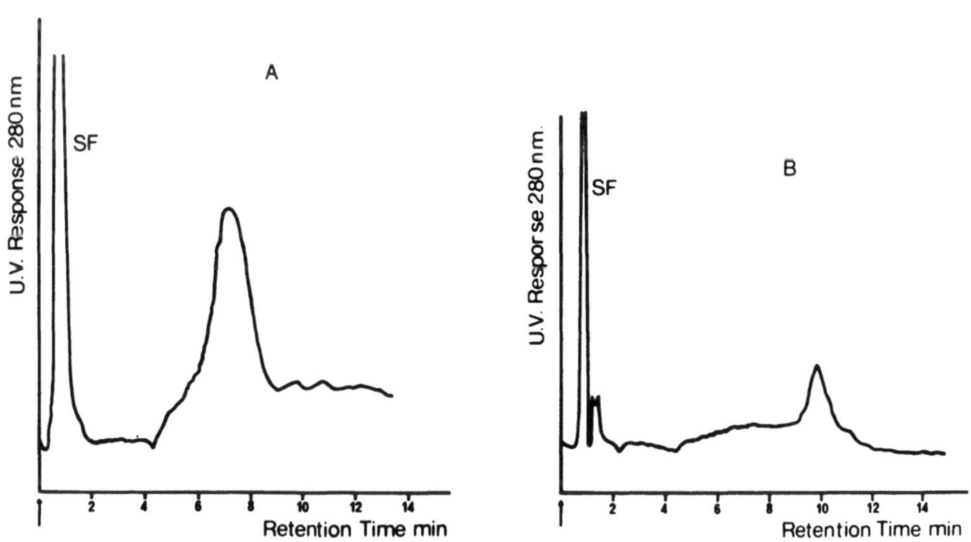

Fig. 3. High-performance liquid chromatographic analysis of biphenyl-carbonyl derivative of brain psychosine. Re-chromatography of the first (a) and second (b) peaks eluted from the column. ↑ represents point of injection. Abbreviation: S.F. = solvent front.

mediates in the biosynthesis of cerebrosides. Only when a genetic metabolic disorder exists, such as in Farber and Gaucher diseases, it was possible to isolate hydroxy fatty acids (HFA) ceramide (Sugita et al., 1973) and glycosyl sphingosine (Raghavan et al., 1973). When labelled ^{3}H sphingosine was injected intracranially (Kanfer & Gal, 1966) only trace amounts of radioactive cerebroside were isolated. Using ^{14}C stearoyl dehydrosphingosine, Kopaczyk & Radin (1965) obtained the same results. These data indicate that the possibility for ceramide to act as a precursor of cerebroside would be remote. Carter & Kanfer (1974) using L-^{14}C serine as precursor demonstrated that the radioactive compound was only incorporated into the non-hydroxy fatty acids (NFA) ceramide. This observation suggests that hydroxy fatty acids (HFA) ceramide may not be the precursor of HFA galactosylceramide. We can speculate that brain HFA cerebrosides may be synthesized by acylation of psychosine.

The fact that we can suggest tentatively the presence of a free psychosine pool in adult rat brain supports the possibility that cerebroside can be synthesized by the psychosine pathway.

CONCLUSIONS

Based on the sensitivity of high performance liquid chromatography and in the fact that compounds with primary amino groups can be converted into the U.V. absorbing biphenylcarbonyl derivatives we developed a method for the analysis of brain psychosine.

Psychosine was extracted from brain. The biphenylcarbonyl derivatives were separated by high performance liquid chromatography into two different peaks that were eluted with the same retention time as the psychosine standards, using dichloromethane:methanol:NH_3 85:8.5:1.5 by vol. as solvent. We assume tentatively that free psychosine is present in brain. According to these results we can propose a possible pathway of cerebroside synthesis by acylation of brain psychosine.

REFERENCES

Brady, R.O. (1962) J. Biol. Chem. 237, 2416-2417.

Carter, T.P. & Kanfer, J.N. (1974) J. Neurochem. 23, 589-594.

Cleland, W.W. & Kennedy, E.P. (1960) J. Biol. Chem. 235, 45-51.

Curtino, J.A. & Caputto, R. (1972) Lipids 7, 525-527.

Curtino, J.A. & Caputto, R. (1974) Biochem. Biophys Res. Commun. 56, 142-147.

Hammarstrom, S. (1971) Biochem. Biophys. Res. Commun. 45, 459-467.

Jungalwala, F.B.; Turel, R.J.; Evans, J.E. & McCluer,R.H. (1975) Biochem. J. 145, 517-526.

Kanfer, J.N. & Gal, A.E. (1966) Biochem. Biophys Res. Commun. 22, 442-446.

Kanfer, J.N. (1969) Lipids 4, 163-168.

Kopaczyk, K.C. & Radin, N. (1965) J. Lipid Res. 6, 140-145.

Radin, N.S. (1974) Lipids 9, 358-360.

Raghavan, S.; Mumford, R.A. & Kanfer, J.N. (1973) Biochem. Biophys. Res. Commun. 54, 256-263.

Sugita, M.; Connolly, P.; Dulaney, J.T. & Moser, H.W. (1973) Lipids 8, 401-406.

III. LIPIDS IN NEURAL TISSUE

(A) BIOSYNTHESIS AND TURNOVER

TRANSFORMATIONS OF THE POLYUNSATURATED FATTY ACIDS IN THE BRAIN

J.F. Mead, G.A. Dhopeshwarkar, and M. Gan Elepano

Laboratory of Nuclear Medicine and Radiation Biology, and Department of Biological Chemistry, School of Medicine, University of California, Los Angeles, California

A great deal of research from several laboratories, including ours, has given a picture, incomplete, to be sure, of the many reactions that combine to determine the composition of the brain lipids. These are outlined in Fig. 1, which provides an overview of the sources and metabolic transformations of the fatty acids that contribute their properties to the major phosphoglycerides and sphingolipids that form the chief structural units of the membranes that are involved in most of the properties of the brain.

It illustrates our finding that fatty acids derived from the diet or synthesized in other locations in the body are readily transported into the brain and rapidly incorporated into the various lipids (Dhopeshwarkar & Mead, 1973). Certain fatty acids can also be synthesized by systems within the brain itself and both the endogenous and exogenous fatty acids are subjected to alterations that render them more suitable for the particular lipids into which they will be incorporated. While these alterations can, and probably do, in part, take place in the liver (Sinclair, 1975), there is ample evidence to prove that all the necessary machinery for the reactions involved is present in the brain and that all types of fatty acid alterations can and do occur in that organ (Dhopeshwarkar & Subramanian, 1975a ; Dhopeshwarkar & Subramanian, 1976). The major uncertainty here is the extent to which the brain or other organs contribute to the final products.

The chief reactions involved are oxidation, desaturation and elongation. Although the brain is not well endowed with the enzymes of β-oxidation and probably derives little of its total energy from the process (Vignais, et al., 1958), it has been demonstrated repeatedly that fatty acids are subjected to partial or

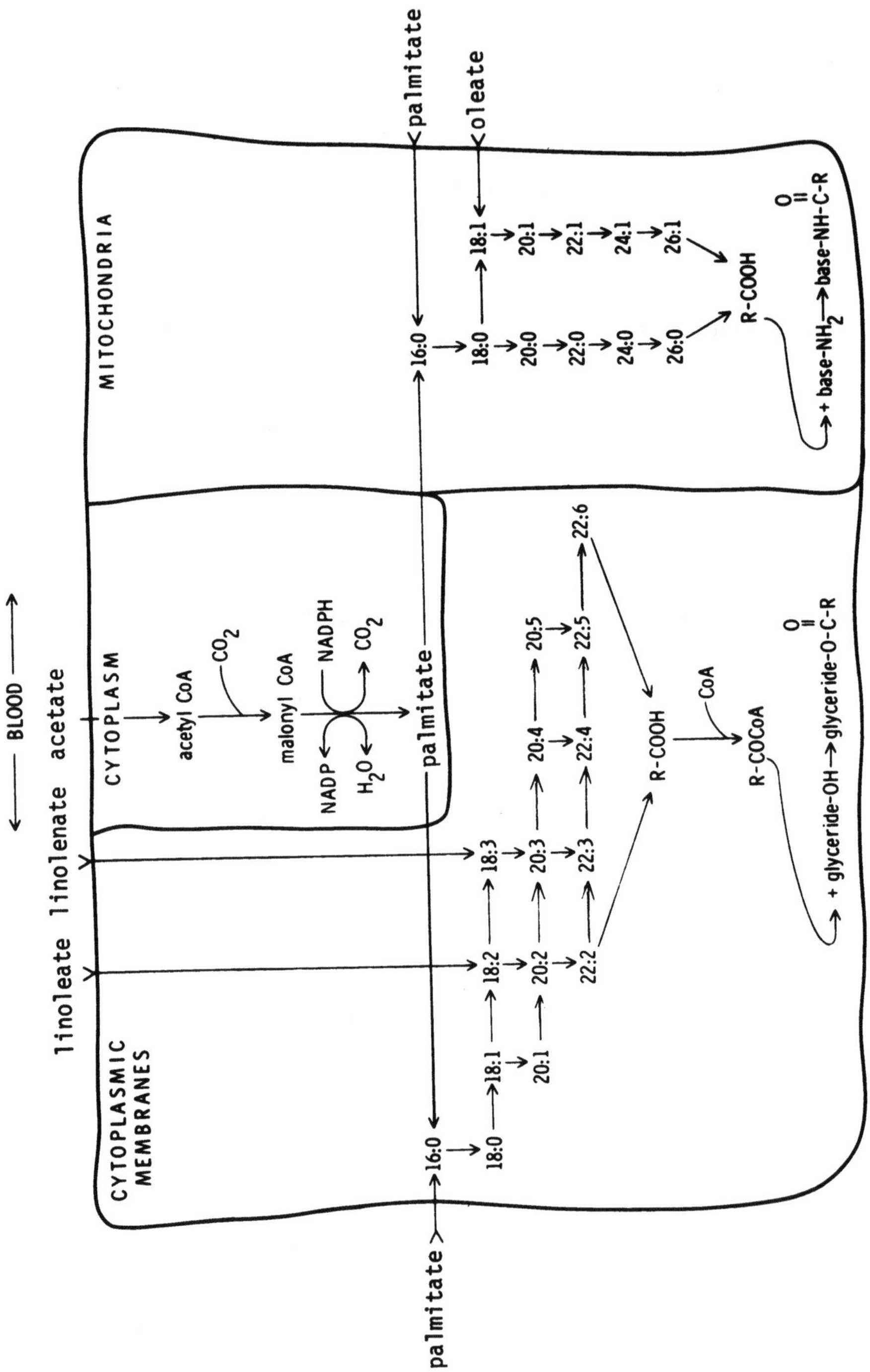

Fig. 1. Fatty acid biosynthesis in brain.

complete degradation in the brain, particularly in suckling animals (Warshaw & Terry, 1976), and that acetyl-CoA derived from such reactions is reutilized for synthesis of fatty acids and cholesterol (Dhopeshwarkar & Subramanian, 1975b).

The various desaturases are also manifestly present in the different cell types of the brain (Yavin & Menkes, 1974; Dhopeshwarkar & Subramanian, 1976). Evidence has been presented for the existence of the Δ5, Δ6 and Δ9 desaturases (Yavin & Menkes, 1974; Dhopeshwarkar & Subramanian, 1976a) and that, as in the case of the liver (Sprecher & Lee, 1975), a Δ8 desaturase appears to be lacking (Dhopeshwarkar & Subramanian, 1976b).

A reaction of special importance in the brain and one concerning which much confusion exits is that of chain elongation. It is through participation of this reaction in the tailoring of the fatty acids derived from the various sources that both the very long-chain saturated and monounsaturated fatty acids of the sphingolipids and the long-chain highly unsaturated fatty acids of the phosphoglycerides are formed. As a matter of fact, an outstanding characteristic of the brain fatty acids in general is their high average chain length. Polyunsaturated, monounsaturated and saturated fatty acids appear to be longer than those of most other tissues indicating that the elongation processes in the brain are either particularly active or are qualitatively different from those of other tissues. It is thus surprising that relatively little is known about the interplay of the brain elongation systems and that until very recently, most of our information for the brain was derived by inference from the liver.

In this latter organ, two fairly well-defined systems are operative (Fig. 1).

The mitochondrial system in the liver is active in the elongation of acyl-CoA derivatives with acetyl-CoA in the presence of NADH and NADPH. The intermediates and enzymes are similar to those of the mitochondrial β-oxidation system but the apparent rate-limiting reaction, NADPH enoyl-CoA reductase, is distinctive (Hinsch & Seubert, 1975).

Although it has been reported (Boone & Wakil, 1970) that this system can carry out the elongation to at least 24 carbons and that it is the sole fatty acid synthesizing system of the heart (Donaldson et al., 1970), several workers in the field have proposed alternative functions for it, such as the transfer of reducing equivalents or 2-carbon fragments into the mitochondria (Hinsch & Seubert, 1975).

The endoplasmic reticular system of the liver appears to be

involved importantly with polyunsaturated fatty acids in that it can effect both elongation and desaturation of their acyl-CoA derivatives. In this case, elongation requires malonyl CoA in addition to reduced pyridine nucleotide. In the presence of exogenous saturated fatty acids the elongation rate of saturated and monounsaturated acids is reduced and elongation of polyunsaturated fatty acids is emphasized (Christianson et al., 1967). The actual binding site is thought to be a pantetheine residue on the elongation enzyme (Podack et al., 1974) and this may be a site for competition.

The relationships between the two elongation systems in liver are not clear, although it has been suggested (Landriscina et al., 1972) that the microsomes are involved primarily with the elongation of 16:0 and 18:0 formed by the *de novo* system as well as in desaturation and elongation of the desaturated products, whereas the mitochondria may carry out the elongation of exogenous fatty acids as well as the transport of acetyl units and reducing equivalents. It may be significant that the microsomal system is under dietary control whereas the mitochondrial system seems to be unaffected by starvation (Donaldson et al., 1970).

In the brain, the lines of demarcation are more obscure. There is some evidence that the roles of the two systems are reversed from those in the liver in that the microsomes are involved with elongation of saturated acids (Goldberg et al., 1973; Bourre, et al., 1975) while the mitochondria are concerned with polyunsaturated acids (Yatsu & Moss, 1969; Aeberhard et al., 1969). In the brain, as in the liver, there is the suggestion that the endoplasmic reticular system may have the function of elongation of endogenously-produced saturated fatty acids, followed by their incorporation into myelin lipids. It is apparently the most active system for C_{16-18} fatty acid elongation in mixed brain homogenates (Cook & Spence, 1974), more active in the myelinating brain (Carey & Parkin, 1975) and, in microsomal preparations, is only slightly stimulated to increased incorporation of malonyl-CoA by added acetyl-CoA (Carey & Parkin, 1975). Thus, there appeared to be a need to determine the location of the active sites of fatty acid elongation in the brain and whether these are different for different fatty acids.

In our laboratory, several fatty acids of different degrees of unsaturation were incubated, under previously determined optimal conditions, with both microsomal and mitochondrial suspensions and the products and the kinetic parameters of their formation were studied. The fatty acids themselves or their CoA derivatives were used and the tracer was present either in the fatty acid or more usually, in acetyl-or malonyl-CoA.

When no fatty acid substrate was added, major radioactive

peaks were seen with radio-GLC only in 22:4 ω6 in mitochondria and in this and 24:4 ω6, 16:0 and 18:0 in microsomes (see Table 1). In both cases, addition of a substrate fatty acid resulted in the formation of an additional peak for a fatty acid two carbons longer than the substrate. The incorporation of radioactivity into products was linear for 15 min with mitochondria and for 30 min with microsomes. Several substrate fatty acids, in particular long-chain saturated or monounsaturated species, were poor substrates for either system.

Examination of the fatty acid composition of brain microsomal and mitochondrial lipids (Table 2) reveals that major fatty acids of both particles are 16:0, 18:1, 20:4 and 22:6. It is not too surprising, then that under the influence of enzyme specificity, endogenous 20:4 should be most readily elongated (to 22:4). How-

Table 1. Relative radioactivity (in percent of total) of fatty acids derived by incubation of brain mitochondrial or microsomal fractions with appropiate cofactors* and with 1-^{14}C acetyl CoA or 2-^{14}C malonyl CoA

Fatty Acid	Mitochondria	Microsomes
14:0	3.1	
16:0	3.5	36.8
18:0	11.8	20.5
18:1	6.3	
18:2	3.4	
20:0 or 18:3	9.6	
20:2	5.5	
22:4	50.4	23.7
24:4	6.4	19.1

* Incubations were carried out as follows: 30 mμ moles of acetyl CoA or 30 mμ moles malonyl CoA, 8 μ moles ATP, 1 μ mole NADPH, 1 μ mole NADH, 60 μ moles of potassium phosphate buffer, pH 6.5 and 1-3 mg mitochondrial or microsomal protein were incubated in a final vol of 0.5 ml, at 37°C. The microsomal malonyl CoA vials were incubated for a period of 30 min and the mitochondrial acetyl CoA vials for 15 min.

Table 2. Fatty acid composition of brain mitochondrial and microsomal fractions from 15-day-old rats (in percent of total)

Fatty Acid	Mitochondria	Microsomes
14:0	0.5	1.2
15:0	1.5	2.9
16:0	16.8	15.4
16:1	3.5	4.5
18:0	12.4	13.0
18:1	13.6	12.4
18:2	4.8	3.9
18:3	0.2	1.1
	0.3	0.5
20:4	17.8	14.5
	4.8	
22:4	4.8	6.7
		1.3
22:6	19.1	22.7

ever, it is not entirely apparent how a relatively minor constituent, as 22:4, could be a major elongation substrate in the microsomes. It is probable that enzyme specificity, substrate concentration and, substrate location or transport are the determining factors in elongation rate. As noted by Carey & Parkin (1975) for 16:0, it appears that endogenous fatty acids are more accesible to the elongation systems than are added substrates. In fact, it is possible that the substrate pool is the 2-position of a phospholipid (the major location of 20:4) and that the action of phospholipase A_2 and acyl CoA synthetase precede the reactions of chain elongation. In this case, exogenous fatty acid might be delayed in reaching the enzymes involved in elongation.

In an attempt to locate such a pool of substrate fatty acid, we followed the incorporation of radioactivity from malonyl-or acetyl-CoA into both lipids and fatty acids of microsomal and mitochondrial preparations. As can be seen in Fig. 2, there is a hint that with the mitochondria the elongation products of the added substrate (18:2) follow the neutral lipids while 22:4, the

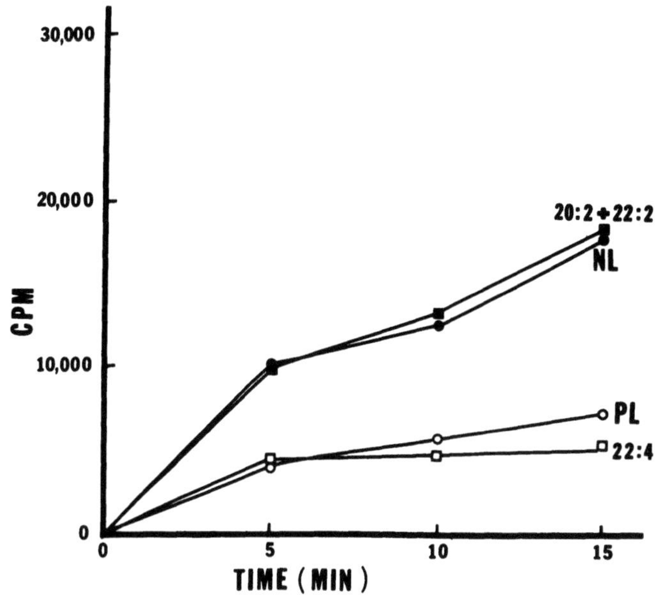

Fig. 2. Incorporation of 18:2 into mitochondrial fatty acids and lipids with time. NL, neutral lipids; PL, polar lipids.

elongation product of a major endogenous fatty acid (20:4), at least during the initial times, follows the phospholipid fraction. Of course, this does not prove that the added substrate is initially incorporated into a neutral lipid or that the differences in the handling of the endogenous and exogenous fatty acids are due to their incorporation into different lipids. Nevertheless, when coupled with information from other laboratories such as the report by Yavin & Menkes (1974) that the activity of fatty acids added to a culture (of brain cells) appears first in the neutral lipid fraction, these results indicate that the pools of endogenous and exogenous fatty acids in the brain may very well be the neutral and phospholipid fractions. A schematic suggestion of these ideas is presented in Fig. 3.

If this is true, it may be a difficult matter experimentally to test the brain elongation systems with added substrates. Nevertheless, kinetic studies have been made using the mitochondrial and microsomal elongation systems with several fatty acids. The K_m for both linoleic and linolenic acids with both microsomes and mitochondria was about 3×10^{-4} M while the V_{max} for both substrates was somewhat greater for mitochondria than for microsomes and greater for 18:3 than for 18:2 (Fig. 4). For palmitate, under the same conditions, K_m was 2×10^{-4} M and V_{max} was again somewhat

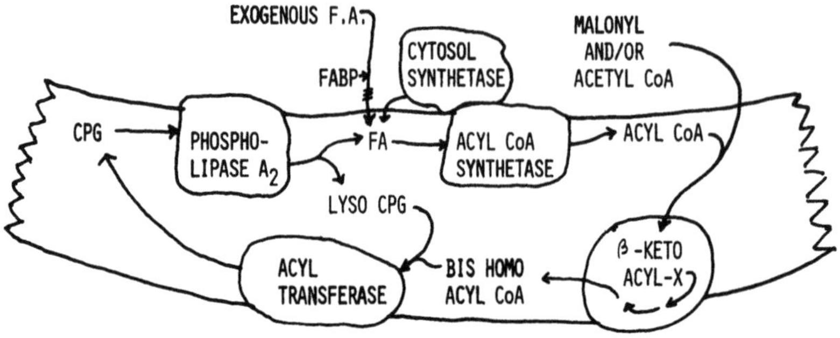

Fig. 3. Fatty acid elongation in a brain membrane system.

greater for mitochondria. It thus appeared that for elongation of a variety of fatty acids by both microsomes and mitochondria, a single enzyme was rate limiting and that rates were dependent on fatty acid structure. If this were true, it might be expected that there would be competition among the different fatty acids

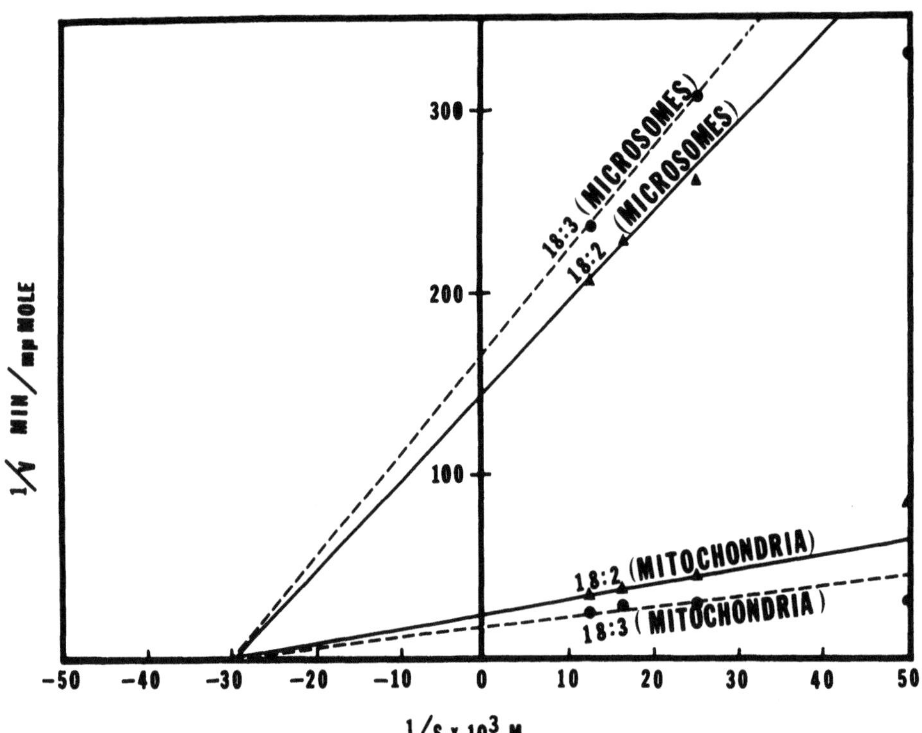

Fig. 4. Double reciprocal plots of the substrate dependence of the rate of elongation of 18:2ω6 and 18:3ω3 by microsomal and mitochondrial preparations. Conditions and components are described in a separate publication.

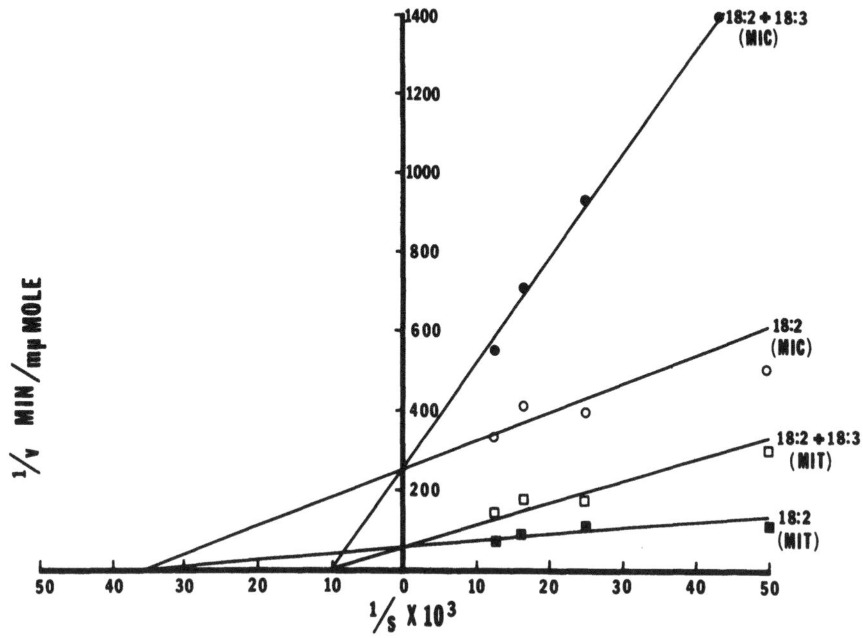

Fig. 5. Competitive inhibition of 18:2 and 18:3.

for the binding site of the enzyme. In Fig. 5, it can be seen that there is competitive inhibition between 18:2 and 18:3 but in Fig. 6, it is evident that 16:0 inhibits the elongation of both the unsaturated fatty acids non-competitively. Since the concentration

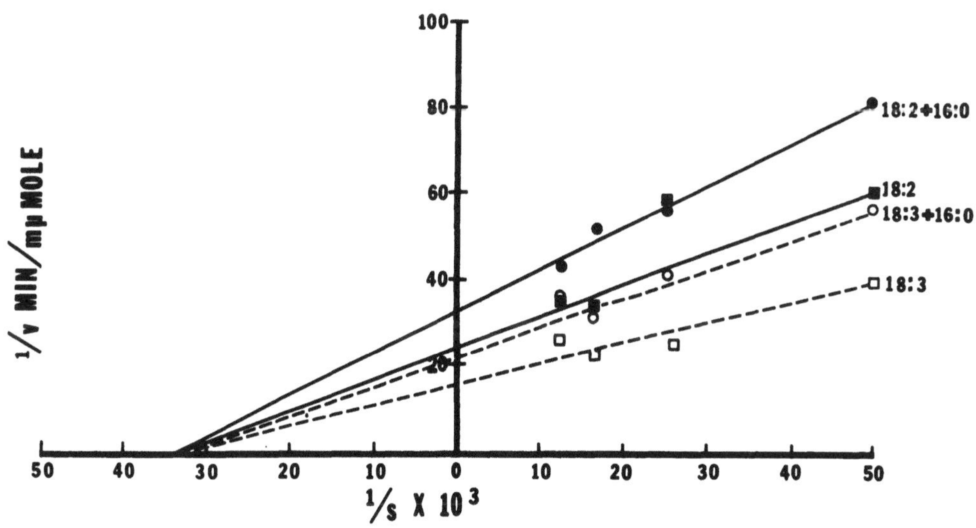

Fig. 6. Non-competitive inhibition of 18:2 or 18:3 whit 16:0.

of acyl CoA found by Carey & Parkin (1975) to be inhibitory was of the order of 2 x 10^{-6}M while the concentration of 16:0 used in these experiments was 3 x 10^{-8}M, the mechanism of this inhibition is not apparent. In any event, the results of these studies appeared to point to a common reaction in the elongation of all sustrates and, although the experiments of Cook & Spence (1974) appear to deny that the acyl CoA synthetase is rate limiting, these results agree with those of Carey & Parkin (1975) and Carey & Cantrill (1975) that this is indeed the common, and probably rate-limiting reaction.

This probability, though it aids in any consideration of the mechanism of fatty acid elongation in the brain, does not unequivocally determine the site of elongation of the different types of fatty acids. Clearly, the information presented so far implicates the mitochondria in the elongation of the polyunsaturated fatty acids since the maximum velocity of the rate-determining step is greatest in these particles. Other factors, however, such as relative amounts of enzyme protein and cofactors in the fractions, must be considered and rates of transport to the enzyme sites may be crucial. For additional information on this subject, the elongation of both palmitate and linoleate was carried out in a mixed preparation of mitochondria and microsomes and the rate of increase in the activity of elongation products was followed in both particles.

It can be seen in Fig. 7 that under our conditions the incorporation of label from added 18:2 in microsomes is an order of magnitude higher than that in mitochondria. Moreover, the formation of the saturated acids, 16:0 and 18:0, is, as would be expected, particularly low in mitochondria (Fig. 8) since a measurement of the percent RCA (relative carboxyl activity) for 14:0 and 16:0 proves them to be formed by a *de novo* process, and, in the presence of exogenous fatty acid, is almost completely inhibited (Fig. 7). It appears that *de novo* synthesis of saturated fatty acids in the endoplasmic reticulum is not strongly inhibited by fatty acids and acyl CoA's whereas the formation of these acids in the mitochondria, probably by elongation, is completely inhibited under these conditions. Thus the appearance of elongation products in the mitochondria in a mixed incubation may very well involve a transfer process. In order to test this probability, microsomes and mitochondria were incubated separately under optimal conditions with radioactive malonyl or acetyl CoA and in the presence or absence of linoleate. They were then washed free of precursors, and the unincubated particles were added, followed by incubation, separation, and fatty acid and lipid analysis. It can be seen in Tables 3 and 4 that there is very rapid exchange of elongation products in both directions despite the absence of the cytosol which presumably contains

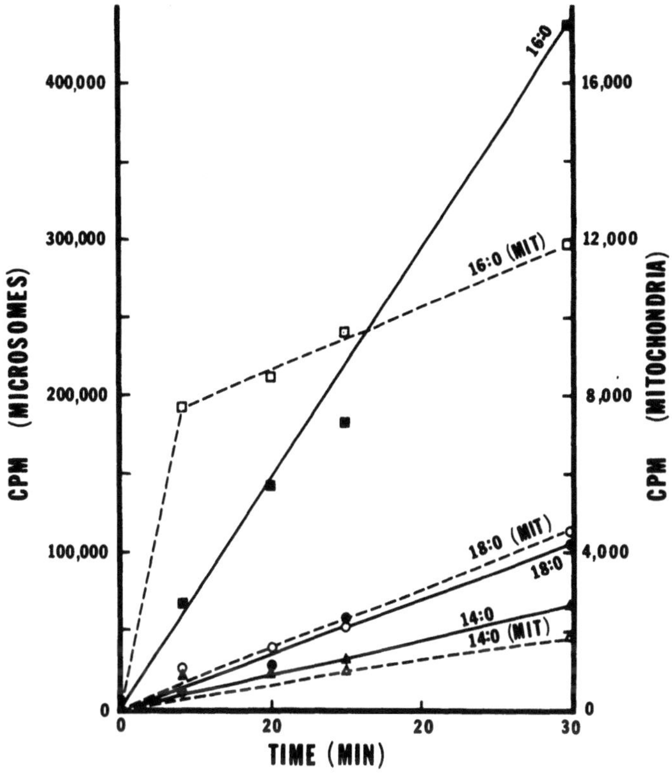

Fig. 7. Elongation of endogenous saturated fatty acids in microsomes plus mitochondria. The incubation mixture contained 30 mμmoles malonyl CoA, 10 mμmoles acetyl CoA, 8 μmoles ATP, 1 μmole NADH, 60 μmoles of potassium phosphate buffer, pH 6.5 and 1-3 mg of homogenate without the supernatant in 0.5 ml vol. 40 mμmoles of 18:2ω6 was added. After each incubation, the mitochondria were separated by layering the whole incubation mixture on a discontinous density gradient consisting of 0.8 and 1.2 M sucrose (3 ml of each/tube) in the SW 27 head of the Beckmann L-50 ultracentrifuge at 50,000 xg for 2 hors. The 2 upper layers were diluted to 0.25 M sucrose and the microsomes were recovered as the pellet when centrifuged at 1000 xg for 30 min.

the fatty acid binding proteins (FABP) (Ockner et al., 1972) and the phospholipid exchange proteins (PEP) (Wirtz et al., 1976). In order to gain some indication of the form in which the fatty acids are transferred between the particles, the lipids of each particle type were analyzed and counted following the combined incubation process (Table 5). From these data it appears that there is exchange of all major classes of lipids.

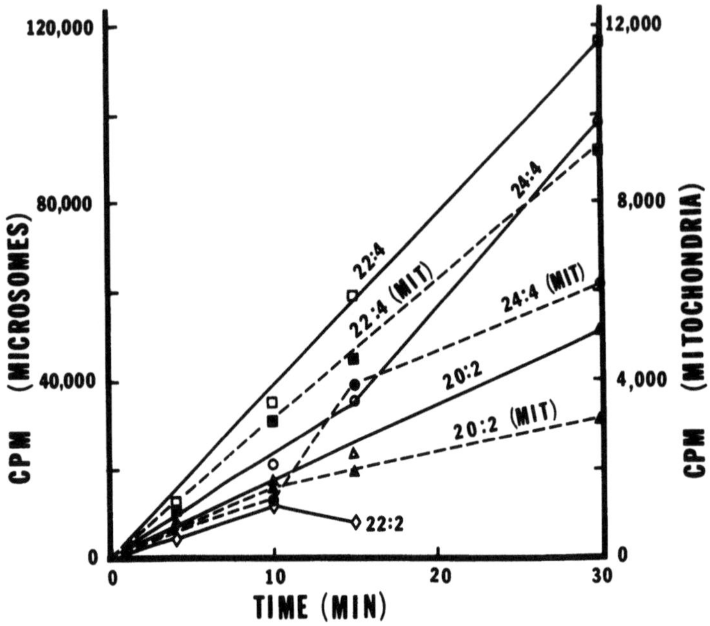

Fig. 8. Elongation of endogenous polyunsaturated fatty acids and added linoleate in microsomes plus mitochondria.

Table 3. Exchange of fatty acids from microsomes to mitochondria

Fatty Acids	Microsomes %	Mitochondria %
Control		
14:0	3.6	3.6
16:0	36.5	27.1
18:0	16.9	14.9
20:0 or 18:3		2.8
20:2		2.5
22:4	28.2	31.7
24:4	14.9	17.4
Plus 18:2		
14:0	4.6	4.2
16:0	29.0	25.3
18:0	7.8	7.6
20:2	11.2	10.8
22:4	25.8	31.5
24:4	21.6	20.7

Table 4. Exchange of fatty acids from mitochondria to microsomes

Fatty Acids	Mitochondria %	Microsomes %
Control		
14:0	5.5	4.7
16:0	7.9	5.6
18:0	11.8	10.6
18:1	8.7	10.3
20:0 or 18:3	17.1	19.0
20:2	17.8	8.5
22:4	31.2	41.3
Plus 18:2		
20:2	70.3	70.2
22:2	9.5	10.4
22:4	20.2	19.4

Table 5. Incorporation of radioactivity into lipids after 30 minutes incubation of mitochondria and microsomes

	Phospholipids	Free Fatty Acids	Triglycerides
Control microsomes	32.9	54.6	12.5
Microsomes + 18:2	20.3	69.7	10.0
Control mitochondria	14.6	76.4	9.0
Mitochondria + 18:2	20.3	70.2	9.5

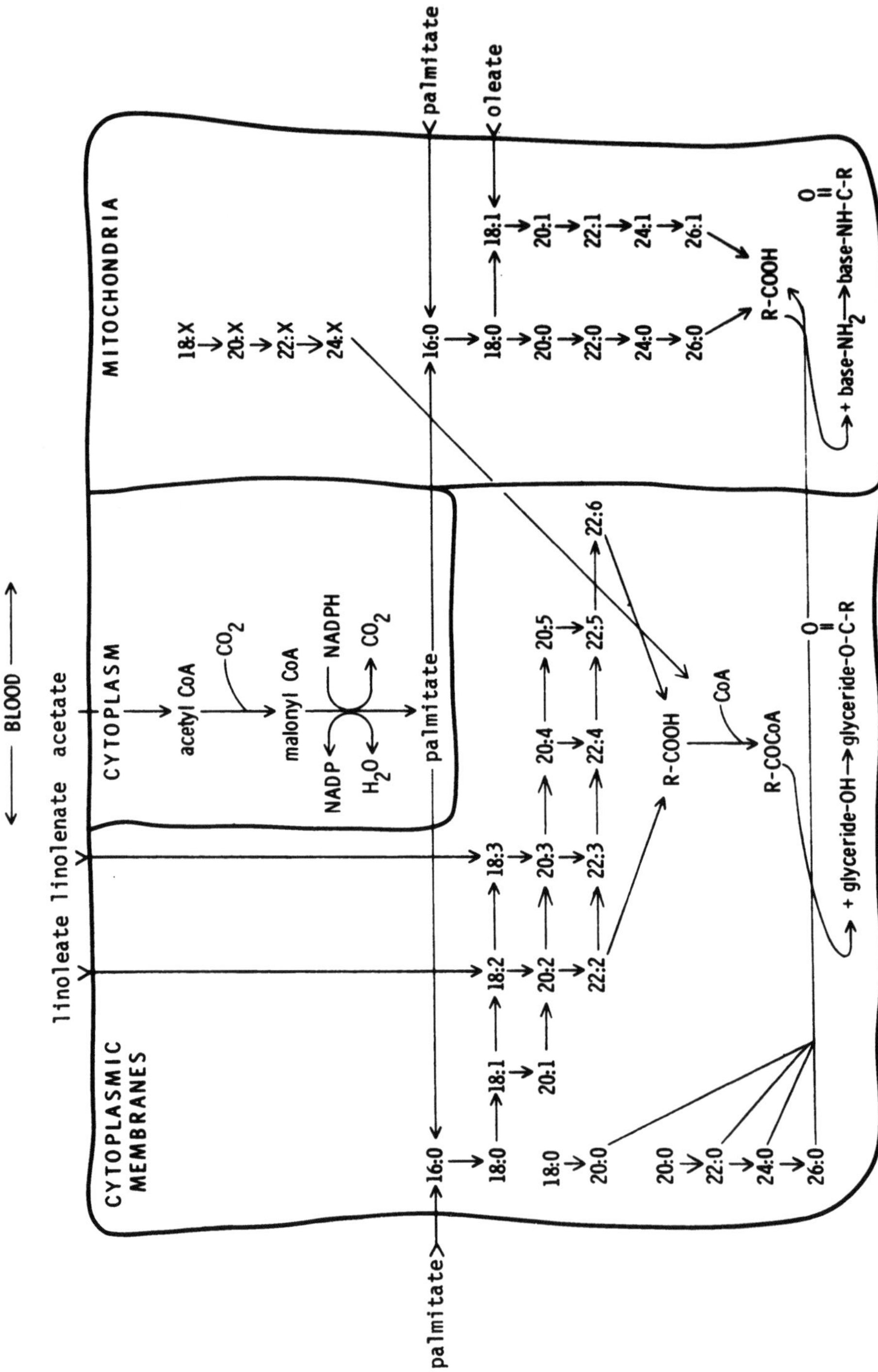

Fig. 9. Fatty acid biosynthesis in brain

As discussed above, conclusions reached under the conditions of these experiments may not be valid for the in vivo situation. However, they give an indication that for the brain (without consideration of the cell types involved) by far the most important site of elongation is the cytoplasmic membrane system and that the products formed in this system are transported rapidly to other membranes. Thus, Fig. 1, which was formulated by inference from the liver, can be refined to some extent, as is shown in Fig. 9 In it is also represented present thinking that the endoplasmic reticulum is also the site of myelin formation, presumably from the fatty acids and lipids formed at the same site.

CONCLUSIONS

The results of these studies have probably given us only hints of the very complex reactions concerned with fatty acid elongation in brain. On this basis we can conclude, with reservations, that the endogenous fatty acid pool is rapidly accesible to the elongation enzymes but that exogenous fatty acids do not readily equilibrate with it. Elongation is preceded by a common step in both microsomes and mitochondria, probably acyl CoA synthesis. Although the mitochondria are eminently capable of elongation, the major production of elongation products is probably in the endoplasmic reticulum in which the fatty acids are incorporated into the various lipids.

Finally, it cannot be claimed that we are even yet in a position of understanding all the reactions contributing to the production of myelin in the developing brain, but at least a start has been made and the painstaking work from many laboratories must ultimately achieve a clear picture of this vital process.

ACKNOWLEDGEMENTS

These studies were supported in part by Contract E(04-1)GEN-12 between ERDA and the University of California, and by U. S. Public Health Service Research Career Award No. GM-K-6-19,177 from the Division of General Medical Sciences, National Institutes of Health.

REFERENCES

AEBERHARD, E., GRIPPO, J. & MENKES, J. H. (1969) Pediat. Res. 3, 590-596.

BOONE, S. C. & WAKIL, S. J. (1970) Biochemistry 9, 1470-1479.

BOURRE, J. M., DAUDU, O. L. & BAUMANN, N. A. (1975) Biochem. Biophys. Res. Comm. 63, 1027-1034.

CAREY, E. M. & CANTRILL, R. C. (1975) J. Neurochem. 24, 807-809.
CAREY, E. M. & PARKIN, L. (1975) Biochim. Biophys. Acta 380, 176-189.
CHRISTIANSON, K., MARCEL, Y., GAN, M. V., MOHRHAUER, H. & HOLMAN, R. T. (1967) J. Biol. Chem. 243, 2969-2974.
COOK, H. W. & SPENCE, M. W. (1974) Biochim. Biophys. Acta 369, 129-141.
DHOPESHWARKAR, G. A. & MEAD, J. F. (1973) in Adv. in Lipid Res., Vol. 11, (Paoletti, R. and Kritchevsky, D., eds.), Academic Press, New York, pp. 109-142.
DHOPESHWARKAR, G. A. & SUBRAMANIAN, C. (1975a) Lipids 10, 238-241
DHOPESHWARKAR, G. A. & SUBRAMANIAN, C. (1975b) Lipids 10, 242-247.
DHOPESHWARKAR, G. A. & SUBRAMANIAN, C. (1976a) Lipids 11, 67-71.
DHOPESHWARKAR, G. A. & SUBRAMANIAN, C. (1976b) J. Neurochem. 26, 1175-1179.
DONALDSON, W. E., WITPEETERS, E. W. & SCHOLTE, H. R. (1970) J. Biol. Chem. 226, 497-509.
GOLDBERG, I., SCHECHTER, I.& BLOCH, K. (1973) Science 182, 497-499.
HINSCH, W. & SEUBERT, W. (1975) Eur. J. Biochem. 53, 437-447.
LANDRISCINA, C., GNONI, G. V. & QUAGLIARIELLO, E. (1972) Eur. J. Biochem. 29, 188-196.
NUGTEREN, D. H. (1965) Biochim. Biophys. Acta 106, 280-290.
OCKNER, R. K., MANNING, J. A., POPPENHAUSEN, R. B. & HO, W. K. L. (1972) Science 177, 56-58.
PODACK, F. R., SAATHOFF, G. & SEUBERT, W. (1974) Eur. J. Biochem. 50, 237-243.
SINCLAIR, A. J. (1975) Lipids 10, 175-184.
SPRECHER, H. & LEE, C. J. (1975) Biochim. Biophys. Acta 388, 113-125.
VIGNAIS, P. M., GALLAGHER, C. H. & ZABIN, I. (1958) J. Neurochem. 2, 283-287.
WARSHAW, J. B. & TERRY, M. L. (1976) Developmental Biology 52, 161-166.
WIRTZ, K. W. A., JOLLES, J., WESTERMAN, J. & NEYS, F. (1976) Nature 260, 354-355.
YATSU, F. M. & MOSS, S (1969) J. Neurochem. 18, 1895-1901.
YAVIN, E. & MENKES, J. H. (1974) J. Lipid Res. 15, 152-157.

THE SIGNIFICANCE OF BASE-EXCHANGE REACTIONS IN LIPID METABOLISM

G. Porcellati, M. Brunetti, G.E. De Medio,
E. Francescangeli, A. Gaiti, and G. Trovarelli

Istituto di Biochimica, Università di Perugia
Perugia, Italy

INTRODUCTION

In recent years a number of investigators have studied, *in vitro*, the Ca^{2+}-stimulated, non-energy-dependent, base-exchange incorporation of ethanolamine, choline and L-serine into corresponding phospholipids (Arienti et al., 1976). The enzymic system, which has been extensively studied in nervous structures (Arienti et al., 1970; Porcellati et al., 1971; Kanfer, 1972; Miller & Dawson, 1972; Goracci et al., 1973; Gaiti et al., 1974), produces the non net synthesis of membrane phospholipids, and is thought to play an interesting role in changing the structure of the polar head groups of membrane lipids (Gaiti et al., 1974). Kinetic data (Porcellati et al., 1971; Gaiti et al., 1974; Bjerve, 1973; Saito et al., 1975), Ca^{2+}requirements (Arienti et al., 1970; Porcellati et al., 1971; Gaiti et al.,1974; Saito et al., 1975) and heat sensitivity measurements (Saito et al., 1975) strongly support the view that more than one enzyme is involved in the exchange reactions in brain (Porcellati et al., 1971; Gaiti et al., 1974; Saito et al., 1975), as proposed also for the liver system (Bjerve, 1973). Very recently, Miura & Kanfer (1976) have brilliantly succeeded in obtaining separate soluble fractions from rat brain which catalyze principally ethanolamine, choline or L-serine incorporation by base-exchange.

Previous reports from this laboratory (Arienti et al., 1976) have already dealt with the nature of the base-exchange reaction and of the influencing parameters, with the cellular compartmentation of the system, and with the nature of the lipid acceptors in the exchange mechanism. The present report deals with additional information on the properties and physiological significance of the system.

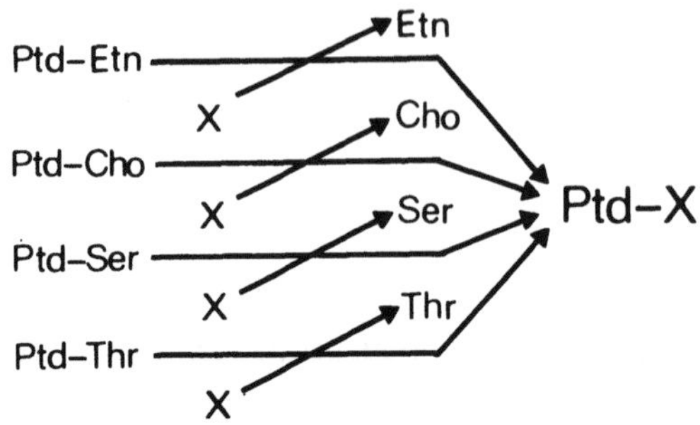

Fig. 1. Possible theoretical exchanges of free nitrogenous bases (X) with membrane-bound endogenous phospholipids (Ptd-Etn, Ptd-Cho, Ptd-Ser, Ptd-Thr). See the text, for abbreviations; X = ethanolamine, choline or L-serine

MEMBRANE POOLS IN THE EXCHANGE REACTION

Little information has been given in the past about the types of phospholipid substrates (Porcellati & di Jeso, 1971; Saito & Kanfer, 1973) that participate in the exchange reaction in brain tissue, although it had been already shown in 1971 that brain microsomal phosphatidyl ethanolamine (Ptd-Etn) is able to exchange with both ethanolamine and serine (Porcellati et al., 1971). As shown in Fig. 1, any of the nitrogenous bases under study, i.e., ethanolamine, choline or L-serine, could theoretically exchange with Ptd-Etn, phosphatidyl choline (Ptd-Cho), phosphatidyl serine (Ptd-Ser) and phosphatidyl threonine (Ptd-Thr), and therefore it would be of extreme interest to examine by experimental approach whether only homologous (for example, ethanolamine with Ptd-Etn) or also heterologous (for example, choline with Ptd-Etn or serine with Ptd-Cho) types of exchange may take place at the membrane level in brain. Fig. 1 takes into account also Ptd-Thr, as an exchangeable phospholipid, not only because of the occurrence of this lipid in animal tissues (Katada et al., 1959) but also because free labelled threonine has been found to be incorporated into brain lipid material by base-exchange (Porcellati et al., 1966). Moreover, monomethyl aminoethanol and dimethyl aminoethanol are also incorporated into the same material by similar processes (Porcellati, 1972).

Fig. 2 shows that brain microsomal Ptd-Cho can act as a membrane-bound substrate for the incorporation of either ethanolamine, L-serine or choline, although with different levels of magnitude. In separate experiments (not shown in the figure) it was also observed that the released tritiated choline, which was displaced by exchange, was quantitatively estimated in the aqueous phase of the incubation mixture.

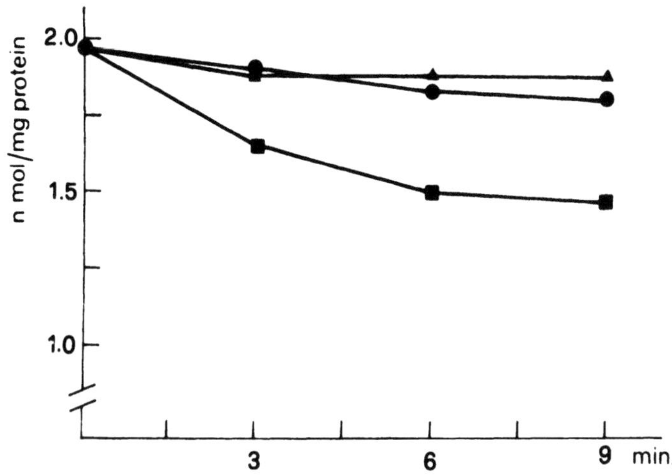

Fig. 2. Displacement of tritiated choline from rat brain microsomal labelled Ptd-Cho by the addition of |^{14}C| labelled ethanolamine (■-■-■), L-serine (●-●-●) and choline (▲-▲-▲). The loss of radioactivity of microsomal Ptd-Cho, expressed as nmol/mg protein, is plotted against minutes of reincubation. Estimations done in quadruplicate. The Ptd-Cho pre-labelling of membranes was done by base-exchange with 3.3 mM |Me-^{3}H| choline for 15 min, followed by washing procedures (Gaiti et al., 1975) and reincubation with either |^{14}C| labelled ethanolamine, L-serine or choline in an exchange medium, as indicated. The experimental data were all corrected for the release of lipid radioactivity due to water alone (*i.e.*, in the absence of any added |^{14}C| base in the reincubation experiments).

Fig. 2 indicates that 25%, 9% and 5% of the tritiated choline in the labelled Ptd-Cho were displaced with |^{14}C| labelled ethanolamine, L-serine and choline, respectively, after 9 min of reincubation in the exchange medium. Now, by estimating the levels of incorporation of the three |^{14}C| labelled bases under the same experimental conditions used in the base-displacement reactions, only 4.6, 2.0 and 0.44 nmol/mg protein/9 min of ethanolamine, L-serine and choline, respectively, were found to be incorporated, and these data, together with those on the percentages of displacement calculated from Fig. 2, indicate that the total available pool of Ptd-Cho for all types of exchange was about 18-20 nmol (8 nmol in the exchange of choline), which represents only a small amount (3-6%) of the total Ptd-Cho (300 nmol) of 1 mg of brain microsomal protein (Biran & Bartley, 1961; Cuzner et al., 1965). It is clear therefore that only a small pool (about 5%) of the total Ptd-Cho is able to act as a substrate for base-exchange reactions, the value of this pool size being strikingly similar to that already found (Gaiti et al., 1975) for exchanging Ptd-Etn molecules (5-6%) in previous

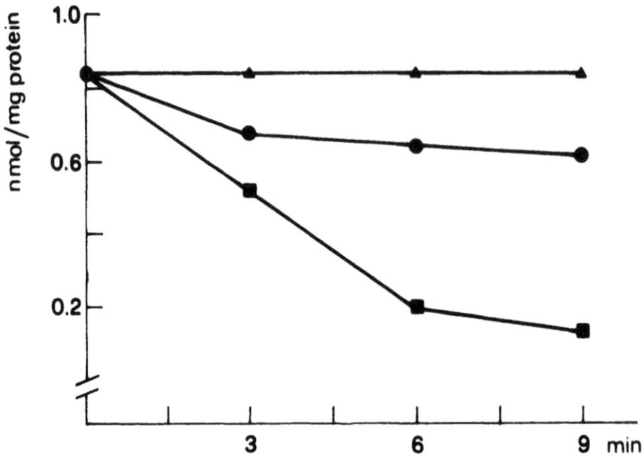

Fig. 3. Displacement of tritiated L-serine from rat brain microsomal labelled Ptd-Ser by the addition of $|^{14}C|$ labelled ethanolamine (■-■-■), L-serine (●-●-●) and choline (▲-▲-▲). The Ptd-Ser pre-labelling of membranes was carried out by base-exchange with 1.95 mM $|3\text{-}^{3}H|$ L-serine for 15 min. For other details, see Fig. 2.

experiments. Incidentally, on this occasion (Gaiti et al., 1975) it was also shown that brain microsomal Ptd-Etn can act as a membrane-bound substrate for the incorporation of either ethanolamine, choline or L-serine.

Fig. 3 shows that after 9 min of incubation, ethanolamine and L-serine displace by 82% and 27%, respectively, the microsomal lipid-serine synthesized by base-exchange in brain, being simultaneously reincorporated during the same intervals of the displacement reaction at a rate of 4.4 and 1.5 nmol/mg protein/9 min, respectively. This result again indicates, as reported also in Fig. 2, that the total pool size of Ptd-Ser available for the exchange with both ethanolamine and L-serine is small (about 5 nmol), i.e., only 6-7% of the total Ptd-Ser (80 nmol) present in 1 mg of microsomal protein (De Medio et al., 1973). These data agree well with those reported for the "active" pools of Ptd-Etn (Gaiti et al., 1975) and Ptd-Cho (Fig. 2), and indicate that small but highly active pools of these lipids are involved in the exchange reactions with free bases in brain. Interestingly, Fig. 3 indicates that choline is not able to displace labelled L-serine from Ptd-Ser, and that therefore this base cannot exchange with membrane-bound Ptd-Ser.

In other experiments, brain microsomal Ptd-Etn or Ptd-Cho were pre-labelled *in vitro* by incubation with CDP-ethanolamine or CDP-choline, respectively, before performing the base-displacement reaction. Table 1 (experiments with Ptd-Etn) indicates that the

Table 1.- The Ca^{2+}-stimulated displacement of rat brain microsomal phosphatidylethanolamine synthesized *in vitro* by the CDP-ethanolamine pathway

Data are reported in nCi/mg protein. Estimations in quadruplicate. Incubation with $|^{14}C|$ labelled CDP-ethanolamine was carried out as reported elsewhere (Porcellati et al., 1970), prior to the reincubation with either unlabelled choline, L-serine or ethanolamine. For other details, see Fig. 2.

Addition	nCi of lipid-bound ethanolamine displaced at the following time intervals of reincubation (min)*		
	3	6	9
Choline	0	0	0
L-Serine	0.12	0	0
Ethanolamine	0.08	0.34	0.56

* nCi of pre-labelled phosphatidylethanolamine at zero time : 12.51 per mg of protein. Correction has been made for release of radio activity by water alone.

displacement by ethanolamine of labelled Ptd-Etn is much lower (4.5%) than that obtained by the same base (41%) on Ptd-Etn synthesized by base-exchange (Gaiti et al., 1975). Table 1 shows in addition that no appreciable displacement by L-serine or choline takes place, whereas these two bases were able to displace respectively by 20% and 10% the Ptd-Etn synthesized by base-exchange (Gaiti et al., 1975). This finding again indicates that only a small pool of the total Ptd-Etn of brain microsomes, of high specific activity, is available to the exchange enzyme; therefore, when the Ptd-Etn molecules are more homogeneously labelled by the Kennedy pathway with CDP-ethanolamine, in place of the exchange reaction, the high degree of base displacement due to the exchange enzyme is no longer observed.

Rather similar data are shown in Table 2, which refers to displacement experiments carried out on microsomal Ptd-Cho pre-labelled *in vitro* from CDP-choline through the Kennedy pathway. Indeed, after 9 min of incubation, L-serine and ethanolamine displace the lipid-choline moiety by only 6 % and 4 %, respectively, and choline by only 2%. These data differ essentially from those described in Fig. 2, which reported the percentages of displacement rates of lipid-choline pre-labelled by base-exchange, and which

Table 2.- Displacement by base-exchange of rat brain phosphatidylcholine synthesized *in vitro* by the CDP-choline-mediated pathway.

Data are reported in nmol/mg protein. Incubation with $|^{14}C|$ labelled CDP-choline was carried out as reported elsewhere (Binaglia et al., 1973), prior to the reincubation with either choline, L-serine or ethanolamine. For other details, see Fig. 2.

Addition	nmol of lipid-bound choline displaced at the indicated min of reincubation*		
	3	6	9
Choline	0.08	0.09	0.10
L-Serine	0.32	0.34	0.32
Ethanolamine	0.22	0.20	0.20

* nmol of pre-labelled phosphatidylcholine at zero time : 5.33 per mg of protein (estimations in quadruplicate). Correction of data has been made for the release of radioactivity by water alone.

were 25%, 9% and 5% for ethanolamine, L-serine and choline, respectively.

The physiological significance of lipid pools at the membrane level is at the moment not well understood. A possibility is that base-exchange reactions are important for the functioning of nervous membranes and that base-exchange is connected to some particular functions. The next sections will try to answer these questions at least partially.

THE INFLUENCE OF BASE-EXCHANGE ON GABA TRANSPORT

It is known from previous work that brain cell plasma membranes are enriched in the base-exchange enzymic system (Goracci et al., 1973). Neuronal perikarya have been found to possess a much higher rate of exchange for both ethanolamine and L-serine than glial cells or synaptosomes (Goracci et al., 1973). In agreement with the results from chick brain (Porcellati et al., 1971), the enzyme activity has been found to be very high in membrane fractions from neuronal perikarya (Goracci et al., 1973).

Little is known, on the other hand, about the functional importance of the base-exchange reaction. Owing to the previous results (Goracci et al., 1973), a study was undertaken therefore to investigate possible effects of the base-exchange system on amino acid transport properties of neuronal membranes. In view of its central position as an inhibitor and a neurotransmitter candidate and of its high-affinity uptake system in the brain and high tissue to medium ratios obtained (Sellström & Hamberger, 1975), γ-amino-*n*-butyric acid (GABA) was chosen to monitor the amino acid uptake. The net GABA uptakes into neuronal cell bodies and glial cells, prepared according to Hamberger et al. (1975), and into synaptosomes, obtained according to Cotman & Mattews (1971), were therefore compared, and the effect of base-exchange on the amino compound uptake examined.

Preliminary experiments (De Medio et al., 1976) have already shown that the preincubation of a crude rabbit brain suspension (Hamberger et al., 1975) with increasing concentrations of ethanolamine in a suitable base-exchange medium (Gaiti et al., 1974), followed by the measurement of GABA transport in a typical uptake medium (Sellström & Hamberger, 1975), causes a significant inhibition of the GABA uptake at 0.2 mM ethanolamine. Upon a further increase of ethanolamine concentration to 2.5 mM, GABA uptake increases by 25% over the control levels, while uptake inhibition is again observed for ethanolamine concentrations over 3 mM. In contrast to this, the incorporation of L-serine by exchange induces an increase of GABA uptake throughout all the range of base concentration used, while the incorporation of choline leads essentially to an inhibition of the process, as was observed after an exchange with ethanolamine.

Fig. 4 shows a slight activation of GABA uptake in rabbit brain synaptosomes (Cotman & Mattews, 1971) after a base-exchange reaction with 0.2 mM L-serine, followed by a very slight inhibition of the uptake or no action at higher base concentration. It must be added, in connection with these results, that a peak of increased GABA uptake of about 15-20% over the control values was observed at 2.5 mM serine (not shown in the figure). Neuronal cell bodies and glial cells behaved very similarly to synaptosomes, although the stimulation of GABA uptake by the serine exchange was considerably higher in neurons than in glia and synaptosomes. When synaptosomes were incubated, without prior incubation for base exchange, with 5 x 10^{-7}M $|^3H|$-GABA in the uptake medium (Sellström & Hamberger, 1975) in the presence of various concentrations of L-serine, the GABA uptake was inhibited by more than 50% at all concentrations tested, starting with 0.2 mM L-serine. This result indicates that the mere presence of various levels of serine in the cells does not cause the effects on GABA uptake shown in Fig. 4, following incubation for base-exchange.

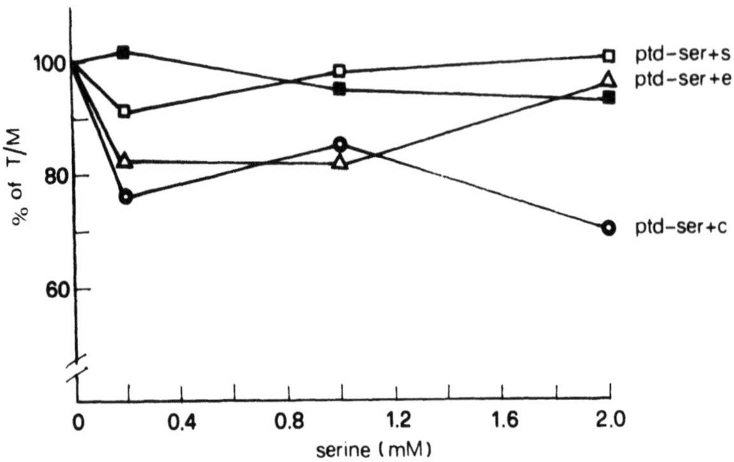

Fig. 4. GABA uptake in synaptosomal fractions of rabbit brain, as a function of base-exchange reaction with increasing concentrations of $|^{14}C|$ L-serine. GABA uptake (% of T/M, see Sellström & Hamberger, 1975) on the left is plotted against L-serine concentration (mM). Experiments were performed by first preincubating for 15 min at 37°C synaptosomes with different amounts of $|^{14}C|$-serine in a base-exchange medium (Gaiti et al., 1974), and then washing them and incubating in a GABA uptake medium (Sellström & Hamberger, 1975) for 10 min at 37°C (■-■-■). In other experiments, the pre-incubation with serine was followed by a displacement reaction (see previous section) of the synaptosomal Ptd-Etn for few min with either unlabelled L-serine (□-□-□), ethanolamine (Δ-Δ-Δ) or choline (o-o-o), prior to GABA uptake measurements. The value of starting control (% of T/M = 100) is taken as the level of GABA uptake obtained by pre-incubating the synaptosomes for 15 min in the absence of any added L-serine or by "displacing" the synaptosomal Ptd-Ser without any added unlabelled base. Each point is the mean of five experiments.

When the synaptosomal Ptd-Ser, produced by base-exchange, was displaced by the three unlabelled bases (ethanolamine, L-serine or choline), as shown in Fig. 4, a general inhibitory effect on GABA uptake was produced, which was well evident when choline or ethanolamine were used as base-displacers. The homologous-type of base-displacement (i.e.,L-serine with Ptd-Ser) produced only a light modification of the basic uptake (Fig. 4).

Fig. 5 shows that the synaptosomal GABA uptake was invariably inhibited after ethanolamine incorporation by exchange, and this result was duplicated in similar experiments performed with neurons and glia (not reported in this work). However, when synaptosomes

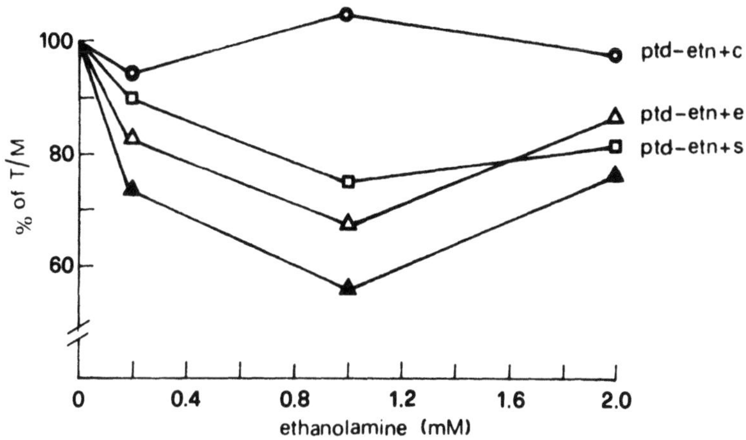

Fig. 5. GABA uptake in synaptosomal fractions of rabbit brain, as a function of base-exchange reaction with increasing |^{14}C|-ethanolamine concentrations. Base-exchange with ethanolamine (▲- ▲- ▲); base-exchange with ethanolamine, followed by displacement of membrane Ptd-Etn by unlabelled ethanolamine (Δ- Δ- Δ), L-serine (□ - □- □) or choline (o-o - o). See Fig. 4 for additional information.

were re-incubated, after the base-exchange, with displacing concentrations of unlabelled choline, L-serine or ethanolamine in order to observe the effect of the displacement by different bases on GABA transport, stimulatory values of uptake were always obtained, as compared to the basic data. It appears therefore that producing additional Ptd-Ser molecules at the expense of Ptd-Etn brings about an activation of the uptake process, and this result is even more pronounced when more Ptd-Cho molecules replace by exchange the Ptd-Etn on the membrane. The homologous type of base-displacement (i.e., ethanolamine with Ptd-Etn) produced less modification of the basic uptake (Fig. 5).

Fig. 6 indicates that choline incorporation induced essentially varying degrees of inhibited GABA uptake in synaptosomes and that the displacement of the Ptd-Cho molecules by external ethanolamine produced an even higher inhibitory effect, while L-serine and choline caused intermediate changes.

SYNAPTOSOMAL POOLS IN THE BASE-EXCHANGE REACTION

To what extent the patterns of variation in GABA uptake, reported in Figs. 4-6, are due to effects of the base-exchange system

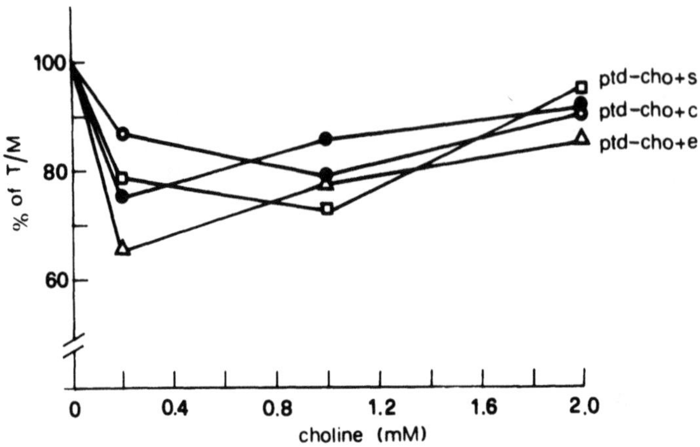

Fig. 6. GABA uptake in synaptosomal fractions of rabbit brain, as a function of base-exchange reaction with increasing $|^{14}C|$-choline concentrations. Base-exchange with choline (● - ● - ●) ; base-exchange with choline, followed by displacement of membrane Ptd-Cho by unlabelled choline (o - o - o), ethanolamine (Δ- Δ - Δ) or L-serine (□ - □ - □). See Fig. 4 for additional information.

on membrane phospholipid structure presumably involved in GABA transport phenomena is hard to establish. It might be possible that, due to the base-exchange, the incorporated as well as the displacing bases may affect, at different structure levels, a possibly small pool of total membrane phospholipids which might be involved in GABA transport, the variation of which might successively change the extent of the uptake. We examined, therefore, the pool size of synaptosomal phospholipid available for exchange, taking into account previous (Gaiti et al., 1975) and present (see first paragraph of this chapter) results, which indicate that only a small pool of membrane phospholipid is active in the exchange reaction with the bases in rat brain microsomes.

These studies were carried out by pre-labelling *in vitro* the synaptosomes from rabbit brain in their Ptd-Etn, Ptd-Ser or Ptd-Cho molecules with $|^{3}H|$ labelled bases in a base-exchange medium (Gaiti et al., 1974) and by chasing successively the membrane-bound radioactivity with $|^{14}C|$ labelled ethanolamine, L-serine or choline, at different time intervals. The degree of displacement of the lipid-bound base moiety of the pre-labelled membrane (which was labelled with the first isotope) due to the addition of a base labelled with a second, different isotope was determined as the loss of the first isotope's radioactivity from the isolated phospholipid. The incorporation of the displacing base through base-exchange was followed under the same conditions in which the displacement of the bases was normally examined, by estimating the increase of radioactivity content of the second isotope in the isolated phospholipid.

Table 3.- Pools of exchanging phosphatidyl-molecules in rabbit brain synaptosomes*

* The pool is expressed as nmol phospholipid/mg protein. The per cent of the total membrane phospholipid content is shown between brackets. Total phospholipid content (Porcellati & Goracci, 1976): Ptd-Etn (220 nmol/mg protein), Ptd-Ser (55nmol), Ptd-Cho (230 nmol). Mean values of four experiments.

Displacing base	Pre-labelled phospholipid		
	Ptd-Etn	Ptd-Ser	Ptd-Cho
Ethanolamine	6.0 (2.7)	6.2 (11.3)	2.1 (1.0)
L-Serine	2.5 (1.1)	5.5 (10.0)	0.50 (0.2)
Choline	0.5 (0.23)	0.83 (1.5)	0.20 (0.1)

It is apparent from Table 3 that only small pools of total membrane phospholipid participate in the exchange reaction in synaptosomes. On using |^{14}C|-ethanolamine as a displacing base, the smallest pool was active when the displaced lipid was Ptd-Cho and the highest when the lipid was Ptd-Ser. Similar data were obtained (Table 3) when the displacing base was |^{14}C|-L-serine or |^{14}C|-choline, although absolute values of the pools were different. Similarly, in the displacement of Ptd-Etn the magnitude of the pool decreased on using ethanolamine, L-serine and choline in this order, the same being true for the other lipids also.

From these data it is concluded that only selected, small pools of synaptosomal phospholipid are available for exchange at membrane level. The size of these pools never exceeds 10% of the total lipid and more often is even less than 0.2%. The involvement of these small pools was at play, of course, also during the experiments on variation of GABA uptake depicted in Figs. 4-6. Now, is there any possible relationship between these pools of base-exchange, their magnitude and the change of GABA uptake caused by the displacement?

It is tempting to speculate whether a large variation in GABA uptake is due to variation of a small pool of exchanging phospholipid or not. The answer is doubtful. This possibility seems in fact to apply when Ptd-Etn is displaced by ethanolamine, L-serine or choline, because these bases interact with gradually decreasing pools of Ptd-Etn in that order (Table 3) and because, conversely, they produce increasing effects on GABA uptake in the same order (Fig. 5). The same phenomenon seems to take place when the displacement of Ptd-Ser is considered (compare Fig. 4 and Table 3), but not when that of Ptd-Cho is under investigation (compare Fig.6 and Table 3). Interestingly, the displacement by homologous bases always produces the smallest changes (Figs. 4-6).

CYCLIC-AMP AND THE BASE-EXCHANGE REACTION FOR PHOSPHATIDYL SERINE SYNTHESIS

The calcium-stimulated exchange reaction of free L-serine with pre-existing membrane phospholipid is known to be the only route of Ptd-Ser synthesis in brain (Arienti et al., 1970; Porcellati et al., 1971). The present work and a previous report (Gaiti et al., 1975) have also shown that small highly active pools of Ptd-Cho and Ptd-Etn are involved at the membrane level in the exchange reaction with L-serine to produce Ptd-Ser in brain.

Ptd-Ser is thought to play a key role for the activity of various membrane-bound enzymes. Recently, this lipid has been demonstrated to take part in the glucagon-responsiveness of adenylate cyclase in liver (Pohl et al., 1971) and heart (Levey, 1971), in the activation of tyrosine hydroxylase in rat brain striatal synaptosomes (Raese et al., 1976), in the increased turnover of dopamine in brain (Toffano & Leon, this volume) and in other functional parameters.

In view of the recent results of Gullis & Rowe (1975), who reported that the turnover of the hydrophobic chains of synaptosomal phospholipids was stimulated *in vitro* by cyclic-AMP and putative neurotransmitters, the exchange of the hydrophylic heads (nitrogenous bases) of membrane phospholipids with externally added L-serine was examined in rat brain synaptosomes and synaptic membranes in the presence of cyclic-AMP or noradrenaline. This work was also prompted by some recent results of Lo & Levey (1976), who showed that the rate of Ptd-Ser synthesis in heart slices, which certainly takes place by base-exchange (Kiss, 1976), is very efficiently stimulated by glucagon.

Table 4 shows, in agreement with preliminary observations (Gaiti et al., 1976), that cyclic-AMP is not able to stimulate *in vitro* the serine base-exchange-mediated incorporation into rat brain synaptosomal and synaptic membrane Ptd-Ser, and that noradrenaline possesses a slight activatory effect at 10^{-4} M. Similar results have been obtained by changing in different ways (Gaiti et al., 1974) buffer composition and/or concentration, pH values, Ca^{2+}-concentration, theophylline amounts and time of observation. Further work is in progress to examine the effect of glucagon on the base-

Table 4.- The effect of cyclic-AMP and noradrenaline on the base-exchange reaction for phosphatidyl serine (Ptd-Ser) synthesis in rat brain synaptosomes*

* Incubation with 1-2 mg synaptosomal protein was carried out at 37°C for 10 min in a base-exchange medium (Gaiti et al., 1974)at pH 8.0 with |^{3}H| L-serine without calcium addition, and increasing amounts of either dibutyryl cyclic-AMP or noradrenaline. Results expressed as nmol Ptd-Ser/mg protein/3 min.

Addition	Concentration (μM)				
	0	100	1	0.1	0.001
Cyclic-AMP	1.52	1.60	1.65	-	1.52
Cyclic-AMP**	1.77	1.62	1.71	-	1.70
Noradrenaline***	1.12	1.29	1.22	1.12	1.11

** 1mM Ca^{2+} added. ***nmol/mg protein/5 min.

exchange reaction for Ptd-Ser synthesis in nervous tissue.

CONCLUDING REMARKS

From the reported data, base-exchange reactions appear to be limited in nervous tissue to small pools of phospholipid molecules at membrane level, which are available for exchange. These pools normally constitute 5% of the total lipid molecules. Ethanolamine exchanges in purified rat brain microsomal membranes with phosphatidyl serine (Ptd-Ser), phosphatidyl ethanolamine (Ptd-Etn) and phosphatidyl choline (Ptd-Cho); L-serine exchanges with Ptd-Etn, Ptd-Ser and Ptd-Cho; and choline only with Ptd-Etn and Ptd-Cho. Similar pools have also been observed at the level of synaptosomal membranes from rabbit brain.

The uptake of γ-amino-*n*-butyric acid (GABA) is affected at the synaptosomal level during a base-exchange reaction. Small pools of synaptosomal phospholipid are involved in the exchange reaction, and probably in the uptake of GABA. Incorporation of L-serine by base-exchange brings about a stimulation of the synaptosomal uptake of GABA whereas that of choline and ethanolamine decreases the rate of transport in this order.

The base-exchange reaction for L-serine incorporation into Ptd-Ser of rat brain synaptosomes and synaptic membranes is not affected *in vitro* by cyclic-AMP addition, whereas noradrenaline induces only slightly stimulatory effects.

ACKNOWLEDGEMENTS

This work has been aided by a grant from the Consiglio Nazionale delle Ricerche, Rome (Contract No. 75.00676.04). Valuable technical assistance was given by Mr. Antonio Boila and Mr. Edmondo Giovagnoli.

REFERENCES

ARIENTI, G., BRUNETTI, M. , GAITI, A., ORLANDO, P. & PORCELLATI, G.: In Function and Metabolism of Phospholipids in the Central and Peripheral Nervous Systems (G.Porcellati, L.Amaducci & C.Galli, eds.), Plenum Press, N. Y., 1976, pp. 63-78.

ARIENTI, G., PIROTTA, M., GIORGINI, D. & PORCELLATI, G. (1970) Biochem. J. 118 : 3P.

BINAGLIA, L., GORACCI, G., ROBERTI, R., PORCELLATI, G. & WOELK, H. (1973) J. Neurochem. 21, 1067.

BIRAN, L. A. & BARTLEY, W. (1961) Biochem. J. 29, 159.

BJERVE, K. S. (1973) Biochim. Biophys. Acta 306, 396.

COTMAN, C. W. & MATTEWS, D. A. (1971) Biochim. Biophys. Acta 249, 380.

CUZNER, M. L., DAVISON, A.N. & GREGSON, M. A. (1965) J. Neurochem. 12, 469.

DE MEDIO, G. E., GAITI, A., GORACCI, G. & PORCELLATI, G. (1973) Trans. Biochem. Soc. 1 , 348.

DE MEDIO, G. E., TROVARELLI, G., HAMBERGER, A., GAITI, A. & PORCELLATI, G.: II National Congress of Biochemistry, Venice, 1-4 October 1976, Abstract No. 170.

GAITI, A., BRUNETTI, M. & PORCELLATI, G. (1975) FEBS Letters 49, 361.

GAITI, A., DE MEDIO, G. E., BRUNETTI, M., AMADUCCI,L. & PORCELLATI, G. (1974) J. Neurochem. 23, 1153.

GAITI, A., FRANCESCANGELI, E., DE MEDIO, G.E., BRUNETTI, M., PALMERINI, C. A. & PORCELLATI, G. : II National Congress of Biochemistry, Venice, 1-4 October 1976, Abstract No. 169.

GORACCI, G., BLOMSTRAND, Ch., ARIENTI, G., HAMBERGER, A. & PORCELLATI,G. (1973) J. Neurochem. 20, 1167.

GULLIS, R. J. & ROWE, C. E. (1975) Biochem. J. 148, 197, 557.

HAMBERGER, A., HANSSON, H. A. & SELLESTROM, A. (1975) Exp. Cell Res. 92, 1.

KANFER, J. N. (1972) J. Lipid Res. 13, 468.

KATADA. M., ZAMA, K. & IGARASHI, H. (1959) Bull. Jap. Soc. Sci. Fish. 24, 735.

KISS, Z. (1976) Europ. J. Biochem. 67, 557.
LEVEY, G. S. (1971) J. Biol. Chem. 246, 7405.
LO, H. & LEVEY, G. S. (1976) Endocrinology 98, 251.
MILLER, E. K. & DAWSON, R.M.C. (1972) Biochem. J. 126, 805.
MIURA, T. & KANFER, J. (1976) Arch. Biochem. Biophys. 175, 654.
POHL, S. L., KRANS, H. M. J., ZOZYREFF, V., BIRNBAUMER, L. & RODBELL, M. (1971) J. Biol. Chem. 246, 4447.
PORCELLATI, G.: In Role of Membranes in Secretory Processes (L. Bolis, R.D.Keynes & W.Wilbrandt, eds.), North-Holland, Amsterdam, 1972, pp. 72-79.
PORCELLATI, G., ARIENTI, G., PIROTTA M. & GIORGINI, G. (1971) J. Neurochem. 18, 1395.
PORCELLATI, G., BIASION, M. G. & PIROTTA, M. (1970) Lipids 5, 734.
PORCELLATI, G. & DI JESO, F.: In Membrane-bound Enzymes (G.Porcellati & F.di Jeso, eds.), Plenum Press, N.Y., 1971, pp. 111-134.
PORCELLATI, G., DI JESO, F. & MALCOVATI, M. (1966) Life Sci. 5, 769.
PORCELLATI, G. & GORACCI, G. :In Lipids, Vol. 1 (R. Paoletti, G. Porcellati & G. Jacini, eds.), Raven Press, N.Y., 1976, pp. 203-214.
RAESE, J., PATRICK, R. L. & BARCHAS, J. D. (1976) Biochem. Pharmacol. 25, 2245.
SAITO, M., BOURQUE, E. & KANFER, J. (1974) Arch. Biochem. Biophys. 164, 420.
SAITO, M., BOURQUE, E. & KANFER, J. (1975) Arch. Biochem. Biophys. 169, 304.
SAITO, M. & KANFER, J. (1973) Biochem. Biophys. Res. Commun. 53, 391.
SELLESTRÖM, A. & HAMBERGER, A. (1975) J. Neurochem. 24, 847.
TOFFANO, G. & LEÒN, A. (This volume).

METABOLISM OF PHOSPHATIDYLCHOLINE IN THE CENTRAL NERVOUS SYSTEM

E.F. Soto, R. Najle, I.F. de Raveglia, and J.M. Pasquini

Departamento de Química Biológica, Facultad de Farmacia

y Bioquímica, Junín 956, Buenos Aires, Argentina

INTRODUCTION

In the present paper we describe the results of our studies on the biosynthesis and turnover of phosphatidyl choline (PC) in rat brain. We will discuss the apparent lack of phospholipid transport mechanisms in cell membranes of the CNS and will show that all brain subcellular fractions seem to have a certain degree of independence for the synthesis of this phospholipid . The presence of rapid and slow turnover pools for PC will be described and possible explanations for these findings will be presented.

EXPERIMENTAL PROCEDURES

The experiments presented in this communication were carried out in Wistar rats. Radioactive choline (^{3}H-methyl or ^{14}C-methyl depending on the type of experiment) was used in all our studies as precursor of PC. For the "in vivo" experiments the radioactive choline was injected intracranially (slightly above and between the eyes) under light ether anesthesia. At pre-established times the brains were removed and subcellular fractions isolated from a total homogenate, following methods currently used in our laboratory (Gómez et al., 1970). Lipids were extracted as described by Folch et al. (1957) and phospholipids were separated and quantitated by TLC according to Seminario et al. (1965). Radioactivity was measured in aliquots of the total lipid extract with a Packard Tri-Carb Spectrometer, using standard procedures (Soto et al., 1972). "In vitro" experiments were done in brain slices incubated at 37°C. The incubating medium (Pasquini et al., 1975) contained radioactive choline or $Na_2{}^{35}SO_4$. At pre-determined times, the slices were removed and washed and subcellular fractions isolated from a total homogenate. Lipid extraction, separation and quantitation of phospholipids by TLC and radioactive counting was done as indicated above.

To study the incorporation of choline into PC of neurons and glial cells, animals were injected as described for the "in vivo" experiments. The cerebral cortex was carefully dissected out from the excised brains and neuronal and glial enriched fractions were obtained following the procedure described by Freysz et al. (1969). Lipids were extracted from the isolated cell suspensions and radioactivity measured as described above.

Incorporation of choline into different molecular species of PC was also carried out in animals injected as for the "in vivo" experiments. Lecithin was isolated by preparative TLC and further separated into mono, tetra and hexaenoic molecular species in Silica gel plates impregnated with $AgNO_3$ as described by Arvidson (1968a). The composition of each fraction was checked by GLC in order to ascertain that the results with brain lecithin were similar to those obtained with liver by this investigator.

To study the incorporation of PC in mitochondria obtained from the cell body and from nerve endings the following procedure was used. Animals were injected with radioactive choline intracranially. The brains were excised at different times and subcellular fractions isolated as described before. The mitochondrial pellet obtained after centrifugation in a discontinous density gradient was taken without further purification as representative of the somatic (cell body) population of mitochondria. Nerve ending fractions were pooled and treated at 0°C for 1 1/2 hours with 6 mM Tris HCl buffer pH 8.1 as described by Cotman & Matthews (1971). This step is essential for the complete osmotic rupture of intact synaptosomes prior to the isolation of intraterminal mitochondria, which was carried out in a sucrose density gradient according to Cotman & Matthews (1971). Purity and morphological structure of both types of mitochondrial fractions was ascertained by electron microscopy and by determination of the activity of succinate dehydrogenase and acetylcholinesterase according to De Robertis et al. (1962).

RESULTS AND DISCUSSION

Turnover of PC in Subcellular Fractions

Incorporation of choline into various subcellular fractions after the intracranial injection of the precursor was studied as a function of time. All fractions except myelin incorporated the label into PC in a similar way, reaching the highest specific radioactivity at 5 hours (Fig. 1). Once the maximum was attained, the specific radioactivity diminished very rapidly for a few hours and continued to decrease thereafter at a slower rate. The only fraction showing very little change in specific radioactivity during the first 48 hours and no clear peak of incorporation was myelin.

Dawson and his group (McMurray & Dawson, 1969; Jungalwala & Dawson, 1970; Miller & Dawson, 1972) postulated that mitochondria have very limited capacity for the synthesis of most phospholipids and suggested that these are synthesized within the endoplasmic reticulum and transferred to other subcellular fractions. If the incorporation of labeled choline takes place in microsomes with subsequent transfer of the entire molecule to other cell fractions one would expect to find a curve of the type precursor-product between microsomes and the other membranes. The results presented in Fig. 1 do not show this type of relationship and are more compatible with the idea that all cell fractions are able to synthesize part of their own PC. The type of decay curve shown by the various subcellular fractions (Fig. 1)

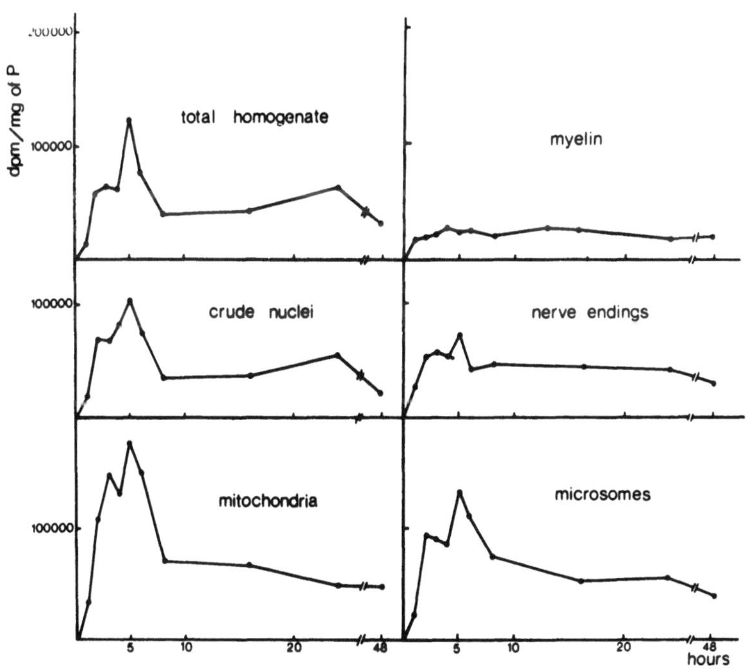

Fig. 1. Incorporation of radioactive choline into subcellular fractions of rat brain. Adult rats were injected intracranially with 2 µCi of choline (^{3}H-methyl). Labelling of PC was studied in subcellular fractions at times between 0-48 h. Results are expressed as dpm/µg of P of phosphatidyl choline. Each point represents the mean value of five different experiments. S.E.M. for each point was 10% or lower. (By permission of Pergamon Press).

indicated the possible presence of two populations of PC with different turnover rates. We performed experiments at times up to 40 days. The maximal specific radioactivity of PC in each fraction was taken as 100% value, and the specific radioactivity at 0 - 2 days and at 2-40 days was calculated for each fraction and plotted separately as percentage of the maximal activity. Regression lines were obtained by the least square method and half lives were calculated graphically from the plots. All fractions except myelin showed decay curves with two completely different half lives. There was a rapid pool with a half life of 1-2 days and a slow pool with a half life of 16 days for microsomes, 23 days for mitochondria and 53 days for nerve endings. In myelin there seemed to be only a slow turning over population with a half life of 44 days. As far as we are aware, the presence of two pools of PC in rat brain with completely different turnover rates was described by us for the first time (Pasquini et al., 1973). Jungalwala (1974) has recently confirmed our findings. Horrocks et al. (1975) on the other hand, studied the possible presence of rapid and slow turnover pools for phosphatidyl ethanolamine and PC using glycerol or ethanolamine as precursors. Their results show half lives for both pools which are coincident with our findings although the precursors used were different. The possible explanations that we gave for our results were: a) differences in the metabolic activity of the various cellular types present in brain (neurons and glial cells), b) different molecular species of PC present in brain could be metabolized at different rates.

Incorporation of Choline into Isolated Neuronal and Glial Enriched Fractions and in Different Molecular Species of PC

Incorporation of radioactive choline into PC of neuronal and glial enriched fractions and turnover of the labelled phospholipids was studied at different times up to 20 days after the intracranial injection of the precursors. The pattern of incorporation and decay was very similar to that found in brain subcellular fractions (Fig. 2). The specific activity in the neuronal fraction was higher than that in the glial enriched fraction. When the results were plotted and half lives were calculated we found that both cell fractions showed the presence of two PC pools with different half lives. The rapid turnover component showed a half life of 26 hours in neurons and 70 hours in glial cells while the slow turnover pool showed a similar half life in both fractions (15-16 days). As the presence in rat brain of PC pools with different turnover rates could not be explained by differences in metabolic activity of two cell types, we decided to explore the possibility that it could be due to differences in the metabolic stability of the various molecular species of PC which are present in the brain. Using the method of Arvidson (1968a) three fractions of PC were clearly separated and isolated, which according to the dominating unsaturated fatty acid present in the molecule were

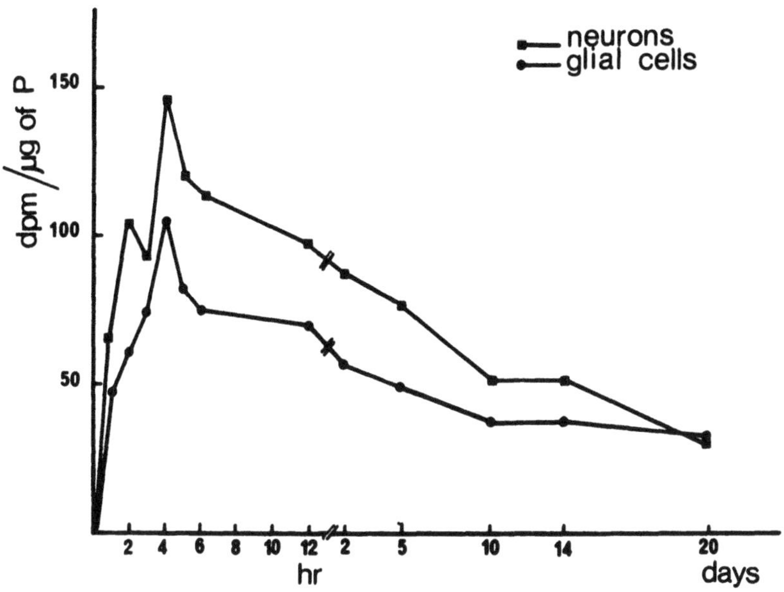

Fig. 2. Incorporation of choline into neuronal and glial enriched fractions. Adult rats were injected intracranially with 2 µCi of radioactive precursor. At the pre-established time, the animals were killed and the cell fractions isolated. Results are expressed as dpm/µg of P of phosphatidylcholine. Each point represent the mean of 4-6 experiments. S.E.M. was 10% or lower.

characterized as mono, tetra and hexaenoic PC_s. The recovery of phosphorus from the regular preparative plates and from the $AgNO_3$ impregnated plates ranged between 85-92%. The distribution of phosphorus among the different fractions (mean of 8 experiments) was 76% for the monoenoic, 16% for the tetraenoic and 8% for the hexaenoic PC_s. This results are quite close to those found by O'Brien &Gieson in a recent study (1974) in spite of the differences in methodology. The dienoic fraction, which is clearly seen in TLC of liver PC, was absent in all our runs.

The pattern of incorporation of labeled choline into total PC showed that peak incorporation occurred at 5 hours in agreement with our previous results. Contrariwise the incorporation of the precursor into different types of PC_s (Fig. 3) showed a very rapid increase in the specific activity of the tetraenoic fraction, which reached a maximum at 2 hours, decreasing quite rapidly thereafter. The incorporation into the monoenoic PC_s attained the maximum 6 hours after the injection of the precursor, while hexaenoic PC_s reached its peak at

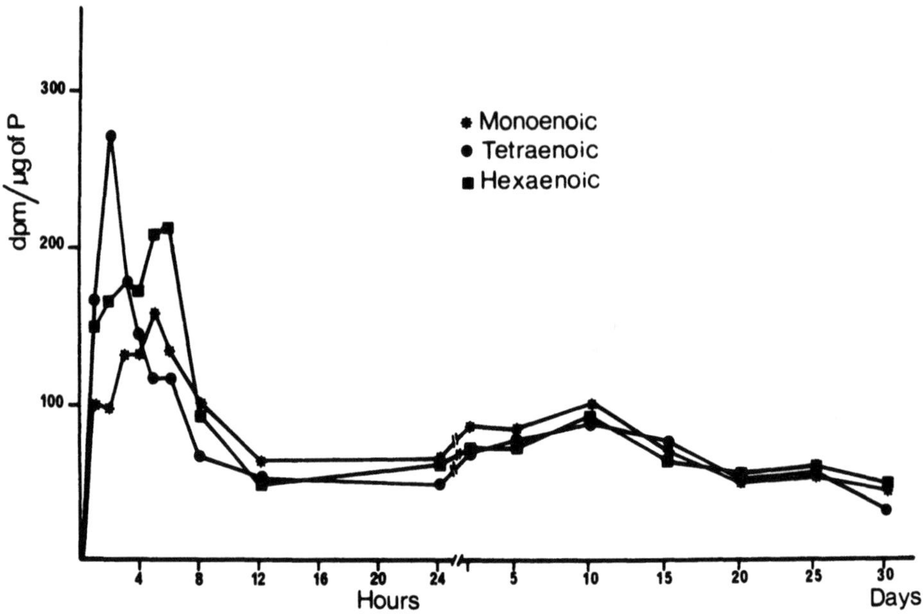

Fig. 3. Incorporation of radioactive choline into different molecular species of PC_s. Adult rats were injected intracranially with 2 uCi of labelled choline. Labelling of monoenoic, tetraenoic and hexaenoic PC_s was studied at different times up to 30 days. Each point represents the mean of 5 different experiments. S.E.M. was 10% or lower.

5 hours. The specific activity decreased in both fractions at quite similar rates. The radioactivity values decreased very slowly in all fractions during the period of time studied (30 days).

Arvidson (1968b) has shown in rat liver the presence of PC_s with different metabolic stability demonstrating that the dienoic, and to a lesser degree the monoenoic species, incorporated radioactive choline and lost the label very rapidly, while tetraenoic PC_s showed a slow and smaller incorporation. The explanation given by this investigator as well as by Van Golde et al. (1969) and by Kanoh (1969) was that the mono-dienoic fraction was synthesized "de novo" through the Kennedy pathway while the tetraenoic derivatives were synthesized by re-acylation of lysophosphatidyl choline. Recent studies by Parkes & Thompson (1973; 1974) however have clearly shown that in guinea pig liver, glycerol was most rapidly incorporated into the polyenoic fraction of PC and phosphatidyl ethanolamine and that this molecular species turned over at faster rates than the less unsaturated fractions. O'Brien & Gieson (1974) showed that the polyenoic PC_s in rat brain were labelled more rapidly after the intracranial injection of ^{14}C-

glycerol, while the diglyceride fraction containing polyunsaturated fatty acid had the lowest specific activity even 4 hours after the injection. McCaman & Cook (1966) found that 1,2 diglyceride from soybean PC, which contained mainly unsaturated fatty acids were more active substrates for brain CDP-choline: 1,2 diglyceride choline phosphotransferase than diglyceride from egg or yeast PC which contained more saturated fatty acids. Among other possible explanations, O' Brien & Gieson (1974) indicate that the most obvious possibility for their findings is that choline phosphotransferase may show specificity towards certain diglycerides species. Thus, the situation in rat brain is opposite to that found in rat liver but is similar to the findings of Parkes & Thompson (1975) in guinea pig liver, raising the possibility that "de novo" synthesis makes a greater contribution to the formation of unsaturated classes of PC both in guinea pig liver and in rat brain. Our findings seem to indicate that in rat brain the tetraenoic species of PC would act as precursor of less (or more) unsaturated components. They would function as "donors" of lysoderivatives for acyl transfer reactions involved in the synthesis of other species of PC. These results however do not allow us to conclude that the presence of rapid and slow turnover pools can be explained by the different metabolic reactivity of PC species. It is clear that the rate of labelling and the specific activity attained by the tetraenoic fraction is different, especially in relation to the monoenoic components, but the type of decay curve of the three species is clearly biphasic, i.e., it shows a rapid and a slow turnover pool.

Horrocks et al. (1975) indicate the various ways in which one can explain the presence of rapid and slow turnover pools. Recycling of the precursor remaining in the acid soluble pool and decaying at rates similar to the labeled phosphoglycerides would lengthen the half lives, explaining the presence of a slow turnover pool. Our findings (Pasquini et al., 1973) and those of Jungalwala (1974) show that the label remaining in the soluble pool after 5 hours is less than 1%, so its contribution to the lengthening of the rate of decay can be considered negligible. The slow turnover pool in our opinion, represents true breakdown of the PC molecule. Horrocks et al. (1975) based on the flux of fatty acids through the diglyceride pool arrive at similar conclusions.

Turnover of the phosphoglycerides by different reactions may also explain, according to the same group of investigators, the presence of two pools. Loss of radioactivity could occur not only by breakdown of the molecule through the action of phospholipases, but also by base exchange, although this type of reactions cannot completely explain the rapid turnover pool because this pool is also detected when glycerol is used as precursor (Horrocks et al., 1975).

The two other reasons put forward by Horrocks et al. (1975) to explain two turnover pools (association of these pools with specific cell types and molecular species of phosphoglycerides being turned over at different rates) have been substantiated by the studies presented in this communication, although it would be important to carry out experiments, in isolated oligodendroglial cells in relation to the other cell types present in brain.

Finally, Horrocks et al. (1975) present a new hypothesis which might explain rapid and slow turnover pools; the rapid turnover pool could be associated with the exchange of the phosphoglycerides between the cytosol carrier proteins and the cytoplasmic side of the membranes. The slow turnover pool would be due to catabolism of membrane segments. A quite similar assumption was made by us in a previous paper (Pasquini et al., 1973).

Jungalwala (1974) has reported the presence of slow and fast turnover pools in myelin. Our own findings however indicate the presence of only a slowly turning over pool in this fraction. If the hypothesis advanced by Horrocks et al. (1975) is true, one would not expect to find a rapid pool in myelin because the cytoplasmic side of the membrane would not be in contact with cytosol carrier proteins, except in young animals during myelination or in limited areas of adult myelin. This means that the slow turnover pool in myelin would represent the true half life of the membrane.

Turnover of PC in Somatic and Intraterminal Mitochondria

Nerve endings are structures containing mitochondria and synaptic vesicles but which have no free ribosomes, or rough endoplasmic reticulum. Except in specialized regions (photoreceptors, neuromuscular junctions), nerve endings are also devoid of smooth endoplasmic reticulum (Pellegrino de Iraldi, personal communication). Intraterminal mitochondria are assumed to be transported by axoplasmic flow from the cell body. They might also originate by division of preexisting mitochondria. Apparently they are similar to those present in the soma.

Although we have shown that there are reasons to assume that brain mitochondria have a certain degree of biosynthetical autonomy for PC (Pasquini et al., 1973) we designed the following experimental set up in order to further reinforce our previous conclusions. If mitochondrial phospholipids are synthesized in the endoplasmic reticulum, one would expect to find that carrying out labelling experiments of PC with a precursor such as choline "in vivo" and measuring the incorporation in mitochondria isolated from nerve endings and cell bodies, one would find at short times, labelling in somatic mitochondria only, because the absence of rough endoplasmic reticulum in the nerve ending would make labelling of intraterminal mitochondria either impossible or very low. At longer time periods, label would increase in this type of mitochondria due to axonal flow of phospholipid containing lipoprotein (carrier proteins?) or perhaps even intact mitochondria, from the cell body towards the nerve endings. On the other hand if mitochondria incorporate label independently, one would expect that a similar pattern of labelling would occur in both types of mitochondria.

The methods of isolation used allowed us to separate both mitochondrial species. Electron microscopy and assay of marker enzymes demonstrated that the isolated fractions were quite pure. Succinate dehydrogenase used as marker of mitochondria showed similar values of specific activity in both types. Ninety to one hundred per cent of the total activity present in the osmotically shocked nerve endings fraction was recovered in the intraterminal mitochondrial fraction, demonstrating that the osmotic shock, as well as the final sucrose density gradient, were effective for the separation of the mitochondria from the rest of the structures present in the synaptic ending. Although our results are somehow preliminary, the assays of AchE show that both mitochondrial fractions have a very low activity of this enzyme, indicating that the contamination by other membrane structures,mainly from the nerve endings or/and microsomes is below 10%.

The incorporation of radioactive choline into PC of both types of mitochondria was studied at various times between 1 and 24 hours, after the intracranial introduction of the precursor in adult rats. The data obtained (Fig. 4) show that both fractions incorporate the label in a similar way, indicating that synthesis of PC takes place in both fractions at the same time, even in the absence of rough endoplasmic reticulum. The higher specific activity found in somatic mitochondria could indicate that in this type of organelles the autonomously synthesized phospholipids could add to those that the mitochondria receives from the microsomes. These results confirm our previous data and give further support to the hypothesis that brain mitochondria are capable of synthesizing part of their own phospholipids.

Transport of Phospholipids and Sulphatides in Brain Cell Membranes

The last aspect of our research which we would like to present is related to the transport of lipid molecules between cell membranes by carrier lipoproteins present in the cytosol, a subject which has received considerable attention in the last few years.

Herschkowitz et al. (1967) studied some aspects of the synthesis of sulphatides in brain and showed that it occurred in the microsomal fraction, the entire lipid molecule being transferred to a soluble lipoprotein present in the supernatant and finally to myelin.

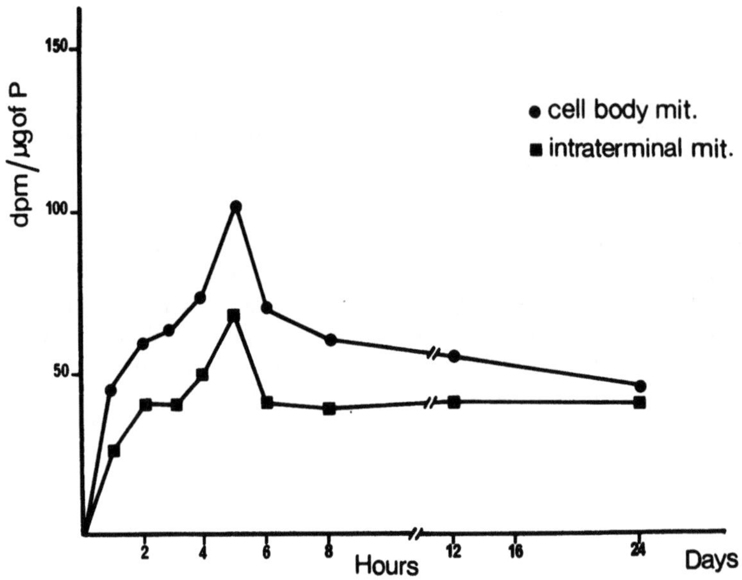

Fig. 4. Incorporation of radioactive choline into PC of somatic and intraterminal mitochondria. The precursor was injected intracranially into adult rats and mitochondria were isolated from the cell body and from nerve endings (see Experimental Procedures). Results are expressed as dpm/µg P of phosphatidyl choline and are the mean of four experiments. S.E.M. was 10% or lower.

Pleasure and Prockop (1972) recently confirmed these findings in peripheral nerve. Negative results obtained by us in preliminary experiments with rat liver, as well as data from Miller & Dawson (1972) and Benjamins & McKhann (1973) related to the presence of such mechanisms for the transport of phospholipids, prompted us to investigate some aspects of this controversial point, using both an "in vivo" and "in vitro" approach (see Experimental Procedures).

After the intracranial injection of radioactive choline in 16 days old rats, we followed the incorporation into PC of brain subcellular fraction at short times, up to 90 min. and at long times between 1 and 6 hours. A group of animals were used to perform chase experiments. They were injected with unlabelled choline 2 hours after the intracranial injection for the long time experiments and 20 min. after, in short time experiments.

Figure 5 shows the incorporation of the precursor into PC of brain subcellular fractions in both situations. The specific radioac-

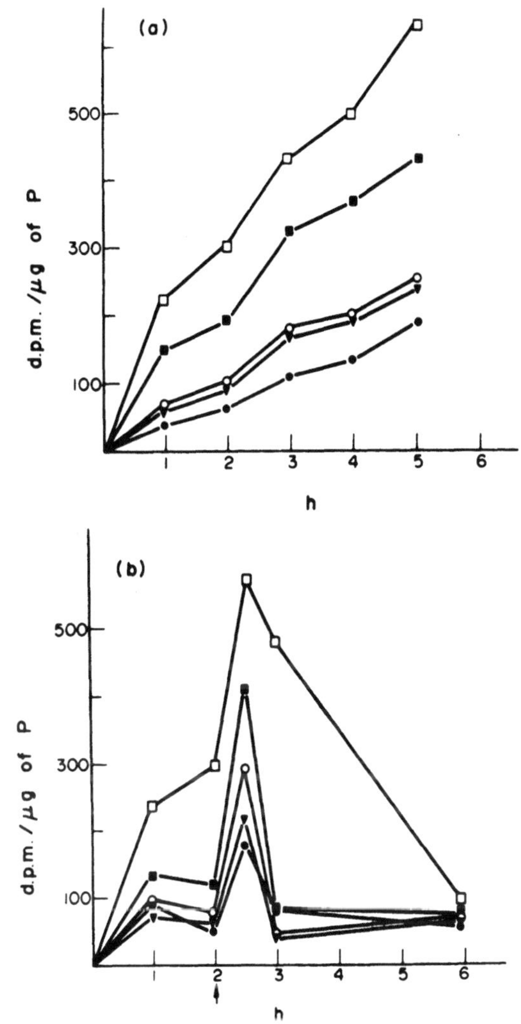

Fig. 5. "In vivo" incorporation of choline (^{3}H-methyl) in subcellular fractions of rat brain. (A) Incorporation into lipids of the different subcellular fractions was measured at various times after the intracranial injection of 2 μCi of the radioactive precursor. (B) Two hours after the intracranial injection of 2 μCi of the radioactive precursor, a chase dose of 500 μg unlabelled choline was given intraperitoneally to each animal (arrow). Incorporation into lipids of the various subcellular fractions was measured at different times. Results in both cases are expressed as dpm/μg of P and are the mean of six experiments. S.E.M. for each point was 10% or lower.

□ Supernatant; ■ mitochondria; ○ microsomes;
▼ nerve endings; ● myelin.

(By permission of Pergamon Press).

tivity of PC increases linearly in all fractions up to five hours, but when the chase dose of choline was used, a rapid decrease in specific radioactivity occurred in all fractions, 45-60 min. after the intraperitoneal injection of unlabelled choline. The radioactivity of the soluble fraction was higher than that found in the other fractions at all times. Mitochondria showed the highest specific activity. In the experiments at short times the pattern of labelling was quite similar. It is clear from these two sets of experiments that the "in vivo" studies did not indicate the presence of a transport mechanism of PC from the microsomes to the other fractions, especially myelin. Although labelled protein-phospholipid complexes are found in the supernatant, their specific radioactivity is higher at all times than that of the other fractions, which in turn, seem to incorporate the label independently. An interpretation of these negative results could be that two phospholipid pools might exist in the microsomal fraction; a small one, with very high specific radioactivity and rapidly exchanging with phospholipids in the cytosol and a large pool of much lower specific radioactivity and of low exchange rate. The large pool would decrease the specific radioactivity of the microsomes, while the exchange of phospholipid molecules of high specific radioactivity would increase the activity of the supernatant. The "in vivo" experiments at short times performed in an attempt to detect changes in the various fractions during the earliest stages of the postulated transport process did not give any clue regarding the possibility just mentioned.

Our "in vitro" experiments were done using brain slices. The incorporation of precursor (either radioactive choline or Na_2SO_4) into brain subcellular fractions was studied up to 120 min. Chase experiments were done washing the slices 15 min. after starting the incubation and re-incubating in the presence of unlabelled precursor. With choline, there is a short initial lag period followed by linear incorporation during the time studied. In chase experiments, the incorporation rapidly reaches a plateau, without decreasing thereafter (Fig. 6).

Experiments carried out with radioactive sodium sulphate however, demonstrated very clearly that in brain slices, the radioactive sulphate is incorporated into sulphatides in the microsomal fraction, it is transferred to the supernatant and finally to myelin (Fig. 7). This finding confirm the "in vitro" results of Pleasure & Prockop (1972) in peripheral nerve and demonstrate that brain slices are a suitable preparation for the exploration of the transport process. It is clear on the other hand that when choline is used as precursor, there is no indication that the phospholipid molecules, probably synthesized in microsomes, are transferred through the supernatant to the myelin or to other subcellular fractions.

Although Miller & Dawson (1972) postulated that transport of phospholipids should occur "in vivo", they could not demonstrate it

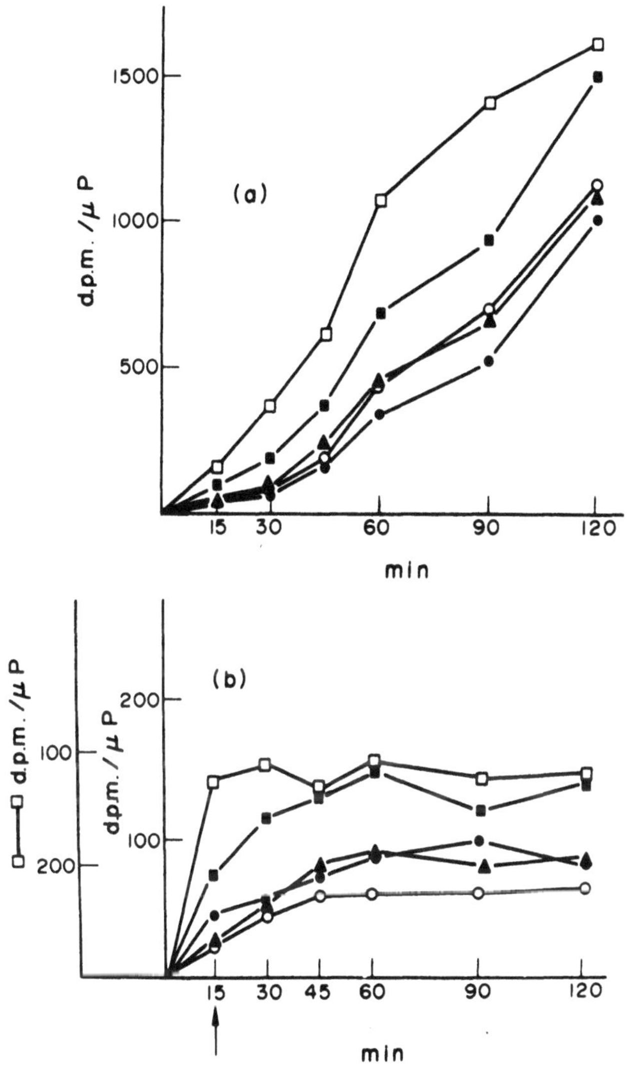

Fig. 6. "In vitro" incorporation of choline (^{3}H-methyl) into rat brain subcellular fractions. (A) Brain slices were incubated with the radioactive precursor. Subcellular fractions were then isolated and incorporation into lipids was measured at different times. (B) Fifteen minutes after starting the incubation with the labelled precursor, the slices were washed twice with fresh medium containing 1% unlabelled choline, and reincubated with this medium. Subcellular fractions were then isolated and incorporation into lipids was measured at different times. Results are expressed as dpm/μg P, and are the mean of six experiments. S.E.M. for each point was 10% or lower. Symbols as in Fig. 5. (By permission of Pergamon Press).

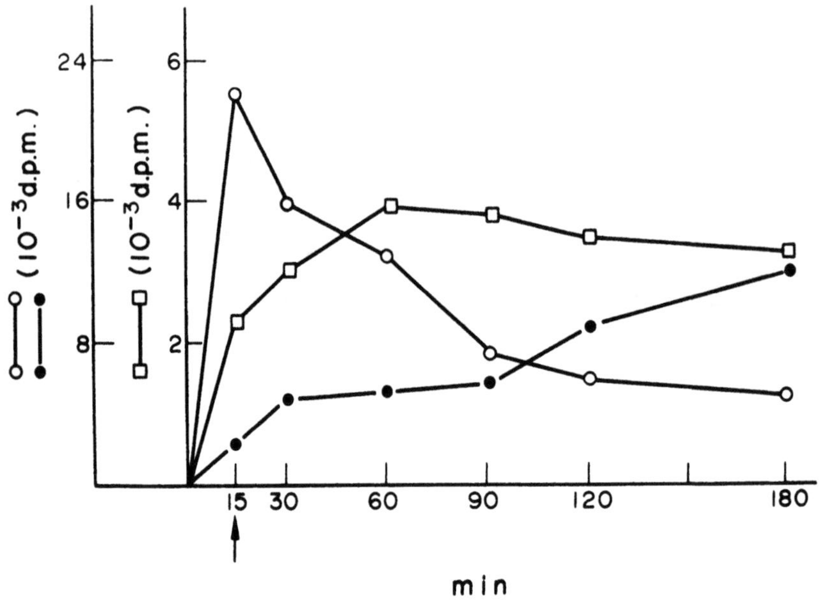

Fig. 7. "In vitro" incorporation of $Na_2{}^{35}SO_4$ into rat brain subcellular fractions. Brain slices were incubated with the precursor for 15 min. and after two washings with fresh medium containing 0.1 mM-Na_2SO_4 they were further incubated for different times. Isolation of subcellular fractions and extraction of lipids was performed as described under "Experimental Procedures". Radioactivity was measured in the total lipid extract. Results are expressed as dpm/brain and are the mean of three experiments. Symbols as in Fig. 5.

"in vitro" using P_{32}. Our findings agree with those of Benjamins & McKhann (1973) who found phospholipid containing lipoproteins with similar properties to those containing sulphatides but who could not show any correlation between lipoproteins and the addition of phospholipids to myelin. One possible explanation for these findings according to Benjamins & McKhann (1973) is that phospholipids might be present in a precursor membrane of myelin, while some of the sulphatides may be added to myelin by a specific cytoplasmic transport system. This would bring the question of whether phospholipids are added to the early or myelin-like fraction by a transport process or if the membrane of the oligodendroglial cell synthesizes its own phospholipids. We have carried out "in vivo" experiments similar to those described above in 10 days old animals (not shown) and have found no evidence that the proposed transport system might be operative at this early stage of development.

These important aspects of membrane formation are currently being explored in our laboratories in order to clarify some of the unexplained events that we have just described.

CONCLUSIONS

We have reviewed our most recent work in the field of metabolism of phospholipids in the Central Nervous System.

Based on our results and in references obtained from the literature we can conclude:

a) All subcellular fractions of the rat CNS, especially mitochondria seem to be capable of synthesizing part of their own phospholipids.

b) The turnover of PC in brain occurs in a bi-phasic way i.e. there is a rapid turnover pool and a slow turnover pool, which is neither due to the presence of different molecular species of PC with different metabolic reactivity, nor to the presence of various cell types.

c) The rapid pool could be due to exchange of phospholipid molecules between the cytosol and the cytoplasmic side of the various membranes in the cell, while the slow pool would be the true expression of the replacement of membrane segments.

d) Mechanisms for the transport of sulphatides between microsomes and myelin, which are mediated by soluble lipoproteins can be demonstrated. We have been unable to find evidences for a similar mechanism involved in the transport of phospholipids.

Acknowledgements

We thank Dr. Amanda P. de Iraldi for the electron microscopic control of our subcellular fractions. The skilful technical assistance of Miss María Teresa Iturregui is acknowledged. This work has been supported by grants from the Consejo Nacional de Investigaciones Científicas y Técnicas, Argentina.

REFERENCES

Arvidson, G.A.E. (1968a) Europ. J. Biochem. 4, 478-486.

Arvidson, G.A.E. (1968b) Europ. J. Biochem. 5, 415-421.

Benjamins, J.A. & McKhann, G.M. (1973) J. Neurochem. 20, 1121-1129.

Cotman, C.W. & Matthews, D.A. (1971) Biochim. Biophys. Acta 249, 380-394.

De Robertis, E.; Pellegrino de Iraldi, A.; Rodriguez de Lorez Arnaiz, G. & Salganicoff, L. (1962) J. Neurochem. 9, 23-35.

Freysz, L.; Bieth, R. & Mandel, P. (1969) J. Neurochem. 16, 1417-1424.

Folch, J.; Lees, M. & Sloane Stanley, G.H. (1957) J. Biol. Chem. 226, 497-509.

Gómez, C.J.; Pasquini, J.M.; Soto, E.F. & De Robertis, E. (1970) J. Neurochem. 17, 1485-1492.

Herschkowitz, N.; McKhann, G.M. & Shooter, E.M. (1967) J. Neurochem. 17, 1485-1492.

Horrocks, L.A.; Toews, A.D.; Thompson, D.K. & Chin, J.Y. (1975), In Function and Metabolism of phospholipids in CNS and PNS (G. Porcellati, ed.) pp 1-18, Plenum Press, London.

Jungalwala, F. & Dawson, R.M.C. (1970) Biochem. J. 117, 481-490.

Jungalwala, F. (1974) Brain Res. 78, 99-108.

Kanoh, H. (1969) Biochim. Biophys. Acta 176, 756-763.

McCaman, R.E. & Cook, K. (1966) J. Biol. Chem. 241, 3390-3397.

McMurray, W.C. & Dawson, R.M.C. (1969) Biochem. J. 126, 805-821.

Miller, E.K. & Dawson, R.M.C. (1972) Biochem. J. 126, 823-835.

O'Brien, J.F. & Geison, R.L. (1974) J. Lipid Res. 15, 44-49.

Parkes, J.G. & Thompson, W. (1973) J. Biol. Chem. 248, 6655-6662.

Parkes, J.G. & Thompson, W. (1975) Canad. J. Biochem. 53, 698-705.

Pasquini, J.M.; Gómez, C.J.; Najle, R. & Soto, E.F. (1975) J. Neurochem. 24, 439-443.

Pasquini, J.M.; Krawiec, L. & Soto, E.F. (1973) J. Neurochem. 21, 647-653.

Pleasure, D.E. & Prockop, D.J. (1972) J. Neurochem. 19, 283-295.

Seminario, L.; Soto, E.F. & Cohan, T. (1965) J. Chromatog. 17, 513-519.

Soto, E.F.; Pasquini,J.M. & Krawiec, L. (1972) Archs. Biochem. Biophys. 150, 362-370.

Van Golde, L.M.G.; Scherphof, G.L. & Van Deenen, L.L.M. (1969) Biochim. Biophys. Acta 176, 635-637.

METABOLIC STUDIES ON ETHANOLAMINE PHOSPHOGLYCERIDES IN NEURONAL AND GLIAL CELLS

H. Woelk, G. Porcellati, L. Binaglia, and G. Goracci

Einheit für Neurochemie, Universitäts-Nervenklinik, Erlangen-Nürnberg, West Germany and Department of Biochemistry, University of Perugia, Italy

Marked differences in phospholipid metabolism have been observed between glial and neuronal cell enriched fractions. Recently, evidence was obtained for a faster turnover of glycerophospholipids in neurons than in glial cells, since neurons contained a considerably higher phospholipase A_1 and A_2 activity when compared to the glial cell enriched fraction (Woelk et al., 1973). Furthermore the neuronal cell bodies were found to possess a much higher rate of exchange of both serine and ethanolamine into the phospholipids than the glial cell enriched fraction (Goracci et al., 1973) and Freysz et al., (1969), showed in their extensive investigation on the kinetics of the biosynthesis of phospholipids in neurons and glial cells, isolated from rat brain cortex, that neuronal phospholipids had a faster turnover than glial phospholipids. Recent results, presented by Woelk et al. (1974), indicated that phosphatidylinositol and phosphatidylcholine had the fastest and ethanolamine plasmalogen the slowest turnover in both cell populations.

Measuring the phospholipid content and composition in fractions enriched in neuronal cell bodies and in glial cells, we found that glial cells contained approx. one-third more phospholipids per unit protein than the neuronal cell bodies (Table 1). The distribution and pattern of phospholipids, relative to the total amount was rather similar in both cell types. However, a slightly larger relative concentration of phosphatidylcholine was observed in the neuronal cell bodies (Table 1). With regard to the content of total phospholipids we obtained somewhat lower figures for the neuronal cell bodies compared to those of Hamberger & Svennerholm (1971). As for the phospholipid content of the glial cell enriched fraction and concerning the relative concentration of the individual phospho-

Table 1. Phospholipid Composition of Neuronal and Glial Cell Enriched Fractions of the Rabbit Cerebral Cortex

	nmol/mg Protein		%P of total lipid-P	
	Neurons	Glia	Neurons	Glia
Phospholipid				
Total phospholipids	395	631	100.0	100.0
Phosphatidylcholine	168	247	42.6	39.2
Phosphatidylethanolamine	65	97	14.2	15.4
Ethanolamine plasmalogen	61	99	15.4	15.8
Phosphatidylserine	25	43	6.5	6.9
Phosphatidylinositol	22	32	5.6	5.1
Sphingomyelin	20	34	5.0	5.5
Diphosphatidylglycerol	19	33	4.8	5.2

lipids in both glia and neurons our figures are in agreement with those published by Hamberger & Svennerholm (1971). Estimates of the exact cellular localization of the phospholipids between glia and neurons are necessary to understand more about their physiological function. In this connection it is interesting that the content of ethanolamine plasmalogen is rather similar in the neuronal and glial cell enriched fractions. The association of the major part of the ethanolamine plasmalogen with the myelin sheath suggested a predominantly glial localization of the ethanolamine plasmalogen, since the myelin sheath probably derives from the plasma membrane of the oligodendrocyte in the central nervous and of the Schwann cell in the peripheral nervous system. The presence of relatively large amounts of ethanolamine plasmalogen in the neuronal cell bodies (Table 1) indicates however, that ethanolamine plasmalogen might also be associated with neuronal function.

Taking into account this localization a study on the enzymic properties of the *de novo* mechanism for phospholipid biosynthesis both in neuronal and glial cell fractions seems to be of value. As a first step we have used neurons and glia to investigate the activity of phosphorylethanolamine diacyl glycerol phosphotransferase (EC 2.7.8.1), which catalyse the last step in the synthesis of phosphatidylethanolamine. For this purpose neuronal or glial homogenates were incubated (ca. 0.2 mg neuronal protein and 0.45 mg glial protein) for 30 min at 39.5 C with 4 mM diacyl glycerol or alkenylacyl glycerol (both prepared from ox heart and 1.2 mM cytidine-5'-diphosphate ethanolamine (specific activity of 1 nCi/nmol). Table 2 shows that, in the absence of any lipid acceptor, the neuronal cell fraction possessed a higher rate of ethanolamine phosphoglyceride (EPG) synthesis, if compared to glia. In terms of specific activity, the neurons displayed a threefold increase of activity for both phosphatidylethanolamine and ethanolamine plasmalogen synthesis. Furthermore, both neurons and glial cells displayed a higher synthesis of diacyl-glycero-3-phosphorylethanolamine (diacyl-GPE) compared to that of alkenylacyl-glycero-3-phosphorylethanolamine (alkenylacyl-GPE). The addition of 4 mM diacyl glycerol caused a 5-fold and a 3-fold stimulation of cytidine-5'-diphosphate

Table 2. Incorporation of Cytidine-5'-Diphosphate Ethanolamine into Phosphatidylethanolamine and Ethanolamine Plasmalogen of Dispersions of Neuronal and Glial Cells from Rabbit Brain in Absence and Presence of Added Diacyl Glycerol or Alkenylacyl Glycerol

Synthesized lipid	Fraction	Type of added diglyceride	Activity[a]	Ratio[b]
Diacyl-GPE	Neurons		1.36	
	Glia		0.43	3.1
Alkenylacyl-GPE	Neurons		0.76	
	Glia		0.28	2.7
Diacyl-GPE	Neurons	Diacyl glycerol	4.00	
	Glia	Diacyl glycerol	1.98	2.0
Alkenylacyl-GPE	Neurons	Diacyl glycerol	1.21	
	Glia	Diacyl glycerol	0.50	2.4
Diacyl-GPE	Neurons	Alkenylacyl glycerol	1.97	
	Glia	Alkenylacyl glycerol	0.68	2.9
Alkenylacyl-GPE	Neurons	Alkenylacyl glycerol	19.0	
	Glia	Alkenylacyl glycerol	5.4	3.5

[a]nmoles/mg protein/30 min, [b]neuronal/glial ratio

ethanolamine (CDPE) incorporation into diacyl-GPE in glia and neurons, respectively, whereas a correspondent 20-fold and 25-fold stimulation of ethanolamine plasmalogen synthesis was observed on adding a similar concentration of alkenylacyl glycerol. On adding diacyl glycerol or alkenylacyl glycerol, the neuronal fraction displayed a noticeably higher activity, compared to glia (2-fold and 3.5-fold increases for diacyl-GPE and alkenylacyl-GPE, respectively). The increased uptake of radioactive CDPE into the two distinct phospholipid moieties following diglyceride addition was not completely specific, because a small but consistent stimulation of alkenylacyl-GPE synthesis was obtained on adding diacyl glycerol, and a smaller one viceversa (Table 2).

Investigation on plasmalogen biosynthesis in intact cells has led to conflicting opinions regarding the aliphatic precursor of the alkenyl moiety. The postmitochondrial fraction of Ehrlich ascites cells and preputial gland tumors can synthesize ethanolamine plasmalogen from the long chain fatty alcohols and dihydroxyacetone phosphate (Snyder et al., 1971) or 1-alkyl-2-acyl-sn-glycero-3-phosphate (Blank et al., 1971). It appears to be well established now that one mechanism by which 1-alk-1'-enyl-2-acyl-sn-glycero-3-phosphorylethanolamine (ethanolamine plasmalogen) can be formed involves dehydrogenation of the 1-alkyl moiety and that cytochrome b5 participates into the reaction (Woelk et al., 1976). It is also possible, however, that diacyl-GPE, alkenylacyl-GPE and alkylacyl-GPE are synthesized independently but by analogous Kennedy's pathways with a supply of appropriate glyceride acceptors (Joffe, 1969). This assumption would implay the presence of the correspondent neutral diglyceride in nervous tissue, a finding which apparently seems to apply only to diacyl glycerol and probably to alkylacyl glycerol (Sun, 1970) but not to alkenylacyl glycerol (Horrocks, 1972). However, it is worth mentioning in this connection that smaller stimulation of diacyl-GPE synthesis was obtained on adding comparable amounts of diacyl glycerol from the same source (ox heart) than that of ethanolamine plasmalogen from the alkenylacyl derivative.

REFERENCES

BLANK M. L., WYKLE R. L. & SNYDER F. (1971) Fed. Eur. Biochem. Soc. Lett. 18, 92-94

FREYSZ L., BIETH R. & MANDEL P. (1969) J. Neurochem. 16, 1417-1424

GORACCI G., BLOMSTRAND C., ARIENTI G., HAMBERGER A. & PORCELLATI G. (1973) J. Neurochem. 20, 1167-1180

HAMBERGER A. & SVENNERHOLM L. (1971) J. Neurochem. 18, 1821-1829

HORROCKS L. A. in "Ether Lipids: Chemistry and Biology", Edited by F. Snyder, Academic Press, New York, N. Y., 1972, p. 177

JOFFE S. (1969) J. Neurochem. 16, 715

SNYDER F., BLANK M. L. & WYKLE R. L. (1971) J. Biol. Chem. 246, 3639-3645

SUN G. Y. (1970) J. Neurochem. 17, 445
WOELK H., GORACCI G., GAITI A. & PORCELLATI G. (1973) Hoppe Seyler's Z. Physiol. Chem. 354, 729-736
WOELK H., KANIG K. & PEILER-ICHIKAWA K. (1974) J. Neurochem. 23, 1057-1063
WOELK H., PORCELLATI G. & GORACCI G. in "Function and Metabolism of Phospholipids in the Central and Peripheral Nervous Systems", edited by G. Porcellati, L. Amaducci & C. Galli, Plenum Publ. Corp., New York, N. Y. 10011, 1976, pp. 55-61

METABOLIC RELATION BETWEEN PHOSPHATIDIC ACIDS, CDP-DIGLYCERIDES, AND PHOSPHOINOSITIDES

W. Thompson

Department of Biochemistry

University of Toronto, Toronto, Canada

The metabolic turnover of phosphoinositides in numerous tissues has received a great deal of attention with particular interest being focused on the polar head groups of the lipids. What has received comparatively little attention, however, is the fatty acid turnover and the mechanisms for regulation of the molecular species of these lipids. In a number of tissues arachidonic acid is the predominant unsaturated fatty acid of each of the inositides and in rat brain this is a unique feature of the inositides. Not only is arachidonate an important fatty acid, but it seems of importance for the cell to position arachidonate in position - 2 of the phosphoinositide molecule.

Our interest has been in the mechanisms by which the molecular species of phosphatidyl inositol are regulated and how the high arachidonate content is achieved. This communication deals with studies of both the de novo pathway and the acyl transferase pathway of inositide formation.

Acyl transferase pathway

Table 1 shows that lysophosphatidyl inositol can be readily acylated by rat brain microsomes with any of a number of acyl CoA thioesters. There is a general preference for unsaturated over saturated thioesters but, more strikingly, there is a high selectivity for arachidonoyl CoA which under certain condition of incubation is some 4 - 8 times more active in acylating the lysolipid than the other thioesters. The eicosatrienoic species used in this series is the precursor of arachidonate with the same arrangement of double bonds at positions 8, 11 and 14. The marked difference in acylation rate between eicosatrienoate and the arachidonate thioester indicates that the fourth double bond between C atoms 5 and 6 may be critical in directing enzyme selectivity for arachidonoyl CoA.

Table 1

Acylation of lysolipids by rat brain microsomes

	nmoles/min/mg protein	
Acyl CoA	1-Acyl GPI	1-Acyl GPC
16:0	2.9	3.7
18:0	3.1	1.5
18:1	4.5	3.7
18:2	8.7	14.6
18:3	5.3	16.8
20:3	12.0	12.2
20:4	41.5	12.9
22:6	5.1	9.0

The second column in Table 1 shows that brain microsomes also acylate lysolecithin but in this case we find no special selectivity for arachidonoyl CoA and, in fact, in brain phosphatidyl choline does not have a high arachidonate content.

The *in vitro* data together with other *in vivo* data (1) lead us to conclude that the deacylation and reacylation sequence must be an important means for establishing arachidonoyl-containing species of phosphatidyl inositol. There are, however, still some elements of the picture to be filled in. For example, there is minimal information about phospholipase A activity towards phosphatidyl inositol and, secondly, the presence of lysophosphatidyl inositol has yet to be demonstrated in fresh tissues which, of-course, is a requisite for the proposed deacylation - reacylation reactions. If lysophosphatidyl inositol does occur in tissues it must undoubtedly be present in very low concentrations.

De-Novo pathway

Our attention was next directed to the *de novo* pathway of synthesis from phosphatidic acid. Although phosphatidic acid was known to have a rather low arachidonate content there was no information about the intermediate, CDP-diglyceride. Detailed knowledge of the de novo pathway required some information about the liponucleotide and it became necessary to attempt its isolation. We have now been able to locate small amounts of CDP-diglyceride in bovine brain (2) and liver (3) and, more recently in rat liver. These products show an interesting fatty acid composition, stearate being the major saturated fatty acid and arachidonate the predominant unsaturated fatty acid. This is a recurring feature in rat liver, bovine liver and bovine brain.

Table 2

Fatty acid composition of CDP-diglyceride of various tissues

	% Composition		
Fatty Acid	Bovine Brain	Bovine Liver	Rat Liver
16:0	4.4	4.4	19.6
18:0	42.4	34.0	22.7
18:1	14.1	16.6	10.5
18:2	5.5	4.5	13.3
18:3	3.4	2.6	0.2
20:3		3.9	2.4
20:4	22.3	21.9	28.7
20:5		0.3	1.5
22:3	5.3	6.4	
22:5	1.2	1.3	0.1
22:6	1.4	3.9	0.8

The liponucleotide has a considerably higher content of arachidonate than its precursor, phosphatidate, and is, in fact, more similar in fatty acid profile to phosphatidyl inositol. Now, does this mean that the de novo pathway of synthesis from phosphatidic acid gives rise to arachidonate-enriched CDP-diglyceride which in turn forms arachidonate-rich phosphatidyl inositol or are there other explanations? A number of hypotheses have been proposed (2) to account for this pattern of fatty acids in CDP-diglyceride and our recent investigations have been directed towards testing one of these hypotheses, namely that the enzyme that synthesises CDP-diglyceride may be selective for arachidonate-containing species of phosphatidic acid.

For the kinetic studies we have used rat liver as it is rather more easy to work with than brain. The left-hand side of Figure 1 shows the labelling of phosphatidic acid, CDP-diglyceride and phosphatidyl inositol following the intraportal injection of ^{3}H-glycerol. Maximum labelling of phosphatidic acid and CDP-diglyceride occurred within 60 seconds after injection and maximum labelling of phosphatidyl inositol within 120 seconds. The radioactivity in the liponucleotide is quite low but it must be turning over very rapidly to account for the large accumulation of radioactivity in phosphatidyl inositol. On the right-hand side of Figure 1 is shown the labelling of these lipids after injection of ^{3}H-glycerol into the jugular vein. The reason for the two different routes of injection will be explained below. In this case the labelling of the liver lipids was lower and slower because of dispersion of tracer in the systemic circulation. The time at which maximal labelling occurred was not determined but is likely to be at points between 1 and 4 hours after injection.

The distribution of radioactivity in the molecular classes of the lipids 15 to 300 seconds after intraportal injection is shown in Table 3. These were examined on the simple premise that if there was random utilisation of species of phosphatidic acid, i.e. no species selectivity, for the synthesis of CDP-diglyceride and phosphatidyl inositol then the labelling pattern in the molecular classes in all three lipids should be similar. The major flux of radioactivity was through the oligoenoic species of phosphatidic acid with one or two double bonds. Labelling of polyenoic species was low in contrast. The labelling of CDP-diglyceride was not identical to that of phosphatidic acid there being somewhat less radioactivity in the monoenoic, some enhancement in the trienoic and about a doubling of radioactivity in the tetraenoic species.

The labelling pattern of the liponucleotide, however, is not only the result of the flow of radioactivity from phosphatidic acid but also from the transfer of radioactivity from the liponucleotide to phosphatidyl inositol (and polyglycerophosphatides which were not measured). The radioactivity in phosphatidyl inositol from the portal vein injections was concentrated largely in the tetraenoic species and to a smaller extent in other polyenoic species with more than 4 double bonds. This pattern was invariant from the earliest time point and it was unlike the pattern from other experiments with brain (4) and liver (5) which had shown an initial high labelling of oligoenoic species of phosphatidyl inositol, followed by a slow redistribution of label to the tetraenoic attributed to deacylation-reacylation reactions. The very high labelling of the tetraenoic species after portal vein injections could also be attributed to rapid uptake and redistribution by acyl transfer mechanisims. Another possibility was that if there was significant back-flow of radioactivity from the highly

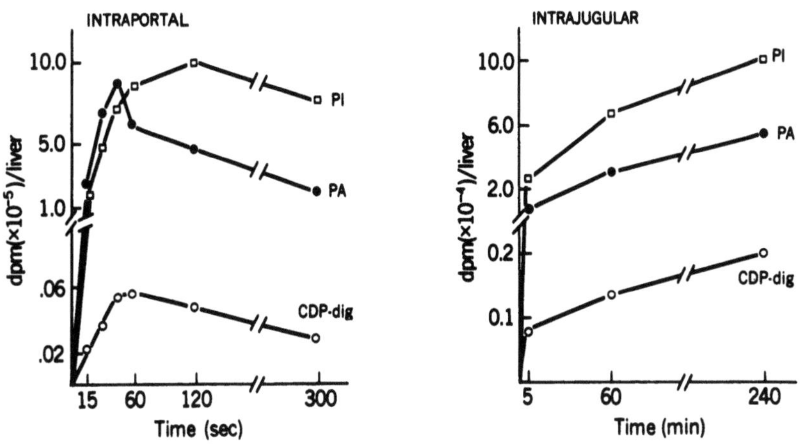

Fig. 1. Labelling of phosphatidic acid, CDP-diglyceride and phosphatidyl inositol following the (a) intraportal and (b) intrajugular injection of ^{3}H-glycerol.

labelled phosphatidyl inositol via the back reaction of CDP-diglyceride: inositol phosphatidyl transferase then this might account for the increased labelling of tetraenoic species of CDP-diglyceride.

To check on these possibilities we resorted to intrajugular injections of ^{3}H-glycerol in which the tracer is utilized more slowly by the liver. With the intraportal injections the time intervals extended up to 300 secs but with the intrajugular injections up to 240 minutes. As shown in Table 4 the radioactivity in phosphatidic acid was again concentrated in the oligoenoic species (Δ1 and Δ2) and again the CDP-diglyceride pattern was different - the oligoenoic species particularly the monoenoic, being less labelled and all polyenoic species (≥ Δ3) being more highly labelled. Again this could suggest some species selectivity.

The labelling pattern of phosphatidyl inositol was quite different from that obtained by intraportal injections. There was an initial concentration of label in the monoenoic but with time this species became less labelled while the tetraenoic species progressively accumulated more label. This redistribution of counts we attribute to deacylation - reacylation reactions in which oligoenoic species, principally the monoenoic, are converted into tetraenoic by substitution of the fatty acids at position 2, and from the redistribution pattern this would appear to be very extensive. Thus, the pattern which takes hours to emerge with

Table 3

Labelling of molecular classes of phosphatidic acid, CDP-diglyceride and phosphatidyl inositol following the intraportal injection of ^{3}H-glycerol

	% distribution of radioactivity								
	PA			CDP-dig			PI		
Molecular class	15"	60"	300"	15"	60"	300"	15"	60"	300"
Δ0	4.1	1.2	2.7	1.8	2.3	2.2	2.0	1.4	0.8
Δ1	29.7	34.3	50.3	16.5	14.9	21.9	5.3	4.8	5.8
Δ2	36.4	35.6	25.2	37.3	42.1	22.0	5.6	4.6	3.8
Δ3	4.9	3.6	4.8	8.9	9.6	9.4	4.3	11.4	10.2
Δ4	14.3	14.5	10.2	28.5	24.4	35.3	61.0	57.9	57.8
>Δ4	10.6	10.8	6.8	7.0	6.7	9.1	21.8	19.9	21.6

Table 4

Labelling of molecular classes of phosphatidic acid, CDP-diglyceride and phosphatidyl inositol following the intrajugular injection of ^{3}H-glycerol

	% distribution of radioactivity								
	PA			CDP-dig			PI		
Molecular class	5'	60'	240'	5'	60'	240'	5'	60'	240'
Δ0	2.9	5.6	4.0	5.8	5.6	7.1	4.2	8.6	3.8
Δ1	42.0	43.5	41.2	24.3	22.6	24.5	49.4	25.6	22.2
Δ2	32.2	27.2	29.6	20.6	22.5	22.2	9.6	9.8	5.4
Δ3	4.7	4.9	5.4	15.6	13.7	15.1	11.2	12.6	12.8
Δ4	7.8	8.9	9.3	17.8	18.4	18.8	16.2	30.5	45.7
>Δ4	10.3	10.0	10.5	16.0	17.2	12.3	9.5	12.8	10.0

intrajugular injection of isotope takes only seconds with intraportal injections.

Since there was no progressive change and redistribution among the species of CDP-diglyceride it can be assumed that the labelling pattern of the liponucleotide was not an expression of the back-flow of radioactivity from phosphatidyl-inositol. When rate constants for the formation of different molecular species of CDP-diglyceride from phosphatidic acid are determined based on the data from Table 4 and a simplified mathematical model then these rate constants do suggest selectivity for trienoic and tetraenoic species of phosphatidic acid, but this can not be claimed as definitive evidence - only as suggestive evidence.

In Vitro Synthesis of CDP-diglyceride

The formation of CDP-diglyceride was also investigated with rat liver microsomes supplemented with CTP and sonicated dispersions of phosphatidic acid labelled with ^{3}H-glycerol. As listed in Table 5, comparison of the molecular classes showed some reduction of label in monoenoic and dienoic classes of liponucleotide (which was somewhat similar to the in vivo data) and some enhancement of label in trienoic and polyenoic classes with more than 4 double bonds. Although these differences were small,

Table 5

Synthesis of molecular classes of CDP-diglyceride from ^{3}H-phosphatidic acid by rat liver microsomes

	% distribution of radioactivity		
Molecular class	Phosphatidic acid	CDP-diglyceride	P value
Δ0	7.7 ± 1.1	8.6 ± 2.4	N.S.
Δ1	23.6 ± 2.0	18.3 ± 3.9	<.001
Δ2	16.7 ± 1.6	14.0 ± 2.3	<.01
Δ3	8.3 ± 2.9	10.7 ± 2.4	<.05
Δ4	20.3 ± 1.5	20.1 ± 2.2	N.S.
>Δ4	23.3 ± 2.4	28.3 ± 3.5	<.001

comparison of means by Student's t test showed them to be statistically significant. There was, however, no significant change in the proportion of radioactivity in the major tetraenoic class of CDP-diglyceride - and in this respect the in vitro and in vivo data were different.

Table 6 compares the relative labelling in vivo and in vitro as a ratio of radioactivity in total polyenoic classes ($\geqslant \Delta 3$) to oligoenoic classes. The relative enhancement of label in polyunsaturated species of CDP-diglyceride in vivo was considerably greater than that in vitro.

Table 6

Comparison of in vivo and in vitro labelling of molecular species of phosphatidic acid and CDP-diglyceride

	Ratio of counts: $\frac{\Delta 3 + \Delta 4 + >\Delta 4}{\Delta 1 + \Delta 2}$	
	Phosphatidic acid	CDP-diglyceride
In vitro	1.28	1.81
In vivo (intrajugular)	0.31	1.10

CONCLUSIONS

While the in vivo data are suggestive of some molecular species selectivity in CDP-diglyceride synthesis the in vitro data provide no convincing evidence for a high enough selectivity to account for the increased level of arachidonoyl species of CDP-diglyceride over that of phosphatidic acid - although it is possible that there is a regulatory mechanism for species selectivity in vivo that was not manifested under these in vitro conditions of incubation.

Other possible explanations for the high level of arachidonate in CDP-diglyceride have been proposed (2) which can be tested experimentally. Undoubtedly, however, the evidence is convincing that the enrichment of arachidonate in phosphatidyl inositol is in large measure achieved by a combination of deacylation and reacylation reactions. The presence in the tissues of CDP-diglyceride with high arachidonate content, whatever its metabolic origins, suggests that arachidonate-rich phosphatidyl inositol can also be derived from the liponucleotide. The formation of this species by both pathways would emphasize the importance of arachidonate-containing phosphatidyl inositol for cellular membrane function.

REFERENCES

Baker R.R. & Thompson W. (1972) Biochim. Biophys. Acta 270, 489.
Thompson W. & MacDonald G. (1975) J. Biol. Chem. 250, 6779.
Thompson W. & MacDonald G. (1976) Eur. J. Biochem. 65, 107.
MacDonald G. Baker R.R. & Thompson W. (1975) J. Neurochem. 24, 655.
Holub B. & Kuksis A. (1971) J. Lipid Res. 12, 699.

NEOBIOSYNTHESIS OF PHOSPHATIDYLINOSITOL AND OF OTHER GLYCEROLIPIDS IN THE ENTIRE CATTLE RETINA

N.G. Bazán, M.G. Ilincheta de Boschero, and N.M. Giusto

Instituto de Investigaciones Bioquímicas, Universidad Nacional del Sur y Consejo Nacional de Investigaciones Científicas y Técnicas, Bahía Blanca, Argentina

INTRODUCTION

The vertebrate retina comprises two different portions, a neural part and a specialized layer of photoreceptor cells. Because the former is an integral part of the central nervous system it contains a large surface of excitable membranes. Also, a very high proportion of the photoreceptor cell mass is made up of tightly folded membranes. Each portion of the retina requires polar lipids for its membranes. However, the lipid composition and metabolism of the two portions differ strikingly. This is in part an assumption since whole photoreceptor cells have not yet been isolated and direct biochemical investigations are thus not possible. However, isolated rod outer segments are currently being intensively studied. They have a high phospholipid content and are very rich in polyenoic acyl chains (Anderson, 1970; Mason et al., 1973; Daemen, 1973). Although retina homogenates and subcellular fractions display several enzymatic activities for phospholipid synthesis and breakdown the rod outer segments seem to be devoid of these pathways (Swartz & Mitchell, 1970, 1973, 1974). The temperature- and light-stimulated continuous photoreceptor membrane renewal points to an active and specialized membrane biosynthetic process in the inner segment. Moreover, Bibb & Young (1974) concluded from autoradiographic analysis that phospholipids formed from (2-^{3}H)-glycerol are concentrated initially in the inner segments, then aggregated at the base of rod outer segments and later appear in the outer segments. Thus an

---

Abbreviations: PA, phosphatidic acid; DG, diacylglycerol; TG, triacylglycerols; IPG, CPG, EPG and SPG, inositol-, choline-, ethanolamine- and serine-phosphoglyceride.

efficient biosynthesis of phospholipids must sustain membrane biogenesis in the retina. The catabolism of photoreceptor membranes takes place in the pigment epithelial cell where active phospholipases A_1 and A_2 were found (Swartz & Mitchell, 1973). Ultrastructural and autoradiographic studies reveal that the membranes are taken up from the apex of the outer segment by the pigment epithelial cell where they are digested in the phagolysosomes (Daemen, 1973; Bibb & Young, 1974).

In our laboratory we have been exploring the composition of endogenous intermediates of lipid metabolism (Aveldaño & Bazân, 1972, 1973, 1974, 1975, 1977) and several aspects of lipid metabolism including the biosynthesis of complex lipids (Bazân et al., 1976a; 1976b; Bazân & Bazân, 1975, 1976, 1977; Giusto & Bazân, 1975, 1977; Aveldaño & Bazân, 1977) in vertebrate retinas. The present paper summarizes studies carried out to understand the neobiosynthetic flux in glycerolipids prevailing in the entire retina *in vitro* and about the factors that might give a clue on the regulation of these pathways.

METHODS

Cattle retinas were incubated in an ionic medium, pH 7.4, containing 2 mg / ml of glucose gassed with 5 % CO_2 in oxygen (Giusto & Bazân, 1977). Radioactive precursors were added after 20 min of preincubation, when drugs were employed they were present throughout the preincubation and incubation periods. Lipids were extracted by grinding with chloroform - methanol 2:1 by volume by means of a motor-driven teflon pestle in a Potter - Elvehjam homogenizer. Diacylglycerol and triacylglycerol were separated by gradient-thickness thin-layer chromatography (Bazân & Bazân, 1975). Monoacylglycerols were isolated as described by Porcellati & Binaglia (1976). Phospholipid classes were separated by two-dimentional TLC (Rouser et al., 1970).

Radioactivity was measured in uneluted TLC chromatographic spots, after mixing with Omnifluor in Toluene, by liquid scintillation counting. When $|U-^{14}C|$glycerol (specific activity 7.4 mCi/mmole) and $1-^{14}C$-palmitic (specific activity 56.91 cCi/mmole) acid were simultaneously used the Silica gel containing each individual lipid was methanolyzed using 14% BF_3 in methanol and partitioned several times with hexane. Then the palmitic acid labeled-acyl groups from the organic phase and the ^{14}C-labeled glycerol backbone from the hydrophilic phase were quantitatively recovered. In a separate study about this procedure, we found less than 1% cross-contamination between the radioactivity contained in these phases (Ilincheta de Boschero & Bazân, in preparation).

RESULTS AND DISCUSSION

Lipid synthesis from ^{14}C-glycerol in the retina

Intact toad (Bazán & Bazán, 1976) or cattle (Giusto & Bazán, 1975; Bazán et al., 1976a) retinas incubated *in vitro* utilize ^{14}C-glycerol as precursor for the glycerol backbone of lipids. Phosphatidic acid is clearly the first diacylated lipid to be formed. The labeling sequence PA-DG-TG opperates as reported in liver (Akeson et al.,1970). The route of ^{14}C-glycerol entrance into PA should still be further investigated to evaluate the contribution, if any, of the acyl dihydroxyacetone phosphate pathway(Agranoff & Hajra, 1971; Pollock et al., 1975).

In the cattle retina there is a slight time-dependent increment in monoacylglycerol labeling from ^{14}C-glycerol (Table 1). The specific activity of this lipid in comparison with that of other glycerolipids reveals that the acylation of glycerol is a minor pathway for the synthesis of diacylglycerols and triacylglycerols. Porcellati & Binaglia (1976) found that after intracerebral injection of ^{14}C-glycerol a steep peak in monoacylglycerol specific activity precedes that of other glycerolipids, thus suggesting the existence of an active monoacylglycerol route in brain. In the retina, likely there is not an earlier peak in monoacylglycerol labeling because under our experimental conditions the precursor was present in the medium throughout the incubation time.

At early incubation times PA synthesis was the higher both in toad and cattle retina; however, a remarkable time-dependent increase in the incorporation of the precursor in triacylglycerol was found. After 90 min of incubation the labeling of the neutral glyceride surpassed that of all other lipid classes (Giusto & Bazán,

Table 1. Time course of specific activities in cattle retina glycerides using ^{14}C-glycerol as precursor

Incubation time (min)	Monoacylated	Diacylated	Triacylated
	(cpm x 10^{-3}/µmole of lipid)		
10	23	385	148
20	86	1034	800
30	129	1345	2033

1975; Bazán et al., 1976a). Of the polar lipids, phosphatidylinositol displayed the highest specific activity. On the other hand, in the toad retina *in vitro* the specific activity of PI was the largest of lipids other than PA. A high rate of PI *de novo* synthesis from ^{14}C-glycerol was found also by using different incubating conditions (Bazán & Bazán, 1977) and in the toad *in vivo* (Bazán & Bazán, 1976). In the latter study it was shown that in the toad brain also takes place an active synthesis of PI, unlike a non-neural tissue, the choroid.

Effects of propranolol or phentolamine on the metabolism of retinal lipids followed by ^{14}C-glycerol and ^{14}C-palmitic acid

Since the glycerolipids are important components of excitable biomembranes we decided to observe the effects of drugs known to affect membrane functioning on retinal lipid formation. Adrenergic blocking drugs at concentrations known to inhibit dopamine-sensitive adenyl cyclase of the retina (Brown & Makman, 1972, 1973) were used. Nevertheless we found that the actions exerted are not related to their antagonistic properties (see Conclusions). A rapid accumulation of ^{14}C-glycerol in PA takes place when cattle retinas are incubated in the presence of 500 μM of either drug (Table 2). The labeling in all lipid classes located in the metabolic sequence beyond the conversion of PA in DG is decreased (Fig. 1). Thus CPG, EPG and TG synthesis are deeply reduced (Fig. 2). IPG and SPG, on the other hand, displayed a large increment (Fig. 1).

Table 2. Changes due to different concentrations of propranolol or phentolamine on ^{14}C-glycerol lipid labeling in retina

Incubation time 5 min. Details as in Methods.

	AP	DG	TG	IPG	CPG
	(% of specific activities respect to control)				
Propranolol					
10 μM	77	231	136	60	156
500 μM	466	17	44	220	20
Phentolamine					
10 μM	94	196	124	—	94
500 μM	370	41	5	520	25

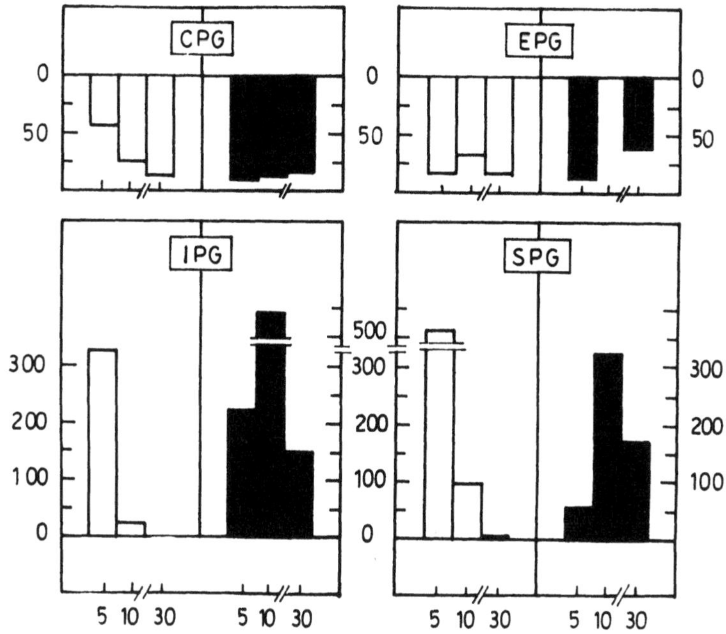

Fig. 1. Changes in ^{14}C-palmitic acid and ^{14}C-glycerol labeling of retina phosphoglycerides by 500 μM propranolol. Values of ordinate: per cent of controls. Abscissa: incubation time in minutes. Black bars, ^{14}C-glycerol; white bars, ^{14}C-palmitic acid.

Phosphatidylserine labeling from ^{14}C-glycerol represents only 1.2, 2.1 and 5% of the labeling of CPG, EPG, and IPG, respectively, after 30 minutes of incubation. Thus the synthesis of phosphatidylserine from ^{14}C-glycerol in the retina is rather low and upon exposure to the drugs an increased labeling takes place. Phosphatidylserine synthesis arises from phosphatidylethanolamine by a Ca^{2+}-stimulated, energy-independent reaction that catalyzes the exchange of ethanolamine for 1-serine (Arienti et al., 1976). Since no enzymatic study has been carried out to date on retina SPG formation we tentatively conclude that the increased SPG synthesis may derive from a small EPG pool, although a drastic reduction in nitrogen-containing phospholipid synthesis is exerted by the drugs (Fig. 1). ^{14}C-glycerol incorporation in EPG still is 10- and 12- fold higher with propranolol and phentolamine respectively. Moreover these drugs may act releasing Ca^{2+} intracellularly, as has been shown to occur in red cell membranes (Porzig, 1975), and consequently bringing about the required concentration of the divalent cation for the base exchange reaction (Arienti et al.,1976).

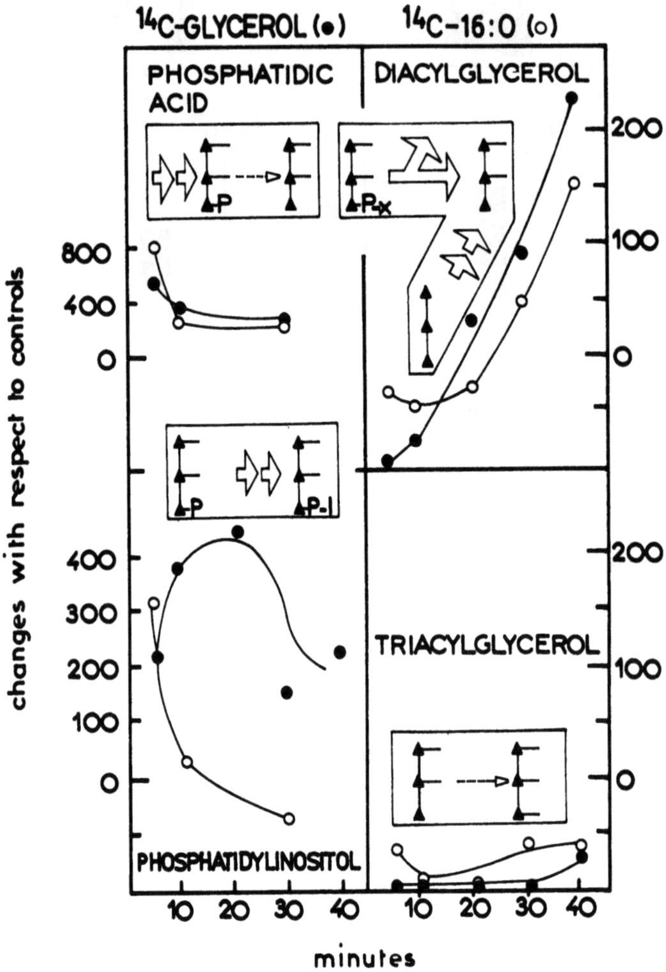

Fig. 2. Propranolol-induced modifications in the synthesis and metabolism of saturated acyl groups of PA, IPG and glycerides. Dashed lines: reduced or inhibited steps and black arrows: increased steps. Open circles: ^{14}C-palmitic acid and black circles: ^{14}C-glycerol. The three joined triangles depict glycerol and the straight lines arising from some of them stand for acyl chains.

A study was performed using cattle retina to inquire about the acylation of ^{14}C-palmitic acid in comparison with the synthesis of lipids from ^{14}C-glycerol. When both precursors were simultaneously added only in PA after 5 min the specific activity of the 3-carbon skeleton was higher than that of the acyl groups. In all the other lipid classes studied, although the incorporation increased for both moieties, the radioactivity of the acyl chains surpassed that of the hydrophilic moiety. ^{14}C-palmitic acid attained the highest specific activity in DG in all the incubation times, followed by PA (Bazán et al., 1976b). Similar labeling profiles by palmitic acid were previously described by Sun & Horrocks (1971) using mouse brain. Retinal CPG and IPG also actively incorporated the labeled fatty acid, by the contrary EPG and SPG were the polar lipids less labeled.

Concentration-dependent effects of propranolol or phentolamine on phosphatidylinositol, glycerides, and nitrogen-containing lipids in the cattle retina

Opposite modifications were encountered in the labeling of retinal lipids by ^{14}C-glycerol when 10 μM and 500 μM concentrations of either drug were added.

At the lower concentration radioactivity in PA and IPG was lower than controls and in diacylglycerols (at 5 min of incubation time) was greatly augmented. TG was slightly enhanced at this low concentration (Table 2).

Dual effect of propranolol or phentolamine on the synthesis of CPG from ^{14}C-choline

The entire cattle retina efficiently utilizes ^{14}C-choline as a precursor for CPG biosynthesis (Ilincheta de Boschero & Bazán, unpublished). Here again, both propranolol or phentolamine produce closely resembling changes. Abdel-Latif (1976) has recently reported that propranolol inhibits the ^{14}C-choline labeling of muscle iris lipids. However this data was obtained after 60 min of incubation. In the cattle retina we were able to show a dual effect of propranolol or phentolamine on the synthesis of phosphatidylcholine from ^{14}C-choline by following the time-course of the lipid labeling. Thus up to 15 min of incubation an increase in CPG synthesis was disclosed whereas afterwards a progresive inhibition of labeling takes place(Ilincheta de Boschero & Bazán, unpublished).

CONCLUSIONS

The effects of propranolol and phentolamine on retinal lipid neosynthesis are similar and likely are unrelated to adrenergic antagonistic properties as previously suggested (Bazán et al., 1976b). Hauser & Eichberg (1975) and Abdel-Latif et al. (1976) reached similar conclusions studying the effect of propranolol on phospholipid metabolism of pineal gland and iris muscles respectively. Brindley & Bowley (1975) using a wide variety of other amphiphilic cationic drugs as well as Allan & Michell (1975), employing liver and lymphocytes respectively, proposed an intracellular site of action (see below).

Dual drug actions take place when a)the drug concentrations are varied, b)the time-course of the diacylglycerol labeling is followed, and c)the incorporation of ^{14}C-choline is surveyed. At higher concentrations than 100 µM an inhibition of PA conversion into DG results. However at concentration of 10 µM a decrease in IPG labeling from ^{14}C-glycerol takes place and an enhanced flow of radioactivity towards CPG, EPG and TG occurs. Thus two different mechanisms are needed to explain these observations. Both stimulatory and inhibitory actions on the phosphatidate phosphatase are suggested to account for the biphasic drug action as well as opposite effects on the conversion of PA in IPG take place (Table 3).

The second dual effect of these drugs was evidenced on diacylglycerol (Fig. 2). During the early stages of the incubation period diacylglycerol labeling from ^{14}C-glycerol was markedly decreased; however, after 15 min of incubation it became increasingly enlarged. The early inhibition is explained by a restriction of the labeled glycerol backbone flow between PA and DG (a, Table 3). After 15 min a clear redirection of the biosynthesis towards IPG with concomitant decreased flux of the label from DG to CPG, EPG and TG occurs. The enhanced radioactivity appearing in DG both from ^{14}C-16:0 and ^{14}C-glycerol may be explained by the presence of at least two pools of diacylglycerols similarly labeled by the precursors, one of which may arise from a step other than the one catalyzed by the PA phosphatase. Either a monoacylglycerol route of diacylglycerol synthesis or a phospholipase C may be involved. However, the close correlation between the labeling by the two precursors does not exclude the possibility of an overflowing from PA to a DG pool unable to become further metabolized (b, Table 3).

The third dual effect was found when ^{14}C-choline labeling was followed. The stimulation observed at early incubation times likely is brought about by a mechanism other than an inhibition of phosphatidate phosphatase (see below). The lowering in ^{14}C-choline incorporation at later incubation times may involve diacylglycerol shortage or a direct action on the route of synthesis from choline.

Table 3. Steps of the neosynthesis of glycerolipids possibly affected by propranolol or phentolamine

Condition or observed change	Steps affected
A - 500 μM of the drugs	PA ⋯→ DG PA → IPG
B - 10 μM of the drugs	PA → DG PA ⋯→ IPG
C - The enhanced DG labeling after 20 min at 250 μM of the drugs may be explained by either of these reactions:	G → PA_1 ⋯a⋯→ DG_1 ⋯→ TG, EPG, CPG; PA_1 —b→ DG_2 G → MG → DG G → X-Glycerolipid (P) → DG + Pi
D - Dual effect on ^{14}C-choline incorporation:	
a) Early incubation times (5 min)	^{14}C-choline → → CPG (DG →)
b) At later incubations times (20 min) either of these changes may occur	^{14}C-choline → → ⋯→ CPG (DG ⋯→) ^{14}C-choline ⋯→ ⋯→ ⋯→ CPG (DG ⋯→)
E - Increased labeling of SPG by drugs (500 μM)	PA ⋯→ DG ⋯→ ⋯→ EPG → SPG

G, glycerol; PA_1, PA_2, DG_1 and DG_2 indicate different pools of PA and DG respectively; Pi, inorganic phosphate.
Dotted lines depict decreased flux through the step and solid lines indicate increased reaction rates.

If these drugs reach the endoplasmic reticulum an altered formation of the diacylglycerol pool specific for the choline-phosphotransferase may explain the inhibition observed. The shift in the biosynthetic flow towards IPG may be due to a rapid inhibition of phosphatidate phosphatase; as a result of this CPG and EPG as well as triacylglycerol synthesis are depressed (Bazán et al., 1976b). This conclusion agrees with the proposal made by Eichberg & Hauser (1974) from studies on the effects of local anesthetics on the ^{32}P labeling of pineal gland phospholipids. Using the same system Eichberg, Hauser and coworkers (1973, 1975) disclosed that propranolol greatly enlarges the ^{32}P labeling of phosphatidic acid, CDP-diglyceride, phosphatidylglycerol and diphosphatidylglycerol. In the retina neither propranolol nor phentolamine were able to stimulate the labeling of diphosphatidylglycerol (unpublished observation). Brindley & Bowley (1975) and Allan & Michell (1975) from studies using other amphiphilic cationic drugs employing non-neural tissues suggested that the negatively charged groups of phosphatidic acid may physicaly interact with the cationic drug preventing catalysis by phosphatidate phosphatase. This mechanism of action explains also the effects observed on retina lipid metabolism at high drug concentrations. However to understand the nature of the changes observed on low drug concentrations one will have to evaluate the possibility of a direct effect on the enzymes or on the membranes where the enzymes are located (Hauser & Eichberg, 1975). Since within 5 min of retina exposure to the drugs the modifications are evident (Ilincheta de Boschero & Bazán, in preparation), they act directly after rapidly entering into the tissue or alternatively they may trigger a change in a cellular metabolite concentration to accomplish the lipid effect. In addition it should still be decided whether or not drug metabolites are partially or entirely responsible for the effects.

Our studies also suggest that a detailed differentiation has to be made between ^{32}P turnover and *de novo* biosynthesis of complex lipids, mainly in the case of IPG and PA. Phosphatidylinositol is a minor component of membranes that has been implicated since long ago in the cell responses to hormones, neurotransmitters, and other agents (Hokin, 1977). In recent years from studies on synaptosomes the proposal has been made that PA phosphatase is stimulated by acetylcholine (Yagihara et al., 1973; Schacht & Agranoff 1973, 1974). It has also been postulated that this strategically located enzyme may be regulable in non-neural tissues (Lamb & Fallon, 1974). Control of PA phosphatase may yield deep changes in membrane lipid composition, mainly when the modification of the biosynthetic pathway affects minor lipids such as IPG (Michell et al., 1976).

Further support for the importance of the neobiosynthetic route of IPG summarized here arises from the finding of a high rate of IPG formation from ^{14}C-glycerol, both *in situ* and *in vitro* that takes place in retina and brain of the toad (Bazán & Bazán, 1976, 1977). Thus the retina, as well as other parts of the nervous system (Baker & Thompson, 1972), is also able to carry on active IPG neobiosynthesis. This pathway may also be modified by light flashes (Bazán & Bazán, 1977) and divalent cations (Bazán & Bazán, 1977; Giusto & Bazán, 1977).

REFERENCES

ABDEL-LATIF, A. A. (1976) in Function and Metabolism of Phospholipids in the Central and Peripheral Nervous Systems (Porcellati, G., Amaducci, L. and Galli, C, eds.) pp. 227-256, Plenum, New York.
ABDEL-LATIF, A. A., OWEN, M. P. & MATHENY, J. L. (1976) Biochem. Pharmacol. 25, 461-469.
AGRANOFF, B. M. & HAJRA, A. K. (1971) Proc. Natl. Acad. Sci. USA 68, 411-415.
AKESON, B., ELOVSON, J. & ARVIDSON, G. (1970) Biochim. Biophys. Acta 210, 15-27.
ALLAN, D. & MICHELL, R. H. (1975) Biochem. J. 148, 471-478.
ANDERSON, R. (1970) Exp. Eye Res. 10, 339-344.
ARIENTI, G., BRUNETTI, M., GAITI, A., ORLANDO, P. & PORCELLATI, G. (1976) in Function and Metabolism of Phospholipids in the Central and Peripheral Nervous Systems (Porcellati, G., Amaducci, L. and Galli, C., eds.) pp. 63-78, Plenum, New York.
AVELDAÑO, M. I. & BAZAN, N. G. (1972) Biochem. Biophys. Res. Commun. 48, 689-693.
AVELDAÑO, M. I. & BAZAN, N. G. (1973) Biochim. Biophys. Acta 296, 1-9.
AVELDAÑO, M. I. & BAZAN, N. G. (1974) FEBS Lett. 40, 53-56.
AVELDAÑO, M. I. & BAZAN, N. G. (1975) J. Neurochem. 23, 1127-1135.
AVELDAÑO, M. I. & BAZAN, N. G. (1977) This volume.
BAKER, R. R. & THOMPSON, W. (1972) Biochim. Biophys. Acta 270, 489-503.
BAZAN, H. E. P. & BAZAN , N. G. (1975) Life Sci. 17, 1671-1678.
BAZAN, H. E. P. & BAZAN , N. G, (1976) J. Neurochem. 27, 1051-1057.
BAZAN, H. E. P. & BAZAN , N. G. (1977) This volume.
BAZAN, N. G. & BAZAN, H. E. P. (1975) in Research Methods in Neurochemistry (Marks, N. and Rodnight, R., eds.) Vol. 3, pp. 309-324, Plenum, New York.
BAZAN, N. G., AVELDAÑO, M. I., PASCUAL de BAZAN, H. E. & GIUSTO, N. M. (1976a) in Lipids (Paoletti, R., Porcellati, G. and Jacini, G., eds.) Vol. 1, pp. 89-97, Raven Press, New York.
BAZAN, N. G., ILINCHETA de BOSCHERO, M. G., GIUSTO, N. M. & PASCUAL de BAZAN, H. E. (1976b) in Function and Metabolism of Phospholipids in the Central and Peripheral Nervous System (Porcellati, G, Amaducci, L. & Galli, C., eds.) pp. 139-148, Plenum, New York.
BIBB, C. & YOUNG, R. W. (1974) J. Cell. Biol. 62, 378-389.
BRINDLEY, D. N. & BOWLEY, M. (1975) Biochem. J. 148, 461-469.
BROWN, J. H. & MAKMAN, M. H. (1972) Proc. Natl. Acad. Sci. USA 69, 539-543.
BROWN, J. H. & MAKMAN, M. H. (1973) J. Neurochem. 21, 477-479.
DAEMEN, F. J. M. (1973) Biochim. Biophys. Acta 300, 255-288.
EICHBERG, J., SHEIN, H. M., SCHWARTZ, M. & HAUSER, G. (1973) J. Biol. Chem. 248, 3615-3622.
EICHBERG, J. & HAUSER, G. (1974) Biochem. Biophys. Res. Commun. 60, 1460-1467.

GIUSTO, N. M. & BAZAN, N. G. (1975) in Abstracts Book of 5th Internat. Meet. Internat. Soc. Neurochem. pp. 378.
GIUSTO, N. M. & BAZAN, N. G. (1977) This volume.
HAUSER, G. & EICHBERG, J. (1975) J. Biol. Chem. 250, 105-112.
HOKIN - NEAVERSON, M. (1977) This volume.
LAMB, R. G. & FALLON, H. J. (1974) Biochim. Biophys. Acta 348, 167-178.
MASON, W. T., FAGER, R. S. & ABRAHAMSON, E. W. (1973) Biochemistry 12, 2147-2150.
MICHELL, R. H., ALLAN, D. & FINEAN, J. B. (1976) in Function and Metabolism of Phospholipids in the Central and Peripheral Nervous Systems (Porcellati, G., Amaducci, L. and Galli, C.,eds.) pp. 3-13, Plenum, New York.
POLLOCK, R. J., HAJRA, A. K. & AGRANOFF, B. W. (1975) Biochim. Biophys. Acta 380, 421-435.
PORCELLATI, G. & BINAGLIA, L. (1976) in Lipids (Paoletti, R., Porcellati G. and Jacini, G., eds.) Vol. 1, pp. 75-88 , Raven Press, New York.
PORZIG, H. (1975) J. Physiol. 249, 27-49.
ROUSER, G., FLEISCHER, S. & YAMAMOTO, A. (1970) Lipids 5, 494-496.
SCHACHT, J. & AGRANOFF, B. W. (1973) Biochem. Biophys. Res. Commun. 50, 934-941.
SCHACHT, J. & AGRANOFF, B. W. (1974) J. Biol. Chem. 249, 1551-1557.
SUN, G. Y. & HORROCKS, L. (1971) J. Neurochem. 18, 1963-1969.
SWARTZ, J. G. & MITCHELL, J. E. (1970) J. Lipid Res. 11, 544-550.
SWARTS, J. G. & MITCHELL, J. E. (1973) Biochemistry 12, 5273-5278.
SWARTS, J. G. & MITCHELL, J. E. (1974) Biochemistry 24, 5053-5059.
YAGIHARA, Y., BLEASDALE, J. E. & HAWTHORNE, J. N. (1973) J. Neurochem. 21, 173-190.

PHOSPHATIDATE, PHOSPHATIDYLINOSITOL, DIACYLGLYCEROLS, AND FREE FATTY ACIDS IN THE BRAIN FOLLOWING ELECTROSHOCK, ANOXIA, OR ISCHEMIA

E.B Rodríguez de Turco, G.D. Cascone, M.F. Pediconi, and N.G. Bazán

Instituto de Investigaciones Bioquímicas, Universidad Nacional del Sur y Consejo Nacional de Investigaciones Científicas y Técnicas, Bahía Blanca, Argentina

INTRODUCTION

The first changes clearly established to occur in membrane lipids of the adult brain in situ at the onset of ischemia or after electroshock are the production of free fatty acids (Bazán, 1970; Bazán & Rascowski, 1970; Bazán, 1976) and of diacylglycerols (Aveldaño & Bazán, 1975). On the other hand, in the CNS of newborn mammalians (Bazán, 1971) and of an adult poikilotherm (Aveldaño & Bazán, 1975) ischemia produces free fatty acids at rates lower by several-fold. Thus it was deemed of interest to study the composition and labeling of lipids in newborn brain during anoxia and in the adult brain after electroshock or ischemia.

METHODS

Adult Balb/c mice of both sexes weighing between 20-24g were used. Ischemia was produced by decapitation at 20-24°C. Electroshock was applied after placing platinum needle electrodes under the skin of the head. Unidirectional rectangular pulses (0.1 ms at 100 V) were applied for 1 s (frequency = 140 Hz). Six µl containing 0.3 µCi of |1- ^{14}C|-arachidonic acid (58 mCi/mM, The Radiochemical Centre) was injected intracerebrally in the area of the lateral ventricle. For the anoxia experiments, mice within 10 hours of birth were placed in a jar containing oxygen-free nitrogen for 50 minutes at 20-24°C. Immediately after, a brain homogenate was made in Krebs-Ringer Bicarbonate pH 7.3, containing 2 mg of glucose per ml under 5% CO_2 in oxygen. The radioactive precursors added were 0.6 µCi of ^{14}C-glycerol (5-10 mCi/mmol, New England Nuclear,

Mass.) or 3 μCi of ^{14}C-arachidonic acid per 400 mg of wet tissue weight.

Lipid extraction in all instances was performed as described elsewhere (Bazán & Bazán, 1975). Lipid classes were isolated by thin layer chromatography (Rouser et al., 1970; Bazán & Bazán, 1975) and radioactivity was determined in uneluted spots from thin-layer chromatograms after mixing with Omnifluor (New England Nuclear Comp., Mass) in toluene and then counting in a Packard Tricarb liquid-scintillation spectrometer. The fatty acid composition was assessed by gas-liquid chromatography (Aveldaño & Bazán, 1975) and PA was separated by the oblique spotting technique (Rodríguez de Turco & Bazán, 1976).

RESULTS

Effects of ischemia and electroshock on ^{14}C-arachidonic acid labeling of brain lipids

A rapid removal of the labeled free 20:4 was observed following intracerebral administration of the fatty acid in agreement with

Table 1. Changes in ^{14}C-arachidonic acid incorporation in lipids of the adult brain by electroshock or ischemia

Figures represent % distribution of radioactivity at 1 and 10 minutes after intracerebral injection of labeled fatty acid. Postdecapitation ischemia or electroshock were carried out 10 seconds after injection of the labeled fatty acid. Experimental details as in Methods.

	Control		Ischemia		Electroshock	
	1	10	1	10	1	10
			(minutes)			
Free fatty acids	70	10	93	85	90	50
Total phospholipids	19	60	5	12	8	37
Triacylglycerols	7	20	0.5	1	1	8
Diacylglycerols	4	10	1	2	2	4
Phosphatidate	4	2	1	1	2	3
Phosphatidylinositol	4	17	1	4	2	11

Yau and Sun (1974). However electroshock or ischemia markedly reduced the labeling of brain lipids by ^{14}C-arachidonic acid. If electrical stimulation is applied 10 seconds after the injection of the fatty acid, 50% of the radioactivity remains in the free fatty acid fraction 10 minutes later. Moreover only 8 and 37% of the labeled fatty acid was present in phospholipids after 1 and 10 min of electroshock, respectively, as compared with 19 and 60% in control animals (Table 1).

Postdecapitation ischemia also deeply inhibits the entrance of ^{14}C-arachidonic acid into brain lipids. Ten minutes after the onset of blood shortage a deep inhibition of lipid labeling was observed and only 12% of radioactivity was in the polar lipids. Ischemia produced a larger decrease in lipid labeling than electroshock; however, in the latter a tendency toward recovery in the incorporation can be seen 10 min after application of the stimulus (Table 1).

Fatty acid composition, ^{14}C-glycerol and ^{14}C-arachidonic acid labeling of newborn brain lipids after anoxia

The greater resistance to oxygen deficiency of the just born brain is accompanied by slower metabolic rates (Thurston & McDougal, 1969) and by lower energy requirements (Duffy et al., 1975) in comparison with the mature brain. Our studies on lipids in relation to anoxia in newborn brain included two aspects. Firstly we observed the fatty acid composition of brain lipid classes and secondly the labeling of brain lipids by ^{14}C-glycerol or ^{14}C-arachidonic acid.

In the present study the newborn mouse was able to survive 40 minutes under the experimental conditions of anoxia employed. Thus a comparison was made between the fatty acids of newborn brain following 40 minutes of anoxia and those of adult mouse brain after 3 minutes of ischemia. This time period of complete circulatory arrest prevents total recovery of CNS functions. Though there are obvious differences between the action of anoxia and ischemia on brain metabolism in general, we found similar changes in the newborn brain fatty acids during anoxia under nitrogen and in previously published data using postdecapitation ischemia (Bazán et al., 1971; Bazán, 1971).

There are striking differences in the resting levels of brain free fatty acids between newborn and adult mice. Free 18:0 and 20:4 in the former are at much lower concentrations. The enhancement in newborn brain free fatty acids during early anoxia is slow, as has been shown before to occur at the onset of ischemia (Bazán, 1971). In the adult brain ischemia produces free fatty acids following a

Table 2. Percent increments in brain free fatty acids after anoxia or ischemia

Newborn mice brain subjected to anoxia as described in Methods were used. The data on ischemia was obtained from adult mouse brain (Aveldaño & Bazán, 1975) following the same procedures used in this study. The selected time periods of anoxia or ischemia were at the time that irreversible impairment of brain function takes place. The figures represent percent increments from controls. In all cases at least three individual samples were analyzed. Each sample of newborn brain was made up of 5-6 brains.

Fatty acids	Anoxia on newborn brain		Ischemia on adult brain	
	40	60	3	10
	(minutes)			
16:0	39	84	62	171
16:1	13	11	6	63
18:0	27	43	82	164
18:1	16	34	57	130
20:4	83	188	198	300
22:6	49	80	244	617
Total	30	63	89	178

steep course during the first 4 minutes of blood shortage (Bazán et al., 1971). In the newborn brain arachidonic acid was produced at higher rates than other fatty acids during N_2 anoxia (Table 2) whereas in the mature brain the larger per cent change between control and 3 minutes of ischemia was accounted for 22:6 (Aveldaño & Bazán, 1975). In the newborn the latter fatty acid increased slightly. Although ischemia also triggers a partial triacylglycerol breakdown (Bazán, 1970), in the newborn brain the neutral glycerides were unchanged (unpublished observations). Furthermore anoxia causes a much slower increase in newborn brain diacylglycerols than was observed in adult brain (Figure 1). Palmitate and oleate were the fatty acids mainly increased in the diacylglycerols, being followed by palmitoleate and docosahexaenoate. Unlike diacylglycerols in adult brain, stearate and arachidonate display only minor modifications.

In agreement with Galli et al. (1976), we found no changes in

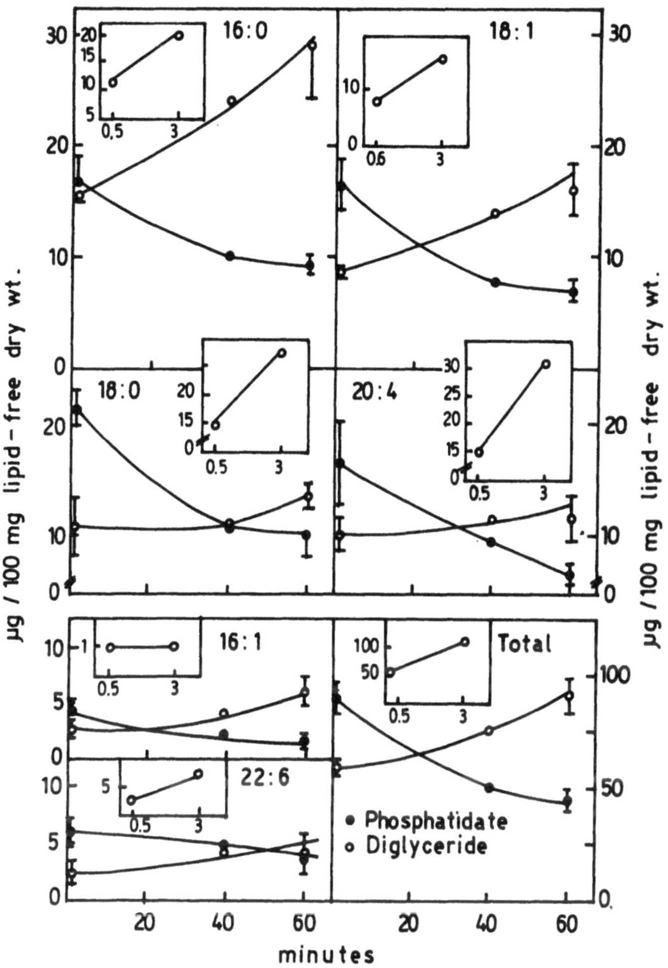

Fig. 1. Modifications in the content of acyl groups in mouse brain diacylglycerols and phosphatidic acid due to anoxia.

Inserts depict the changes in diacylglycerols from adult mouse brain determined by the same procedural outline (Aveldaño & Bazán, 1975). The newborn samples were pools of 5-6 brains. The points without standard derivation are average of at least two samples and the others are averages of four different samples.

total PA nor in their fatty acid composition in adult brain undergoing 3 minutes of ischemia (unpublished observations). In the newborn brain, however, a 50% decrease in PA was found after 60 minutes of anoxia. The acyl groups lost from PA, with the exception of stearate and arachidonate, seem to be responsible for the increase in diacylglycerols (Fig. 1).

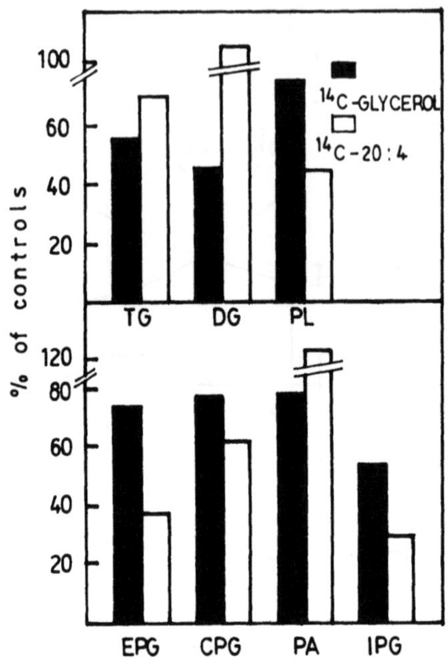

Fig. 2. Changes in the labeling of lipids by ^{14}C-glycerol or ^{14}C-arachidonic acid in brain homogenates of anoxic newborn mice. Anoxia time was 50 minutes as described in Methods. TG : triacylglycerol; DG: diacylglycerol; PL: total phospholipids; EPG: phosphatidylethanolamine; CPG: phosphatidylcholine; PA: phosphatidic acid and IPG: phosphatidylinositol.

Brain homogenates, from anoxic newborn mice, incubated under aerobic conditions scarcely incorporated ^{14}C-glycerol or ^{14}C-arachidonic acid. Phosphatidylinositol reached the lower incorporation in both cases, that is 54 and 29% of controls for ^{14}C-glycerol and ^{14}C-arachidonic acid. Although the labeling by the fatty acid was reduced by less than 50% in total phospholipids, a slight enhancement in PA labeling was found (Fig. 2). The incorporation of the precursor in diacylglycerols following anoxia was not different from controls.

DISCUSSION

Similar changes take place in newborn brain free fatty acids after anoxia or ischemia. Since arachidonic acid is the most rapidly produced fatty acid after prolonged anoxia, a relationship may exist with the tissue capacity to biosynthesize prostaglandins. In mature CNS an active phospholipase A_2 releases arachidonic acid during the first few minutes of ischemia (Bazán, 1970; Bazán, 1976) or after electroshock (Bazán & Rakowski, 1970). A comparatively slow enzymatic system is present in the newborn brain, thus likely the process of releasing 20:4 for prostaglandin synthesis is different. Wolfe and coworkers (1967) early recognized in a non-neural tissue the relationship between nerve stimulation, phospholipase A_2 activation and prostaglandin formation from released arachidonic acid.

We have also evidenced in this paper that ischemia, besides stimulating the production of free fatty acids, causes a rapid inhibition in the incorporation of labeled arachidonic acid into brain lipids, likely by lowering the arachidonoyl thioquinase step due to ATP shortage. Following electroshock a similar effect was disclosed although it is reversible and less pronounced.

The increment in diacylglycerol triggered by anoxia in the newborn brain differs from that observed in the mature ischemic brain (Aveldaño & Bazán, 1975) in that 18:0 and 20:4 only increase very slightly. Thus the lost phosphatidate during anoxia is not quantitatively accumulated in diacylglycerol. Lost 20:4 and 18:0 do not appear in diacylglycerols. Likely anoxia activates phosphatidate breakdown through a phosphatase and some molecular species are accumulated as diglycerides and others are further metabolized (e.g. to monoglycerides).

The stability of mature brain phosphatidate during ischemia appears to be a consequence of a developmental change and points to an interesting peculiarity in the membrane lipid metabolism of the newborn brain. Another feature of PA is that the incorporation of ^{14}C-20:4 is slightly enhanced in homogenates of anoxic newborn brain while in most lipids the labeling is markedly diminished. Diacylglycerol shows a similar tendency, whereas in phosphatidyl inositol a more sharply reduced incorporation has been observed.

The sluggish production of free fatty acids in the newborn brain in response to anoxia and ischemia (Bazán, 1976) may be related to the greater resistance of the newborn CNS to sustaining longer periods under such conditions. Conversely, in the mature brain the rapid production of free fatty acids due to blood shortage may be involved in the irreversible impairment of brain function (Bazán, 1976). It is interesting that not only phospholipase A_2 activity but also a rapid reduction of the incorporation of the fatty acid into the lipids contributes to the accumulation of free arachidonic acid.

REFERENCES

AVELDAÑO, M. I. & BAZAN, N. G. (1974) J. Neurochem. 23, 1127-1135.
AVELDAÑO, M. I. & BAZAN, N. G. (1975) Brain Res. 100, 99-110.
BAZAN, N. G. (1970)Biochim. Biophys. Acta 218, 1-10.
BAZAN, N. G. & RAKOWSKI, H. (1970) Life Sciences 9 : 501.
BAZAN, N. G. (1971) Acta Physiol. Lat. Amer. 21, 15-20.
BAZAN, N. G., BAZAN, H. E. P., KENNEDY, W. P. & JOEL, C. D. (1971) J. Neurochem. 18, 1387-1393.
BAZAN, N. G. & BAZAN, H. E. P. (1975) in Research Methods in Neurochem. (Marks, N. & Rodnight, R. eds.) Vol. 18 pp. 309-329, Plenum Press, New York.
BAZAN, N. G. (1976) in Function and Metabolism of phospholipids in the central and peripheral Nervous Systems. (Porcellati, G., Amaducci, L. & Galli, C. eds.) 317-335, Plenum Press, New York.
CENEDELLA, R. J., GALLI, C. & PAOLETTI, R. (1975) Lipids 10, 290-293.
COCEANI, C., PACE-ASCIAK, C., VOLTA, F. & WOLFE, L. S. (1967) Amer. J. of Phisiol. 213, 1056-1063.
DUFFY, T. E., KOHLE, S. J. & VANNUCCI, R. C. (1975) J. Neurochem. 24, 271-276.
GALLI, C. & SPAGNUOLO, C. (1976) J. Neurochem. 26, 401-404.
RODRIGUEZ de TURCO, E. B. & BAZAN, N. G. (1977) J. Chromatog. (in press).
ROUSER, R., FLEISCHER, S. & YAMAMOTO, A. (1970) Lipids 5, 494-496.
THURSTON, J. H. & Mc DOUGAL, D. B. Jr. (1969) Amer. J. Physiol. 216, 348-352.
YAU, T. M. & SUN, G. Y. (1974) J. Neurochem. 23, 99-104.

ACYL GROUPS, MOLECULAR SPECIES, AND LABELING BY ^{14}C-GLYCEROL AND ^{3}H-ARACHIDONIC ACID OF VERTEBRATE RETINA GLYCEROLIPIDS

M. I. Aveldaño de Caldironi and N. G. Bazán

Instituto de Investigaciones Bioquímicas, Universidad Nacional del Sur and Consejo Nacional de Investigaciones Científicas y Técnicas, Bahía Blanca, Argentina

INTRODUCTION

The toad retina diacylglycerol content is high and enriched in docosahexaenoate, both in comparison with toad brain and mammalian retina (Aveldaño & Bazán, 1973, 1974) as well as with mammalian brain diacylglycerols (Sun, 1970; Keough et al., 1972; Aveldaño & Bazán, 1975) where arachidonic acid predominates. This suggested that diacylglycerols in the toad retina may be uniquely engaged in the metabolism of 22:6-containing membrane lipids. Here we report an analysis of acyl groups and molecular species showing further peculiarities in toad retina lipids. In addition the cattle and toad retina glycerolipid labeling by ^{14}C-glycerol and ^{3}H-arachidonic acid is surveyed.

MATERIALS AND METHODS

Toad (Bufo arenarum, Hensel) brains and retinas were excised immediately after decapitation. Cattle eyes obtained from a local slaughterhouse were kept within crushed ice and used within 2 hours. Cattle and toad retinas were incubated at 37°and 23°C respectively under 5% CO_2 in oxygen in the medium of Ames & Hastings (1956) containing 2 mg/ml glucose. 5 µCi per cattle retina (or per 17 toad retinas) of both |U-^{14}C|-glycerol (8.75 mCi/mmol) and |5,6,8,9,11,12,14,15-^{3}H|-free arachidonic acid (80 Ci/mmol) from New England Nuclear were added to the media.

Abbreviations used: DG, diacylglycerol; TG, triacylglycerol; PA, phosphatidic acid; PC, PE, PI and PS, phosphatidyl -choline, -ethanolamine, -inositol and -serine; TLC, thin-layer chromatography.

Lipid extracts were prepared and washed according to Folch et al. (1957), then were dried under N_2 and applied as oblique bands on preparative two-dimensional TLC plates (Rodríguez de Turco & Bazán, 1977). The lipids were scraped off and eluted according to Arvidson (1968). Neutral lipids were rechromatographed as described elsewhere (Aveldaño & Bazán, 1974).

For the analysis of molecular species, PC and PE from toad brain and retina were subject to phospholipase C treatment (Parkes & Thompson, 1973, 1975). The diacylglycerols were purified by boric acid-impregnated TLC, acetylated and resolved by argentation TLC in two solvent systems (Fig. 2).

Fatty acid analysis was carried out by gas-liquid chromatography after transmethylation with 14% BF_3 in methanol. Quantitation of neutral lipids was performed by this method, and that of phospholipids by phosphorus measurement (Rouser et al., 1970). Radioactivity was determined in a Packard Tri-Carb spectrometer.

RESULTS AND DISCUSSION

The fatty acid composition of toad brain and retina glycerolipids is comparatively presented in Fig. 1. The toad brain lipids, although being more unsaturated as expected for a poikilotherm, share the general features known to be present in mammalian brain: high concentration of arachidonate in PI and PE, and of docosahexaenoate in PS, PC on the other hand being poor in polyenoic constituents.

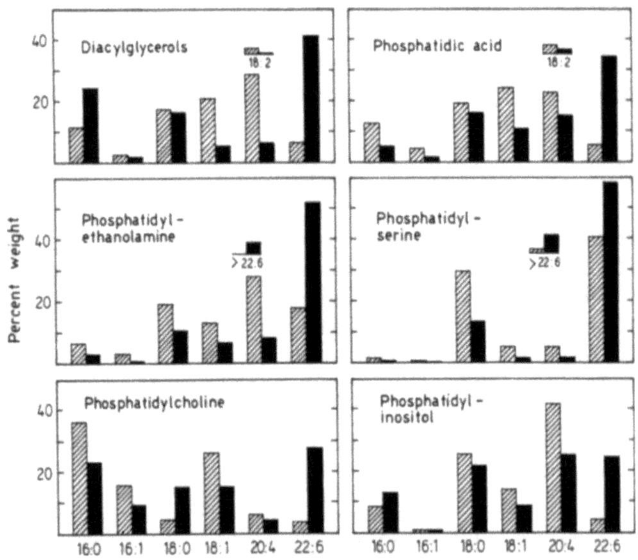

Fig. 1. Fatty acid distribution in toad brain (▨) and retina (■) glycerolipids.

Toad brain phosphatidic acid shows a similar distribution but higher percentages of arachidonic and docosahexaenoic acids than those found in rat (Baker & Thompson, 1972) and mouse brain (Rodríguez de Turco & Bazán, unpublished). Diacylglycerols are as in these species (Sun, 1970; Keough et al., 1972; O'Brien & Geison, 1974; Aveldaño & Bazán, 1975; Porcellati & Binaglia, 1976) predominantly tetraenoic (Figs. 1 and 2). Thus DG and PA fatty acid composition slightly differ, unlike the pattern encountered in rat brain (Baker & Thompson, 1972 and Keough et al., 1972; Porcellati & Binaglia, 1976).

Large quantities of docosahexaenoate are present in all retina glycerolipids, due to the contribution of lipids from the outer segments of photoreceptors, which are known to be enriched in this fatty acid. However, the same holds for diacylglycerols and phosphatidic acid, which likely are not concentrated in these structures. The unusual composition of these intermediates suggests that an important proportion of toad retinal lipid metabolism may be involved in supporting photoreceptor lipid composition. The concentration of both DG and PA is 4-5 fold higher in the toad than in the cattle retina (Aveldaño & Bazán, 1974 and unpublished).

The sum of 22:6 *plus* 20:4 amounts to 50% both in PA and DG,

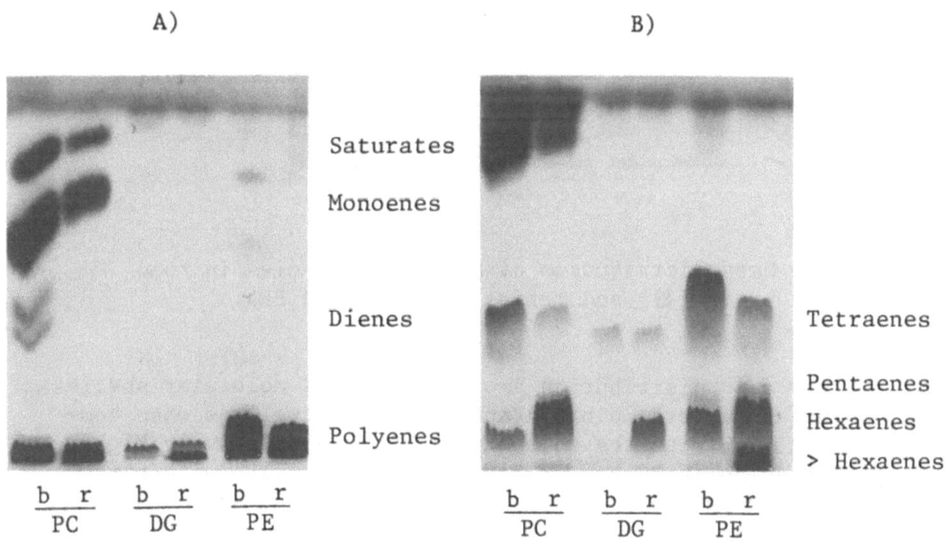

Fig. 2. Molecular species distribution in toad brain (b) and retina (r) phosphatidylcholines, diacylglycerols and phosphatidylethanolamines. Equal aliquots of acetylated diacylglycerols were spotted on each plate (silica gel G-20% $AgNO_3$) and developed in chloroform-methanol A) 99:1 and B) 94:6

indicating that the bulk of these lipids are made up by tetra and hexaenoic molecular species. Fig. 2 shows this is the case in retina diacylglycerols. The molecular species distribution in toad brain and retina PC and PE is also illustrated in this figure and their relative amounts are shown in Fig. 3. Whereas in brain PC the monoenoic species predominates, as does in rat (O'Brien & Geison, 1974; MacDonald et al., 1975; Porcellati & Binaglia, 1976) and rabbit brain (Bräuning & Gercken, 1976), in retina the major amount corresponds to the hexaenoic and then to the monoenoic. PEs also show a remarkably contrastant species distribution in both tissues: tetraenes preponderate in the brain, while hexaenes and a species even more unsaturated prevail in retina.

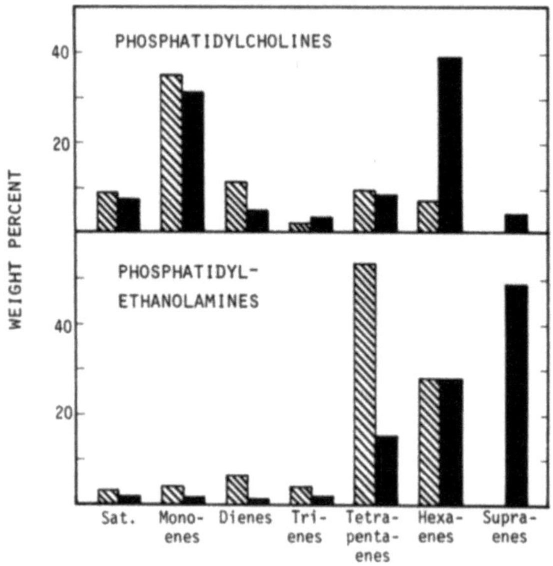

Fig. 3. Mass distribution of molecular species in toad brain (▧) and retina (■) PCs and PEs.

The fatty acid distribution among individual molecular species showed a similar pattern in brain and retina, as well as when comparing PCs and PEs, with the sole exception of a relatively higher amount of C16 acids in PC and of C18 acids in PE. The main fatty acids in the species shown in Fig. 3 can be arranged as follows: saturates: 14:0, 16:0, 18:0 (70-91%); monoenes: 16:0, 18:0, 16:1, 18:1 (80-96%); dienes: 16:0, 18:2, 16:1, 18:1 (64-83%); tetraenes: 16:0, 18:0, 20:4 (80-84%); hexaenes: 16:0, 18:0, 22:6 (67-90%). In the "supraene" species, present only in retina, 22:6 alone accounted for 60-65% of the total acyl groups, the rest being formed by several long-chain highly unsaturated fatty acids such as 20:4, 22:4, 22:5 and 24:6. An outstanding characteristic of this species is that it virtually lacks saturated fatty acids, indicating that both positions of the glycerol backbone are occupied by polyenoic acids. Positional analysis of fatty acids in unfractionated PE from frog rod outer segment fragments revealed a 42% of polyunsaturated fatty acids among the components of the 1-position (Anderson & Risk, 1974).

Cattle retina glycerolipids also contain high percentages of polyenoic constituents (Table 1). However the distribution of acyl groups shows a more definite tendency than in the toad retina. Thus as found by Anderson et al., (1970) in the whole cattle retina, PS is enriched in 18:0-22:6 and PI in 18:0-20:4 pairs respectively (Table 1) whereas in toad retina 22:6 is also an important component of PI (Fig. 1). Although in DPG 18:1 and 18:2 predominate, it seems a peculiarity of retina the high content of polyenoic acids in this lipid. No studies have been made on retina cardiolipin. Anderson et al., (1975) were unable to find it even in the mitochondrial fraction, perhaps due to the two-dimensional TLC system employed. Phosphatidic acid shows an intermediate composition, standing out the relatively high level of docosahexaenoate. When comparing the fatty acid distribution in this lipid and its products, it becomes an attractive idea to assign a strong specificity of biosynthetic enzymes towards certain molecular species of PA. Specificity towards tetraenoic species in the biosynthesis of PI is being currently studied (Thompson, this volume). In addition several lines of evidence have indicated the importance of acyl and base exchange reactions in determining the composition of phosphoglycerides.

Table 1. Fatty acid composition of cattle retina glycerolipids.

Fatty acid, %	PA	PI	PS	DPG**	DG*	PC	PE	TG*
16:0	19.0	9.3	0.8	6.0	18.6	37.7	6.2	24.6
16:1	1.8	0.5	0.1	2.9	1.3	1.4	0.3	3.9
18:0	26.1	33.9	34.5	2.5	27.1	14.8	23.9	16.9
18:1	11.4	6.2	6.9	43.5	8.2	16.6	4.9	17.6
18:2	4.0	0.5	0.1	17.4	1.1	1.1	0.7	1.9
20:4	9.6	37.0	1.9	6.7	27.2	4.6	7.7	5.3
22:5	1.8	1.2	3.6	1.2	1.0	1.4	2.5	1.6
22:6	16.7	5.9	39.3	8.4	6.6	17.2	42.0	17.2

** diphosphatidylglycerol; *Data taken from Aveldaño & Bazán, 1974.

In cattle retina diacylglycerols 18:0-20:4 predominate (Table 1) a combination which does not appear in any of the lipids derived from them but in PI, as shown in the rat brain (Keough et al., 1972). Based on this fact and in the presence of active phosphoinositide phosphodiesterases, these authors proposed a relationship between arachidonoyl diglycerides and the phosphoinositides in the central nervous system. The rapid production of diacylglycerols enriched in 18:0-20:4 that takes place in the mammalian brain during ischemia (Aveldaño & Bazán, 1975), may be related to this possibility. By analogy we suggested that such a metabolic link might also be present in the mammalian retina (Aveldaño & Bazán, 1974).

To inquire about the relations among some glycerolipids in the cattle retina we undertook a time-course study on the simultaneous uptake of ^{14}C-glycerol and ^{3}H-arachidonic acid. A comparison is made with the toad (Table 2). The specific activities of cattle retina lipids show a rapid labeling by glycerol of PA at 5 min incubation, followed by a steadily increasing accumulation of label in the other lipids, particularly diacyl- and triacylglycerols.

Table 2. Specific activities of cattle and toad retina lipids during incubation in the presence of ^{14}C-glycerol and ^{3}H-arachidonic acid

Retina	Incubation time (min)	PC	PE	PS	PI	PA	TG	DG
		(cpm/μmole of lipid)						
^{14}C-glycerol								
	5	60	98	33	1339	64833	1808	23224
Cattle	15	1170	1430	224	13805	160560	58300	364837
	30	6593	4325	259	29492	272200	614789	928653
Toad	30	282	320	93	10088	16333	6384	8230
^{3}H-arachidonic acid								
	5	14	787	13	522	22611	11510	46467
Cattle	15	514	1427	77	3597	54960	24777	155563
	30	1809	2283	65	8942	76300	179873	314173
Toad	30	99	1015	48	2465	7250	3309	23395

With arachidonate on the other hand, the highest specific activities are found in diacylglycerols. The marked TG labeling with both precursors is consistent with previous results in retina (Bazán et al., 1976) and brain (MacDonald et al., 1975), pointing to the high metabolic activity of these lipids in neural tissues. The large ^{3}H/^{14}C ratio in TG at early incubation times stresses the extreme lability of arachidonoyl-TG in retina and the comparison with that found in DG clearly indicates that they turn over independently of *de novo* biosynthetic mechanisms. MacDonald et al. (1975) demonstrated that 20:4 is mainly located in position 1 (3) of the glycerol backbone of brain TG, in contrast with DG, where it is almost exclusively at the 2- position. Based on the highly unsaturated nature of TG (Table 1) we suggested that they might function as acyl donors in neural tissues (Aveldaño & Bazán, 1974).

PE is also more heavily labeled with arachidonate than with glycerol at early incubation times, in evident contrast with PI. Taking into account the contribution of 20:4 to the total mass of both lipids (Table 1) a more active turnover of tetraenoic PE as compared to PI becomes apparent. The high $^{3}H/^{14}C$ ratio in PE seems to be a character of retina, since it is also present in the toad retina (Tables 2 and 3). The slow labeling of PI with arachidonate in cattle retina points to an interesting contrast with the results obtained in the rat brain *in vivo* (Baker & Thompson, 1972). The differences in the rates of arachidonate uptake and $^{3}H/^{14}C$ ratios in PI and DG at early incubation times suggest that arachidonoyl-DG are actually involved in biosynthetic routes. Thus the relationship between DG and PI in brain and retina cannot be excluded nor sustained without proposing a compartmentation of diglyceride pools.

Table 3. Distribution of radioactivity from ^{14}C-glycerol and ^{3}H-arachidonic acid among retina glycerolipids after 30 min incubation (%)

Lipid	^{14}C-glycerol		^{3}H-arachidonate	
	cattle	toad	cattle	toad
PC	25.5	20.0	20.3	6.3
PE	10.5	19.4	14.0	49.3
PI	11.0	36.2	9.9	8.4
PA	7.7	8.8	7.6	3.4
TG	18.2	4.0	16.9	1.8
DG	26.6	9.9	28.3	25.1

As shown by Bazán & Bazán (1976), the *de novo* biosynthetic pathway in the toad retina favours the route PA⟶PI (Table 2). Diglyceride labeling by ^{14}C-glycerol is relatively poor after 30 min incubation, whereas they attain the highest specific activity of retina lipids with ^{3}H-arachidonate. This is an interesting finding in view of the low contribution of tetraenoic DG to total diacylglycerols (Figs. 1 and 2). Triglyceride synthesis is not such an active process in the toad as in the cattle retina, as seen by ^{14}C-glycerol and ^{3}H-arachidonate incorporation (Tables 2 and 3). These contrasts may be related to the presence in the toad choroid, but not in the cattle, of a large and highly unsaturated TG pool (Aveldaño & Bazán, 1974). These lipids take up *in vivo* as much radioactivity from ^{14}C-glycerol as do polar lipids in this tissue, whereas in retina most of the label is concentrated in polar lipids (Bazán & Bazán, 1976, and Table 3).

The distribution of radioactivity among retina lipids after 30 min incubation shows that in the toad retina the preferred flow of glycerol is towards PI, whereas arachidonate gets predominantly into PE (Table 3). By the contrary, the label from both precursors in cattle retina lipids is similarly distributed.

REFERENCES

AMES, A. III & HASTINGS, A.B. (1956) J. Neurophysiol. 19, 201-212

ANDERSON R.E., FELDMAN L.S., & FELDMAN, G.L. (1970) Biochim. Biophys. Acta 202, 367-373

ANDERSON, R.E., MAUDE, M.B. & ZIMMERMAN, W. (1975) Vision Res. 15, 1087-1090

ANDERSON, R.E. & RISK, M. (1974) Vision Res. 14, 129-131

ARVIDSON, G.A.E. (1968) Eur. J. Biochem. 4, 478-486

AVELDAÑO, M.I. & BAZÁN, N.G. (1973) Biochim. Biophys. Acta, 296, 1-9

AVELDAÑO, M.I. & BAZÁN, N.G. (1974) J. Neurochem. 23, 1127-1135

AVELDAÑO, M.I. & BAZÁN, N.G. (1975) J. Neurochem. 26, 919-920

BAKER R.R. & THOMPSON, W. (1972) Biochim. Biophys. Acta 270, 489-503

BAZÁN, H.E.P. & BAZÁN, N.G. (1976) J. Neurochem. 27, 1051-1057

BAZÁN, N.G., AVELDAÑO, M.I., BAZÁN, H.E.P. & GIUSTO, N.M. in *Lipids*, (R. Paoletti, G. Porcellati & G. Jacini Eds.) Raven Press, N. York, pp. 89

BRÄUNING, C & GERCKEN G. (1976) J. Neurochem. 26, 1257-1261

FOLCH, J., LEES, M. & SLOANE STANLEY, G.H. (1957) J. biol. Chem. 226,497-509

KEOUGH, K.M.W., MACDONALD, G. & THOMPSON, W (1972) Biochim. Biophys. Acta 270, 337-347

MACDONALD, G., BAKER, R.R. & THOMPSON, W. (1975) J. Neurochem. 24, 665-661

O'BRIEN, J.F. & GEISON, R.L. (1974) J. Lipid Res. 15, 44-49

PARKES, J.G. & THOMPSON, W. (1973) J. biol. Chem. 248, 6655-6662

PARKES, J.G. & THOMPSON, W. (1975) Can. J. Biochem. 53, 698-705

PORCELLATI, G. & BINAGLIA, L. (1976) in *Lipids*, (R. Paoletti, G. Porcellati & G. Jacini, Eds.), Raven Press, N. York, pp 75.

RODRÍGUEZ DE TURCO, E.B. & BAZÁN, N.G. (1977) J. Chromatog. (in press)

ROUSER, G., FLEISCHER, S. & YAMAMOTO, A. (1970) Lipids 5, 494-496

SUN, G.Y. (1970) J. Neurochem. 17, 445-446

III. LIPIDS IN NEURAL TISSUE

(B) FUNCTIONS

EFFECT OF PHOSPHOLIPID LIPOSOMES ON THE REGULATION OF CEREBRAL METABOLISM

G. Toffano, A. Leon, G. Savoini, and P. Orlando*

-F.I.D.I.A. Res. Lab.- Abano Terme, Italy

* Institute of Pharmacology, University S. Cuore, Rome, Italy

INTRODUCTION

In studying the pharmacological effects of a sonicated dispersion of bovine brain phospholipids on CNS, the following observations have been made: a) in animals, increase of both the conditioning avoidance response and motor activity (Toffano et al., 1976b) increase of brain glucose (Bruni et al., 1976a; 1976b) and of brain phospholipid synthesis (Orlando et al., 1976); and b) in men, decrease of prolactin secretion (Polleri et al., unpublished).

All the above parameters are affected by Dopamine or by dopamine-like drugs. Brown et al. (1973) suggested that DA is involved in the behavior performances. Tyce & Owen (1973)showed that L-dopa increases brain glucose and brain/blood glucose ratio, while Shwartz et al. (1975) reported that 6-OH-DA, which disrupts catecholaminergic neurons, decreases brain glucose uptake. As showed by many authors (Friedel et al., 1974; Hokin, 1970; Abdel-Latif et al., 1974) catecholamines affect phospholipid synthesis, in vitro and in vivo. Moreover, serum prolactin level is known to be under the inhibitory control of DA along the hypothalamic-pituitary axis (McLeod & Lehmeyrer, 1974). Hence an attempt has been made: (i) to investigate the possible effects of exogenous phospholipids on brain dopamine and on its related enzymatic activities; (ii), to determine the active components of total phospholipid mixture. It appears that exogenous bovine brain phospholipids affect dopaminergic system and that phosphatidylserine is the most active component of phospholipid mixture.

EXPERIMENTS

Liposome preparation

The composition of phospholipid mixture from bovine cerebral cortex (BC-PL) and the preparation of single phospholipid classes, to identify the active components of lipid mixture, were previously reported (Bruni et al., 1976a; Maniero et al., 1973). The phospholipid composition of BC-PL mixture is the following: 6.6 % phosphatidic acid, 20.2 % phosphatidylethanolamine, 18.7 % phosphatidylserine, 15 % lysophosphatidylethanolamine (serine), 29.9 % phosphatidylcholine and 9.4 % sphyngomielin. Liposomes were obtained by sonication at 0°C of a phospholipid solution suspended in 50mM Tris-HCl, pH 7.5, in a MSE apparatus for 8 minutes. Suspensions were injected intravenously in albino male mice or Sprague-Dowley rats weighing 20-25 g and 120-150 g respectively.

Dopamine and homovanillic acid assay

Three pooled brains of decapitated mice were homogenized with five volumes of cold 0.4N $HClO_4$ with 0.1% $Na_2S_2O_5$ and centrifuged at 10,000 x g for 15'. DA (Dopamine) was assayed in 3 ml of the clear supernatant according to Chang (1964), while HVA (homovanillic acid) was assayed in 3 ml of the same supernatant according to Korf et al. (1971).

Cyclic AMP

Mice were killed by decapitation so as to allow the head to fall immediately into liquid nitrogen. Brain hemispheres were removed and powedered under liquid nitrogen together with 1 ml 3% trichloroacetic acid. Cyclic AMP was separated according to Krishna et al. (1968) and assayed according to Kuo & Greengard (1970).

Adenylate Cyclase

Adenylate cyclase assay was carried out into mice brain crude mitochondrial fraction by procedure based on the method of Krishna et al.(1968).

Tyrosine hydroxylase

Rats were sacrified by decapitation, the striata removed and gently homogenized in 50 mM Tris-acetate buffer, pH 6.0, containing 0.2% Triton X-100. Tyrosine hydroxylase was tested into the supernatant according to Zivkovic et al.(1975)

Protein

Protein was measured by the method of Lowry et al. (1951) using bovine serum albumin as a standard.

RESULTS

1) Effect of BC-PL on Dopaminergic System and on its Related activities

Fig. 1 shows that, in normal mice, liposomes of cerebral phospholipids induce a rapid fall of brain DA. The parallel increase of HVA suggests that the decrease of DA is due to an increased DA release and not to an inhibition of its synthesis.

To study DA turnover, BC-PL were injected into mice pretreated with α-methyl-p-tyrosine (α-MT) (Fig. 2). α-MT induces catecholamine disappearance by means of an inhibitory effect on tyrosine hydroxylase activity (Anden et al., 1972; Nybäch & Sedwall, 1968). Immediately after the injection, BC-PL potentiate the disappearance of DA, whereas after a certain period of time, they reverse the α-MT effect. The first effect is interpreted as an increased release, neuroleptics behave in the same manner (Nybäch & Sedwall, 1968), while the second effect may be interpreted in terms of an increased synthesis.

The effect of BC-PL on tyrosine hydroxylase (TH) of rat striatum is shown in Fig. 3 which reports a double reciprocal plot of the rate of tyrosine hydroxylase activity against cofactor concentration. The apparent affinity for cofactor, $DMPH_4$, is increased by BC-PL treatment: the Km for cofactor changes from 0.8mM to 0.3mM. BC-PL do not affect the apparent Vmax as well as the apparent Km for tyrosine (data non reported).

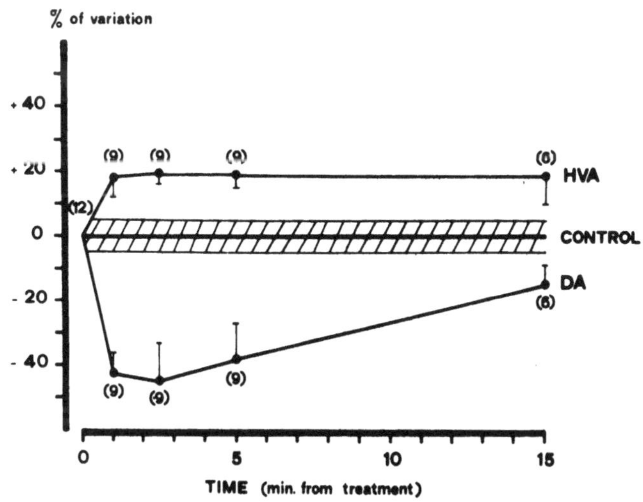

Fig. 1. Effect of sonicated BC-PL (150 mg/kg, i.v.) on cerebral DA and HVA content in mice. Values are mean ± S.E.M. (n). Normal values: DA = 0.75 γ/g, HVA = 0.25 γ/g wet weight. From Toffano et al. (1976a).

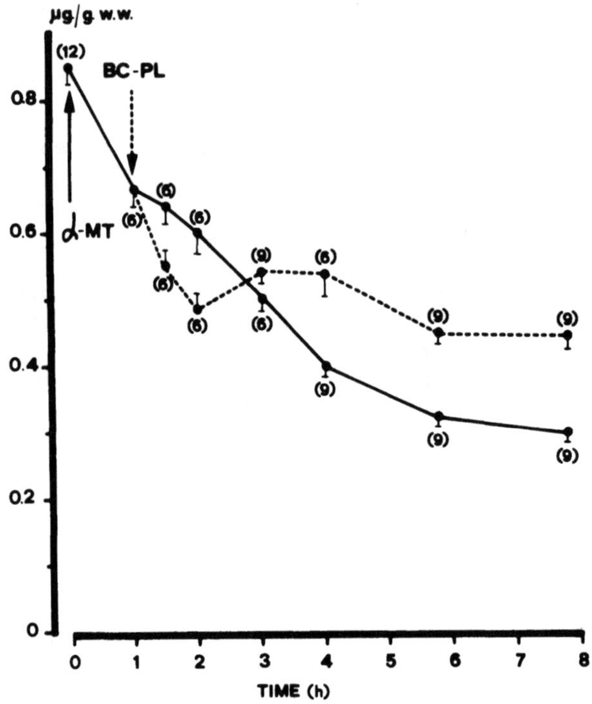

Fig. 2 . Effect of BC-PL (100 mg/kg i.v.) on cerebral DA in mice pretreated with α-methyl-p-tyrosine (α-MT: 200 mg/kg i.p.). Values are mean ± S.E.M.(n).

The phospholipid-induced increase of both DA release and synthesis is of interest in view of the connection between dopaminergic transmission and cAMP formation (Iversen, 1975; Iversen et al., 1975). BC-PL produce cAMP accumulation into mice brains (Fig. 4).

Low doses of BC-PL (50 mg/kg, i.v.) induce a significant increase of cAMP a few minutes after the injection, and a 3-fold increase occurs after 10 minutes. High doses of BC-PL (150 mg/kg i.v.) produce a 5 fold cAMP accumulation after 10 minutes. In 1 mg/kg haloperidol-treated mice, 150 or 50 mg/kg BC-PL do not produce cAMP accumulation.

It is known that cAMP levels depend on neurotransmitter availability and on the combined activity of two enzymes, namely phosphodiesterase and adenylate cyclase. Fig. 5 shows the effect of BC-PL on cerebral adenylate cyclase.

BC-PL stimulate the DA-sensitive adenylate cyclase activity (Zivkovic et al., 1975; Lowry et al., 1951), while they do not affect the basic activity. Above 100 mg/kg, BC-PL inhibit the

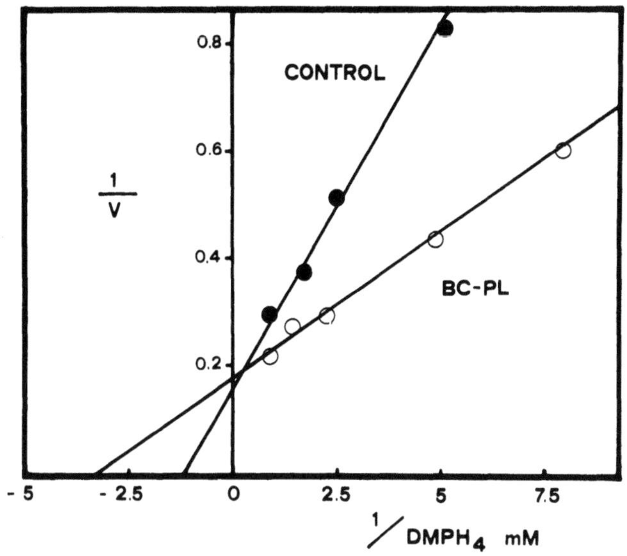

Fig. 3. Effect of BC-PL (100 mg/kg, i.v.) on Tyrosine hydroxylase of rat striatum. Double reciprocal plot of the rate of tyrosine hydroxylase against cofactor concentration. Control group Km = 0.8mM, BC-PL treated group Km = 0.3mM for $DMPH_4$.

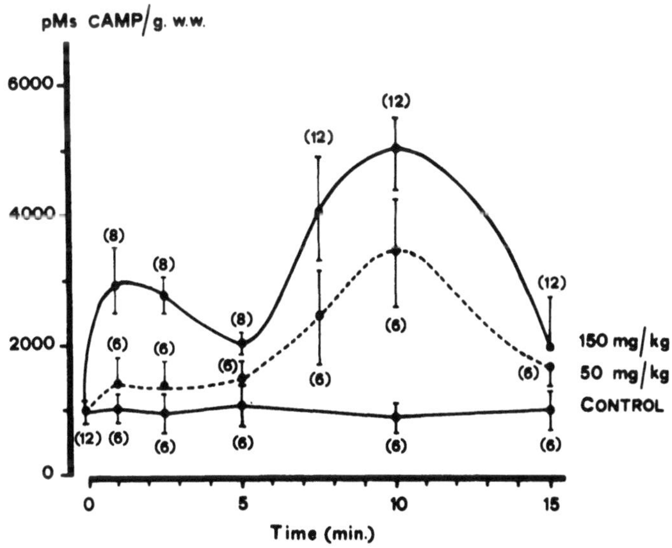

Fig. 4. Effect of BC-PL on cyclic AMP in mice brains. Mice were sacrificed in liquid nitrogen. cAMP was removed through a Dowex-x4 column and assayed by means of a cAMP-protein kinase dependent reaction. Values are mean ± S.E.M. (n).

response to dopamine, thus duplicating the effect of high amounts of apomorphine and catecholamines on the same enzyme (Kebabian et al., 1975). In haloperidol (1mg/kg i.v.) treated mice, BC-PL fail to stimulate the DA-sensitive adenylate cyclase. PDE activity is not inhibited in vivo by BC-PL treatment (data not presented).

Fig. 6 shows that BC-PL are able to affect in men prolactin secretion and spinal lumbar HVA suggesting an effect on dopaminergic system also in humans. Phospholipids induce a decrease of serum prolactim and an increase of HVA in the lumbar spinal fluid. Serum prolactin is believed to be under the inhibitory control of DA along the hypothalamic pituitary axis (Mc Leod & Lehmeyrer, 1974), while lumbar HVA may be indicative of DA turnover (Curzon, 1975). Furthermore, BC-PL reverse the chlorpromazine effect on serum prolactim. The effect of phenothiazines on prolactin secretion seems to depend on a block of dopamine receptors and it is counteracted by the administration of L-dopa and of dopamine agonists (Frantz et al., 1972).

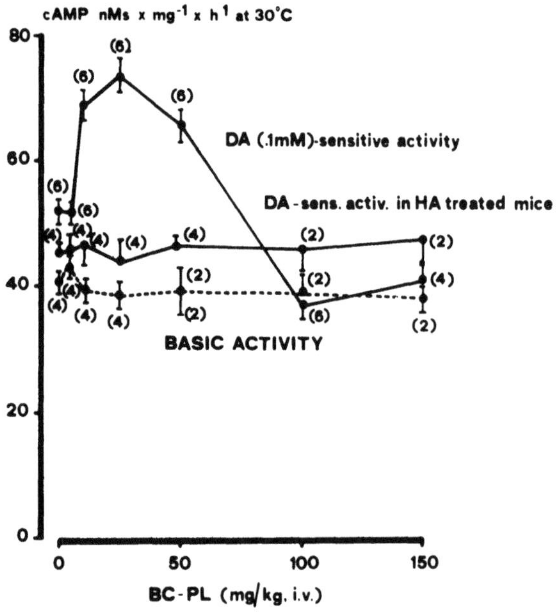

Fig. 5. Effect of BC-PL on adenylate cyclase. Enzymatic activity was assayed 10 min after treatment. In haloperidol (HA) treated mice (1mg/kg, i.v.) BC-PL were injected 1h after. Basic activity means the normal activity assayed according to Krishna et al.,(1968) DA-sensitive activity means the activity assayed in the presence of 0.1mM DA.

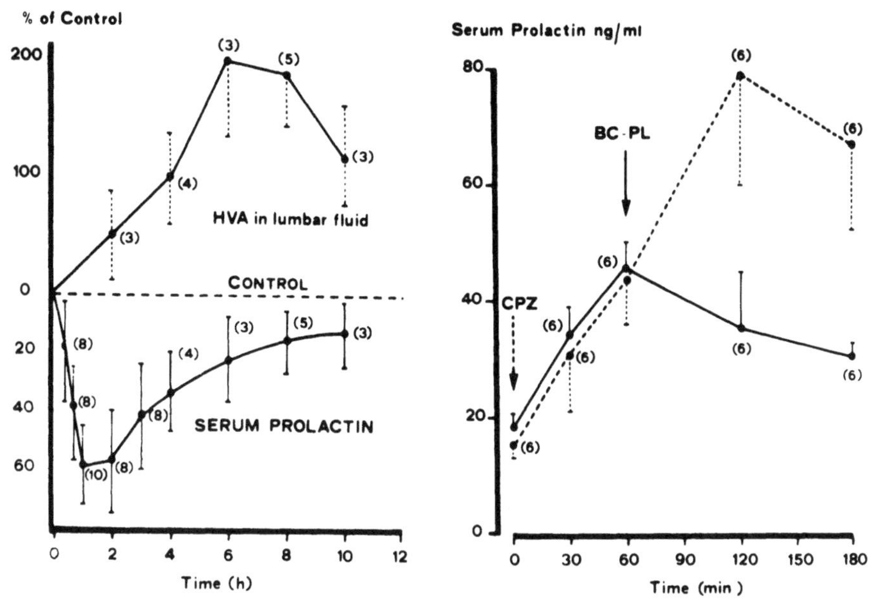

Fig. 6. Effect of BC-PL (200 mg, i.v.) on lumbar HVA and on serum prolactin levels in normal and chlorpromazine (CPZ, 50 mg i.m.) treated subjects. From Polleri et al. (1976) and Massarotti et al. (1976). Serum Prolactin control=9-16 ng/ml; HVA control=0.045±0.002 µg/ml.

The reversal effect of BC-PL on the chlorpromazine-increased prolactin may be interpreted as the result of an effect on a dopaminergic mechanism mediated by their effect on DA as indicated by the parallel increase of lumbar HVA observed in normal subjects.

2) Active Component of BC-PL Mixture

Single phospholipid classes were separated to identify the active components of lipid mixture. Non-lipid contaminants were removed on Sephadex G_{25}. Lipids were the passed through a silicic acid column, washed with chloroform to eliminate neutral lipids and fatty acids, and phospholipids fractioned with various amount of methanol in chloroform. Single purified phospholipid classes were obtained from these enriched fractions by mean of one dimentional TLC on 20 x 20 cm Silica Gel G plates in chloroform-methanol-water (70:30:5), taken to dryness under nitrogen stream and dissolved in 50mM Tris-HCl pH 7.5.

Phosphatidylserine is the most active component in inducing increased DA turnover and adenylate cyclase stimulation (Table 1).

Table 1. Effect *in vivo* of single phospholipid classes on DA release and on DA-sensitive adenylate cyclase in mouse brain

a = DA-release was measured assaying the ratio DA/HVA, 2.5 min after treatment with 75 mg/kg phospholipids (Toffano et al., 1976a).

b = nMoles cAMP x mg^{-1} x h^{-1} formed in presence of 0.1mM DA. Animals were injected with 50 mg/kg of phospholipids and sacrificed 10 min later (Leon et al., unpublished).

Values are mean ± S.E.M. of 6-10 determinations.

Treatment	DA/HVA ratio	DA-sensitive Adenylate cyclase
None	2.8±0.21[a]	53.2±2.6[b]
Phosphatidylserine	1.4±0.20	86.7±3.5
Phosphatidylethanolamine	2.2±0.19	78.4±1.7
Lyso-phosphatidyl-ethanolamine	2.4±0.25	68.8±2.9
Phosphatidylcholine	2.4±0.26	56.9±3.6

DISCUSSION

The data reported above indicate that a sonicated preparation of bovine brain phospholipids affects dopaminergic system in brain of animals and men. The effect on DA system at pre-and post-synaptic level (Fig. 7).

At the presynaptic level, BC-PL induce an increased DA release as indicated by the fall of DA and by the simultaneous increase of its HVA catabolite (Fig. 1). At the same time BC-PL produce an activation of the tyrosine hydroxylase (Fig. 3). The activation is due to an increased affinity of the enzyme toward cofactor without modification of both the Km for tyrosine and the Vmax. From a physiological point of view this may be of interest, because, normally, tyrosine in the brain is present in an amount sufficient to saturate the enzyme, while the cofactor is present in an amount insufficient to saturate the enzyme.

At the post-synaptic level BC-PL produce a 5 fold cAMP accumulation (Fig. 4). Apomorphine or amphetamine in vivo are able to produce only about a 2 fold increase of cAMP. The phospholipid effect on cAMP seems to depend on two factors. The first is the

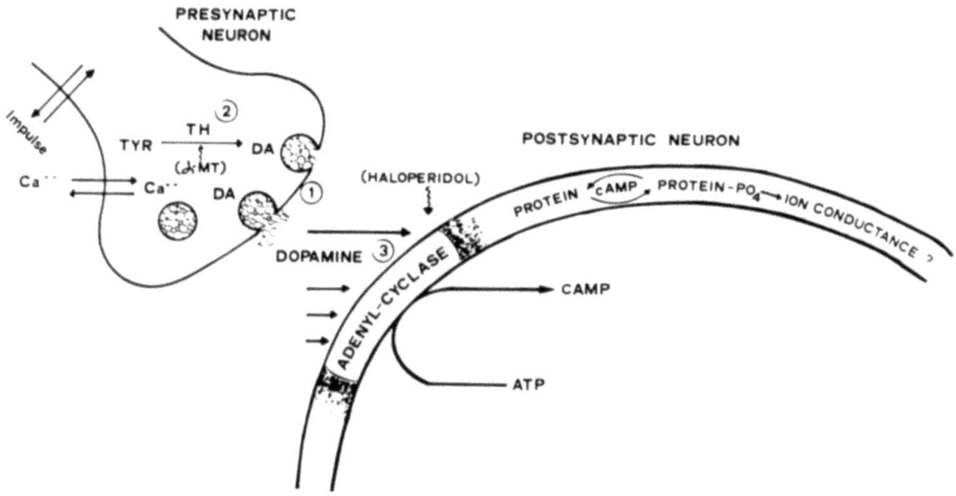

Fig. 7. Possible effect of BC-PL on Dopaminergic transmission.

1) Release of dopamine into intersynaptic side as indicated by the increased HVA (Fig. 1).

2) Activation of tyrosine-hydroxylase activity by means of an increased affinity toward cofactor (Fig. 3).

3) Increase of DA-sensitive adenylate cyclase sensibility to DA (Fig. 5), followed by an increased cAMP accumulation at post-synaptic neuron (Fig. 4).

increased DA availability, due to the increased release (Fig. 1), the second is the increased response of adenylate cyclase to DA released by neuron terminals (Fig. 5). In fact in animals pre-treated with BC-PL the response of the DA-sensitive adenylate cyclase to DA, added in vitro, is increased. In animals pretreated with haloperidol, BC-PL fail to produce cAMP accumulation and to increase the response of DA-sensitive adenylate cyclase to DA. This suggests a specific effect on DA receptors.

Greengard and his collegues (Kebabian & Greengard, 1971; Kebabian et al., 1972; Mc Afee & Greengard, 1972), showed that dopaminergic transmission operates through the formation of cAMP. cAMP regulates some protein kinase activities which in turn regulate the phosphorylation of specific proteins in neuronal cells thus modifying membrane excitability and metabolic function of the cells (Greengard, 1976; Ueda et al., 1973). Therefore the induced increase of DA release and synthesis, the activation of the DA-sensitive adenylate cyclase and the cAMP accumulation may account for the modifications in animals of motor activity and conditioning

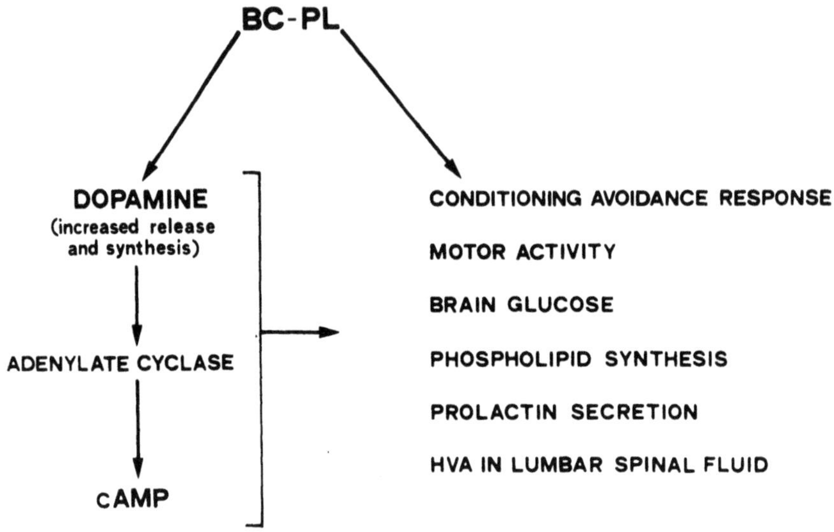

Fig. 8. Correlation between the effect of BC-PL on dopaminergic system and other pharmacological effects.

avoidance responce, of brain glucose and of brain phospholipid synthesis, and for the effect on prolactin secretion observed in men (Fig. 8).

Among individual phospholipids, phosphatidylserine was the most active component (Table 1).

The same pattern of specificity was found by Goth et al. (1971) on histamine release from isolated mast-cells during antigen-antibody interaction; by Pepeu et al. (1976) on the Ach-release from rat cortex; by Bruni et al. (1976b) on the increase of brain glucose. Lloyd & Kaufman (1974) found that PS specifically stimulates tyrosine hydroxylase activity in bovine caudate nucleus, while Raese et al., (1976) found that PS stimulates the tyrosine hydroxylase also in rat striatum.

This peculiar property of phosphatidylserine as pharmacological active phospholipid may be related to its capacity to induce extensive cell fusion in cell cultures in the presence of calcium (Papahadjopoulos et al., 1973). By analogy with the proposed mechanism for the phosphatidylserine-induced liberation of histamine in vitro (Goth et al., 1971; Foreman & Mongar, 1975) and the role of Ca^{++} in the process of Ach (Babel-Guerin, 1974) and cathecolamine release (Blanstein et al., 1972) and of adenylate cyclase stimulation

(Lynch et al., 1976), it is possible that exogenous acidic phospholipids induce modification of the Dopaminergic system by affecting Ca^{++} movement.

REFERENCES

ABDEL-LATIF, A. A., YAN, S. J. & SMITH, J. P. (1974) J. Neurochem., 22, 383.

ANDEN, N. E., CORRODI, H. & FUXE, K. (1972) J. Pharm. Pharmacol., 24, 177.

BABEL-GUERIN, E. (1974) J. Neurochem., 23, 525.

BLAUSTEIN, M. P., JOHNSON, E. M. & NEEDLEMAN, P. (1972) Proc. Natl. Acad. Sci. U.S.A., 69, 2237.

BROWN, R., DAVIS, J.N. & CARLSSON, A. (1973) J. Pharm. Pharmacol., 25, 412.

BRUNI, A., LEON, A. & BOARATO, E.: Advances in experimental medicine and biology, 72:271 (Porcellati G. Amaducci L. and Galli C., eds) Plenum Press, N. Y. (1976a).

BRUNI, A., TOFFANO, G., LEON, A. & BOARATO, E. (1976b) Nature, 260, 331.

CHANG, C. C.(1964) Int. J. Neuropharmacol., 3, 643.

CURZON, G. (1975) Advan. Neurol., 9, 349.

FOREMAN, J. C. & MONGAR, J. L.: In "Calcium transport in contraction and secretion", (Carafoli E., Clementi F., Drabikowski W.& Margreth A., eds.) pag. 175-184, North-Holland Publ. Co., (1975).

FRANTZ, A. G., KLEINBERG, D. J. & NOEL, G. L. (1972) Rec. Progr. Horm. Res., 28, 527.

FRIEDEL, R. O., BERRY, D. E. & SCHANDERG, S. M. (1974) J. Neurochem. 22, 873.

GOTH, A., ADAMS, H. R. & KNOOHUIZEN M. (1971) Science, 173, 1034.

GREENGARD, P. (1976) Nature, 260, 101.

HOKIN, M. R. (1970) J. Neurochem., 17, 357.

IVERSEN, L. L. (1975) Science, 188, 1084.

IVERSEN, L. L., HORN, A. S. & MILLER, R. J. (1975) Advan. Neurol., 9, 197.

KEBABIAN, J. W. & GREENGARD, P. (1971) Science, 174, 1346.

KEBABIAN, J. W., PETZOLD, G. L. & GREENGARD, P. (1972) Proc. Natl. Acad. Sci. U.S.A., 69, 2145.

KEBABIAN, J. W., CLEMENT-CORNIER, Y. C., PETZOLD, G. L. & GREENGARD, P.(1975) Advan. Neurol., 9, 1.

KORF, J., OTTEMA, S. & VAN DER VEEN, I. (1971) Anal. Biochem., 40, 187.

KRISHNA, G., WEIS, B. & BRODIE, B. B. (1968) J. Pharmacol. Exp. Ther., 163, 379.

KUO, J. F. & GREENGARD, P. (1970) J. Biol. Chem., 245, 4067.

LLOYD, T. & KAUFMAN, S. (1974) Biochem. Biophys. Res. Commun., 59, 1262.

LOWRY, O. H., ROSEBROUGH, N. J., FARR, A. L. & RANDALL, R. J. (1951)

J. Biol. Chem., 193, 265.
LYNCH, T. J., TALLANT, E. A & CHEUNG, W. Y. (1976) Biochem. Biophys. Res. Commun., 68, 616.
MANIERO, G., TOFFANO, G., VECCHIA, P. & ORLANDO, P. (1973) J. Neurochem., 20, 1401.
MANTOVANI, P., PEPEU, G. & AMADUCCI, L. : Advances in experimental medicine and biology, 72, 285 (Porcellati G., Amaducci L. & Galli C., eds.) Plenum Press, N. Y. (1976).
McAFEE, D. A. & GREENGARD, P. (1972) Science, 170, 310.
McLEOD, R. M. & LEHMEYRER, J. E. (1974) Endocrinology, 94, 1077.
NYBÄCH, H. & SEDVALL, G.(1968) J. Pharmacol. Exp. Ther., 162, 294.
ORLANDO, P., CERRITO, F. & PORCELLATI, G. : Advances in experimental medicine and biology, 72, 79 (Porcellati G., Amaducci L. & Galli C., eds.) Plenum Press, N. Y. (1976).
PAPAHADJOPOULOS, D., POSTE, G. & SCHAEFFER, B. E. (1973) Biochim. Biophys. Acta, 323, 23.
POLLERI, A., ROLANDI, E., BARRECA, T., GIANROSSI, R., MASTURZO, P., APORTI, F. & TOFFANO, G. (unpublished).
RAESE, J., PATRICK, R. L. & BARCHAS, J. D. (1976) Biochem. Pharmacol., 25, 2245.
SCHWARTZ, W. J., SHARP, F. R., GUNN, R.H. & EVARTS, E. V.(1975) Nature, 261, 155.
TOFFANO, G., LEON, A., BENVEGNÙ, D. & AZZONE, F. (1976a) Pharm. Res. Commun. 8, 581.
TOFFANO, G., LEON, A., BENVEGNÙ, D. & CERRITO, F. (1976b) Atherosclerosis, (in press).
TYCE, G. M. & OWEN, C. A. Jr. (1973) J. Neurochem., 20, 1563.
UEDA, T., MAENO, H. & GREENGARD, P. (1973) J. Biol. Chem., 248, 8295.
ZIVKOVIC, B., GUIDOTTI, A. & COSTA, E. (1975) J. Pharm. Pharmacol., 27, 359.

METABOLISM OF PHOSPHATIDIC ACID AND PHOSPHATIDYLINOSITOL IN RELATION TO TRANSMITTER RELEASE FROM SYNAPTOSOMES

J. N. Hawthorne and M. R. Pickard

Department of Biochemistry, University Hospital and

Medical School, Nottingham NG7 2UH, England

Phosphatidylinositol accounts for about one-twentieth of the total phospholipid fraction of nervous tissue. Phosphatidic acid, the precursor of phosphatidylinositol and a key intermediate in lipid metabolism, occurs in smaller amounts. The possible involvement of these phospholipids in synaptic transmission was first suggested by the work of Hokin & Hokin (1954) showing that acetylcholine increased the incorporation of labelled phosphate into lipids of brain slices. Phosphatidic acid and phosphatidylinositol were the compounds chiefly affected. Diphosphoinositide and triphosphoinositide, the phosphorylated derivatives of phosphatidylinositol, did not respond to acetylcholine.

Turning from brain slices to subcellular fractions, the acetylcholine effect was localised in the synaptosomal fraction by Durell & Sodd (1966). Synthesis of phosphatidic acid from [^{32}P]ATP was sensitive to acetylcholine, but not synthesis from [^{32}P] glycerophosphate. This indicated that reaction 2 rather than reaction 1 was involved (Hokin & Hokin, 1959). Confirmation comes from the finding that diacylglycerol kinase is about ten times as

$$\text{glycerol 3-phosphate} + 2\ \text{acyl-CoA} \rightarrow \text{phosphatidic acid} + 2\ \text{CoASH} \qquad (1)$$

$$\text{1,2 diacylglycerol} + \text{ATP} \rightarrow \text{phosphatidic acid} + \text{ADP} \qquad (2)$$

active as the acylation system in brain (Lapetina & Hawthorne, 1971).

The synthetic reactions, however, may be part of a recovery

process. In pancreas (Hokin-Neaverson, 1974) and parotid gland (Jones & Michell, 1974) hydrolysis of phosphatidylinositol seems to be the initial response to cholinergic agonists. The most likely hydrolysis route is to diacylglycerol and an inositol phosphate, initially the cyclic 1,2-phosphate and then D-inositol 1-phosphate. Permanent loss of phosphatidylinositol will damage synaptosomal membranes, so a cycle of breakdown and resynthesis has been proposed (Fig. 1).

The aim of our recent work has been to determine whether specific synaptosomal membranes are involved in these phospholipid changes and in what way the phospholipid effects are related to synaptic events.

LABELLING STUDIES WITH SYNAPTOSOMES

In these experiments synaptosomes prepared from guinea-pig brain cortex were incubated in media containing $^{32}P_i$ so that the effect of acetylcholine/eserine on phospholipid labelling could be studied (Yagihara et al., 1973). Sub-synaptosomal membranes were subsequently prepared by osmotic shock and density-gradient centrifugation. The labelled phospholipids from the various sub-synaptosomal fractions were extracted and separated by two-dimensional thin-layer chromatography. After incubation for an hour, phosphatidic acid was the most highly labelled phospholipid and the bulk of its radioactivity was either in the synaptic

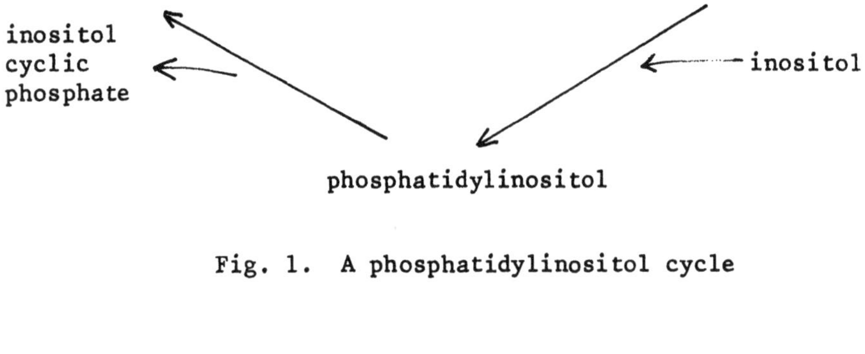

Fig. 1. A phosphatidylinositol cycle

vesicle fraction or in fractions containing mitochondria, as shown by the presence of succinate dehydrogenase. The mitochondrial pool was unaffected by acetylcholine but the specific radioactivity of the vesicle phosphatidic acid was more than doubled. Phosphatidylinositol was also well labelled but the increases produced by acetylcholine were less reproducible than those seen in phosphatidate. The phosphatidylinositol changes were seen in fractions rich in rotenone-insensitive NADH: cytochrome c reductase, an indication that membranes from the endoplasmic reticulum were present.

LABELLING OF ELECTRICALLY STIMULATED SYNAPTOSOMES

The major action of acetylcholine is at post-synaptic membranes, so it is surprising that this transmitter affects phospholipid metabolism in isolated synaptosomes. As Fig. 2 indicates, a fragment of post-synaptic membrane is usually attached to synaptosomes, but this will not have access to the metabolic pathways involved in phospholipid synthesis. Transmission of the nerve impulse involves phospholipid changes in both pre- and post-synaptic membranes. L. E. Hokin (1969) concluded from pre-ganglionic denervation experiments with sympathetic ganglia that there was a pre-synaptic phosphatidic acid effect but that the phosphatidylinositol changes induced by acetylcholine were largely post-synaptic.

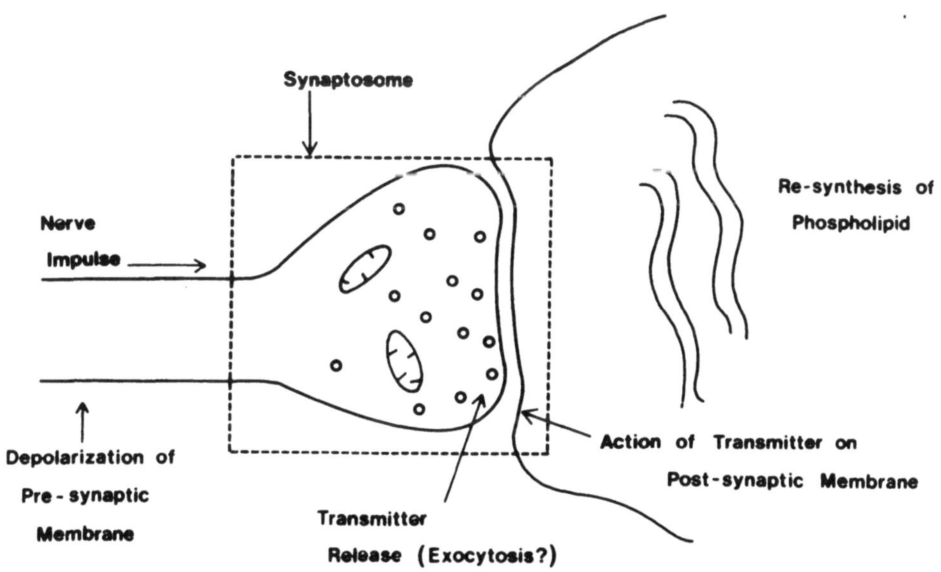

Fig. 2. Sketch of synaptic membranes and associated events

Studies of the action of cholinergic agonists on synaptosomes from brain cortex can be criticised on at least two grounds. First, only a minority of the nerve endings will be cholinergic and second, release of transmitter is more important at pre-synaptic endings than response to it. We have therefore turned to electrical stimulation of synaptosomes (De Belleroche & Bradford, 1972) as a model of the pre-synaptic events in nerve impulse transmission. Bradford and his colleagues have shown that depolarization in this way or with high external potassium ion concentration causes specific release of transmitters when calcium ions are available in the surrounding medium. At the same time there is increased respiration and only a minor leakage of lactate dehydrogenase, indicating that the synaptosomes remain intact.

A 10 min period of electrical stimulation in the presence of $^{32}P_i$ led to increased labelling of synaptosomal phosphatidate (Bleasdale & Hawthorne, 1975) but no consistent increase in phosphatidylinositol specific radioactivity. The labelling of ATP was unaffected. When sub-synaptosomal membrane fractions were prepared, the increased phosphatidate labelling was seen to be associated with the synaptic vesicle fraction, as in the acetylcholine experiments. Further work (Hawthorne & Bleasdale, 1975) showed that electrical stimulation only increased the labelling of phosphatidate in the vesicle fraction when the medium contained calcium ions. This and the time course of labelling (changes seen 2 min after the onset of stimulation) suggest that the metabolism of phosphatidate is closely associated with the process of transmitter release.

ELECTRICAL STIMULATION OF SYNAPTOSOMES LABELLED *IN VIVO*

As mentioned above, there is evidence that the initial phospholipid change on excitation is hydrolysis of phosphatidylinositol (Fig. 1). Hydrolysis of phosphatidate to diacylglycerol has also been postulated to occur in synaptosomes incubated with acetylcholine (Schacht & Agranoff, 1974). These hydrolytic changes would not be detected in experiments measuring incorporation of labelled inorganic phosphate into synaptosomal phospholipids. We have therefore labelled synaptosomal phospholipids with ^{32}P *in vivo* and studied the loss of label caused by electrical stimulation.

Guinea pigs received an intraventricular injection of carrier-free $^{32}P_i$ and were killed two hours later. At this time there was maximal labelling of phosphatidate and phosphatidylinositol in the synaptosomal fraction (Hawthorne *et al.*, 1976). Other phospholipids were only poorly labelled. Synaptosomes were prepared from the brain cortex and washed in medium free from

^{32}P. They were then incubated in similar medium for electrical stimulation as described above. In some experiments 2 mM 2,4-dinitrophenol and 10 mM 2-deoxyglucose were added to block further synthesis of phospholipids. Electrical stimulation for 10 min at 100 Hz produced a marked loss of labelled phosphatidate and phosphatidylinositol from the synaptosomes. In the presence of the metabolic inhibitors more than 90% of the phosphatidate radioactivity was lost. The corresponding figure for phosphatidylinositol was 60%.

Preparation of sub-synaptosomal fractions showed that after labelling _in vivo_, phosphatidate of Fraction E which contains marker enzymes for endoplasmic reticulum and plasma membrane (but not in higher concentration than some other fractions) had the highest specific radioactivity (Table 1). The most active phosphatidylinositol was in the synaptic vesicle fraction. Electrical stimulation provoked loss of these particular pools of phospholipid, while there was little effect on either phospholipid in the remaining fractions, as can be seen from the Table.

CDP-diacylglycerol could also be isolated from the labelled synaptosomes by thin-layer chromatography. Electrical stimulation increased the labelling of this compound (Table 2), suggesting that some of the labelled phosphatidate may have reacted with CTP to form CDP-diacylglycerol. The figures in Table 2 refer to experiments without added dinitrophenol/deoxyglucose. Similar results were obtained in the presence of the inhibitors, the only important difference being that electrical stimulation then produced a significant increase in the chemically measurable loss of phosphatidate. Preliminary results indicate that diacylglycerol kinase is much more active in Fraction E than in the other sub-synaptosomal fractions of Table 1. There was also activity in Fractions O and D, but very little in Fractions F to I. The enzyme was assayed in the presence of deoxycholate by the method of Lapetina & Hawthorne (1971).

CONCLUSIONS

Breakdown and resynthesis of phosphatidylinositol and phosphatidic acid are closely associated with the events following depolarization of isolated synaptosomes. The most important of these events is transmitter release and the weight of the present evidence favours the view that exocytosis is the mechanism of release. This is supported by our finding that major phospholipid effects are seen in the synaptic vesicle fraction. Work with pre-labelled synaptosomes showed that stimulation led to loss of phosphatidylinositol from this fraction. The enzyme converting it to diacylglycerol is most likely to be involved. Entry of calcium

TABLE 1. EFFECT OF ELECTRICAL STIMULATION ON ^{32}P-LABELLED PHOSPHOLIPIDS OF SUB-SYNAPTOSOMAL MEMBRANES

Synaptosomes were prepared from labelled brain as outlined in the text. Stimulation *in vitro* is described by Hawthorne & Bleasdale (1975).

	Specific radioactivity (c.p.m./µmol)			
	Phosphatidate		Phosphatidylinositol	
Fraction	Control	Stimulated	Control	Stimulated
D, synaptic vesicle	180	190	965	250
E, microsomal	625	60	150	140
F, disrupted membrane	110	120	220	245
G, membranes and synaptosome ghosts	190	250	120	200
H, synaptosome ghosts	100	160	385	310
I, damaged and whole synaptosomes	485	440	180	170

TABLE 2. EFFECTS OF ELECTRICAL STIMULATION ON ATP AND PHOSPHOLIPIDS OF ^{32}P-LABELLED SYNAPTOSOMES

Figures are means with S.D. from the number of analyses given in parenthesis. Metabolic inhibitors were not added. P values refer to differences between lines 2 and 3 in each case. N.S., no significant difference.

	ATP	Phosphatidate	CDP-diacylglycerol	Phosphatidylinositol
Specific radioactivity (c.p.m./nmol)				
After 15 min (3)	278±29	48.3±14.2	12.6±2.4	5.7±1.0
After further 10 min stimulation (5)	307±19	37.9±7.5	19.7±1.9	3.8±0.7
After 25 min without stimulation (3)	286±23	54.2±6.3	11.9±2.7	5.8±0.9
Significance (P)	N.S.	<0.001	<0.001	<0.01
Concentration (nmol/mg protein)				
After 15 min (3)	2.93±0.72	0.78±0.13	1.68±0.21	14.6±2.7
After further 10 min stimulation (5)	3.41±0.27	0.52±0.18	1.23±0.19	13.8±1.3
After 25 min without stimulation (3)	3.26±0.18	0.85±0.11	1.71±0.26	14.0±1.9
Significance (P)	N.S.	<0.05	<0.01	N.S.

ions when the nerve ending is depolarized could activate this enzyme. That diacylglycerol is produced in the vesicle membrane is indicated by the stimulation experiments with $^{32}P_i$. These showed that radioactive phosphatidate was formed in the vesicle fraction, presumably by the diacylglycerol kinase reaction.

The loss of labelled phosphatidate from the 'microsomal' Fraction E (Table 1) is harder to understand. Some of it is converted to CDP-diacylglycerol for resynthesis of phosphatidylinositol but some could be converted to diacylglycerol by phosphatidate phosphohydrolase. Fraction E contains plasma membrane fragments as well as endoplasmic reticulum and so the phosphatidate changes could reflect synthetic reactions (endoplasmic reticulum) or hydrolysis to diacylglycerol (plasma membrane, Cotman et al., 1971).

If this is a correct interpretation of the labelling changes the key event could be the production of diacylglycerol in the vesicle membrane, if not the plasma membrane as well. Production of this lipid can lead to membrane fusion processes in the erythrocyte membrane (Allan et al., 1976). We suggest therefore that production of diacylglycerol in the synaptosome promotes the membrane fusion required for exocytosis and transmitter release. These effects should be clearly distinguished from phosphatidylinositol changes following from the activation of post-synaptic receptors.

ACKNOWLEDGEMENTS

This work was supported by the Medical Research Council.

REFERENCES

Allan, D., Watts, R. and Michell, R.H. (1976) Biochem. J. 156, 225-232

Bleasdale, J.E. and Hawthorne, J.N. (1975) J. Neurochem. 24, 373-379

Cotman, C.W., McCaman, R.E. and Dewhurst, S.A. (1971) Biochim. Biophys. Acta 249, 395-405

De Belleroche, J.S. and Bradford, H.F. (1972) J. Neurochem. 19, 1817-1819

Durell, J. and Sodd, M.A. (1966) J. Neurochem. 13, 487-491

Hawthorne, J.N. and Bleasdale, J.E. (1975) Mol. Cell. Biochem. 8, 83-87

Hawthorne, J.N., Bleasdale, J.E. and Pickard, M.R. (1976) Adv. Exp. Med. Biol. 72, 199-209

Hokin, L.E. (1969) in Structure and Function of Nervous Tissue, vol. III (Bourne, G.H., ed.) pp.161-184. Academic Press, New York

Hokin, L.E. and Hokin, M.R. (1959) J. Biol. Chem. 234, 1387-1390

Hokin, M.R. and Hokin, L.E. (1954) J. Biol. Chem. 209, 549-558

Hokin-Neaverson, M.R. (1974) Biochem. Biophys. Res. Commun. 58, 763-768

Jones, L.M. and Michell, R.H. (1974) Biochem. J. 142, 583-590

Lapetina, E.G. and Hawthorne, J.N. (1971) Biochem. J. 122, 171-179

Schacht, J. and Agranoff, B.W. (1974) J. Biol. Chem. 249, 1551-1557

Yagihara, Y., Bleasdale, J.E. and Hawthorne, J.N. (1973) J. Neurochem. 21, 173-190

METABOLISM AND ROLE OF PHOSPHATIDYLINOSITOL IN ACETYLCHOLINE-STIMULATED MEMBRANE FUNCTION

M. HOKIN-NEAVERSON

Departments of Psychiatry and Physiological Chemistry

University of Wisconsin, Madison, WI 53706, U.S.A.

I should like to discuss some of the work from this laboratory on the nature and function of stimulus-induced changes in phosphatidylinositol metabolism. Such changes have long been known to occur in many types of cells in response to stimulation by some neurotransmitters, hormones, and various other agents. In the early work, the changes were characterized as an increased turnover of the hydrophilic part of the phosphatidylinositol molecule-phosphate and inositol - without a concomitant increase in the turnover of the hydrophobic, diglyceride part of the molecule. Tissues which show such a response include, among others, neural tissue and some exocrine and endocrine glands.

In neural tissue, the neurotransmitter which has been the most studied is acetylcholine, and it appears to evoke two rather different types of phosphatide responses. In brain slices and in postsynaptic neurons of sympathetic ganglia, changes in phosphatidylinositol turnover occur which seem to be analogous to the changes which occur in response to acetylcholine in innervated gland cells (Hokin, M.R. et al., 1960; Hokin, L.E., 1966; Larrabee & Leicht, 1965). In preganglionic nerve endings the response involves primarily an increased turnover of the phosphate of phosphatidic acid, with little involvement of phosphatidylinositol. So far, only this response in nerve endings has been found to be amenable to study in cell-free preparations (Hokin, L.E. & Hokin, M.R., 1958; Durell & Sodd, 1966; Schacht & Agranoff, 1972, 1973, 1974a, b; Yagihara & Hawthorne, 1972).

The changes in phosphatidylinositol metabolism in brain and ganglia are of considerable interest as molecular events which are associated in some way with the actions of neurotransmitters. However, the great heterogeneity of cell types in the nervous system, and the profound complexity of structure and activity, make an analysis of the nature of the changes and of the role which they may play in membrane function particularly difficult in neural tissue. For this reason my colleagues and I have chosen to study in some depth responses to acetylcholine in two tissues which, while they are of interest in their own right, can also be regarded as model systems for the possible postsynaptic actions of acetylcholine on phosphatidylinositol metabolism in brain and ganglia. We hope to derive from these simpler systems information which can be used for further investigation of the role or roles of neurotransmitter-induced changes in phosphatidylinositol metabolism in the nervous system.

The two tissues I shall discuss are the avian salt gland and the exocrine pancreas. An advantage which these two glands offer as models is that each is a highly specialized tissue which responds to acetylcholine with well-defined functional activities which can be monitored. Each has one major membrane system, which is associated with the specialized function of the cell and which contains the bulk of the total phosphatidylinositol of the cell. The specialized functions which are regulated by acetylcholine in the two tissues are different, but both glands show similar and large changes in phosphatidylinositol metabolism in response to acetylcholine.

Breakdown of phosphatidylinositol in response to acetylcholine

Although there is an increased turnover of phosphatidylinositol in stimulated tissue, the primary action of acetylcholine is to cause a net breakdown of this phosphatide. The breakdown was first observed as a loss of radioactivity from prelabeled phosphatidylinositol in response to acetylcholine in the avian salt gland (Hokin, M.R. & Hokin, L.E., 1964; Hokin, M.R., 1965, 1967) and later in the mouse pancreas (Hokin, M.R., 1974). The extent of breakdown of prelabeled phosphatidylinositol suggested that there was probably a net change in the level of phosphatidylinositol in the stimulated tissue. Phosphatidylinositol is a relatively minor component of the total phospholipids of animal tissues. Its separation and quantitative estimation by early methods was difficult and somewhat unreliable. In early experiments with albatross salt gland slices, phosphatidylinositol levels were measured using a relatively crude method. When mean values were calculated from all the individual observations on tissue from four birds, the value for the level of phosphatidylinositol in acetylcholine-stimulated tissue was lower than that for the controls, but the difference between the two means was not statistically significant (Hokin, L.E. & Hokin, M.R., 1963). However, if the percentage difference between values for control and acetylcholine-stimulated tissue for each animal is taken, the mean

Table 1. Net decrease in level of phosphatidylinositol in response to acetylcholine in albatross salt gland and mouse pancreas

Tissues were incubated at 38°C in Krebs-Henseleit glucose-bicarbonate medium with 95% O_2 + 5% CO_2 as the gas phase.

Tissue	Phosphatidylinositol µmoles/g fresh tissue						
	Control		+ ACh*		% loss		
	mean	SE	mean	SE	mean	SE	P<†
Albatross salt gland	3.94	±0.68	2.25	±0.28	-40%	±5	0.01
Mouse pancreas	1.81	±0.16	1.06	±0.15	-41%	±7	0.01

*ACh -acetylcholine. Salt gland, 0.1-100µM ACh; pancreas 10-100µM ACh. In each case 0.1mM eserine was added with the acetylcholine. †P values derived from % loss by paired t test; for each set of pairs, N=4.

percentage difference indicates that there is a significant decrease of approximately 40% in the level of phosphatidylinositol in response to acetylcholine in the albatross salt gland (Table 1). More accurate methods for the separation and quantitative estimation of phosphatidylinositol were developed and were used to study the levels in mouse pancreas under various conditions (Hokin, M.R., 1974; Hokin-Neaverson, 1974). The results indicate that acetylcholine produces a net decrease in the level of phosphatidylinositol in mouse pancreas. In pancreas tissue incubated with 10-100µM acetylcholine, there is a loss of approximately 40% of the tissue phosphatidylinositol (Table 1). Jones and Michell (1974) have observed a breakdown of |^{32}P|phosphatidylinositol and a decrease in the ratio of phosphatidylinositol to other phospholipids in response to acetylcholine in rat parotid gland fragments.

Formation and turnover of stearoyl,arachidonoyl phosphatidic acid

In both salt gland and pancreas, the breakdown of phosphatidylinositol is accompanied by the formation of phosphatidic acid. In mouse pancreas incubated with 100µM acetylcholine there is an increase in the level of phosphatidic acid which is approximately equal to the decrease in phosphatidylinositol level (Hokin-Neaverson, 1974).

The fact that phosphatidylinositol has a fatty acid composition which is very different from that of the other phospholipids of pancreas was used to test whether the newly-formed phosphatidic acid in the stimulated tissue was derived from phosphatidylinositol (Geison et al., 1976). The structure of phosphatidylinositol from other mammalian tissues has been shown to be 1-stearoyl, 2-arachidonoyl-sn-glycero-3-phosphorylinositol (Holub & Kuksis, 1971; Baker & Thompson, 1972). We find that this fatty acid composition is also typical of pancreas phosphatidylinositol. Lipid-soluble products of phosphatidylinositol which retain the diacylglycerol moiety should therefore contain stearic and arachidonic acids in equal proportions. In pancreas tissue incubated with acetylcholine, there is stoichiometry between the amounts of stearic and arachidonic acids which are lost from the phosphatidylinositol fraction and the amounts which appear in the phosphatidic acid fraction. The newly-formed phosphatidic acid appears therefore to have the same fatty acid composition as the phosphatidylinositol which is broken down. It presumably contains the 1-stearoyl, 2-arachidonoyl glycerol moiety which was originally in phosphatidylinositol. I shall use the prefix (18:0,20:4) to denote lipids which contain this moiety. The (18:0,20:4)phosphatidic acid which is formed in stimulated tissue is a novel species; there is very little stearic acid and arachidonic acid in the phosphatidic acid from unstimulated pancreas.

1-Stearoyl, 2-arachidonoyl glycerol

In both salt gland (Hokin, M.R. & Hokin, L.E., 1967) and pancreas (Hokin, M.R., 1968a), phosphatidic acid formed in response to acetylcholine undergoes continuous turnover of its phosphate group, as measured by ^{32}P incorporation, without a concomitant turnover of its glycerol group. (18:0,20:4)diglyceride is presumably therefore an intermediate in this turnover. We have found a great increase in total diglyceride in response to acetylcholine in mouse pancreas (Banschbach et al., 1974). However, the composition of fatty acid in the newly appearing diglyceride is very different from the composition of fatty acids in phosphatidylinositol; it resembles very much more the fatty acid composition of triglycerides in the tissue. In pancreas incubated in the presence of 10μM acetylcholine, at 10 minutes there is no significant difference in the levels of stearic and arachidonic acids in diglyceride from control and stimulated tissue; at 30 minutes there is a small but significant increase in the levels of these two fatty acids in diglyceride in the stimulated tissue (Geison, R.L., Banschbach, M.W. & Hokin-Neaverson, M., unpublished). From this it appears that, although the main effects of acetylcholine on the bulk changes in the level of diglycerides in pancreas are not closely related to the breakdown of phosphatidylinositol, a small component of (18:0,20:4)diglyceride could have been derived from (18:0,20:4)phosphatidylinositol or from newly-formed (18:0,20:4)phosphatidic acid. This could be an intermediate in the stimulated turnover of the phosphate of (18:0, 20:4)phosphatidic acid by a cycle involving reactions catalyzed by the enzymes phosphatidate phosphohydrolase and diglyceride kinase.

Resynthesis of (18:0,20:4)phosphatidylinositol from (18:0,20:4)phosphatidic acid during reversion to the unstimulated state

Atropine blocks acetylcholine receptors in the salt gland and pancreas; it can be used to cause acetylcholine-stimulated tissue to revert to the unstimulated state. In both salt gland and pancreas, $|^{32}P|$phosphatidic acid formed in response to acetylcholine disappears when atropine is added to the tissue. $|^{32}P|$Phosphatidylinositol is formed; and there is also a great increase in $|^{3}H|$inositol incorporation into phosphatidylinositol (Hokin, M.R. & Hokin, L.E., 1964; Hokin, M.R., 1974). Addition of atropine to acetylcholine-stimulated pancreas gives changes in the levels of phosphatidic acid and phosphatidylinositol which are the reverse of those which occurred on stimulation. There is a net synthesis of phosphatidylinositol back to the control level and this is accompanied by a net loss of a stoichiometric amount of phosphatidic acid (Hokin-Neaverson, 1974). The phosphatidic acid which disappears is specifically the stearoyl, arachidonoyl species (Geison et al., 1976). It is presumably the substrate for the resynthesis of phosphatidylinositol by the cytidine nucleotide pathway, with formation of CDP-diglyceride as an intermediate.

The $|^{32}P|$-phosphatidic acid formed in the stimulated state in pancreas is lost from the phosphatidic acid fraction, together with the specific disappearance of (18:0,20:4)phosphatidic acid. This indicates that the (18:0,20:4)phosphatidic acid which is formed from phosphatidylinositol in the stimulated state is the only species of phosphatidic acid which undergoes increased turnover of its phosphate group.

Summary of overall lipid changes

To summarize these changes which I have discussed, stimulation of salt gland and pancreas with acetylcholine leads to a breakdown of phosphatidylinositol. Under some conditions, the (18:0,20:4)-diglyceride moiety of this phosphatidylinositol is converted almost quantitatively to (18:0,20:4)phosphatidic acid. During the stimulated state, this novel species of phosphatidic acid undergoes continuous turnover of its phosphate group, presumably with (18:0, 20:4)diglyceride as an intermediate. It is specifically used for the resynthesis of phosphatidylinositol when stimulated tissue reverts to the unstimulated state (Fig. 1).

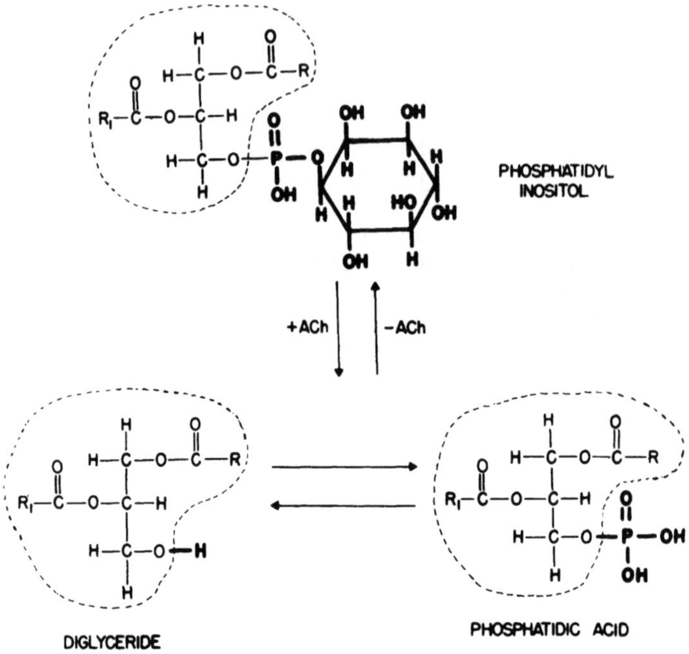

Fig. 1. Lipid interconversion which occur in response to acetylcholine and its removal. Broken lines circle the relatively stable 1-stearoyl(R), 2-arachidonoyl(R_1)glycerol moiety. Heavy type denotes the polar head groups which undergo turnover.

The enzyme pathway of stimulated phosphatidylinositol breakdown

When acetylcholine-stimulated breakdown of phosphatidylinositol was first observed in the avian salt gland, it seemed logical to assume that this would be by cleavage of phosphatidylinositol to form diglyceride and inositol phosphate (Hokin, M.R. & Hokin, L.E., 1964). A Ca^{2+}-dependent enzyme with phospholipase C type activity which catalyzed this reaction had been demonstrated by Kemp et al. (1961). More recently, Dawson et al. (1971) showed that the products are inositol 1,2-cyclic phosphate, inositol 1-phosphate and diglyceride. There is an enzyme with these properties in mouse pancreas. The enzyme is not activated by acetylcholine in soluble extracts, and we have found no evidence that it is concerned in acetylcholine-stimulated breakdown of phosphatidylinositol in the intact cell, nor that the function of phosphatidylinositol breakdown is to form inositol 1,2-cyclic phosphate, as has been suggested by Michell & Lapetina (1972).

When we examined the water-soluble products of stimulated phosphatidylinositol breakdown in pancreas we did not find any increase in the levels of inositol 1,2-cyclic phosphate or inositol 1-phosphate either during or after the breakdown period. Under all conditions studied, the water-soluble product of stimulated phosphatidylinositol breakdown was a stoichiometric amount of free inositol (Hokin-Neaverson et al., 1975).

The production of free inositol from phosphatidylinositol suggests the possibility that a phospholipase D type of activity might be involved in this system. The occurrence of phospholipase D in animal tissues was not reported until recently, when an enzyme which cleaves phosphatidylcholine to give choline and phosphatidic acid was found in brain tissue (Saito & Kanfer, 1975). It is not known whether phosphatidylinositol can act as its substrate in an analogous reaction.

Another possibility is that phosphatidylinositol is degraded via the back reaction of the CDPdiglyceride:inositol transferase; this also would give free inositol as a breakdown product:

$$\text{CDPdiglyceride + Inositol} \underset{}{\overset{Mn^{2+}}{\rightleftharpoons}} \text{Phosphatidylinositol + CMP}$$

CDPdiglyceride:inositol phosphatidyltransferase

This enzyme is part of the cytidine nucleotide pathway for phosphatidylinositol synthesis, and it is present in animal tissues. In their early work on this pathway, Petzold & Agranoff (1965) had some evidence which they thought might indicate that CDPdiglyceride was formed from phosphatidylinositol. Thompson & MacDonald (1975, 1976) have isolated and characterized CDPdiglyceride from bovine liver and brain. They find 1-stearoyl,2-arachidonoyl CDPdiglyceride to be a major species in these tissues. Because this fatty acid composition is the same as that of phosphatidylinositol in these tissues, and is different from that of the bulk of the phosphatidic acid of the tissue, these workers also have raised the possibility that CDPdiglyceride may be formed from phosphatidylinositol by the back reaction of CDPdiglyceride:inositol transferase. We have recently obtained preliminary evidence that this back reaction can occur. |^{3}H|CDPdiglyceride is formed from |^{3}H|CMP in dialyzed microsomal preparations from mouse pancreas. Its formation is greatly increased by the addition of phosphatidylinositol prepared from liver, which contains predominantly the 1-stearoyl,2-arachidonoyl species. The reaction requires Mn^{2+}, as does the forward reaction of CDPdiglyceride:inositol transferase, and it is inhibited by the addition of either myo-inositol or CDPdiglyceride, both of which would be products of the back reaction (Table 2).

The lipid-soluble product of stimulated phosphatidylinositol breakdown can appear as (18:0,20:4)phosphatidic acid or (18:0,20:4) diglyceride in stimulated tissue. It follows that if the back reaction of CDPdiglyceride:inositol transferase is the mechanism of phosphatidylinositol breakdown, the CDPdiglyceride formed must in turn be split either to phosphatidic acid or to diglyceride. Cleavage of CDPdiglyceride to phosphatidic acid or diglyceride has not been reported in animal tissues. A nucleotide pyrophosphohydrolase which splits CDPdiglyceride to give phosphatidic acid and CMP has been found in membranes from E. coli. (Raetz et al., 1972).

Non-random distribution of acetylcholine-responsive phosphatidylinositol

The hormone pancreozymin (cholecystokinin-pancreozymin) stimulates the protein secretory cycle in the exocrine pancreas (Harper & Raper, 1943); it also elicits phosphatidylinositol breakdown, formation of phosphatidic acid, and increased turnover of the polar headgroups of these phosphatides. These effects are essentially the same as those observed in response to acetylcholine (Hokin, M.R.,

Table 2. Formation of CDPdiglyceride from |^{3}H|CMP and 1-stearoyl,2-arachidonoyl phosphatidylinositol

Basic incubation system - 20mM tris buffer, pH 8.4; 3mM $MnCl_2$; 3mM |5-^{3}H|cytidine-5'-monophosphate, sp.act. 10 Ci/mole; pancreas microsome fraction, dialysed for 17 hours at 4°C against 200 volumes of 10mM tris HCl buffer, pH 8.4, approx. 0.8 mg. protein; total volume, 250μl. Incubated at 38°C for 60 min. |^{3}H|CDPdiglyceride was separated by co-chromatography with added carrier CDP-diglyceride.

Incubation conditions	|^{3}H|CDPDG formed nmoles/mg protein	
	No added PI	+ 3mM PI*
Basic system	1.54	19.1
+ 3mM myo-inositol	0.12	0.86
+ 3mM CDPdiglyceride†	1.90	2.75
- Mn^{2+}	0.37	0.62

* Phosphatidylinositol from pig liver (Serdary Research Labs, Inc., London, Ontario, Canada)
† CDP1,2-dipalmitoyl-sn-glycerol (Sigma Chemical Co., St. Louis, Mo., U.S.A.

1968b, 1974). However, unlike the responses to acetylcholine, the responses to pancreozymin are not blocked or reversed by atropine. Because of this, pancreas tissue can be put through a stimulation-reversion-stimulation sequence by addition of these three agents to the incubation medium at appropriate time intervals in the sequence: acetylcholine-atropine-pancreozymin. As discussed above, addition of atropine to acetylcholine-stimulated tissue leads to a resynthesis of phosphatidylinositol during the time that the tissue is reverting to the unstimulated state, and, in the presence of $|2\text{-}^3H|$-inositol, there is a great increase in the incorporation of ^3H into phosphatidylinositol. If tissue in which phosphatidylinositol has been labeled in this way with $|2\text{-}^3H|$inositol is then restimulated with pancreozymin, there is a breakdown specifically of this newly-formed $|^3H|$phosphatidylinositol (Fig. 2). This is some evidence that the same phosphatidylinositol molecules that are synthesized after addition of atropine to acetylcholine-stimulated tissue are

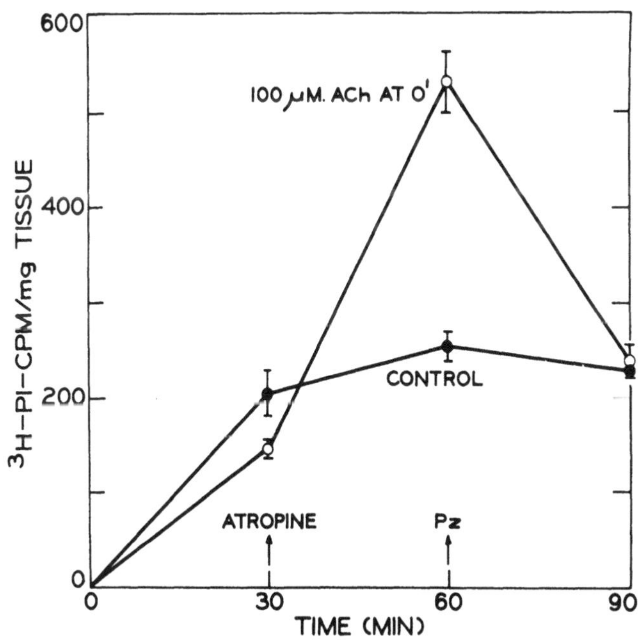

Fig. 2. Incorporation of $|2\text{-}^3H|$inositol into phosphatidylinositol during reversion from the acetylcholine-stimulated state, and its loss on restimulation with pancreozymin. Mouse pancreas was incubated in the presence of $|2\text{-}^3H|$inositol without and with acetylcholine (plus eserine). Atropine (10μM) and pancreozymin, (10 units per ml.) were added in sequence, as indicated, to both control and acetylcholine-stimulated tissues.

Table 3. Non-random distribution of newly-synthesized phosphatidyl-inositol

Mouse pancreas was incubated in the presence of $|^{3}H|$glycerol, with additions made in sequence, as indicated. Values are means±SE, N=3.

Specific Activity of Phosphatidylinositol cpm/nmole			
0 - 30 min	30 - 60 min	60 - 80 min	
Acetylcholine 10μM	Atropine 1μM	Pancreozymin 10U/ml	
5.05 ± 0.56 ⟶	22.3 ± 1.6 ⟶	14.5 ± 0.3	P< 0.01*
Control	Atropine 1μM	Pancreozymin 10U/ml	
3.80 ± 0.07 ⟶	5.05 ± 0.20 ⟶	5.54 ± 1.14	n.s.

*t test - significance of difference between means before and after pancreozymin.

broken down again in response to pancreozymin, and that these molecules do not mix with the other phosphatidylinositol molecules of the tissue.

A stimulation-reversion-stimulation experiment in which phosphatidylinositol was labeled with $|^{3}H|$glycerol in the relatively stable backbone has provided further evidence that there is not random distribution of stimulus-responsive phosphatidylinositol. In pancreas tissue incubated in the presence of $|^{3}H|$glycerol, there is some incorporation of $|^{3}H|$glycerol into phosphatidic acid. In acetylcholine-stimulated tissue to which atropine is added, some of this $|^{3}H|$glycerol-labeled phosphatidic acid is used for the resynthesis of phosphatidylinositol when the tissue reverts to the unstimulated state. The specific activity of phosphatidylinositol rises dramatically, to become only slightly less than that of phosphatidic acid. The rise in phosphatidylinositol specific activity does not occur in response to atropine if the tissue has not previously been exposed to acetylcholine. This confirms that the rise is not a direct response to atropine, but is due to resynthesis of phosphatidylinositol after acetylcholine-induced breakdown. In pancreas tissue which has been labeled in this manner by an acetylcholine-atropine sequence, addition of pancreozymin causes a significant fall in the specific activity of the $|^{3}H|$glycerol-labeled phosphatidylinositol (Table 3). If there had been random distribution of the newly-labeled molecules, the breakdown of phosphatidylinositol in response to the stimulus would not result in a decrease in the specific activity. Such a decrease can only occur if molecules of higher than average specific activity are selectively degraded. The observed fall indicates that stimulus-responsive phosphatidylinositol of higher than average specific activity was selectively degraded in response to restimulation with pancreozymin. The results show clearly that newly synthesized phosphatidylinositol formed during reversion to the unstimulated state does not mix randomly with the other phosphatidylinositol in the tissue.

The non-random distribution of newly-synthesized phosphatidylinositol is evidence against a scheme recently proposed by Michell (1975). He has postulated that phosphatidylinositol molecules associated with cell-surface receptors in the plasma membrane are degraded, and diglyceride or phosphatidic acid molecules derived from them are transferred by lipid transfer-protein from the plasma membrane to the endoplasmic reticulum for resynthesis of phosphatidylinositol, followed by random distribution of the newly-synthesized phosphatidylinositol molecules via lipid transfer protein to the different membranes of the cell. The observation that stimulation by pancreozymin leads to the selective breakdown of phosphatidylinositol of high specific activity, which was formed during the recovery from the response to acetylcholine, is further evidence against Michell's scheme. The receptors for acetylcholine and pancreozymin in the pancreas are presumably different since atropine blocks the actions of acetylcholine, but not those of pancreozymin. In the scheme put forward by Michell, the phosphatidylinositol associated with pancreozymin receptors should be either unlabeled or randomly labeled under the conditions of our experiments, since acetylcholine receptors, not pancreozymin receptors, were stimulated to give the first breakdown of phosphatidylinositol. The results show a selective breakdown of phosphatidylinositol of higher-than-average specific activity in response to pancreozymin, rather than a breakdown either of unlabeled phosphatidylinositol or of randomly distributed labeled phosphatidylinositol. These experiments argue against a direct association of the responsive phosphatidylinositol molecules with cell-surface receptors and receptor function per se in the plasma membrane. Rather, they suggest that the stimulus-responsive phosphatidylinositol molecules in the pancreas are associated with a function common to the actions of both acetylcholine and pancreozymin, and that they can respond to a signal generated by at least two different types of receptor.

The non-random distribution of the responsive (18:0,20:4)phosphatidylinositol and its derivatives implies that these responsive molecules are not freely transported from membrane to membrane by lipid exchange protein and that they do not have free lateral movement in the lipid phase of the membranes in which they occur. They appear to remain bound in a membrane complex, possibly by hidrophobic association of the fatty acids with hydrophobic areas of membrane proteins.

Subcellular location of phosphatidylinositol breakdown in the pancreas

Since the stimulus-responsive phosphatidylinositol molecules do not distribute randomly in the tissue, but appear to have restricted movement in the membrane, the subcellular location of phosphatidylinositol loss should be meaningful and the membrane systems in which it occurs should give some indication of the possible functional roles of phosphatidylinositol breakdown. When we first attempted to study the subcellular location in the pancreas, we found that the methods of Meldolesi et al. (1971a,b) for the isolation of membrane fractions from guinea pig pancreas resulted in an extensive loss of phosphatidylinositol and other lipids from all subcellular fractions. Modifications of these methods were developed for the separation of subcellular fractions with minimal lipid losses from mouse pancreas (Harris & Hokin-Neaverson, unpublished). Mouse pancreas was prelabeled with ^{32}P in vivo and was then incubated for 10 minutes in vitro without or with 10μM acetylcholine; the following subcellular fractions were prepared: plasma membrane, zymogen granules, mitochondria, roug endoplasmic reticulum, smooth endoplasmic reticulum, and postmitochondrial supernatant fraction. When the results were expressed as level of phosphatidylinositol or $|^{32}P|$phosphatidylinositol per gram of original tissue, the rough endoplasmic reticulum was the only fraction which showed a significant intrapair percentage loss of phosphatidylinositol or $|^{32}P|$phosphatidylinositol in response to acetylcholine. Of the total phosphatidylinositol recovered in the fractions, approximately 80% was recovered in the rough endoplasmic reticulum, and the loss of phosphatidylinositol and $|^{32}P|$phosphatidylinositol from this fraction accounted for over 90% of the sum of the differences in all fractions. The results indicate that the bulk of the phosphatidylinositol which is broken down in response to acetylcholine in the pancreas is located in the rough endoplasmic reticulum (Harri & Hokin-Neaverson, 1977).

The second messenger problem

If the major site of acetylcholine-stimulated phosphatidylinositol breakdown in the pancreas is the rough endoplasmic reticular membranes rather than at the cell surface, then it is necessary to postulate that some second messenger system is needed to carry the information from the receptor at the cell surface to the site of the response. The nature of this is not known. The dibutyryl derivatives of cyclic AMP and cyclic GMP do not give rise to phosphatidylinositol breakdown in the pancreas; nor does the influx of Ca^{2+} ion brought about by the ionophore A23187 (Harris & Hokin-Neaverson, unpublished). It is relevant perhaps in this context that acetylcholine-activated phosphatidylinositol breakdown in pancreas is inhibited by 2,4-dinitrophenol (Hokin, M.R., 1974). This suggests that a step which requires metabolic energy is involved in the overall process. A possibility which could account for this is that energy is required for the synthesis of a second messenger.

The function of stimulated phosphatidylinositol breakdown

The acinar cell of the exocrine pancreas is highly specialized for the synthesis and secretion of export proteins - the zymogens, or digestive enzymes. The course of synthesis, storage and extrusion of export protein has been detailed in work from Palade's laboratory (Jamieson & Palade, 1967a,b). Synthesis of export protein takes place on the membrane-bound ribosomes of the rough endoplasmic reticulum. As the proteins are being synthesized, the polypeptide chains pass through the membrane into the cisternae. The proteins are transported via small vesicles from the cisternae to the Golgi complex, where they are packaged into condensing vacuoles which eventually become zymogen granules. Extrusion of export protein occurs by exocytosis i.e. by fusion of zymogen granule membrane with the apical plasma membrane and extrusion of the zymogen granule contents.

Acetylcholine and pancreozymin both stimulate zymogen extrusion in the pancreas. However, many lines of evidence have indicated that the phosphatidylinositol response is not closely related to the process of exocytosis in the pancreas or in other glands. One point of difference between stimulation of exocytosis and stimulation of phosphatidylinositol breakdown in mouse pancreas is that maximum extrusion of enzyme occurs in vitro in response to 0.1μM acetylcholine; stimulation of $|^3H|$phosphatidylinositol breakdown increases linearly between concentrations of acetylcholine of 0.1μM to 100μM (Hokin, M.R. 1974).

The location of the major phosphatidylinositol breakdown in the rough endoplasmic reticulum in the pancreas suggests a relationship between phosphatidylinositol breakdown and protein synthesis in this tissue. Although there have been some conflicting reports on the effects of cholinergic agents on protein synthesis in the pancreas, we have observed in mouse pancreas a consistent stimulation by acetylcholine of protein synthesis, as measured by the incorporation of $|^3H|$phenylalanine and $|^{14}C|$leucine into total protein, and into amylase (Hokin-Neaverson & Williams, unpublished). With 100μM acetylcholine, the incorporation of labeled amino acid is approximately 50% higher than in control tissue. The increase is regulated at the translational rather than the transcriptional level. The percentage increase in the incorporation of $|^3H|$phenylalanine into protein in the pancreas in response to different concentrations of

Table 4. Comparison of the stimulation by acetylcholine of protein synthesis and of phosphatidylinositol breakdown in mouse pancreas

ACh concentration μM	Δ* |^{3}H|Phenylalanine incorporation % of maximum	Δ* |^{3}H|PI breakdown % of maximum
0.1	28	25
1	73	50
10	95	71
100	(100)	(100)

*Δ - difference between control and acetylcholine-stimulated tissue.

acetylcholine shows a curve which is quite similar to the acetylcholine concentration curve for the percentage increase in |^{3}H|phosphatitylinositol breakdown (Table 4). This, together with the location of the major breakdown of phosphatidylinositol in the rough endoplasmic reticulum, is suggestive evidence that the function of stimulated phosphatidylinositol breakdown in the pancreas may be related to the translational control of the rate of protein synthesis on the membrane-bound polysomes. One possibility is that phosphatidylinositol breakdown in the membrane may allow a faster rate of transint of the newly-forming proteins through the membrane into the cisternae, which may in turn lead to a faster rate of formation of the polypeptide chains.

The functional activity of the avian salt secreting gland is very different from that of the exocrine pancreas. The gland is an extrarenal excretory organ. It secretes a fluid which contains concentrations of NaCl which range, in different species, from 0.5M to 1M. Secretory activity involves the activity of NaK-ATPase, the Na^+ pump enzyme, and it is intermittent rather than continuous. It is regulated in an on-off manner by the liberation of acetylcholine at nerve endings in the gland (Fänge et al., 1958). Ultrastructure studies have shown that the bulk of the membranous material of the cell is in two structural elements - the plasma membrane, which is packed into numerous infoldings to give a very large secretory surface, and the mitochondria, which are present in very high concentration; there is only a very small amount of endoplasmic reticulum (Doyle, 1960; Fawcet, 1962; Komnick, 1962; Komnick & Niprath, 1970). Early cell fractionation studies indicated that the responsive phosphatidylinositol, and the new phosphatidic acid which is formed as a result of phosphatidylinositol breakdown, are located primarily in the "microsomal" fraction (Hokin, L.E.& Hokin,M.R.,1960).

Table 5. Comparison of acetylcholine stimulation of ouabain-sensitive respiration and of phosphatidylinositol breakdown in goose salt gland

ACh concentration μM	Δ* $|^{14}C|$glucose oxidation % of maximum	Δ* PI† % of maximum
0.1	85	82
1	98	99
10	(100)	(100)

†calculated from the amount resynthesized after atropine, as measured by the incorporation of radioisotope.
*Δ - difference between control and acetylcholine-stimulated tissue.

This "microsomal" fraction from salt gland consists largely of smooth membrane fragments which appear to be derived from the extensive plasma membrane of cell (Slautterback, Hokin, L.E. & Hokin, M.R., unpublished). The major site of acetylcholine-stimulated phosphatidylinositol breakdown in the salt gland appears therefore to be this extensive plasma membrane, which contains the NaK-ATPase and carries out the secretory function of the gland.

The NaK-ATPase activity which is initiated in response to acetylcholine can be monitored by measurement of the 3-fold increase in the rate of respiration which it evokes. This increase in the respiratory rate is blocked by ouabain, which specifically blocks NaK-ATPase activity, and by atropine, which specifically blocks acetylcholine receptors. In the salt gland, an almost maximal increase in secretory NaK-ATPase activity occurs in response to 0.1μM acetylcholine; and in this tissue, in contrast to the pancreas, the breakdown of phosphatidylinositol, as measured by the amount resynthesized after addition of atropine to stimulated tissue, also occurs almost maximally at this same low concentration of acetylcholine (Table 5). These observations lead me to propose that in the salt gland cell, the function of stimulated phosphatidylinositol breakdown, and of its resynthesis during the reversion to the unstimulated state, is to exert on-off control of the activity of the secretory NaK-ATPase molecules. These are "turned on" when this tissue is stimulated, and are "turned off" when the tissue reverts to the non-secreting state. A simple mechanism for this might be that when the responsive lipid molecules are in the phosphatidylinositol form, an active site of the secretory NaK-ATPase is buried in a hydrophobic area of the membrane, and that when the phosphatidylinositol is broken down, this results in a membrane change which allows the active site to be exposed across the lipid-aqueous interface, where it can interact with its water-soluble ligand.

The stimulus-responsive lipid interconversions are remarkably similar in the exocrine pancreas and in the salt gland. The membrane functions with which they appear to be related are markedly different. To account for this I suggest, as a unifying hypothesis, that the responsive (18:0,20:4)phosphatidylinositol and its derivatives, together with the associated enzymes that carry out the interconversions, form a general mechanism for the control of some membrane functions, and that the changes can affect different types of proteins in the various membrane systems of differentiated cells. I envisage, tentatively, that the membrane changes which result form phosphatidylinositol breakdown might affect the degree to which the relevant membrane proteins can penetrate or span the lipid bilayer and be exposed to the lipid-aqueous interface. This could give different end effects, depending on the nature of the protein in the membrane.

As a final speculative comment, I shall suggest that, by analogy with the pancreas and the salt gland, part of the neurotransmitter-induced change in phosphatidylinositol in neural tissue may be associated with regulation of protein synthesis on the highly developed rough endoplasmic reticulum of the neuron cell body, and part may be associated with the regulation of transport processes in the neuronal plasma membrane. I must point out, however, that if, as proposed above, the phosphatidylinositol system is a general control mechanism, it opens up the possibility that the phosphatidylinositol response in neural tissue may be regulating membrane functions other than the ones which have been discussed here.

ACKNOWLEDGEMENTS

This work has received support from the National Institutes of Health, U.S. Public Health Service (grants NS-06745 and GM-00302) and the University of Wisconsin Graduate School. I acknowledge with gratitude the expert technical assistance of Mr. Kenneth Sadeghian.

REFERENCES

BAKER R.R. & THOMPSON W. (1972) Biochim.Biophys. Acta 270, 489-503

BANSCHBACH M. W., GEISON R. L. & HOKIN-NEAVERSON M. (1974) Biochem. Biophys. Res. Commun. 58, 714-718

DAWSON R. M. C., FREINKEL N., JUNGALWALA F. B. & CLARKE N. (1971) Biochem J. 122, 605-607

DOYLE W. L. (1960) Exptl. Cell Res. 21, 386-393

DURELL J. & SODD M. A. (1966) J. Neurochem. 13, 487-491

FÄNGE R., SCHMIDT-NIELSEN K. & ROBINSON M. (1958) Amer. J. Physiol. 195, 321-326

FAWCETT D. W. (1962) Circulation 26, 1105-1125
GEISON R. L., BANSCHBACH M. W., SADEGHIAN K. & HOKIN-NEAVERSON M. (1976) Biochem. Biophys. Res. Commun. 68, 343-349
HARPER A. A. & RAPER H. S. (1943) J. Physiol. 102, 115-125
HARRIS D. W. & HOKIN-NEAVERSON M. (1977) Federation Proc. (In press)
HOKIN L. E. & HOKIN M. R. (1958) J. Biol. Chem. 233, 822-826
HOKIN L. E. & HOKIN M. R. (1960) J. Gen. Physiol. 44, 61-85
HOKIN L. E. & HOKIN M. R. (1963) Federation Proc. 22, 8-18
HOKIN L. E. (1966) J. Neurochem. 13, 179-184
HOKIN M. R., HOKIN L. E. & SHELP W. D. (1960) J. Gen. Physiol. 44, 217-226
HOKIN M. R. & HOKIN L. E. (1964) in Metabolism and Significance of Lipids. (Dawson R.M.C. & Rhodes D.N., eds.) pp. 423-434, John Wiley, London.
HOKIN M. R. (1965) in Funktionelle und morphologische Organisation der Zelle; Sekretion und Excretion. (Wohlfarth-Bottermann, K.E.,ed) pp. 283-288, Springer-Verlag, Berlin, Heidelberg and New York.
HOKIN M. R. (1967) Neurosci. Res. Prog. Bull. 5, 32-36
HOKIN M. R. & HOKIN L. E. (1967) J. Gen. Physiol. 50, 793-811
HOKIN M. R. (1968a) Arch. Biochem. Biophys. 124, 271-279
HOKIN M. R. (1968b) Arch. Biochem. Biophys. 124, 280-284
HOKIN M. R. (1974) in Secretory Mechanisms of Exocrine Glands. (Thom N.A. & Petersen O.N., eds.) pp. 101-112, Munksgaard, Copenhagen.
HOKIN-NEAVERSON M. (1974) Biochem. Biophys. Res. Commun. 58, 763-768
HOKIN-NEAVERSON M., SADEGHIAN K., MAJUMDER A. L. & EISENBERG F. (1975) Biochem. Biophys. Res. Commun. 67, 1537-1544
HOLUB B. J. & KUKSIS A. (1971) J. Lipid Res. 12, 699-705
JAMIESON J. D. & PALADE G. E. (1967a) J. Cell Biol. 34, 577-596
JAMIESON J. D. & PALADE G. E. (1967b) J. Cell Biol. 34, 597-615
JONES L. M. & MICHELL R. H. (1974) Biochem. J. 142, 583-590
KEMP P., HUBSCHER G. & HAWTHORNE J. N. (1961) Biochem. J. 79,193-200
KOMNICK H. (1962) Protoplasma 55, 414-418
KOMNICK H. & KNIPRATH E. (1970) Cytobiologie 1, 228-247
LARRABEE M. G. & LEICHT W. S. (1965) J. Neurochem. 12,1-13
MELDOLESI J., JAMIESON J. D. & PALADE G. E. (1971a) J. Cell Biol. 49, 109-129
MELDOLESI J., JAMIESON J. D. & PALADE G. E. (1971b) J. Cell Biol. 49, 130-149
MICHELL R. H. & LAPETINA E. G. (1972) Nature 240, 258-260
MICHELL R. H. (1975) Biochim. Biophys. Acta 415, 81-147
PETZOLD G. L. & AGRANOFF B. W. (1965) Federation Proc. 24, 426
RAETZ C. R. H., HIRSCHBERG C. B., DOWHAN W., WICKNER W. T. & KENNEDY E. P. (1972) J. Biol. Chem. 247, 2245-2247
SAITO M. & KANFER J. (1975) Arch. Biochem. Biophys. 169, 318-323
SCHACHT J. & AGRANOFF B. W. (1972) J. Biol. Chem. 247, 771-777
SCHACHT J. & AGRANOFF B. W. (1973) Biochem. Biophys. Res. Commun. 50, 934-941
SCHACHT J. & AGRANOFF B. W. (1974a) J. Biol. Chem. 249, 1551-1557

SCHACHT J. & AGRANOFF B. W. (1974b) in Neurochemistry of Cholinergic Receptors. (de Robertis, E. & Schacht, J., eds.) pp. 121-129, Raven Press, New York.
THOMPSON W. & MacDONALD G. (1975) J. Biol. Chem. 250, 6779-6785
THOMPSON W. & MacDONALD G. (1976) Eur. J. Biochem. 65, 107-111
YAGIHARA Y. & HAWTHORNE J. N. (1972) J. Neurochem. 19, 355-367

THE POSSIBLE INVOLVEMENT OF PHOSPHATIDYLINOSITOL BREAKDOWN IN THE MECHANISM OF STIMULUS-RESPONSE COUPLING AT RECEPTORS WHICH CONTROL CELL-SURFACE CALCIUM GATES

R. H. Michell, S. S. Jafferji, and L. M. Jones

Department of Biochemistry, University of Birmingham

P.O. Box 363, Birmingham B15 2TT, U.K.

Although the general contributions of phospholipids to membrane structure and function are now becoming reasonably well understood, there are still relatively few situations in which one can confidently point to a correlation between a specific metabolic or structural characteristic of an individual lipid and a specific membrane function. However, one of the few very clear indications that individual phospholipids can display highly individual behaviour was provided many years ago by Hokin and Hokin (1) when they discovered that in the pancreas acetylcholine and pancreozymin greatly stimulated the incorporation of ^{32}Pi into phosphatidylinositol (PI), a quantitatively minor anionic membrane phospholipid. Since that time a similar 'phosphatidylinositol response' has been observed in cells exposed to some, but not all, of the extracellular stimuli that interact with receptors on cell surfaces. Despite this, there is still no single accepted view on the function of this response in the overall pattern of events brought about by cell stimulation. We have recently developed a hypothesis which proposes that the PI response is intimately involved in the functioning of a variety of cell-surface receptor systems which control cell-surface permeability to Ca^{2+} ions (2-5). This paper will review the evidence relating to this hypothesis and will speculate briefly on how PI breakdown might be implicated in the control of cell surface Ca^{2+} gates.

HISTORICAL BACKGROUND

The Hokins quickly realised that the PI response was often observed in exocytotic secretory cells (including nerve cells) when they were exposed to secretogogues and they therefore suggested that it had some essential role in the membrane fluxes needed for the secretory cycle (6). However, it was clear by about 1970 that this was not correct: the PI response occurs in several non-secretory tissues and it can be dissociated from secretion in secretory cells (for reviews, see 2,3,7). Since that time an increasing amount of attention has been given to the possibility that the PI response is involved in some way in the function of receptor systems. The first carefully argued case was presented in 1969 by Durell and his colleagues, who proposed a role for PI breakdown in cholinergic, and maybe other, receptors in the brain (8,9). These views were not widely accepted for two main reasons: (a) the experimental studies on which they were based were rather ill-defined and were not reproduced in other laboratories, and (b) the receptor model which was offered included features of nicotinic cholinergic receptor function whereas the characteristics of the PI response indicated the involvement of muscarinic cholinergic receptors (see 2). By 1973, therefore, it appeared that activation of cells by stimuli which interact with cell surface receptors was a common feature which united all PI responses(7): in particular, it appeared that many stimuli which provoked a PI response did not stimulate adenylate cyclase, and vice-versa(2). The mechanisms of adenylate cyclase-independent receptors were not, and still are not, fully understood, and it seemed worthwhile to see whether further detailed investigations of the PI response might provide an insight into the molecular mechanisms of these receptors.

IDENTIFICATION AND CONTROL OF THE STIMULATED REACTION

An early observation by the Hokins, and one which has subsequently been confirmed several times, was that renewal of the inositol and phosphate moieties of the PI molecule was more rapid during stimulation, but there was not an equivalent increase in the incorporation of glycerol into PI (see 2). The obvious conclusion from this is that a cycle of reactions occurs in which PI is both broken down and resynthesized, but that the glycerol backbone of the molecule is conserved throughout this process. There is still some dispute over the exact nature of this cycle of reactions, which will be discussed later.

Thus the first major question still unresolved in 1973 was to find out which reaction in PI metabolism was controlled by the stimulus. Earlier work by the Hokins (10,11) on avian salt gland had suggested that it was probably PI breakdown, and Durell had also presented strong arguments for this view (8,9). Confirmation that PI breakdown is indeed the controlled step then came independently from our laboratory (12-14) and from that of Hokin (15-17). In these experiments it was shown that stimulation provoked a loss of labelled phosphate and inositol from prelabelled PI and, more important, that stimulation could cause a marked decrease in the tissue PI concentration. Table 1 summarises the current status of the evidence that PI breakdown can be stimulated by a variety of stimuli in their target tissues: it seems likely that this will turn out to be the reaction that initiates the PI response in all situations (see ref. 2 for a review). The increased labelling of PI that follows upon stimulation, and which has normally been the

Table 1

A summary of the evidence that PI breakdown is the reaction controlled by receptor activation

			Evidence for PI breakdown		
		Fall in concn.	loss of ^{32}P	loss of ^{3}H-inos	Refs.
Pancreas	cholinergic	yes	yes	yes	15-17
	pancreozymin	yes	yes	yes	
Parotid gland	cholinergic	yes	yes		12-14
	adrenergic	yes			
Salt gland	cholinergic		yes		10
Ileum smooth muscle	cholinergic	yes			18
	histamine	yes			unpub. data
Iris smooth muscle	adrenergic	yes			19
Vas deferens smooth muscle	adrenergic	yes			unpub. data
Islets of Langerhans	gluc. or gal. (not mann.)			yes	20,21
Cerebral Cortex	cholinergic		yes		22
Synaptosomes	electrical	no	yes		23
Lymphocytes	phytohaema-agglutinin		yes		24
Fibroblast	serum factors, pop.density, viral trans.		yes	yes	25-27

effect which has allowed detection of the PI response, is therefore a metabolic consequence of PI breakdown rather than a direct quantitative measure of its occurence. As a result, the PI labelling response can now be regarded as a useful pointer to the prior occurrence of a PI breakdown response, and maybe also as a semi-quantitative indication of its magnitude.

PI breakdown in cell-free systems from several tissues involves the cleavage of the glycerol-phosphate bond, with the release of 1,2-diacylglycerol and a mixture of inositol 1,2-cyclic phosphate and inositol 1-phosphate (see 2). Initially we suggested (7,28) that the significance of the stimulation of PI breakdown in cells was likely to lie in the production of inositol 1,2-cyclic phosphate within these cells, and we proposed this molecule as a putative second messenger associated with receptors which trigger a PI response and do not control adenylate cyclase. As mentioned above, the nature of the PI breakdown reaction in stimulated cells is still a matter of some disagreement to which we shall return later. Even if we ignore this, however, little evidence has been adduced in support of the idea that inositol 1,2-cyclic phosphate is a second messenger (2), even thought it might have some modulatory function in controlling the responses of stimulated cells (29). As a result, we have thought further about possible effects of a local breakdown of phosphatidylinositol in the plasma membranes of target cells.

PI BREAKDOWN IS CONTROLLED BY RECEPTORS WHICH CONTROL CELL-SURFACE CALCIUM PERMEABILITY

We then realised that the receptors which trigger PI breakdown appeared to share three common features: (a) they did not stimulate adenylate cyclase, (b) they brought about an increase in cell surface Ca^{2+} permeability, and (c) they caused an increase, at least transiently, in the intracellular cyclic GMP concentration (2,30,31). Of the latter two effects, there was evidence from studies involving Ca^{2+} deprivation that the primary event was the change in Ca^{2+} permeability (2,31). It therefore seemed that the role of the PI breakdown response should be sought either in the mechanisms through which these receptors controlled cell-surface Ca^{2+} permeability or among the cell responses brought about by an increase in the intracellular Ca^{2+} concentration (2). The latter alternative was then eliminated by experiments which showed that stimulated PI breakdown was Ca^{2+}-independent (see later). Thus attention became focussed on the idea that PI breakdown might have a key role in control of cell-surface Ca^{2+} permeability (2-5).

It is usually considered that receptor systems at cell surfaces must include three essential elements: (1) a receptor site which recognises and binds the extracellular signal, (2) some form of transduction mechanism which recognises a change in configuration in the receptor and transmits this information on to (3) an amplifier which usually translates the received message into an increase in the intracellular concentration of a chemical species (the second messenger) (see, for example, refs 30-33). The second messenger can then diffuse away from the membrane into the cytosol and there exert control over a wide variety of intracellular processes. The best understood receptor systems, and those which are most easily studied, are those which control enzymes that are responsible for synthesis of second messengers. At present, all known receptor systems of this type exert their effects through control of adenylate cyclase, although there have been suggestions that guanylate cyclase may sometimes be subject to the same sort of direct control (e.g. ref. 34).

Another mode of receptor-controlled information transmission, and one which need not necessarily involve any covalent enzymic reaction, depends upon the normal existence of metabolically sustained ion gradients across the cell surface. Any change in cell permeability to particular ions can therefore lead to a change in the intracellular ionic environment: in these circumstances the ions can act as second messengers or the ion movements can perturb the electrical characteristics of the membrane. In the best known system of this type, the nicotinic cholinergic receptor, the ion fluxes are extremely rapid (implying that an enzymic reaction is probably not involved) and they bring about a membrane depolarisation that causes an action potential which then propagates along the plasma membrane and carries the receptor message to remote parts of the target cell.

Those receptors whose activation brings about a PI response belong to a relatively large group which also act through a change in intracellular ionic environment, but in this case the key change is an increase in intracellular Ca^{2+} concentration. This is brought about, at least in part, by an increase in cell surface Ca^{2+} permeability that allows Ca^{2+} to enter cells by flowing down the substantial gradient that normally exists at the plasma membrane (concentration about 10^3-10^6 fold lower inside cells than outside) (see refs. 30,31 for reviews). Since the arguments that follow rest entirely on the validity of the claim that the same receptors both stimulate PI breakdown and increase intracellular Ca^{2+} concentrations, it is worthwhile to briefly summarise here the types of evidence that have been used to assign to these receptors a stimulus-response coupling mechanism based wholly or in part on an increase in cell surface Ca^{2+} permeability. This information has normally been obtained using several of the following experimental approaches which, taken together, can be regarded as making a fairly rigorous assessment of the involvement of changes in cell surface Ca^{2+} permeability (see ref. 30,31,35-37) in rather the same way that 'Sutherland's criteria' (see ref. 32) provide a reference set of tests with which to check for involvement of cyclic AMP in a receptor-controlled response. (a) Cells lose responsiveness to stimuli when extracellular Ca^{2+} is depleted or removed and regain it when the extracellular Ca^{2+} supply is restores: examples of such responses would include secretion of macromolecules, contraction, K^+ efflux, elevation of intracellular cyclic GMP concentration, etc. (b) Responses are inhibited by Ca^{2+} antagonistic drugs (e.g. D-600, nifedipine, cinnarizine, local anaesthetics) or ions (e.g. La^{3+} and other lanthanides, Mn^{2+} or H^+) and this inhibition is relieved by elevation of the extracellular Ca^{2+} concentration. (c) Responses can be elicited by Ca^{2+} introduced directly into the cell by either microelectrodes or ionophores. (d) There is electrophysiological evidence of an inward Ca^{2+} current during stimulation. (e) Studies with $^{45}Ca^{2+}$ show an increase in cell-surface Ca^{2+} permeability and tissue Ca^{2+} content, and this can be blocked by Ca^{2+}-antagonistic ions or drugs. (f) Evidence from studies with intracellular indicators (e.g. aequorin) shows that intracellular $|Ca^{2+}|$ rises on stimulation.

If we are now to proceed with the assumption that all receptors which fulfil the criteria listed above may be members of a single family in which a common fundamental mechanism always underlies the controlled Ca^{2+}-gating phenomenon, then it is appropriate to examine whether this is a proposition consonant with the available experimental information. It would certainly seem reasonable on general grounds of biological unity to expect such similarities, and the limited evidence appears to support this idea. The prototype for biological Ca^{2+} gates is the potential-sensitive gate in squid axon (38): this is a fairly slow responding system which is inhibited by the drugs and ions mentioned above and which is inactivated by prolonged stimulation. Similar potential-sensitive Ca^{2+} gates exist in some, but not all, mammalian tissues (see, for example, 39), and in ileum smooth muscle activation of this Ca^{2+} gate is accompanied by a stimulation of PI turnover (40). Information on most situations in which Ca^{2+} gates are controlled by receptors is more fragmentary, but it is known that the gates controlled by muscarinic cholinergic stimulation respond rather slowly to stimulation (41), that their action is blocked by Ca^{2+} antagonistics drugs and ions (42,42a), and that they are inactivated by maintained exposure to high levels of stimulation (43). Thus they appear very similar to the potential-sensitive systems in several respects. Finally, there is a phenomenon in which exposure of a tissue to a high concentration of one type of Ca^{2+}-mobilising stimulus subsequently desensitises the tissue to stimulation by another stimulus which acts through a different type of receptor (e.g. exposure to acetylcholine will reduce the sensitivity of ileum smooth muscle to subsequent stimulation by histamine, and vice-versa): this is most easily understood as consequence of the sharing of a sensitive common component (the Ca^{2+} gate and/or its control mechanism?) by the two receptor systems (44).

In Table 2a are listed a number of receptors which both produce a PI response and, by the criteria mentioned above, seem to produce their major effects on target tissues through effects on the intracellular Ca^{2+} concentration. Most have been listed and reviewed previously (2), but thyrotrophin deserves some special comment since it is a hormone whose effects are usually ascribed mainly to activation of adenylate cyclase (see 32). Although this is undoubtedly correct, there is also some evidence that certain effects of thyrotrophin, particularly stimulation of glucose oxidation and of protein iodination, are triggered by a Ca^{2+} influx into thyroid cells (45). It is therefore especially interesting that this is the only receptor which appears to control both adenylate cyclase and a PI response (unless, by analogy with adrenaline or histamine, it transpires that thyrotrophin exerts two types of control over cells through interaction with two independent sets of receptors (2)).

Several additional receptor systems which seem to exert control through their effects on the intracellular Ca^{2+} concentration are also listed in Table 2b. These are some systems for which there is evidence, often quite limited, that they exert at least some of their effects through control of cell surface Ca^{2+} permeability, but for which there is not yet reliable evidence as to whether they elicit a PI response. Future investigations of these systems will obviously be useful in determining how general is the suggested association between stimulated PI breakdown and controlled Ca^{2+} permeability.

PI BREAKDOWN MAY BE INVOLVED IN THE COUPLING BETWEEN RECEPTORS AND CALCIUM GATES

When it first became apparent that the PI response was brought about by those receptors which control physiological cell responses through an increase in intracellular Ca^{2+}, the obvious inference was that this was also likely to be the route by which the PI response was triggered. However, at that time Trifaró had already published very clearcut data on the adrenal medulla which appeared to show that the cholinergic PI response of that tissue was quite unaffected by the removal of extracellular Ca^{2+}(59), despite the fact that this was the tissue with which the classic studies of Douglas and his colleagues on the Ca^{2+}-dependence of cholinergically-stimulated catecholamine secretion had been performed (35,36). We and the laboratory of Selinger then undertook independent studies of the effects of Ca^{2+} deprivation on the cholinergic and adrenergic PI responses of the parotid gland. These responses were not diminished by removal of extracellular Ca^{2+} and, in addition, they were not elicited when Ca^{2+} was introduced into the parotid acinar cells with an ionophore (13,14,60). More recently, we have also shown that a variety of calcium-antagonistic drugs do not diminish the PI response to either cholinergic or histaminergic stimuli in ileum smooth

Table 2

(a) Stimuli which control both cell surface Ca^{2+} permeability and PI breakdown (and the tissues affected)*

Muscarinic cholinergic (various)	Pancreozymin (or caerulein) (pancreas)
α-adrenergic (various)	Thyrotrophin (thyroid)
H_1-histamine (brain and smooth muscle)	Corticotrophin-releasing factor (anterior pituitary)
5-Hidroxytryptamine (brain and smooth muscle)	Phytohaemagglutinin and other mitogens (lymphocytes)
Mitogenic serum components (fibroblasts)	D-glucose or D-mannose (islets of Langerhans)
ADP, thrombin or collagen (platelets)	Phagocytizable particles (polymorphonuclear leucocytes)

Depolarization, electrical or by extracellular K^+ (various)

(*for further information, see reference 2).

(b) Some stimuli for which evidence points to control of cell surface Ca^{2+} permeability, but for which there are no definitive studies of effects on PI metabolism

Receptor	Evidence for Ca^{2+}-mediated responses
Opiates (morphine etc)	Ca^{2+}-dependent rise in cyclic GMP (46-48) and contraction of vascular smooth muscle (49).
Mast cell activators (Antigens, ATP etc)	Ca^{2+}-dependent secretion (e.g. 36)
Angiotension II	Ca^{2+}-dependent smooth muscle contraction (50) and cyclic GMP↑ (50a)
Touch (Paramecium) or Light (Chlamydomonas)	Ca^{2+}-dependent changes in ciliary motility (51,52)
Substance P, (and the related peptides physalaemin and uperolein)	Excitatory putative neurotransmitter with smooth muscle contracting ability (53)
Prostaglandin $F_{2\alpha}$	Smooth muscle contraction and elevation of cyclic GMP (54-56)
Oxytocin and vasopresin	Uterine muscle contraction and elevation of cyclic GMP (54,57)
Bradykinin	Smooth muscle contraction (58)

muscle (61). We were therefore faced with an apparent paradox: the PI response was a response to stimuli which effect cell responses via changes in the intracellular Ca^{2+} concentration, but we had good evidence that the PI response did not itself conform to any of the first three criteria that we consider diagnostic of responses controlled by Ca^{2+}-mobilising receptors. Two alternative explanations were then considered by both Selinger and his colleagues and ourselves (2,60): either the PI response was controlled in a manner fundamentally different from other cell responses or it must precede Ca^{2+} influx in the sequence of events which constitute stimulus-response coupling at the effective receptors. The latter idea appeared to us to be both more interesting and more parsimonious and we therefore formulated, and undertook studies to test, the general hypothesis that PI breakdown is an intrinsic and essential event in the coupling between receptor activation and the opening of cell surface Ca^{2+} gates (2-5,61).

Since that time we have tested this hypothesis in a variety of ways and it has not been disproved. Studies of the receptors involved in triggering the PI response were undertaken to ascertain whether previously untested Ca^{2+}-gating receptors (5-hydroxytryptamine and H_1-histamine) gave a PI response. They did (62). Experiments were done to test whether the cholinergic response occured in a previously untested tissue (ileum smooth muscle) and whether, contrary to evidence in the literature, the response of sympathetic ganglia was pharmacologically of the muscarinic type. In both cases the answer was positive (63,64). It seemed possible that PI breakdown might also be involved in the activity of potential-sensitive Ca^{2+} gates, so we tested for responses to elevated extracellular K^+ concentrations in tissues which possess (ileum smooth muscle) and do not possess (pancreas) potential-sensitive Ca^{2+} gates: only the smooth muscle showed increased PI turnover when exposed to high extracellular K^+ (40). Finally, as mentioned above, we used calcium antagonistic drugs to check the Ca^{2+}-insensitivity of the PI response (61). As a result of these investigations, the probability that PI breakdown is in some way implicated in the mechanisms of action of a number of different Ca^{2+}-mobilising receptor systems at the cell surface now seems high, and the remainder of this discussion will be devoted to the analysis of this idea in greater depth.

However, before attempting to discuss the possible manner in which an enzymic reaction such as PI breakdown might be implicated in the control of cell-surface Ca^{2+} gates, it is necessary to check whether such a role would be compatible with the known characteristics of the candidate enzyme reaction and of the appropriate receptor systems. Discussion of these points will here be confined to the muscarinic cholinergic receptor, since this is the receptor for which the most detailed information is available.

When one compares the dose-response curve for a physiological response, such as contraction or secretion, with the curve describing the interaction between the receptor and an agonist of high efficacy it is immediately apparent that far higher agonist concentrations are needed to saturate all receptor sites than to elicit the majority of physiological responses (3,4,65). This is because maximum cell responses can brought about by the activation of only a small fraction of the total muscarinic cholinergic receptor population of sensitive target tissues. In contrast, the production of a maximum PI response appears to require a very high degree of receptor occupation. This suggests that PI breakdown is closely coupled to the activated receptors, and that no amplification stage intervenes between the activated receptor and the activation of the enzyme responsible for PI breakdown (3,4). This type of behaviour has only previously been described for adenylate cyclase (66-68).

The kinetic characteristics of a receptor system may also be able to cast some light on whether an enzyme reaction is likely to be implicated in its mechanism. Very fast responses, such as those at the nicotinic cholinergic receptor, seem unlikely to involve an enzymic reaction as an essential step, whereas slower responses may often involve such a reaction. For example, those receptors which control adenylate cyclase do not usually evoke appreciable effects in their target cells in less than a second or two: the fastest known effects are those seen in the heart, where the cyclic AMP concentration oscillates appreciably within the 1-2 sec period of a single beat (69,70). The most sensitive method for detecting rapid effects of stimuli on cells is probably by using microelectrodes to analyse changes in the electrical characteristics of the plasma membrane of the stimulated cells. Fortunately, this can be done with some cells that respond to muscarinic cholinergic stimuli, and such experiments have always revealed an appreciable delay (about 0.1-0.2 sec) between the application of an agonist and the earliest detectable change in the electrical characteristics of the cell surface (41). Care was taken in these studies to ensure that this effect was not due to any artefact, such as the diffusion of agonist, being rate-limiting. It therefore seems clear that this appreciable delay between stimulus and electrophysiological response reflects some intrinsic property of the stimulus-response coupling mechanism within the receptor system: an enzyme-mediated reaction might well be a cause of this latent period.

WHAT IS THE NATURE OF THE PI BREAKDOWN REACTION?

In thus appears that the muscarinic cholinergic receptor and other Ca^{2+}-mobilising receptors might well involve an enzymic step, and PI breakdown appears to be an excellent candidate for this role. Some possible ways in which PI breakdown might be implicated in a stimulus-response coupling sequence at the receptor will be discussed later, but first one needs to be certain of the precise mechanism and intracellular site of the reaction which causes PI breakdown. Here there is still some lack of both information and agreement. By 1974 it was agreed that the reaction controlled by receptor activation is PI breakdown and that a considerable proportion of the diacylglycerol units released during PI breakdown are recycled back to PI. It had also been widely agreed that this cycle of reactions probably consisted of the removal of the phosphorylinositol headgroup of PI by a PI-specific phospholipase C, followed by the resynthesis of PI via phosphatidate (Scheme 1 of Fig.1, see ref.2). However, Hokin-Neaverson and her colleagues recently presented and alternative hypothesis which suggests that the cycle of reactions involved consists of a reversible removal and replacement of the inositol group of PI, maybe involving either a reversal of the final two steps of the biosynthesis of PI or the activity of a PI-specific phospholipase D (Scheme 2 of Fig. 1, see ref. 17). This new model was designed to reconcile three experimental observations in the stimulated pancreas which were not be expected as the most probable results of the previous model, even thought they were all compatible with it. (a) There was no detectable accumulation of a diacylglycerol with the fatty acid composition expected of that released by PI breakdown (71); (b) there was no detectable accumulation of either inositol cyclic phosphate or inositol 1-phosphate (17); and (c) the disappearance of PI was accompanied by appearance of a stoichiometrically equivalent amount of phosphatidate rich in stearate and arachidonate (15-17, 71a).

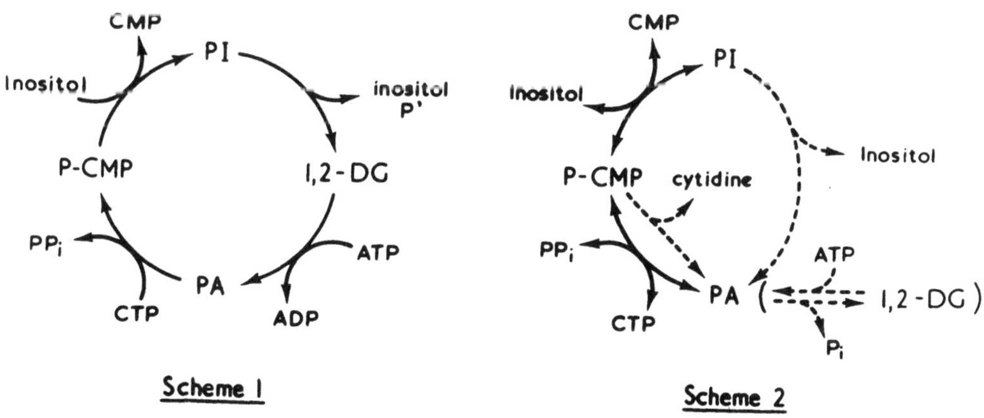

Fig. 1. Two alternative schemes proposed to explain the stimulated breakdown and labelling of PI. Abbreviations: PA, phosphatidate; 1,2-DG, 1,2-diacylglycerol; P-CMP, CDP-diacylglycerol.

Despite these difficulties, however, the weight of evidence still favours the original idea that PI is degraded to diacylglycerol and that this is then converted to phosphatidate and thence back to PI. The reasons for this view are as follows. (a) The increase in PI labelling with ^{32}P which has been for over twenty years the standard way of detecting the PI response would not occur as a result of the removal and renewal of the inositol group of PI. (b) Experiments both on the stimulated labelling and the stimulated breakdown of PI have usually given results which indicate that the turnover of phosphate and of inositol are stimulated to a similar extent: there has been no evidence for preferential renewal of the inositol portion of the headgroup (see, for example, 16). (c) No enzyme has ever been reported that catalyses a direct PI to phosphatidate interconversion, but enzymes which catalyse the breakdown of PI to diacylglycerol are widespread and highly active in animal tissue (d) Reversal of the biosynthetic steps leading from phosphatidate to PI is highly unlikely on energetic grounds, since it would entail the synthesis of CTP from CMP and PPi. If this were to occur, it would be more likely in energy-depleted than in healthy tissue, but it has been suggested that the stimulated PI breakdown is, if affected at all, prevented by energy depletion (16). (e) The fact that diacylglycerol derived from PI does not accumulate is not specially surprising, since it has been shown recently that diacylglycerol produced at the plasma membrane of healthy cells(erythrocytes and lymphocytes) is very rapidly converted to phosphatidate by the diacylglycerol kinase of the plasma membrane (73,74). (f) Although the inositol phosphates that would be produced by the loss of the phosphorylinositol headgroup of PI in a stimulated cell were not detected either by Hokin-Neaverson in the pancreas (17) or by Freinkel and Dawson in Islets of Langerhans (74), this does not exclude the possibility that they were produced but had a very short lifetime in the cytoplasm. This latter view is supported by the recent demonstration by Clements and Rhoten (21) that about one quarter of the ^{3}H-myoinositol released from PI in glucose-stimulated islets of Langerhans is found in the medium as either inositol 1-phosphate or inositol 1,2-cyclic phosphate, and that these esters are present in about the same ratio as they are generated by the widespread PI-specific phospholipase C of animal tissues.

Thus it would appear that there is no existing experimental information which is incompatible with the view that a primary event brought about stimuli is the removal of the phosphorylinositol group of PI to release diacylglycerol, which then has a short lifetime (probably seconds rather than minutes) before being phosphorylated to phosphatidate. There are, however, a number of facts which are incompatible with the alternative proposal that simply envisages a cyclic interconversion of PI and phosphatidate. Obviously, one can add to the latter model a phosphatidate/diacylglycerol cyclic interconversion (bracketed in Scheme 2 of Fig. 1) and thus accomodate additional experimental data satisfactorily, but there is at present no good reason for rejecting the simpler and more widely accepted view. A definitive result that could decide the issue in favour of direct PI/phosphatidate interconversion (Scheme 2) would be a demonstration that conservation of ^{32}P occurred in the breakdown of PI and generation of phosphatidate in a stimulated tissue that had been pulse-labelled with ^{32}P: this type of result has not been reported in any of the published studies.

There is, however, a need at this stage to sound a further note of caution relating to mechanisms of stimulated PI breakdown. It is quite possible that situations will emerge, particularly in tissues where endogenous prostaglandin production in response to stimuli is important, in which stimulated PI breakdown may occur via a reaction catalysed by a phospholipase A. Indeed, Haye and his colleagues have proposed just such a mechanism to explain their observation of thyrotrophin-stimulated liberation from PI of arachidonate for prostaglandin biosynthesis in thyroid homogenates (57-77). A similar effect has also recently been reported in human platelets, where thrombin provoked release of arachidonate from PI and phosphatidylcholine (78). Lapetina (79) has extended this observation to horse platelets, where he found that about 60% of the PI was degraded during 5 min of stimulation with thrombin, whether measured as a disappearance of labelled arachidonate-labelled PI or as a net decrease in the PI content of the cells: this effect was accompanied by a marked increase in the incorporation of ^{32}Pi into PI. These results, with the exception of the ^{32}P labelling, are most simply explained as effects of a stimulated PI-specific phospholipase A. However, they would also be satisfactorily explained if some of the diacylglycerol produced during the normal process of stimulated PI breakdown were to be attacked by a plasma membrane diacylglycerol lipase (e.g. 80,81). It is to be hoped that further information that will resolve this uncertainty will soon be available.

POSSIBLE MECHANISMS FOR INVOLVEMENT OF PI BREAKDOWN IN CONTROL OF CALCIUM GATES

How might a receptor-stimulated loss of the phosphorylinositol headgroup from PI at the plasma membrane, followed by rapid rephosphorylation of the released diacylglycerol, be involved in the control of a Ca^{2+}-gating process at the plasma membrane? A detailed answer to this question cannot become available until PI breakdown has been brought under consistent control in cell-free systems: although such control has been claimed in the past (8,9,82), it has never yet been either characterised adequately or independently confirmed (83-86). Until such experiments allow us to analyse the control of PI breakdown in detail, we must therefore confine ourselves to speculation on possible mechanisms. These appear to be of three fundamentally different types. Two would involve a direct role for anionic lipids (either PI or phosphatidate) as crucial binding sites for Ca^{2+}. In one of these it would be envisaged that PI molecules provide binding sites at the entrance to Ca^{2+} channels, and that when these PI molecules are broken down the Ca^{2+} is released and the channels are also opened for a subsequent flow of Ca^{2+} ions into the stimulated cells. In the second alternative PI, a lipid that is not an ionophore for Ca^{2+}, is converted into phosphatidate, a molecule that is an excellent Ca^{2+} ionophore (87). Third, PI and the other lipids may play no direct role in the recognition or carriage of Ca^{2+}, but the local change in lipid microenvironment caused by PI breakdown may cause a change in the functional state of a Ca^{2+}-gating structure so that it is switched from a closed to an open configuration.

There is at present little information that can help to distinguish between these three possibilities. The problem that arises most obviously with both of the models in which a lipid provides a binding site for Ca^{2+} lies in the fact that the known ion-binding characteristics of PI and phosphatidate have not so far suggested any very great specificity for this ion over other multivalent cations (see 2). However, this may not be a serious problem, since there is evidence which suggests that the 'slow Ca^{2+} channels' of membranes may not discriminate greatly between different divalent cations: they may act physiologically as Ca^{2+} channels simply because there is no great gradient of Mg^{2+}, the only other major physiological divalent cation, at the cell surface (88). A test of the rather specific idea that the Ca^{2+} gate might be a reflection of phosphatidate acting as an ionophoric diffusing carrier for Ca^{2+} might be possible by using 'calcium antagonists' to determine whether the 'Ca^{2+} gate' is equally susceptible to inhibition at either surface of the membrane and is therefore a functionally symmetrical system. In addition, it must be remembered that the ionophoric properties of phosphatidate have so far only been demonstrated in very artificial situations (87), and that it is not known how good an ionophore this molecule will be in the environment of a plasma membrane. In this context, we have some preliminary evidence which suggests that a substantial rise in the phosphatidate content of the human erythrocyte membrane may not lead to a substantial rise in its Ca^{2+} permeability (D. Allan and R.H. Michell, unpublished).

Possibly the most likely way in which to synthesise a hypothetical model of a Ca^{2+} gate from the known events, irrespective of whether lipids act as ion-binding entities or as modulators of a protein structure, is to suggest that the conversion of PI to diacylglycerol, the reaction initiated by the stimulus, serves to open a gate which is normally closed, and that the subsequent rapid phosphorylation of diacylglycerol closes the gate. This would allow for rapid opening of gates followed by their fairly rapid closure. Since stimuli seem normally to control the breakdown only of PI, and not of phosphatidate, the closed phosphatidate-associated configuration of the gate would be unable to reopen until its phosphatidate had been exchanged for PI, presumably by lateral diffusion of lipid molecules in the plane of the membrane. Thus, this type of model could immediately offer molecular mechanisms both for the opening and the closing of gates and also for the refractory period of receptor 'desensitisation' which invariably follows intense stimulation of cells. This is especially interesting, since recent studies (65) have shown that receptor desensitisation at the muscarinic receptor, unlike that at the adenylate cyclase-linked β-adrenergic receptor (67,68), must be achieved by some mechanism subsequent to the agonist-receptor interaction. Further study will obviously be necessary to see whether this type of speculative model has any basis in reality.

CONCLUSIONS

A variety of physiological stimuli which cause the opening of cell-surface Ca^{2+} gates also stimulate phosphatidylinositol breakdown. Studies of various receptors and tissues have shown that the breakdown of phosphatidylinositol is not caused by an increase in the intracellular Ca^{2+} concentration. It seems most likely that the function of stimulated PI breakdown lies in the coupling between activated cell-surface receptors and the opening of membrane Ca^{2+} gates, and some possible mechanisms for this coupling have been briefly discussed.

ACKNOWLEDGEMENTS

We are grateful to the Medical Research Council for financial support, to Dr. E.G. Lapetina for access to unpublished information and to Dr. Allan for helpful discussions.

REFERENCES

1. HOKIN, M.R. & HOKIN, L.E. (1953) J. Biol. Chem. 203, 967-977.
2. MICHELL, R.H. (1975) Biochim. Biophys. Acta. 415, 81-147.
3. MICHELL, R.H., JONES, L.M. & JAFFERJI, S.S. (1976) in Stimulus-Secretion Coupling in the Gastrointestinal Tract (Case, R.M. & Goebell, H., eds.) pp 88-105, MTP Press, Lancaster.
4. MICHELL, R.H., JAFFERJI, S.S. & JONES, L.M. (1976) FEBS Lett. 69, 1-5.
5. MICHELL, R.H., JONES, L.M. & JAFFERJI, S.S. (1977) Biochem. Soc. Trans. in press.
6. HOKIN, L. E. (1968) Intern. Rev. Cytol. 23, 187-208.
7. LAPETINA, E.G. & MICHELL, R.H. (1973) FEBS Lett. 31, 1-10.
8. DURELL, J., GARLAND, J.T. & FRIEDEL, R.O. (1969) Science 165, 862-866.
9. DURELL, J. & GARLAND, J.T. (1969) Ann. N.Y. Acad. Sci. 165, 743-754.

10. HOKIN, M.R. & HOKIN, L.E. (1964) in Metabolism and Physiological Significance of Lipids (Dawson , R.M.C. & Rhodes, D.N.) pp 423-434, John Wiley, New York.
11. HOKIN, M.R. (1967) Neurosci. Res. Program Bull., 5, 32-36.
12. JONES, L.M. & MICHELL, R.H. (1974) Biochem. J. 142, 583-590.
13. JONES, L.M. & MICHELL, R.H. (1975) Biochem. J. 148, 479-485.
14. JONES, L.M. & MICHELL, R.H. (1976) Biochem. J. 158, 505-507.
15. HOKIN-NEAVERSON, M.R. (1974) Biochem. Biophys. Res. Commun. 58, 763-768.
16. HOKIN-NEAVERSON, M.R. (1974) in Secretory Mechanisms of Exocrine Glands (Thorn, N.A. & Petersen, O.H. eds.) pp 701-712,Munksgaard, Copenhagen.
17. HOKIN-NEAVERSON, M.R., SADEGHIAN, K., MAJUMBER, A.L. & EISENBERG, F. (1975) Biochem. Biophys. Res. Commun. 67, 1537-1544.
18. JAFFERJI, S.S. & MICHELL, R.H. (1976) Biochem. J. 154, 653-657.
19. ABDEL-LATIF, A.A., OWEN, M.P. & MATHENY, J.L. (1976) Biochem. Pharmacol. 25, 461-469.
20. FREINKEL, N., El YOUNSI, C. & DAWSON, R.M.C. (1975) Eur. J. Biochem. 59, 245-252.
21. CLEMENTS, R.S. & RHOTEN, W.B. (1976) J. Clin. Invest. 57, 684-691-
22. LUNT, G.A. & PICKARD, M.R. (1975) J. Neurochem. 24, 1203-1208.
23. HAWTHORNE, J.N., BLEASDALE, J.E. & PICKARD, M.R. (1976) in Function and Metabolism of Phospholipids in the Central and Peripheral Nervous Systems (Porcellati, G., Amaducci, L. & Galli, C., eds.) pp 199-209, Plenum Press, New York.
24. FISCHER,D.B. & MUELLER, G.C. (1968) Proc. Natl. Acad. Sci. 60, 1396-1402.
25. RISTOW, H.-J., FRANK, W. & FROHLICH, M. (1973) Z. Naturforsch. B. 28, 188-194.
26. DIRINGER, H. & KOCH, M.A. (1973) Biochem. Biophys. Res. Commun. 51, 967-971.
27. KOCH, M.A. & DIRINGER, H. (1973) Biochem. Biophys. Res. Commun. 55, 305-311.
28. MICHELL, R.H. & LAPETINA, E.G. (1972) Nature New Biol. 240, 258-260.
29. LAPETINA, E.G. & ZIEHER, L.M. (1976) in Function and Metabolism of Phospholipids in the Central and Peripheral Nervous Systems (Porcellati, G., Amaducci, L. and Galli, C., eds) pp 257-263, Plenum Press, New York.
30. RASMUSSEN, H., GOODMAN, D.B.P. & TENENHOUSE, A. (1972) Crit. Rev. Biochem. 1, 95-148.
31. BERRIDGE, M.J. (1975) Adv. Cyclic Nucleotide Res. 6, 1-98.
32. ROBISON, G.A., BUTCHER, R.W. & SUTHERLAND, E.W. (1971) Cyclic AMP, Academic Press, New York.
33. CUATRECASAS, P. (1974) Ann. Rev. Biochem. 43, 327-356.
34. RUDLAND, P.S., GOSPODAROWICZ, D. & SEIFERT, W. (1974) Nature 250, 741-744.
35. DOUGLAS, W.W. (1968) Brit. J. Pharmacol. 34, 451-474.
36. DOUGLAS, W.W. (1974) Biochem. Soc. Symp. 39, 1-28.

37. TRIGGLE, D.J. (1971) Adv. Memb. Surf. Sci. 5, 267-331.
38. BAKER, P.F. (1972) Prog. Biophys. Mol. Biol. 24, 177-223.
39. RINK, T.J. & BAKER, P.F. (1975) in Calcium Transport in Contraction and Secretion (Carafoli, E. et al., eds) pp 235-242, North-Holland, Amsterdam.
40. JAFFERJI, S.S. & MICHELL, R.H. (1976) Biochem. J. 160, 397-399.
41. PURVES, R.D. (1976) Nature 261, 149-151.
42. TICKU, M.K. & TRIGGLE, D.J. (1976) Gen. Pharmacol. in press.
42a. PINTO, J.E.B. & TRIFARO, J.M. (1976) Br. J. Pharmacol. 57, 127-132.
43. CHANG, K.-J. & TRIGGLE, D.J. (1973) J. Theoret. Biol. 40, 155-172.
44. BOWN, F., GRAHAM, J.D.P. & TAHA, S.A. (1973) J. Pharmacol 22, 64-74.
45. GRENIER, G., Van SANDE, J., GLICK, D. & DUMONT, J.E. (1974) FEBS Lett. 49, 96-99.
46. GULLIS, R., TRABER, G. & HAMPRECHT, B. (1975) Nature 256, 57-59.
47. BRANDT, M., GULLIS, R.J., FISCHER, K., BUCHEN, C., HAMPRECHT, B. MORODER, L. & WUNSCH, E. (1976) Nature 262, 311-313.
48. MINNEMAN, K. & IVERSEN, L.L. (1976) in Opiates and Endogenous Opioid Peptides (Kosterlitz, H. W., ed) pp 137-142. North-Holland, Amsterdam.
49. LEE, C.-H. & BERKOWITZ, B.A. (1976) J. Pharmacol. Exp. Ther. 198, 347-356.
50. PAGE, I.H. & BUMPUS, F.M. (eds) (1974) Handbook of Experimental Pharmacology, Vol 37, Angiotensin, Springer-Verlag, Berlin.
50a. BUONASSISI, V. & VENTER, J.C. (1976) Proc. Natl. Acad. Sci. 73, 1612-1616.
51. BROWNING, J.L., NELSON, D.I. & HANSMA, H.G. (1976) Nature 259, 491-494.
52. SCHMIDT, J.A. & ECKERT, R. (1976) Nature 262, 713-715.
53. OTSUKA, M., KONISHI, S. & TAKAHASHI, T. (1975) Fed. Proc. 34, 1922-1928.
54. GOLDBERG, N.D. et al. (1974) in Cyclic AMP, Cell Growth and the Immune Response (Braun, W. et al., eds.) pp 247-262, Springer-Verlag, Berlin.
55. JIMENEZ De ASUA, L., CLINGAN, D. & RUDLAND, P.S. (1975) Proc. Natl. Acad. Sci. 72, 2724-2728.
56. NAHORSKI, S.R., PRATT, C.N.F.W. & ROGERS, K.J. (1976) Br. J. Pharmacol. 56, 445P-446P.
57. MUNSICK, R.A. (1960) Endocrinol. 66, 451-457.
58. WALLASZEK, E.J. (1970) in Handbook of Experimental Pharmacology, Vol 25, Bradykinin, Kallidin and Kallikrein (Erdos, E.G. ed) pp 421-429, Springer-Verlag, Berlin.
59. TRIFARÓ, J.M. (1969) Mol. Pharmacol. 5, 424-427.
60. ORON, Y., LOWE, M. & SELINGER, Z. (1975) Mol. Pharmacol. 11, 79-86.
61. JAFFERJI, S.S. & MICHELL, R.H. (1976) Biochem. J. 160, 163-169.
62. JAFFERJI, S.S. & MICHELL, R.H. (1976) Biochem. Pharmacol. 25,

1429-1430.
63. JAFFERJI, S.S. & MICHELL, R.H. (1976) Biochem. J. 154, 653-657.
64. LAPETINA, E.G., BROWN, W.E. & MICHELL, R.H. (1976) J. Neurochem. 26, 649-651.
65. BIRDSALL, N.J.M. & HULME, E.C. (1976) J. Neurochem. 27, 7-16.
66. MICKEY, J., TATE, R. & LEFKOWITZ, R.J. (1975) J. Biol. Chem. 250, 5727-5729.
67. KEBABIAN, J.W., ZATZ, M., ROMERO, J.A. & AXELROD, J. (1975) Proc. Natl. Acad. Sci. 72, 3735-3739.
68. MUKHERJEE, C., CARON, M.G. & LEFKOWITZ, R.J. (1976) Endocrinol. 99, 347-357.
69. BROOKER, G. (1973) Science 182, 933-934.
70. WOLLENBERGER, A., BABSKII, E.B., KRAUSE, E.-G., GENZ, S., BLOHM, D. & BOGDANOVA, E.V. (1973) Biochem. Biophys. Res. Commun. 55 446-452.
71. HOKIN-NEAVERSON, M.R. (1975) personal communication.
71a. GEISON, R.L., BANSCHBACH, M.W., SADEGHIAN, K. & HOKIN-NEAVERSON, M.R. (1975) Biochem. Biophys. Res. Commun. 68, 343-349.
72. ALLAN, D., LOW, M.G., FINEAN, J.B. & MICHELL, R.H. (1975) Biochim. Biophys. Acta 413, 309-316.
73. ALLAN, D., WATTS, R. & MICHELL, R.H. (1976) Biochem. J. 156, 225-232.
74. FREINKEL, N. & DAWSON, R.M.C. (1973) Nature 243, 535-537.
75. HAYE, B., CHAMPION, S. & JACQUEMIN, C. (1973) FEBS Lett. 30, 253-260.
76. HAYE, B., CHAMPION, S. & JACQUEMIN, C. (1974) FEBS Lett. 41, 89-93.
77. CHAMPION, S., HAYE, B. & JACQUEMIN, C. (1974) FEBS Lett. 46, 289-292.
78. BILLS, T.K., SMITH, J.B. & SILVER, M.J. (1976) Biochim. Biophys. Acta 424, 303-314.
79. LAPETINA, E.G. (1976) personal communication.
80. MICHELL, R.H., COLEMAN, R. & FINEAN, J.B. (1973) Biochim. Biophys. Acta 318, 306-312.
81. VYVODA, O.S. & ROWE, C.E. (1973) Biochem. J. 132, 233-248.
82. CANESSA De SCARNATTI, O. & RODRIGUEZ De LORES ARNAIZ, G. (1972) Biochim. Biophys. Acta 270, 218-225.
83. SCHACHT, J. & AGRANOFF, B.W. (1973) Biochem. Biophys. Res. Commun. 50, 934-941.
84. SCHACHT, J. & AGRANOFF, B.W. (1974) J. Biol. Chem. 249, 1551-1557.
85. LAPETINA, E.G. & MICHELL, R.H. (1973) Biochem. J. 131, 433-442.
86. LAPETINA, E.G. & MICHELL, R.H. (1974) J. Neurochem. 23, 283-287.
87. TYSON, C.A., ZANDE, H.V. & GREEN, D.E. (1976) J. Biol. Chem. 251, 1326-1332.
88. FLECKENSTEIN, A., NAKAYAMA, K., FLECKENSTEIN-GRUN, G. & BYON, Y. K. (1975) in Calcium Transport in Contraction and Secretion (Carafoli, E. et al., eds) pp 555-556, North-Holland, Amsterdam.

THE BIOSYNTHESIS OF PROSTAGLANDINS AND THROMBOXANES BY NERVOUS TISSUE

L. S. Wolfe, J. Marion, and K. Rostworowski

Montreal Neurological Institute, McGill University

Montreal, Quebec, H3A 2B4 Canada

Prostaglandins $F_{2\alpha}$ and E_2 are rapidly synthesized and released *in vivo* from various regions of the central nervous system and from autonomic nerves during stimulation or after trauma (Wolfe, 1975). They alter the effects of neurotransmitters at pre- and postsynaptic sites. PG's appear in increased amounts in human CSF following epileptic seizures, trauma, infections and stroke (Wolfe & Mamer, 1975). Measurement of PG biosynthesis by cerebral cortex and other brain regions *in vitro* shows considerable biosynthetic capacity from endogenous precursors. Catabolic enzyme activities however are exceedingly low. A metabolically stable pool of arachidonic acid appears in brain tissue after animal death which saturates the brain fatty acid cyclo-oxygenase (Wolfe et al., 1976a; 1976c) localized in microsomal membrane elements (Baker, R. R. & Wolfe, L. S., unpublished results).

The labelling of rat brain lipids *in vivo* with ^3H arachidonic acid and their subsequent analysis at different times postmortem indicates that the released arachidonic acid originates from certain phospholipids. In whole brain at all time interval studies, the specific activity of the liberated arachidonate is intermediate between that of phosphatidylinositol and phosphatidylcholine but higher than that of phosphatidylethanolamine and much lower than that of the neutral glycerides. A similar situation was found in the microsomal fraction.

Although the phospholipase A_1 activity of brain is found mostly in the microsomal fraction and the phospholipase A_2 activity occurs mainly in the mitochondrial fraction (Woelk & Porcellati, 1973), we found that the level of the saturated and unsaturated free fatty acids in both fractions increased with postmortem time. A similar observation has also been reported by Lunt & Rowe (1968). The different subcellular fractions probably generate free arachidonic acid independently.

If the tetraenoic species of phospholipids are preferentially hydrolysed, as suggested by the high rate of release of arachidonic acid (Bazán, 1971; Cenedella et al., 1975), phosphatidylinositol, 75% of which is tetraenoic species, is a likely precursor of the arachidonate (Mc Donald et al., 1975). Phosphatidylinositol did not show a significant decrease in radioactivity postmortem. However, the loss in radioactivity expected from phosphatidylinositol or phosphatidylcholine, if either served as exclusive precursor for arachidonic acid, would be at most of the order of 10-15% which is at the limit of the variability of our present experiments.

Biogenic amines greatly stimulate $PGF_{2\alpha}$ biosynthesis likely through activation of the conversion of the endoperoxide PGG_2 to PGH_2. Recently, it was found that the stable prostaglandin endoperoxide metabolite, thromboxane B_2, is formed in equal or even greater amounts than the PG's from endogenous arachidonic acid by short incubations of cerebral cortex. Noradrenaline also stimulates thromboxane formation. Cerebellar tissues from the rat and guinea pig show very low capacity for thromboxane B_2 formation. This is of considerable interest since cerebellar tissues show the highest levels of PGE_2 formation in brain (Wolfe et al., 1976b). A hypothesis has been presented in which the labile thromboxane A_2 activates calcium ionophores to affect actin filaments at postsynaptic sites following interaction of neurotransmitters with their specific receptors.

ACKNOWLEDGEMENTS

The research was supported by a grant from the Medical Research Council of Canada.

REFERENCES

BAZAN, N. G. (1971) J. Neurochem. 18, 1379-1385.

CENEDELLA, R. J., GALLI, C. & PAOLETTI, R. (1975) Lipids 10, 5, 290-293.

LUNT, G. G. & ROWE, C. E. (1968) Biochim. Biophys. Acta 152, 681-693.

MacDONALD, G., BAKER, R. R. & THOMPSON, W. (1975) J. Neurochem. 27, 655-661.
WOELK, H. & PORCELLATI, G. (1973) Hoppe Seyler's Z. Physiol. Chem. 354, 90-100.
WOLFE, L. S. (1975) in Advances in Neurochemistry, Vol. I (Agranoff, B. W. & Aprison, M. H., eds.) pp. 1-49. Plenum Publishing Corp., New York.
WOLFE, L. S. & MAMER, O. A. (1975) Prostaglandins 9, 183-192.
WOLFE, L. S., PAPPIUS, H. M. & MARION, J. (1976a) in Advances in Prostaglandin and Thromboxane Research, Vol. 1 (Samuelsson, B. & PAOLETTI, R., eds.) pp. 345-355. Raven Press, New York.
WOLFE, L. S., ROSTWOROWSKI, K. & MARION, J. (1976b) Biochem. Biophys. Res. Comm. 70, 907-913.
WOLFE, L. S., ROSTWOROWSKI, K. & PAPPIUS, H. M. (1976c) Can. J. Biochem. 54, 629-640.

THE RELATIONSHIP BETWEEN CHOLINERGIC PROTEOLIPID AND PROTEODETERGENT IN *TORPEDO* ELECTROPLAX MEMBRANES

E. De Robertis, S. Fiszer de Plazas, and M. C. Llorente de Carlin

Instituto de Biología Celular, Facultad de Medicina, Universidad de Buenos Aires, 1121 Buenos Aires, Argentina

INTRODUCTION

The two methodological approaches used for the separation of the cholinergic receptor from electric tissue: i.e. organic solvents (De Robertis et al., 1967, La Torre et al., 1970) or detergent extraction (Changeux et al., 1970; Miledi et al., 1971) have led to the isolation of hydrophobic intrinsic proteins from the electroplax membranes. Because of the intimate relationship of these proteins with lipids, in one case, or detergents, in the other, here they will be called "cholinergic proteolipid" and "cholinergic proteodetergent". The problem which will be considered is that of the identity or non-identity of these proteins; in other words if they are totally different or if they have some similarities. Because both the proteolipid and the proteodetergent have high affinity binding for cholinergic agonists and antagonists of nicotinic type and could be separated from acetylcholinesterase (De Robertis and Fiszer de Plazas, 1970), we were inclined to think that they were similar or at least partially related (Fiszer de Plazas and De Robertis, 1972). Later on, since most of the work on the proteodetergent was based on α-bungarotoxin we studied the binding on this α-toxin to the proteolipid from the electroplax. We found that α-bungarotoxin displaced the site of high affinity for the cholinergic fluorescent probe dansylcholine (De Robertis and Barrantes, 1972). Furthermore, it was possible to demonstrate the binding of labeled α-bungarotoxin to the proteolipid by the use of Sephadex LH 20 chromatography. In experiments done on the electroplax membranes extracted with organic solvents, the membranes lost the binding capacity for α-bungarotoxin, while there was binding to the extracted proteolipid (Fiszer de Plazas and De Robertis, 1972). The specificity of the α-bungarotoxin binding was further demonstrated by experiments using a partition method (Weber et al., 1971). It was observed that the α-toxin could displace completely the binding of |^{14}C|-acetylcholine and of |^{3}H|-decamethonium from the cholinergic proteolipid (Table 1).

The fact that both the cholinergic proteodetergent (Olsen et al., 1972; Karlsson et al., 1972; Schmidt and Raftery, 1972; Klett et al., 1973; Eldefrawi and Eldefrawi, 1973) and the proteolipid (Barrantes 1973; Barrantes et al., 1975b) could be isolated and purified by affinity chromatography also favored the existence of similarities between these proteins.

Affinity Labeling and Extractability of the Cholinergic Proteolipid

A more definite way to solve the issue was the use of affinity labeling for the cholinergic receptor. This method is based on the use of N-maleimido benzyl (or phenyl) trimethylammonium (i.e. MBTA or MPTA) which binds to the anionic site of the receptor and makes a covalent link with a reduced S-S group in its vicinity (Karlin and Cowburn, 1973). With this method and SDS gel electrophoresis

Table 1. Competition between α-bungarotoxin and other cholinergic ligands studied by partition

The cholinergic proteolipid from *Electrophorus* (16 μg protein) in 2 ml of chloroform was submitted to binding by partition (Weber et al., 1971) for 24 hours, with 50 μg of α-bungarotoxin in 2 ml of water saturated with chloroform at pH 8.0 in the upper phase. Then this phase was removed and replaced by water saturated with chloroform containing 100 mM Tris-HCl buffer (pH 7.2) and the radioactive ligand. The partition, with stirring of the two phases, was carried out for 4 hours. In the control there was no previous addition of α-bungarotoxin.

Labeled ligand	Blank	nmoles of ligand in lower phase: control	α-bungarotoxin treated	% Inhibition
|^{14}C| acetylcholine (5×10^{-6}M)	0.25	0.49	0.26	100
|^{3}H| decamethonium (2×10^{-6}M)	0.02	0.17	0.02	100
	0.02	0.24	0.02	100

it was found that the protein binding the affinity label (i.e. $|^3H|$-MBTA) had a molecular weight of about 40,000 dalton in the Electrophorus (Karlin, 1974). In the cholinergic receptor isolated by affinity chromatography from Torpedo californica several protein subunits were detected (i.e. 64,000, 58,000, 48,000 and 39,000 dalton) of which only the one of 39,000 dalton carried the $|^3H|$-MBTA.

Karlin (1974) and Barrantes et al., (1975a) did experiments of $|^3H|$-MBTA or $|^3H|$-MPTA binding to electroplax membranes, followed by extraction with chloroform-methanol. Since they found that the label was not extractable they concluded against the identity of the proteolipid with the proteodetergent. Both groups failed, however, in making a quantitative and qualitative analysis of the proteolipid extracted after affinity labeling. In a preliminary work we demonstrated that the drastic treatment with S-S and SH reagents involved in the affinity labeling with $|^3H|$-MPTA resulted in the almost complete insolubility of the cholinergic proteolipid (De Robertis et al., 1976). In other words, after affinity labeling the receptor is no longer extractable with organic solvents although it can be solubilized with detergents.

In our experiments of affinity labeling of Torpedo electroplax membranes we confirmed the results of Karlin (1974) and Barrantes et al., (1975a) as far as that, after chloroform-methanol extraction, most of the label remained in the residue. As shown in Table 2 in two experiments done with different concentrations of $|^3H|$-MPTA the radioactivity extracted from affinity labeled membranes only

Table 2. Extraction of $|^3H|$-MPTA with chloroform-methanol (2:1) from Torpedo membranes

Membranes of Torpedo electroplax enriched with cholinergic receptor (Changeux et al., 1970) were labeled with $|^3H|$-MPTA (Karlin and Cowburn 1973). In experiment #1, 0.3 nmoles and in #2, 6 nmoles of $|^3H|$-MPTA were used for labeling the membranes.

Experiment	$|^3H|$ -MPTA in dpm		% Extraction
	Residue	Extracted	
1	62,780	4,216	6.3
2	864,800	146,800	14.0

Table 3. Extraction of cholinergic proteolipid after treatment with S-S and -SH reagents and affinity labeling with |^{3}H|-MPTA

Electrophorus membranes were treated with dithiothreitol and mercaptoethanol as used in the affinity labeling. *Torpedo* membranes were affinity labeled as indicated in Table 2.

	Total proteolipid mg/Kg		% reduction	Cholinergic proteolipid mg/Kg		% reduction
	Control	Treated		Control	Treated	
Electrophorus	21.6	12.0	45	5.2	2.0	61
Torpedo (1)	49.2	25.0	53	23.0	7.0	70
Torpedo (2)	55.1	25.5	54	24.9	5.0	80

represented 6.3 % and 14 % of the total. However as shown in Table 3 the little extraction of the radioactivity may be explained by a change in the solubility of the proteolipid which is no longer extracted from the membrane. The treatment of *Electrophorus* membranes with DTT and mercaptoethanol, reagents used in the affinity labeling, leads to a reduction of 45 % in the extraction of total proteolipids. In *Torpedo* membranes, after affinity labeling, the reduction in total proteolipids reached 53 and 54 %. Much more impressive is the effect of these reagents on the cholinergic proteolipid. The presence of this proteolipid was assayed by the binding with 3×10^{-7}M |^{14}C|-acetylcholine, followed by chromatography on small columns of Sephadex LH 20 (15 x 0.8 cm) equilibrated with chloroform. The protein peak eluting in chloroform together with the radioactivity was considered to correspond to the cholinergic receptor proteolipid. As shown in Table 3, in membranes of *Electrophorus* electroplax the treatment with DTT and mercaptoethanol causes a 61 % reduction in the extraction of the cholinergic proteolipid. On the other hand in *Torpedo* membranes in two experiments of affinity labeling with |^{3}H|-MPTA the reduction reached 70 and 80 %. In other words, the amount of cholinergic proteolipid that may be extracted represents only 30 or 20 % of that found in control membranes. It should be stressed here that because of the lack of cold MPTA and the small supply of |^{3}H|-MPTA we could not reach a higher concentration of the label, which probably could have insolubilized even more the cholinergic proteolipid. Another interesting finding shown in Table 2 is that

the Torpedo membranes contain about twice the amount of total proteolipids found in Electrophorus membranes. This difference is much higher when referred to the cholinergic proteolipid. While in Electrophorus membranes there is about 5 mg of protein per Kg tissue, in the Torpedo membranes this amounts to about 24 mg/kg. It is interesting that by affinity chromatography Meunier et al., (1974) reported the extraction of 2.9 mg/Kg of cholinergic receptor from the Electrophorus. In the original paper of La Torre et al., (1970) is already stressed the higher amount of proteolipid present in the Torpedo electroplax as compared with the Electrophorus.

TRANSFER OF CHOLINERGIC PROTEOLIPID INTO DETERGENT

We have transferred the cholinergic proteolipid from Electrophorus or Torpedo into a buffer solution (1 mM Tris-HCl at pH 7.3)

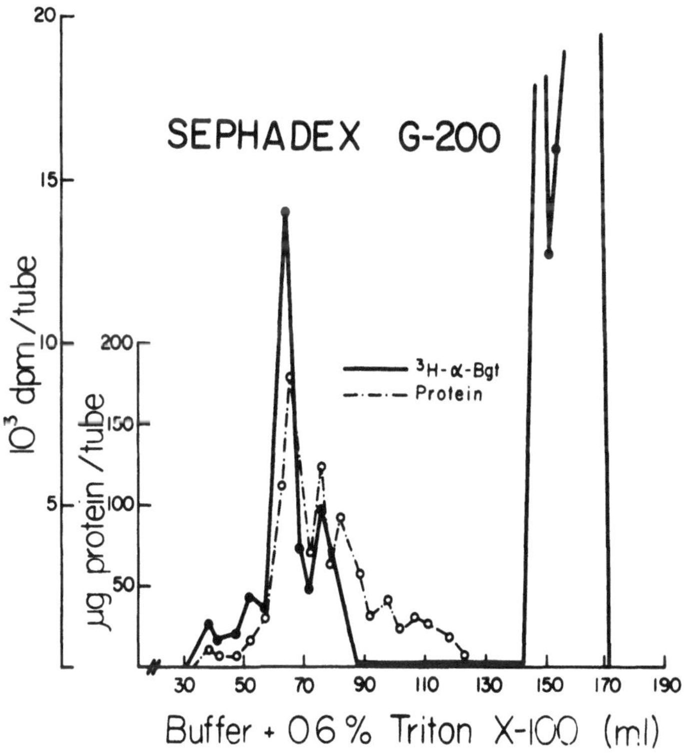

Fig. 1. Column chromatography in Sephadex G-200 of a proteolipid separated from Electrophorus electroplax and bound with $|^3H|$-α-bungarotoxin. The bound and the free toxin are clearly separated (see the description in the text).

containing 0.6% Triton X-100. This transfer is achieved by first evaporating in a vaccum the proteolipid peak into a small volume, with repeated additions of chloroform to reduce the methanol. This is followed by addition of the buffer solution and evaporation with N_2 until all the organic solvent is eliminated. At this point the solution becomes completely clear and can be treated as a water soluble proteodetergent.

In Fig. 1 it may be observed the result of one experiment in which the cholinergic proteolipid from Electrophorus was isolated and treated with |^{3}H| α-bungarotoxin. The extract was loaded on a G-200 Sephadex column (40 x 2.5 cm) and eluted with buffer containing 0.6% Triton X-100. Between 30 and 90 ml the bound radioactivity was eluted in coincidence with two main peaks of protein. The free counts of the α-toxin start to be eluted at about 140 ml. These findings demonstrate that after transfer into detergent the proteolipid is capable of binding α-bungarotoxin in an aqueous medium.

TRANSFER OF PROTEODETERGENT INTO ORGANIC SOLVENTS

Experiments were carried out to obtain the transfer of the

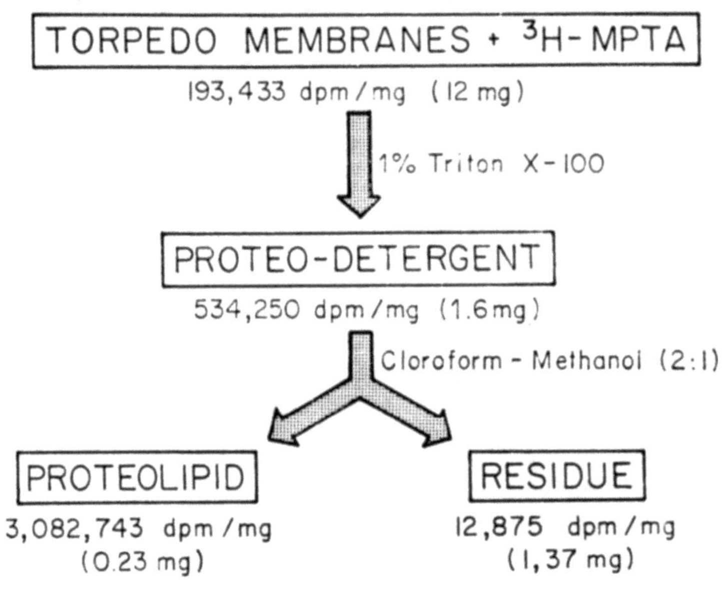

Fig. 2. Experiment in which Torpedo electroplax membranes were affinity labeled with |^{3}H|-MPTA and then submitted to extraction with 1% Triton X-100. The reextraction of the proteodetergent by chloroform-methanol(2:1) is also indicated (see the description in the text).

detergent extracted receptor into chloroform-methanol. From 32.5 g of Torpedo electroplax extracted with 1 % Triton X-100, according to the method of Meunier et al., (1974), we obtained 86.5 mg of protein in the detergent. When this extract was lyophilized and reextracted with chloroform-methanol (2:1), 4 mg of protein (i.e. 4.6 % of the total) together with 4.7 mg of phospholipids, went into the organic solvent. This amount of protein corresponds to 125 mg/Kg of electroplax.

In view of the previously described failure to extract the proteolipid from affinity labeled Torpedo membranes (Table 3) we made first an extraction with detergent. Torpedo membranes were labeled with 3 nmoles of |^{3}H|-MPTA according to Karlin and Cowburn (1973) and then extracted with 1 % Triton X-100. In Fig. 2 it is observed that the detergent extracted 13.3 % of the total protein, together with about 36 % of the radioactivity bound to the membrane. The increase in specific radioactivity in the proteodetergent was 3.3 fold. After reextraction with chloroform-methanol (2:1), only 14.4 % of the protein but 97.6 % of the radioactivity were found in the organic solvent. In this extract the specific radioactivity is 5.6 times higher than in the detergent and 19.1 times higher than in the membranes. The labeled proteolipid was submitted to chromatography on small Sephadex LH-20 columns and the radioactivity appeared in coincidence with a single proteolipid peak. These findings demonstrate that the |^{3}H|-MPTA bound to the membrane, which was extracted by the detergent, is fully extractable, as a proteolipid, by the organic solvent.

SDS GEL ELECTROPHORESIS OF TORPEDO MEMBRANES

To investigate further the problem of the relationship between the cholinergic proteolipid and the proteodetergent, receptor-enriched membranes from Torpedo electroplax were submitted to SDS polyacrylamide gel electrophoresis using the discontinuous buffer system of Laemmli (1970) as modified for slab gels by Studier (1973).

In experiments carried out with |^{3}H|-MPTA labeled membranes it was possible to confirm the finding of Karlin (1974) in the sense that only the 39,000 dalton band was labeled. This was demonstrated by the use of fluorography (Bonner and Laskey, 1974) on prefogged films according to Laskey and Mills (1975).

Other membranes were extracted with chloroform-methanol (2:1) or with the same solvent but containing 1 % desoxycholate (DOC). As shown in Fig. 3 in the extracted membranes we observed the disappearance of the 39,000 dalton band. With DOC in the organic solvent the proteolipid extraction was larger and three bands at about 30,000, 39,000 and 77,500 dalton were missing in the extracted membranes. Since these results, for unknown reasons, were not completely reproducible we used the two dimensional electrophoresis,

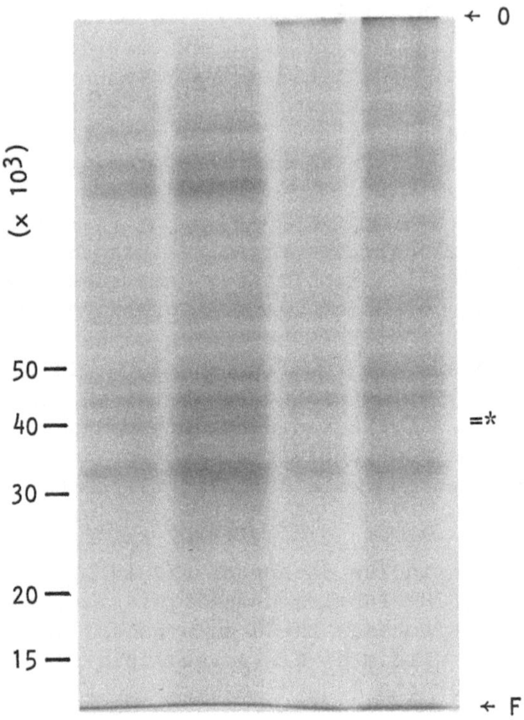

Fig. 3. Polyacrylamide slab gel of the proteins present in Torpedo electroplax membranes. C, control; E, membranes extracted with chloroform-methanol (2:1); 0, origin; F, front. The extraction of the 39,000 dalton band is indicated.

based on isoelectric electrofocusing and SDS gel electrophoresis, described by O'Farrel (1975). This technique has a much higher resolution because it separates the various proteins both by electrical charge and molecular weight. In these gels, after chloroform methanol extraction of the Torpedo membranes, we observed the disappearance of a protein spot in the 40,000 dalton range and having an isoelectric point slightly below pH 6.0. Another protein spot of higher isoelectric point and molecular weight was also missing.

While these results should be considered still preliminary they suggest that the organic solvent is capable of extracting the subunit of 39,000 dalton which according to Karlin (1974) is the only one containing the cholinergic binding site.

DISCUSSION AND CONCLUSIONS

Based on the binding of α-bungarotoxin in 1972 we carried out an extensive investigation on the relationship between the cholinergic proteolipid from Electrophorus electroplax and the proteodetergent α-bungarotoxin complex isolated by other (Fiszer de Plazas and De Robertis, 1972). From this work and the new experiments presented here the following points, which suggest the existence of similarities between the two proteins, can be stressed:

I. Hydrophobic properties. Both the proteolipid and the proteodetergent are extracted with drastic methods used to separate intrinsic proteins from membranes. Triton X-100, one of the most used detergents, is a highly branched octyphenol (hydrophobic portion) containing ten oxyethylene groups (hydrophilic portion). This detergent can induce the formation of lipid micells and substitute the phospholipid interaction by detergent-protein associations (Gitler, 1972). Both organic solvents and detergents can thus extract rather hydrophobic proteins from the membrane. However the amount and number of proteins extracted by detergents is much higher than the more hydrophobic ones that are extracted by organic solvents.

II. Separation by affinity chromatography. Both the cholinergic proteodetergent and the proteolipid can be separated by affinity chromatography using a specific cholinergic group to recognize the receptor protein. Our affinity column, that uses p-phenyltrimethyl ammonium as the active end, can recognize both nicotinic and muscarinic cholinergic proteolipids (Barrantes et al., 1975b).

III. Binding properties. Both types of proteins show high affinity binding for a variety of cholinergic ligands of nicotinic type which include among others: acetylcholine, decamethonium, d-tubocurarine, α-bungarotoxin and MBTA. With α-bungarotoxin we found a single type of binding sites corresponding to a molecule of about 40,000 dalton and a similar molecular weight was calculated from the high affinity binding for acetylcholine (De Robertis et al, 1971). On the proteolipid α-bungarotoxin was able to displace the high affinity binding site of the fluorescence cholinergic drug dansylcholine (De Robertis and Barrantes, 1972) and, as shown in Table 1, it inhibited completely the high affinity binding for acetylcholine and decamethonium. Similar binding properties have been described for the cholinergic proteodetergent.

IV. Transfer of Proteolipid into Proteodetergent. Methods were developed to transfer the cholinergic proteolipid into a water solution by the addition of detergents and the binding of α-bungarotoxin to the proteolipid in aqueous media could be demonstrated (Fig. 1).

V. Transfer of Proteodetergent into Proteolipid. The affinity labeling of Torpedo membranes with |^{3}H|-MPTA, followed by detergent extraction, has permitted to demonstrate that practically all the radioactivity present in the proteodetergent is extractable by the organic solvent and can be found in coincidence with a peak of proteolipid protein (Fig. 2). The failure to extract directly the radioactivity from the membranes with the organic solvents (Karlin, 1974; Barrantes et al., 1975a; De Robertis et al., 1976) may be explained by the fact that the -S-S and SH reagents used in the affinity labeling, cause a change in the solubility properties of the proteolipid while present within the membrane. On the other hand the extractability of the receptor from the membrane by detergents after the affinity labeling is less impaired (Fig. 2).

VI. Extraction of the 39,000 subunit. The results of the SDS polyacrylamide gel electrophoresis suggest that the subunit of 39,000, which carries the binding site, is extracted by the organic solvent (Fig. 3). Since the cholinergic proteodetergent has a much larger molecular weight, with four different subunits (Karlin 1974), we may conclude that the proteolipid probably represents only a part of the total receptor molecule. According to our reconstitution studies this highly hydrophobic portion of the receptor contains not only the cholinergic binding site but also the ionophoric portion of the receptor. This is suggested by the experiments carried out with the cholinergic proteolipid from Electrophorus on lipid bilayers (De Robertis, 1975) and the more recent ones on acetylcholine membrane noise in which the proteolipid from Torpedo, isolated by affinity chromatography, was used (Schlieper and De Robertis, unpublished).

ACKNOWLEDGEMENTS

We would like to express our gratitude to Dr. E. De Robertis Jr. from the MRC Molecular Biology Laboratory (Cambridge, England) for making the fluorography and the two dimentional gel electrophoresis by the method of O'Farrel (1975).

This work was supported by grants from the Burroughs Wellcome Laboratory (USA) and the CONICET (Argentina).

REFERENCES

BARRANTES F. J. (1973) Proc. 9th Internat. Congress Biochem. Stockholm, p. 443

BARRANTES F. J., CHANGEUX J. P., LUNT G. G. & SOBEL A. (1975a) Nature 256, 325-327

BARRANTES F. J., OCHOA E. L. M., ARBILLA S., de CARLIN M. C. L. & DE ROBERTIS E. (1975b) Biochem. Biophys. Res. Commun. 63, 194-200

BONNER W. M. & LASKEY R. A. (1974) Eur. J. Biochem. 46, 83-88
CHANGEUX J. P., KASAI M., HUCHET M. & MEUNIER J. C. (1970) C.R. Acad. Sci. Paris 270, 2864-2867
DE ROBERTIS E. (1975) Synaptic Receptors. Isolation and Molecular Biology, Dekker, New York
DE ROBERTIS E. & BARRANTES F. J. (1972) Eur. J. Pharmacol. 17, 303-305
DE ROBERTIS E., FISZER de PLAZAS S. & SOTO E. F. (1967) Science 158, 928-929
DE ROBERTIS E. & FISZER de PLAZAS S. (1970) Biochem. Biophys. Acta 219, 388-397
DE ROBERTIS E., FISZER de PLAZAS S. & de CARLIN M. C. L. (1976) Nature 259, 605-606
DE ROBERTIS E., LUNT G. S. & LA TORRE J. L. (1971) Mol. Pharmacol. 7, 97-103
ELDEFRAWI M. E. & ELDEFRAWI A. T. (1973) Arch. Biochem. Biophys. 159, 362-373
FISZER de PLAZAS S. & DE ROBERTIS E. (1972) Biochim. Biophys. Acta 274, 258-265
GITLER C. (1972) Ann. Rev. Biophys. Bioeng. 1, 51-125
KARLIN A. (1974) Life Sciences 14, 1385-1415
KARLIN A. & COWBURN D. (1973) Proc. Natl. Acad. Sci. U. S. A. 70, 3636-3640
KARLSSON E., HELBRONN E. & WINDLUND L. (1972) FEBS Letters 28, 107-111
KLETT R. P., FULPIUS B. W., COOPER D., SMITH M., REICH E. & POSSANI L. (1973) J. Biol. Chem. 248, 6841-6853
LAEMMLI V. K. (1970) Nature 227, 680-685
LASKEY R. A. & MILLS A. D. (1975) Eur. J. Biochem. 56, 335-341
LA TORRE J. L., LUNT G. S. & DE ROBERTIS E. (1970) Proc. Natl. Acad. Sci. U. S. A. 65, 716-720
MEUNIER J. C., SEALOCK R., OLSEN R. & CHANGEUX J. P. (1974) Eur. J. Biochem. 45, 371-394
MILEDI R., MOLINOFF P. & POTTER L. T. (1971) Nature 229, 554-557
O'FARRELL P. H. (1975) J. Biol. Chem. 250, 4007-4021
OLSEN R. W., MEUNIER J. C. & CHANGEUX J. P. (1972) FEBS Letters 28, 96-100
SCHMIDT J. & RAFTERY M. A. (1972) Biochem. Biophys. Res. Commun. 49, 572-578
STUDIER F. W. (1973) J. Mol. Biol. 79, 237-248
WEBER G., BORRIS D. P., DE ROBERTIS E., BARRANTES J. F., LA TORRE J. L. & LLORENTE de CARLIN M. C. (1971) Mol. Pharmacol. 7, 530-537

EFFECTS OF DIVALENT CATIONS, K^+, AND X-537A ON GLYCEROLIPID METABOLISM IN THE CATTLE RETINA

N.M. Giusto and N.G. Bazán

Instituto de Investigaciones Bioquímicas, Universidad Nacional del Sur y Consejo Nacional de Investigaciones Científicas y Técnicas, Bahía Blanca, Argentina

INTRODUCTION

The retina besides containing the specialized photoreceptor cell layer is a portion of CNS grey matter comprising a large surface of excitable membranes. Thus it is endowed with a high content and wide variety of polar lipids.

^{14}C-glycerol can be used as a precursor to follow the neosynthesis of retinal lipids, and propranolol and phentolamine are able to redirect the biosynthetic flow (Bazán et al., 1976a; 1976b; Bazán et al., 1977). Propranolol as well as other amphiphilic cationic drugs exert a similar action in lymphocytes (Allan & Michell, 1975) and in rat liver (Brindley & Bowley, 1975). The susceptibility of the pathway to being shifted by these drugs suggests that in the retina in situ may operate a controlling mechanism of the different metabolic branches to insure the provision of specific lipids to membranes. This may notably be case in the steps leading from phosphatidic acid to diglycerides and from phosphatidic acid to phosphatidylinositol through CDP-diglycerides (Bazán et al., 1977).

When cattle retinas are incubated in an ionic medium that resembles cerebrospinal fluid in its electrolyte composition (Ames & Hastings, 1956; Ames & Gurian, 1963) a heightened rate of ^{14}C-glycerol flux towards triacylglycerol is disclosed (Bazán et al.,

---

Abbreviations used: DG, diacylglycerols; TG, triacylglycerols; TPL, total phospholipids; PA, phosphatidic acid; IPG, CPG and EPG, inositol-choline-, and ethanolamine- phosphoglycerides; TLC, thin layer chromatography.

1976a, 1976b, 1977). Since divalent cations modify the activity of phosphatidic acid phosphatase when studied in cell-broken preparations (Jamdar & Fallon, 1973; Mc Caman et al, 1965) it was of interest to survey the effects of changes in the ionic environment. This paper reports preliminary studies varying the ion concentrations, notably that of divalent cations and the effect of adding the ionophoric antibiotic X-537A on the ^{14}C-glycerol labeling of cattle retina lipids.

METHODS

Cattle retinas brought to the laboratory in light-tight containers packed within crushed ice were used no later than three hours after slaughter. In all cases a preincubation of 10 min was followed by the addition of 5 µCi of (U-^{14}C)glycerol per retina. (New England Nuclear Corp. Mass., specific activity 8.75 mCi/mmol). Incubations were performed in the medium of Ames and Hastings (1956) containing 2 mg of glucose per ml and is referred to in the text as control medium. The concentrations of Ca^{2+} and Mg^{2+} were varied as indicated in the Tables. The osmolarity was kept constant by adding choline chlorhydrate when the cation concentration was lowered. The concentration of X-537A (bromo analogue) was 40 µM. When Ca^{2+} was omitted 0.2 mM EGTA was added. Thereafter at different time intervals the retinas were removed from the incubation media and homogenized with chloroform-methanol by means of a Potter-Elvehjem homogenizer (Folch et al., 1957). Phospholipid classes were isolated by two-dimensional TLC (Rouser et al., 1970) and neutral glycerides by gradient-thickness TLC (Bazán & Bazán, 1975). Aliquots of each sample were simultaneously run on two plates, in the case of phospholipids and following visualization by iodine vapor from one plate the spots were scrapped off for P determination and from the other for counting by liquid scintillation.

RESULTS AND DISCUSSION

Effects of high concentrations of K^+

Preincubation of cattle retinas during 10 min with depolarizing concentrations of K^+ (Table 1), although it does not change the ^{14}C-glycerol labeling of phosphatidic acid at 10 min, slightly decreases the incorporation thereafter. A small lowering in the specific activity of IPG and DG at early incubation times was apparent, whereas TG labeling was deeply decreased.

45 mM K^+ reduces by 50% the ^{14}C-glycerol labeling of cattle retina lipids after 30 min of incubation with the sole exception of phosphatidylinositol and DG which approach control values.

Table 1. Changes in glycerolipid synthesis from ^{14}C-glycerol due to high concentration of K^+

		Control	45 mM K^+
		(cpm / µmol of lipid) x 10^{-3}	
10 min	PA	281	280
	IPG	34	21
	DG	656	320
	TG	202	32
20 min	PA	410	308
	IPG	46	48
	DG	790	750
	TG	922	148
30 min	PA	486	369
	IPG	86	81
	DG	1.296	1.182
	TG	2.071	475

CPG, EPG and TG incorporation is inhibited by 60, 40 and 80% respectively. The changes produced by the high K^+ concentration in the neuronal membranes likely diminished the uptake of the precursor or lowered its entrance into the lipid pathway. In any event since the synthesis of TG is markedly slowed as compared with the inhibition in the labeling of the nitrogen-containing phospholipids two different effects may underlie such actions. The former may be brought about by an inhibition of the diacylglyceride acyltransferase and the later by a reduction in the conversion resulting from the cytidyltransferases.

Effects of extracellular Ca^{2+} and Mg^{2+}

The precursor uptake was enhanced by increasing Ca^{2+} and Mg^{2+} concentration from 0.1 mM to 3 mM. The highest labeling of neutral glycerides was found at 1.15 and 1.2mM of Ca^{2+} and Mg^{2+} respectively, that is, in Ames and Hastings medium (Table 2). Phosphatidic acid phosphatidylinositol and phosphatidylcholine, on the other hand, attain highest labeling in the presence of 3 mM Mg^{2+} plus 0.1 mM Ca^{2+}. Under the former condition TG radioactivity was observed to be the lowest.

Table 2. Effects of varying the divalent cations concentration on cattle retina lipids labeling by ^{14}C-glycerol

Condition	Time (min)	PA	IPG	CPG	EPG	TPL	DG	TG
			(cmp / 100 mg		of protein)	x 10^{-5}		
Ca^{2+} 0.1mM Mg^{2+} 0.1mM	10	0.8	0.2	0.2	0.2	1.4	0.7	0.2
	20	1.4	0.8	0.8	0.7	3.8	2.6	1.5
	30	1.2	1.0	1.7	1.1	5.1	2.9	4.4
Ca^{2+} 1.15mM, Mg^{2+} 1.2 mM *	10	0.9	0.3	0.3	0.2	1.7	1.4	0.6
	20	1.3	1.0	1.2	0.9	4.5	3.7	3.5
	30	1.4	1.6	2.9	1.7	7.6	4.8	8.8
Ca^{2+} 3 mM Mg^{2+} 0.1mM	10	1.1	0.4	0.3	0.3	2.1	1.5	0.5
	20	1.0	0.7	1.1	0.7	3.6	2.0	2.3
	30	1.8	1.7	2.5	2.3	8.3	3.8	6.1
Mg^{2+} 3 mM Ca^{2+} 0.1mM	10	1.2	0.5	0.3	0.3	2.3	1.3	0.4
	20	1.2	1.0	1.0	0.7	4.0	3.1	2.5
	30	2.0	2.1	3.0	1.8	9.0	3.6	4.1

* = Control medium.

The addition of either Ca^{2+} or Mg^{2+} above the ion concentrations of Ames and Hastings medium augments the ^{14}C-glycerol incorporation in total phospholipids, as shown also in toad retina (Bazán & Bazán, 1977). This is mainly due to an increased labeling of PA, IPG, CPG and EPG. On the other hand, both diacylglycerol and triacylglycerol markedly diminish. If the retinas are incubated in a medium low in both Ca^{2+} and Mg^{2+} a striking reduction in the labeling of polar lipids (33%) as well as of neutral glycerides (62%) is observed. These changes suggest that the very high triacylglycerol formation taking place when incubation is carried out in Ames & Hastings medium proceeds as such because the required precise concentration of divalent cations is met. The step when diacylglycerols are further acylated to yield triacylglycerols may be the sensible site of these ions.

Effects of X-537A when Ca^{2+}, Mg^{2+}, or both are omitted

After 30 min of incubation in the basal medium 37% of the

radioactivity was found in total phospholipids and 22 and 40% in DG and TG respectively (Table 3). If the ionophoric antibiotic X-537A is added a drastic inhibition in lipid labeling takes place, the neutral lipids being mainly affected. When Mg^{2+} was omitted the inhibition in phospholipids was more pronounced. However if both cations are absent the antibiotic inhibits the labeling by 75%. It is known that X-537A produces alterations unrelated to the properties of forming lipid-soluble complexes with divalent cations to facilitate their entrance through biomembranes. Thus X-537A seems to have an effect of its own on retinal lipid labeling. This may be related to the ionophoric action for monovalent cations (Pfeiffer et al., 1974; Pressman, 1976), to the decrease in ATP levels described for other ionophoric antibiotics

Table 3. Modifications in ^{14}C-glycerol incorporation in retina lipids due to divalent cations and X-537A

Condition	Time (min)	TPL	DG	TG
		(cpm / 100mg of protein) x10^{-5}		
minus X-537A	10	2.68	2.35	0.87
plus Ca^{2+} 1.15mM	20	4.10	2.83	3.97
plus Mg^{2+} 1.2 mM*	30	7.51	4.65	8.96
plus X-537A	10	1.13	0.51	0.21
plus Ca^{2+} 1.15mM	20	1.99	0.66	0.48
plus Mg^{2+} 1.2 mM*	30	2.66	0.85	0.82
plus X-537A	10	0.61	0.30	0.17
plus Ca^{2+} 1.15mM	20	1.10	0.64	0.44
minus Mg^{2+}	30	1.62	0.85	1.07
plus X-537A	10	1.04	0.43	0.18
plus Mg^{2+} 1.2mM	20	2.38	1.01	0.53
minus Ca^{2+}	30	2.93	1.18	1.16
plus X-537A	10	0.73	0.33	0.09
minus Ca^{2+}	20	1.83	0.88	0.37
minus Mg^{2+}	30	3.07	1.33	0.94

* = Control medium.

(Kirkpatrick et al., 1975; Allan & Michell, in press), or to a new action.

Effects of X-537A in the presence of Ca^{2+} and Mg^{2+}

In Table 4 are depicted further experiments with X-537A in the presence of Ca^{2+} and Mg^{2+}. The most marked inhibition of PA labeling was observed at 3 mM of either of these cations, although this phospholipid was affected to a lesser extent. After 10 min of incubation at low ion concentration the inhibition in IPG was 42%, whereas at 3 mM concentration it was more reduced (70-75%).

Table 4. Percent of inhibition in retina lipid labeling from ^{14}C-glycerol due to X-537A

The changes represent the reduction in incorporation calculated from retinas individually incubated under the same condition in media lacking X-537A.

Condition	Time (min)	PA	IPG	TPL	DG	TG
Ca^{2+} 0.1mM, Mg^{2+} 0.1mM	10	24	42	28	41	-
	20	22	62	43	76	68
	30	22	79	62	73	89
Ca^{2+} 1.15mM, Mg^{2+} 1.2 mM*	10	25	45	35	63	68
	20	19	79	56	82	86
	30	18	81	66	82	91
Ca^{2+} 3 mM, Mg^{2+} 0.1mM	10	39	75	53	73	69
	20	33	78	52	67	76
	30	37	85	71	81	87
Mg^{2+} 3 mM, Ca^{2+} 0.1mM	10	49	70	55	76	77
	20	30	78	60	87	90
	30	50	85	76	88	90

* = control medium.

At 30 min there were not great differences of the inhibitory action of X-537A in the presence of various cation concentrations. In total phospholipid labeling a similar trend of changes was seen (Table 4).

The drastic inhibitory effect on the labeling of TG and DG brought about by varying the divalent cation concentration suggests a reduction in the biosynthetic route towards the glycerides. In this conection it is of interest that in brain homogenates Mc Caman et al. (1965) have reported that Ca^{2+} and Mg^{2+} inhibit phosphatidate phosphohydrolase. Mg^{2+} in adipose tissue activates this enzyme at low concentration, whereas at high concentration it produces an inhibitory action (Jamdar & Fallon, 1973).

CONCLUSIONS

The aim of the present work was a preliminary assessment of the influence that ions may have on the functioning of the complex lipid biosynthetic pathway as a whole. Since these studies were carried out using the entire retina only tentative statements can be made about the enzymes involved. However due to the clear-cut effects exerted by propranolol or phentolamine in redirecting the "de novo" synthesis of lipids towards phosphatidylinositol in cattle retina (Bazân et al., 1976b ; 1977) and following the alterations by divalent cations it is tempting to suggest that in the retina are present heterogeneous pools of metabolicaly very active lipids. The shifts in ion concentration may alter the lipid labeling by either a secondary effect or a direct action on the metabolic steps. In either case, the early time-course of the labeling suggests that the precursor can enter different pools of retina lipids and that different branches of the "de novo" biosynthesis of retina glycerolipids are very sensitive to shifts in the ionic environment. Further experiments with lower concentrations of X-537A as well as with another carboxylic acid lipid soluble ionophore will likely help in clarifying the dual effect shown here to be exerted by this antibiotic.

REFERENCES

ALLAN, D. & MICHELL, R. H. (1975) Biochem. J. 148, 472-478.
ALLAN, D. & MICHELL, R. H. (1977) Biochem. J. (in press).
AMES, A. & HASTINGS, B. (1956) J. Neurophysiol. 19, 201-212.
AMES, A. & GURIAN, B. S. (1963) J. Neurophysiol. 26, 617-634.
BAZAN, N. G. & BAZAN, H. E. P. (1975) in Research Methods in Neurochemistry (Marks, N. and Rodnight, R., eds.), Vol. 3, pp.309-324.

Plenum, New York.
BAZAN, N. G., AVELDAÑO, M. I., PASCUAL de BAZAN, H. E. & GIUSTO, N. M. (1976a) in Lipids (Paoletti, R., Porcellati, G. and Jacini G., eds.) Vol. 1, pp. 89-97. Raven Press, New York.
BAZAN, N. G., ILINCHETA de BOSCHERO, M. G., GIUSTO, N. M. & PASCUAL de BAZAN, H. E. (1976b) in Function and Metabolism of Phospholipids in the Central and Peripheral Nervous Systems (Porcellati, G., Amaducci, L. & Galli, C., eds.) pp. 139-148. Plenum, New York.
BAZAN, H. E. P. & BAZAN, N. G. (1977) This volume.
BAZAN, N. G., ILINCHETA de BOSCHERO, M. G. & GIUSTO, N. M. (1977) This volume.
BRINDLEY, D. N. & BOWLEY, M. (1975) Biochem. J. 148, 461-469.
FOLCH, J., LEES, M. & SLOANE STANLEY, G. H. (1957) J. Biol. Chem. 226, 497-509.
JAMDAR, S. C. & FALLON, H. J. (1973) J. Lipid Res. 14,517-524.
KIRKPATRICK, F. H., HILLMAN, D. G. & LA CELLE, P. L. (1975) Experienta 31, 653-654.
Mc CAMAN, R. E., SMITH, M. & COOK, K. (1965) J. Biol. Chem. 240, 3513-3517.
PFEIFFER, D. R., REED, P. W. & LARDY, H. A. (1974) Biochemistry 19, 4007
PRESSMAN, B. C. (1976) Annu. Rev. Biochem. 45, 501-530.
ROUSER, G., FLEISHER, S. & YAMAMOTO, A. (1970) Lipids 5, 494-496.

EFFECTS OF TEMPERATURE, IONIC ENVIRONMENT, AND LIGHT FLASHES ON THE GLYCEROLIPID NEOSYNTHESIS IN THE TOAD RETINA

H.E. Pascual de Bazán and N.G. Bazán

Instituto de Investigaciones Bioquímicas, Universidad Nacional del Sur y Consejo Nacional de Investigaciones Científicas y Técnicas, Bahía Blanca, Argentina

INTRODUCTION

Several extracellular stimuli are known to affect the turnover of membrane lipids, particularly that of phosphatidylinositol (Michell, 1975). However, similar effects on the neosynthesis of phospholipids and neutral glycerides are not clearly known. Nevertheless it appears that this pathway must be under metabolic control to ensure the provision of individual lipids for biogenesis, partial renewal, or repair of biomembranes.

An active glycerolipid synthesis from ^{14}C-glycerol, predominantly phosphatidylinositol formation, takes place in the toad retina both in vivo and in vitro (Bazán & Bazán, 1976). To further study this process a survey of different incubating conditions was carried out. These included variations in temperature, the use of different ionic environments, and the action of a divalent ionophore. In addition it was of particular interest to inquire about the effect of the physiological stimulus of the retina, the light, on the de novo biosynthesis of lipids from ^{14}C-glycerol in the entire toad retina.

EXPERIMENTAL

The retinas were excised from Bufo arenarum Hensel of both sexes just prior to incubation. Incubation was performed in the media of Ames-Hastings (1956), pH 7.33, or Sickel (1965), pH 7.8, each containing 2 mg per ml of glucose and saturated with 5% CO_2 in oxygen. Ten μCi of (U-^{14}C)-glycerol (specific activity 7.4 μCi/μmol, New England Nuclear, MA) per 18 retinas was added to the incubation media. The phospholipids and neutral lipids were separated by two dimensional TLC (Rouser et al., 1970) and by gradient-thickness TLC (Bazân & Bazân, 1975) respectively. Radioactivity was measured in uneluted spots by liquid scintillation spectrometry (Bazân & Bazân, 1976).

RESULTS

Lipid labeling from ^{14}C-glycerol in cattle and toad retinas yielded different profiles (Bazân et al., 1976). In both instances the glycerophosphate-phosphatidate-diacylglycerol route operates, however only in the former a very high labeling of triacylglycerol was found. This lipid in the toad retina, on the other hand, was labeled to a small extent whereas phosphatidylinositol attained the highest incorporation of ^{14}C-glycerol. Since incubation in these experiments was performed in Ames-Hastings medium (1956) at 37°C, we decided to inquire if the high rate of phosphatidylinositol de novo biosynthesis observed in the toad retina was due to the incubating conditions used. Thus we have examined a) the effects of an ionic medium devised for amphibian retina (Sickel, 1965); b) the effect of temperature ; c) the action of divalent cations and d) the effects of light flashes and of darkness during incubation.

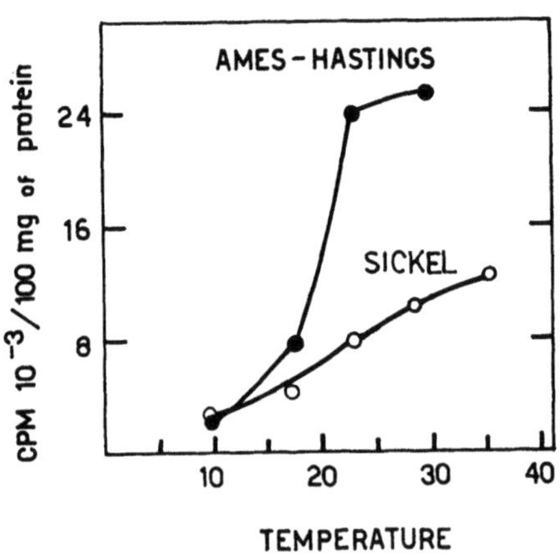

Fig. 1. Dependance of ^{14}C-glycerol labeling of retina lipids on temperature. Each point represents an incubation flask containing 8 retinas in 2.8 ml of medium and 4 µCi of ^{14}C-glycerol. After 10 min of incubation the retinas were lipid extracted.

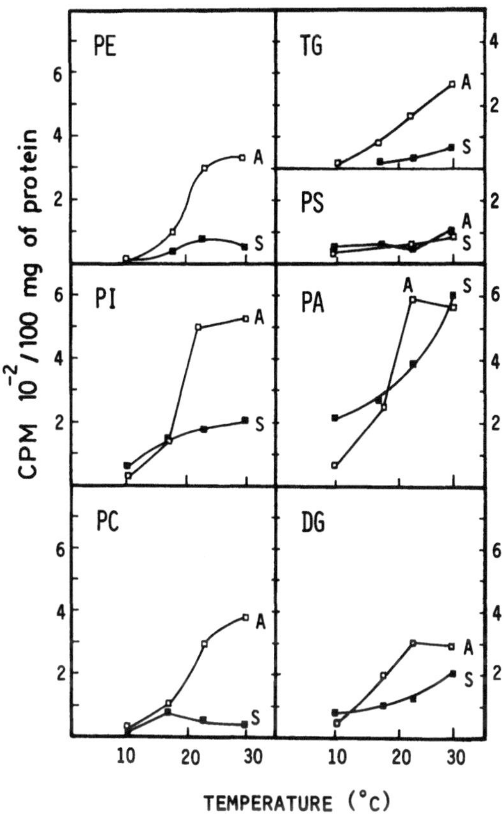

Fig. 2. Effects of temperature on the ^{14}C-glycerol labeling of retina lipid classes. PE: phosphatidylethanolamine ; PI: phosphatidylinositol; PC: phosphatidylcholine; TG: triacylglycerols; PS: phosphatidylserine; PA: phosphatidate and DG: diacylglycerols; A: Ames-Hastings medium and B: Sickel medium.

Although the media used make widely different ionic environments, similar lipid labeling profiles were found when the early time-course of ^{14}C-glycerol uptake was followed. In both cases a temperature-dependent incorporation, higher for the Ames-Hastings medium, is evident. The precursor uptake increases up to 23°C. Hereafter all the experiments were carried out at this temperature (Fig. 1). Phosphatidic acid and phosphatidylinositol were the highest labeled lipids under all these conditions (Fig. 2).

Next, we decided to observe the effect of varying calcium concentrations on the ^{14}C-glycerol uptake. The labeling of total phospholipids increased in a Ca^{2+} concentration - dependent fashion in agreement with a similar observation made in our laboratory using cattle retina (Giusto & Bazán, 1977). Again phosphatidic acid

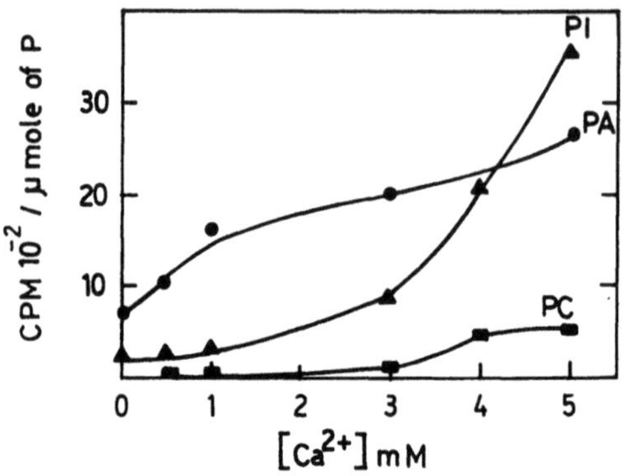

Fig. 3. Calcium-dependent changes in ^{14}C-phospholipid specific activities of retinas. Abbreviations as in Fig. 2. Phospholipid P was measured according to Rouser et al. (1970).

and phosphatidylinositol were the phospholipids mainly affected by Ca^{2+} (Fig. 3).

Since it is known that the ionophore X-537A increases the membrane permeability to divalent cations we incubated toad retinas in Ames-Hastings medium during 20 minutes in darkness in the presence of different calcium concentrations; here again the uptake is Ca^{2+}-dependent. When 27 µg/ml (40 µM) of ionophore was added, the incorporation in total lipids increased and the higher labeling was found in PA and PI.

To observe the effects, if any, of Mg^{2+} on lipid synthesis in the entire retina, the tissue was incubated in Mg^{2+}-depleted Ames-Hastings medium. The total ^{14}C-glycerol incorporation disminished but again the higher specific activities corresponded to PA and PI. The ionophore X-537A produces an enhancement of the effect (Table 1).

The changes documented here, as well as others described elsewhere (Bazán et al., 1976), in the drugs' action on membrane lipid synthesis in the retina pose the question of whether or not they are related to the normal functioning of the central nervous system. The test of such a possibility was carried out by applying to the entire retina its natural stimulus and comparing it with retinas kept in darkness.

In both media light flashes stimulated polar lipid labeling (Fig. 4). The effect was clearly observed after 20 minutes of incubation. The largest increments produced by light were found in PI and PA.

Table 1. Lipid specific activity after 30 min of incubation with ^{14}C-glycerol in Ames-Hastings medium lacking Mg^{2+}

Specific activity: CPM 10^{-3}/ μmole of lipid.
The media isotonicity was mantained by adding chlorhydrate of choline, the ionophore X-537A was disolved in 100 μl of ethanol/7 ml of incubation media and the same quantity of ethanol was added to the controls. Abbreviations as in Fig. 2.

	Without ionophore		With ionophore	
	Specific activity	%	Specific activity	%
PA	9.9	64	27.1	64
PI	1.3	9	6.6	16
PC	0.1	0.7	0.1	0.2
DG	2.4	16	5.2	12
TG	1.7	11	3.6	8

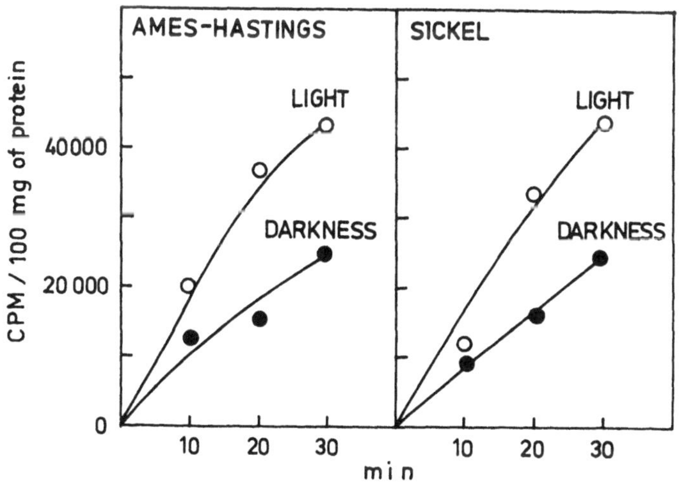

Fig. 4. Effect of light-flashes on the labeling of retina total polar lipids by ^{14}C-glycerol in different media. The light flashes were of one second duration every 30 seconds. The light source was a projector lamp located 30 cm below the incubation bath. Within the flasks 40 ft candle flashes were measured. Each sample contains 18 retinas. Total proteins were determined by Lowry et al. (1951).

DISCUSSION

The ^{14}C-glycerol uptake in toad retina lipids is temperature-dependent, and in the two media employed until 23°C there is a linear incorporation, so we selected this temperature for further experiments. Such optimal temperature was expected for a poikilotherm. At higher temperatures there are not significant changes. In Sickel medium phosphatidic acid and diacylglycerols after 23°C show a rise in the precursor uptake as temperature is augmented. A similar pattern of change followed triacylglycerols in Ames-Hastings medium. The temperature activates the biosynthetic pathway. However similar trends of alteration in the labeling of individual phospholipids of the retina were seen, unlike the different 32P incorporation of nitrogen-containing phospholipids reported to occur when the incubating temperature is varied in Ehrlich ascites tumor cells (Baranska & Banskalieva, 1976).

The incorporation is affected also by the ionic environment and is higher in Ames-Hastings that in Sickel media. This is in agreement with the larger content of divalent cations in the former. When the intracellular Ca^{2+} concentrations were increased by means of the ionophore X-537A an increase in the specific activity of phosphatidic acid and phosphatidylinositol was observed. In media lacking Mg^{2+} the ionophore again produced a higher uptake in the two phospholipids.

The mechanism through which Ca^{2+} ions exert their action is not apparent; however, there are several intracellular events affected by the divalent cation that might bear a relationship with the phospholipid effect here described. Ca^{2+} causes neurotransmitter release from nerve terminals (Del Castillo & Katz, 1954; Pressman 1972), plays an important role in visual excitation (Hagins, 1972; Hagins & Yoshikami, 1975), and influences several other cellular reactions (Carafoli et al., 1976). In addition a direct influence of the divalent cation on the biosynthetic pathway might take place.

The effect exerted by the ionophoric antibiotic seems to be due to an enhanced entrance of Ca^{2+} into the tissue; however, the involvement of other effects of the ionophore cannot be ruled out at the present time.

The light-sensitive biosynthetic pathway of membrane lipids here described operates in two different ionic media and may not be located in the outer segments of the photoreceptor cells. This conclusion is in agreement with the presence of the biosynthetic machinery in the inner portion of the photoreceptor cell. Very likely the light effect takes place in the neural portion and is related to normal neural activity.

These results are different from the data presented by Urban et al. (1973), who failed to show any effect after illumination on the ^{32}P phospholipid turnover. However, they used a different precursor, longer incubation time periods, and different animal species.

REFERENCES

AMES, A. III & HASTINGS, A. B. (1956) J. Neurophysiol. 19, 201-212.
BARANSKA, J. & BANSKALIEVA, V. B. (1976) Febs Letters 65, 24-29.
BAZAN, N. G. & BAZAN, H. E. P. de (1975) in Research Methods in Neurochemistry (Marks, N. & Rodnight, R. eds.), pp. 309-324, Plenum Press, New York.
BAZAN, H. E. P. de & BAZAN, N. G. (1976) J. Neurochem. 27, 1051-1057.
BAZAN, N. G., AVELDAÑO, M. I., BAZAN H. E. P. de & GIUSTO, N. M. (1976) in Lipids Vol 1 (Paoletti, R., Porcellati, G. & Jacini, G. eds.) pp. 89-97, Raven Press, New York.
CARAFOLI, E. MALMSTRÖM, K., SIGEL, E. & CROMPTON, M. (1976) Clin. Endocrinol 5, Suppl., 49s-59s.
DEL CASTILLO, J. & KATZ, B. (1954) J. Physiol. 124, 560-573.
GIUSTO, N. M. & BAZAN, N. G. (1977) This volume.
HAGINS, W. A. (1972) Ann. Rev. Biophys. Bioeng. 1, 131-158.
HAGINS, W. A. & YOSHIKAMI, S. (1975) Ann. New York Acad. Sci., 264, 314-325.
LOWRY, O. H., ROSENBROUGH, N. J., FARR, A. L. & RANDALL, R. J. (1951) J. Biol. Chem. 193, 265-275.
MICHELL, R. H. (1975) Biochim. Biophys. Acta 415, 81-147.
PRESSMAN, B. C. (1972) in The role of membranes in metabolic regulation, (Mehlman, M. A. & Hanson, R. W., eds.) pp. 149-164.
ROUSER, R., FLEISCHER, S. & YAMAMOTO, A. (1970) Lipids 5, 494-496.
SICKEL, W. (1965) Science 148, 648-651.
URBAN, P. F., DREYFUS, H., NESCOVIC, C. N. & MANDEL, P. (1973) J. Neurochem. 20, 325-335.

MUSCARINIC CHOLINERGIC STIMULATION OF PHOSPHATIDYL INOSITOL TURNOVER IN THE CNS

O. Canessa de Scarnatti, M. Sato, and E. De Robertis

Instituto de Biología Celular, Facultad de Medicina

Universidad de Buenos Aires, 1121 Buenos Aires, Argentina

INTRODUCTION

A large variety of cells undergo a marked increase in turnover of PI under the influence of different physiological and pharmacological stimuli (for reviews, see Lapetina & Michell, 1973, Michell, 1975). The features in common of the so-called PI effect are (1) it is probably mediated by receptor sites at the cell surface and (2) it involves only the phosphorylinositol moiety of PI.

In the CNS, ACh was found to stimulate the PI turnover of brain slices (Hokin & Hokin, 1955, 1958), homogenates (Hokin & Hokin, 1958; Redman & Hokin, 1964), and fractions enriched in nerve endings (i.e. synaptosomes) (Durrell & Sodd, 1966; Yagihara & Hawthorne, 1972; Schacht & Agranoff, 1973). The PI effect was also found in electrically stimulated brain slices (Pumphrey, 1969) and synaptosome beds (Bleasdale & Hawthorne, 1974).

It is well known that the CNS has both muscarinic and nicotinic receptor sites (for a review, see De Robertis, 1975); thus the PI effect stimulated by ACh could be mediated by one or the other type of receptors. Scattered observations favor the view that the muscarinic receptors are related to the metabolism of PI. In particulate fractions Hokin & Hokin (1958) found that the PI effect was blocked by atropine, and Schacht & Agranoff (1973), in a synaptosomal fraction, observed that the ACh stimulation was insensitive to d-tubocurarine. To investigate this problem further, in addition to using appropriate cholinergic agonists and antagonists, an approach would be to take into consideration the regional distribution of both types of receptors in the CNS. For example, while in the entire mouse brain the cholinergic receptor sites are almost equally divided between nicotinic and muscarinic (Schleifer & Eldefrawi, 1974), in the n. caudatus there is a predominance of muscarinic binding sites (Hiley & Burgen, 1974; Yamamura et al., 1974). Recently Saraceno & De Robertis (1976) found that the cholinergic receptor proteolipid protein isolated by affinity chromatography from bovine caudate nucleus probably contained 90 % muscarinic sites and only 10 % nicotinic.

In this work we have taken advantage of the regional distribution of receptors by using the bovine caudate nucleus as an essentially muscarinic tissue, in comparison with the rat cerebral cortex in which there is a mixture of the two types of receptor sites. With this approach and the use of appropriate cholinergic blocking agents we have reached the conclusion that the PI effect stimulated by ACh in the CNS is mediated only by muscarinic receptors.

RESULTS AND DISCUSSION

1) Phospholipid composition of bovine caudate nucleus. The total

Table 1. Phospholipid composition of bovine caudate nucleus

Lipids were extracted according to Folch-Pi et al. (1957) with chloroform-methanol (2:1, v/v) and partitioned with chloroform-methanol-water (3:48:47). Phospholipids were separated by TLC (Skipski et al., 1964) and after scraping the various bands (Lunt & Lapetina, 1970) phosphate (Chen et al., 1956) and radioactivity were determined.

The results are the mean of 3 experiments ± S. D.

Phospholipid	% Distribution
Sphingomyelin	4.8 ± 0.2
Phosphatidylcholine	35.8 ± 0.8
Phosphatidylinositol	6.4 ± 0.5
Phosphatidylserine	9.7 ± 0.7
Phosphatidylethanolamine	30.6 ± 3.0
Phosphatidic acid + cardiolipin	10.2 ± 2.0
Origin	2.5

phospholipid content of the n. caudatus of the cow is 46.5 ± 3.5 mg/g fresh tissue. In Table 1 it is shown that quantitatively the major species are phosphatidylcholine (PC) and phosphatidylethanolamine (PE), while PI represents only 6.4 ± 0.5 % of the total phospholipids. The value for phosphatidylserine (PS) was obtained by subtracting the amount of PI obtained by paper chromatography (Kai & Hawthorne, 1966), from that of the two phospholipids together resulting from the TLC.

2. Stimulation of phosphatidylinositol turnover by ACh. As shown in Table 2 the slices of caudate nucleus actively incorporate ^{32}Pi into the various phospholipids; the highest specific radioactivity being achieved by the fraction containing PI plus PS. When the slices were incubated in presence of 10 mM ACh, there was a

Table 2. Action of ACh upon the incorporation of ^{32}Pi into phospholipids of bovine caudate nucleus

Transverse slices of approximately 0.5 mm thickness were cut, and about 200 mg of such slices were put into individual flasks containing 10 ml of Krebs-Ringer bicarbonate. This solution was gassed with a mixture of 95% O_2 and 5% CO_2 for 1 h prior to the addition of 11 mM glucose. 150 µCi of ^{32}Pi (carrier free, Argentine Atomic Energy Commission) was added to each flask and the incubation was carried out at 37°C for 2 h under constant shaking of the samples. The incubation was stopped by the addition of cold buffer and the slices were washed three times with the same buffer and centrifuged; the pellets were used for lipid extraction. When required, ACh was added to the incubation medium at a final concentration of 10 mM (Hokin & Hokin, 1955; Lapetina & Michell, 1972); 1 mM eserine sulfate was used to inhibit acetylcholinesterase.

Results are the mean ± S.D. of three different experiments. In the control there were no additions, while in the experimental 10 mM ACh and 1 mM eserine were added to the incubation medium.

Phospholipid	Specific radioactivity (d.p.m./µg) Control	Experimental	Mean % increase
Phosphatidylinositol + phosphatidylserine	152 ± 20	360 ± 35	237
Phosphatidylcholine	50 ± 10	59 ± 9	—
Phosphatidyl-ethanolamine	85 ± 23	115 ± 17	—

considerable increase in the incorporation of ^{32}Pi in this fraction while there was no stimulation of PC and PE. Using paper chromatography to separate PS from PI, it was demonstrated that the PI is the only phospholipid in which the ^{32}Pi incorporation is stimulated by ACh. After the two hour period of incubation there is no change in the absolute content of the various phospholipids.

3. Effect of muscarinic and nicotinic blocking agents on ^{32}Pi incorporation into PI. Figure 1 shows that the incorporation of ^{32}Pi into the PI of the caudate nucleus increases more than 100 % after stimulation with ACh and that this effect is completely blocked by 1 μM atropine, a muscarinic antagonist; on the other hand it is not inhibited by d-tubocurarine, a nicotinic blocking agent.

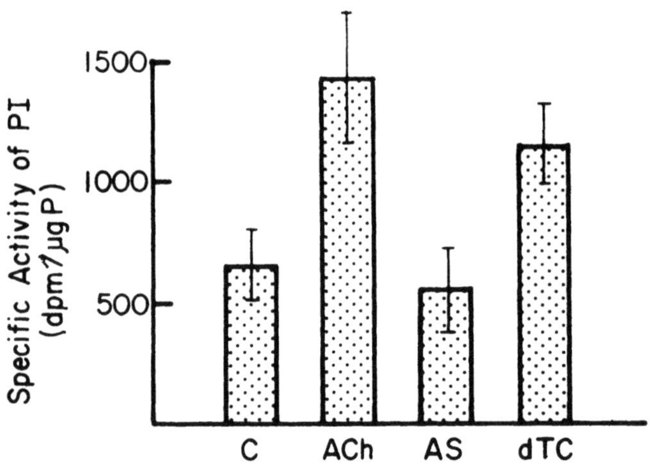

Fig. 1. Action of ACh and of cholinergic blocking agents on the ^{32}Pi incorporation into phospholipids of bovine caudate nucleus. C, control; ACh, 10 mM ACh; AS, 10 mM ACh + 1 μM atropine; dTC, 10mM ACh + 0.1 mM d-tubocurarine. In all the experimental samples 1 mM eserine was added. The results are the mean ± S.D. of three different experiments. The small differences between ACh and dTC are not significant by the "t" test ($P < 0.2$).

Similar results are shown in Table 3 for the cerebral cortex of the rat. In the control condition the incorporation is much higher in PI than in the other phospholipids. Under the action of ACh there is an activation of 174 % in PI turnover, while in the other phospholipids there is no change. The results with the cholinergic blocking agents show that while atropine completely blocks the activation by ACh, there is no change with d-tubocurarine.

Table 3. Action of ACh upon the incorporation of ^{32}Pi in phospholipids of rat cerebral cortex and action of cholinergic blocking agents

Slices of the rat cerebral cortex were incubated as indicated in Table 2 for the caudate nucleus. The results are the mean ± S.D. of three experiments.

	Specific radioactivity (d.p.m./μg)			
Phospholipid	Control	1 mM ACh	1 mM ACh 1 μM atropine	1 mM ACh 0.1 mM d-tubocurarine
Phosphatidyl-inositol	280 ± 26	490 ± 52	255 ± 32	486 ± 51
Phosphatidyl-serine	53 ± 3	47 ± 5	45 ± 8	45 ± 5
Phosphatidyl-ethanolamine	26 ± 9	16 ± 3	13 ± 3	14 ± 2
Phosphatidyl-choline	75 ± 18	55 ± 17	49 ± 14	61 ± 23

The results presented here confirm, both in the caudate nucleus and the cerebral cortex, that the ACh stimulation of the phospholipid turnover is only confined to PI. Even in the control conditions the incorporation of ^{32}Pi is highest in this phospholipid. The finding that in a predominantly muscarinic structure, such as the bovine caudate nucleus (Saraceno & De Robertis, 1976) there is a marked stimulation of PI turnover by ACh suggests that this effect is mediated by muscarinic receptors. This interpretation is strongly supported by the fact that both in the caudate nuclei and rat cerebral cortex the ACh stimulation of PI turnover is blocked by atropine and not by d-tubocurarine (Fig. 1 and Table 3). These findings are in line with the previous observations of Hokin & Hokin (1958) in subcellular fractions in which the PI effect was blocked by atropine and those of Schacht & Agranoff (1973) showing the lack of effect of d-tubocurarine on a synaptosomal fraction. This interpretation is also borne out by results in other tissues. For example the PI effect in rat parotid gland is evoked by acetyl-β-methyl choline, a muscarinic agonist (Jones & Michell, 1974). An apparent exception to this rule is the finding that the PI effect induced in sympathetic ganglia by electrical stimulation is blocked by d-tubocurarine (Larrabee & Leicht, 1965); however some data mentioned by Michell (1975) in superior cervical ganglion suggest that the PI effect stimulated by ACh is insensitive to d-tubocurarine.

Previous studies from this laboratory have shown that in the isolated cholinergic proteolipid from the cerebral cortex (De Robertis et al., 1967, 1969a) there is considerable concentration of PI. In fact, while this phospholipid represents only 5-6 % of the total in the rat cerebral cortex, in the cholinergic proteolipid fraction it reaches 30-33 % (Lunt et al., 1971). From the point of view of the interpretation of the mechanism of the PI effect at the molecular level, of even greater interest was the finding that the only PI pool stimulated by ACh was the one that is tightly associated with the cholinergic receptor proteolipid (Lunt et al., 1971). It should be emphasized here that this protein not only binds d-tubocurarine (De Robertis et al., 1967, 1969a) but also atropine (De Robertis et al., 1969b), indicating the presence of both muscarinic and nicotinic binding sites. The presence of these two types of receptor sites has been recently confirmed in the cholinergic proteolipid isolated from the cerebral cortex by affinity chromatography (Saraceno et al., unpublished).

While for the moment we can conclude that the acetylcholine stimulated PI turnover in the CNS is only mediated by muscarinic receptors, further studies on the isolated receptor proteins are needed to interpret the molecular mechanism involved in the association between the receptor and the PI effect.

ACKNOWLEDGEMENTS

This work was supported by a grant of the Consejo Nacional de Investigaciones Científicas y Técnicas (Argentina) and the Burroughs Wellcome Laboratory (USA).

REFERENCES

CANESSA de SCARNATTI O. E. & LAPETINA E. G. (1974) Biochem. Biophys. Acta 360, 298-305

CHEN P. S., TORIBARA T. Y. & WARNER H. (1956) Analyt. Chem. 28, 1756-1758

BLEASDALE J. E. & HAWTHORNE J. N. (1974) Biochem. Soc. Trans. 2, 261-262

DE ROBERTIS E., FISZER S. & SOTO E. F. (1967) Science 158, 228-229

DE ROBERTIS E., FISZER S., PASQUINI J. & SOTO E. F. (1969a) J. Neurobiol. 1, 41-52

DE ROBERTIS E. (1975) in Synaptic Receptors, Isolation and Molecular Biology, Dekker, New York
DURRELL J. & SODD M. A. (1966) J. Neurochem. 13, 487-491
FOLCH-PI J., LEES M. & SLOANE-STANLEY G. H. (1957) J. Biol. Chem. 226, 497-509
HILEY G. R. & BURGEN A. S. V. (1974) J. Neurochem. 22, 159-163
HOKIN M. R. & HOKIN L. E. (1955) Biochim. Biophys. Acta 18, 102-110
HOKIN L. E. & HOKIN M. R. (1958) J. Biol. Chem. 233, 822-826
JONES L. M. & MICHELL R. H. (1974) Biochem. J. 142, 583-690
KAI M. & HAWTHORNE J. N. (1966) Biochem. J. 98, 62-67
LAPETINA E. G. & MICHELL R. H. (1972) Biochem. J. 126, 1141-1147
LAPETINA E. G. & MICHELL R. H. (1973) FEBS Lett. 31, 1-10
LARRABEE M. G. & LEICHT W. S. (1965) J. Neurochem. 12, 1-13
LUNT G. G. & LAPETINA E. G. (1970) Brain Res. 18, 451-459
LUNT G. G., CANESSA O. E. & DE ROBERTIS E. (1971) Nature New Biol. 230, 187-190
MICHELL R. H. (1975) Biochim. Biophys. Acta 415, 81-147
PUMPHREY A. M. (1969) Biochem. J. 112, 61-70
REDMAN C. M. & HOKIN M. R. (1964) J. Neurochem. 11, 155-163
SARACENO H. & DE ROBERTIS E. (1976) Biochem. Biophys. Res. Commun. 69, 555-565
SCHACHT J. & AGRANOFF B. W. (1973) Biochem. Biophys. Res. Commun. 50, 934-991
SLEIFER L. S. & ELDEFRAWI M. E. (1974) Neuropharmacol. 13, 53-63
SKIPSKI V. P., PETERSEN R. F. & BARCLAY M. (1964) Biochem. J. 90, 374-378
YAGIHARA Y. & HAWTHORNE J. N. (1972) J. Neurochem. 19, 355-367
YAMAMURA M. I., KUHAR M. J., GREENBERG D. & SNYDER S. H. (1974) Brain Res. 66, 541-546

SELECTIVE LIPID ALTERATIONS DURING EXPERIMENTAL ALLERGIC ENCEPHALOMYELITIS -- AN INTERPRETATION OF THE CHANGES

F. A. Cumar, B. Maggio, and G. A. Roth

Departamento de Química Biológica. Facultad de Ciencias Químicas, Universidad Nacional de Córdoba, Córdoba, Argentina

Experimental allergic encephalomyelitis (EAE) is an autoimmune, paralytic disease that has been widely used as a model of human demyelinating diseases (Paterson, 1969). We have described selective changes of the glycosphingolipid content and presence of increased amounts of esterified cholesterol in the central nervous system (CNS) of rats in which EAE was induced by whole CNS homogenates. A striking initial observation in these studies was that, depending on the species or strain of animals in which EAE was induced, some of the lipid changes were not necessarily present simultaneously to the paralytic symptoms (Maggio et al., 1972; Maggio & Cumar, 1974; Vasan & Bachhawat, 1971). It is well known that the basic protein of CNS myelin is the specific antigen for inducing a paralytic disease essentially indistinguishable from EAE produced by injection of whole CNS and that other constituents of myelin possess antigenic properties (Rapport & Graf, 1965). Our approach was, therefore, to study the CNS lipid composition of animals sensitized with purified myelin basic protein (BP) as well as with other myelin components.

The studies briefly reviewed here further support the possibility that the CNS alterations of the induced animals can result from specific disruptive responses towards membrane constituents elicited independently by individual components of the CNS (Maggio & Cumar, 1974; 1975).

Animals injected with whole CNS homogenized in Freund's complete adjuvant showed major changes in the level of sulphatides, cerebrosides and esterified cholesterol (Table 1). Animals in which EAE was induced with purified BP showed alterations of the sulphatide content only. The changes in the level of cerebrosides were produced by the presence of CNS lipids in the injection mixture. Esterified

Table 1.- Brain lipid content in EAE and non-EAE guinea pigs. Each group comprised five animals and the numbers represent percentage differences with respect to the values of control animals injected with Freund's adjuvant (see Maggio & Cumar, 1975). Animals injected with: whole CNS (group CNS); purified myelin basic protein (group BP); purified BP plus CNS lipids (group BPL); purified apoprotein of Folch-Lees proteolipid (group AFP); serum albumin (group Ab); serum albumin plus CNS lipids (group AbL); poly-L-lysine, M. W. 41000 (group Py); poly-L-lysine plus CNS lipids (group PyL).

Group	Sulphatides	Cerebrosides	Total Cholesterol	Esterified Cholesterol [c]
EAE				
CNS	-55*	-46*[b]	0.5	detectable
BP	-31*	- 1	0.1	not detectable
BPL	-51*	-52*	4	not detectable
Non-EAE				
AFP[a]	-38*	- 8	-19**	detectable
Ab	-17	5	2	-
AbL	-53*	-33*	- 4	-
Py	-43*	1	6	-
PyL	-66*	-25*	- 1	-

Data are significant at $p<0.01$* or $p<0.02$**. [a] Data obtained from rats. [b] A similar difference was found in EAE rats durin the recovery period but not in the acute stage (see Maggio et al., 1972). [c] See Fig. 1.

cholesterol appeared in increased amounts in animals sensitized with the purified apoprotein of the Folch-Lees proteolipid (Fig. 1). Moreover, these studies clearly showed that lipid alterations can be present without the concomitant appearance of the classical neurological symptoms of EAE (see groups AFP, AbL, Py and PyL in Table 1).

On the basis of the previous results it was proposed that EAE induced with whole CNS could be interpreted as a composite result of different individual effects elicited independently by different components of the CNS (Maggio & Cumar, 1975). On the other hand, variations of the activity of the arylsulphatase A, an enzyme involved in the metabolism of sulphatides, could not be readily correlated

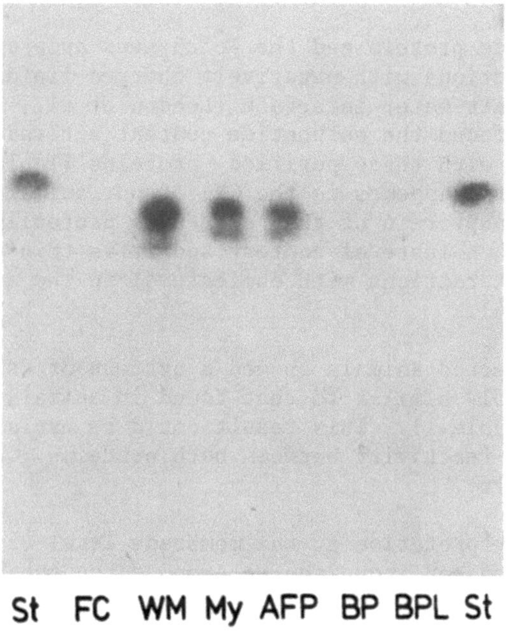

Fig. 1.- Thin layer chromatography of esterified cholesterol. Cholesterol esters were purified according to Horning et al. (1960) and run on TLC with heptane-ethyl ether (24:1). Spots were visualized with Liebermann-Bouchard reagent. Cholesterol oleate standard (St). Material from animals injected with: Freund's adjuvant (FC); white matter (WM); myelin (My); apoprotein of Folch's proteolipid (AFP); basic protein (BP); basic protein plus CNS lipids (BPL).

to the changes in the concentration of this lipid (Maggio et al., 1973)

The stability of biological membranes is a result of carefully balanced intermolecular forces. If these are modified as a consequence of interactions of some of its constituents with external agents, the membrane structure will probably have to bear rearrangements that may have important effects on the metabolism of the membrane components. The myelin membrane is not outside this generalization (Matus et al., 1973). This structure has a unique lipid and protein composition. Lipid constituents such as cerebrosides, sulphatides and cholesterol are present in high amounts. The protein composition of CNS myelin is relatively simple, the three major components being the basic (encephalitogenic) protein, the Folch-Lees proteolipid and the acidic Wolfgram proteins. There are evidences for assuming an asymmetrical distribution of some of these constituents in the myelin membrane (Kirschner & Gaspar, 1972; Poduslo & Braun, 1975) where the lipids seem to be immobilized to a marked degree (Williams et al., 1973), with cholesterol probably keeping the assembly in a fluid-like liquid-crystalline structure (Chapman, 1975).

The myelin basic protein and the Folch-Lees apoprotein show preferential interactions with negatively charged lipids such as sulphatides at the air-water interface (London et al., 1973). Coincidentally, we found the sulphatide content altered in the CNS of animals injected with these purified proteins (Table 1). Esterified cholesterol appears in the CNS of our animals sensitized with the purified apoprotein of the Folch-Lees proteolipid toghether with changes in the cholesterol content and it is this protein that shows remarkable interactions with cholesterol in the model membrane (London et al., 1974).

Polylysine injected animals showed a pattern of CNS lipid alterations remarkably similar to that found in animals sensitized with purified BP (Table 1). This result could be explained by the immunological cross-reactivity between both proteins (Chelmicka-Szorc & Arnason, 1975).

A possible interpretation at the membrane level of the lipid changes briefly discussed above can be proposed on the basis of these observations. Humoral or cellular immunological agents directed towards membrane proteins or lipid haptens may establish new, or modify previous, intermolecular interactions in localized domains of the membrane structure leading to disturbances of the metabolic steady-state. The lipid (and protein ?) components of the perturbed membrane could thus be loosened from specific structural restraints and made available to normal enzymatic activities present in the damaged tissue or membrane. From this point of view, an enzyme activity, which was previously controlled by particular interactions or structural arrangements, need not necessarily be substantially altered in the pathological state (Maggio et al., 1973) but merely be given the possibility of functioning unrestrained. This structure-dynamics orientated interpretation would seem more plausible and in accordance to all the data reviewed than one in which several enzymatic activities should have to be specifically and independently altered as a response to each of the components injected.

Different immunological responses are involved in EAE such as lymphocytes sensitized to BP (Paterson, 1969), capable of passively transfer the disease, and antibodies against cerebrosides that have demyelinating effects in cord tissue cultures (Fry et al., 1974). Disturbances in the membrane structure may occur by several mechanisms involving immunological agents. Aggregation of cell surface receptors can occur in myelin (Matus et al., 1973) and patching phenomena induced by antigen-antibody interactions (Bretscher & Raff, 1975) involving membrane components of both a sensitized cell an a target membrane might lead to endocytotic processes that could selectively remove certain membrane constituents. Alterations of a lipid bilayer

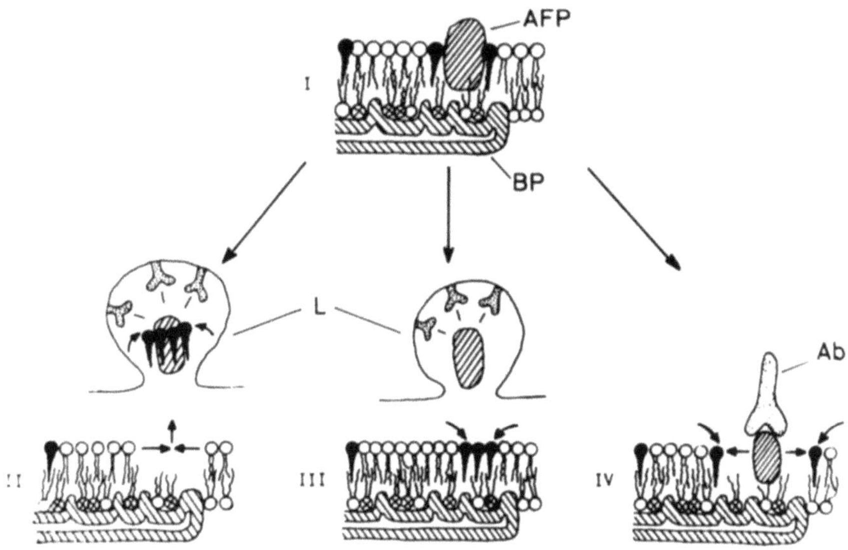

Fig. 2.- Summarized interpretation of possible events leading to selective lipid alterations in EAE.
(I) Simplified structure of part of a normal myelin membrane. The myelin basic protein (BP) is shown in the conformation proposed by London et al. (1973), interacting preferentially with sulphatides ( ). The Folch-Lees proteolipid (AFP) is shown interacting preferentially with cholesterol ( ). (II) Altered membrane after interaction with a sensitized lymphocyte (L) removing AFP and the associated cholesterol. (III) Same as before but only the AFP is being removed. (IV) The association of the AFP with cholesterol is perturbed by interactions of a humoral antibody with the proteolipid. Curved arrows point to the lipid now available for metabolic processes.

structure could also occur by alternative mechanisms mediated by humoral immunological agents (Kinsky, 1972; Henkart & Blumental, 1975). A summarized picture of this interpretation is given in Fig. 2. For simplicity, only some of the possible mechanisms have been considered, involving the AFP; similar effects may occur with other membrane components (cerebrosides, basic protein) as well as with alternative immunological agents (i.e. complement, antibody-dependent phagocytosis).

Several human demyelinating diseases show lipid abnormalities closely related to those seen in EAE. Among other changes there is a general coincidence in reporting alterations of the glycosphingolipid content and increased amounts of esterified cholesterol in the diseased CNS. However, the type or magnitude of the glycosphingolipid alterations have not been always coincidental (Davison & Wajda, 1972;

Clausen & Hansen, 1970). Antiglyco-sphingolipid (Yokoyama et al., 1962) and demyelinating (Appel & Bornstein, 1964) antibodies have been found in the serum of patients with multiple sclerosis. The lipid alterations might be interpreted, and several inconcistencies explained, by considering that some mechanisms similar to those proposed here for the EAE alterations could also participate in human demyelinating diseases. Differing lipid changes in alternative regions of the CNS may depend on the type of antigen exposed to the immunocompetent system by different causative agents, the time elapsed since this has occurred and the responsiveness of the particular patient to specific autoantigens of the CNS.

Apart from the interest that membrane alterations may have in pathological processes, the specificity of immunological reactions involving particular membrane markers might be exploited as a tool for studying the organization of membranes " in situ " in experimental animals.

CONCLUSIONS

Preferential changes in the level of some of the CNS lipids occur during EAE which depend on the presence of particular constituents of myelin in the injection mixture. Animals induced by whole CNS showed changes in sulphatides, cerebrosides and esterified cholesterol. Only sulphatides were altered in animals induced by the encephalitogenic protein. Lipid components of myelin in the sensitizing mixture were responsible for the induction of changes in the cerebroside and sulphatide contents. The Folch-Lees proteolipid is responsible for the appearance of esterified cholesterol. The alterations of CNS lipids were not directly related to the presence of neurological symptoms. The observations may be interpreted as a consequence of specific disruptions of particular lipid-protein interactions by cellular or humoral immunological agents directed towards some of the interacting molecules or associations in myelin. The phenomena are viewed as a primary change in the myelin membrane structure which would subsequently lead to the metabolic alterations.

This work was undertaken with financial aid from the Consejo Nacional de Investigaciones Científicas y Técnicas, Argentina. B.M. is a Career Investigator and G.A.R. is a recipient of a scholarship from the above institution.

REFERENCES

APPEL,S.H. & BORNSTEIN, M. B. (1964) J.Exp.Med. 119, 303-312.
BRETSCHER, M.S. & RAFF, M.C. (1975) Nature (London) 258, 43-49.

CHAPMAN, D. (1975) Quart. Rev. Biophys. 8, 185-235.
CHELMICKA-SZORC, E. & ARNASON, B. G. W. (1975) Clin. Exp. Immunol. 22, 539-545.
CLAUSEN, J. & HANSEN, I. B. (1970) Acta Neurol. Scand. 46, 1-17.
DAVISON, A. N. & WAJDA, M. (1972) J. Neurochem. 9, 427-432.
FRY, J. M., WEISSBARTH, S., LEHRER, G. M. & BORNSTEIN, M. B. (1974) Science 183, 540-542.
HENKART, P. & BLUMENTAL, R. (1975) Proc. Natl. Acad. Sci. U.S.A. 72, 2789-2793.
HORNING, M. G. , WILLIAMS, E. A. & HORNING, E. C. (1960) J. Lipid Res. 1, 482-485.
KINSKY, S. C. (1972) Biochim. Biophys. Acta 265, 1-23.
KIRSCHNER, D. A. & CASPAR, D. L. D. (1972) Ann. N. Y. Acad. Sci. 195, 309-320.
LONDON, Y., DEMEL, R.A., GEURTS VAN KESSEL, W. S. M., VOSSEMBERG, F. G. A. & VAN DEENEN, L. L. M., (1973) Biochim. Biophys. Acta 311, 520-530.
LONDON, Y., DEMEL, R. A., GEURTS VAN KESSEL, W. S. M., ZAHLER, P. & VAN DEENEN, L. L. M. (1974) Biochim. Biophys. Acta 332, 79-84.
MAGGIO, B., CUMAR, F. A. & MACCIONI, H. J. (1972) J. Neurochem. 19, 1031-1037.
MAGGIO, B. & CUMAR, F. A. (1974) Brain Res. 77, 297-307.
MAGGIO, B. & CUMAR, F. A. (1975) Nature (London) 253, 364-365.
MAGGIO, B., MACCIONI, H. J. & CUMAR, F. A. (1973) J. Neurochem. 20, 503-510.
MATUS, A., DE PETRIS, S. & RAFF, M.C. (1973) Nature New Biol. 244, 278-280.
PATERSON, P. Y. (1969) Annu. Rev. Med. 20, 75-100.
PODUSLO, J. F. & BRAUN, P. E. (1975) J. Biol. Chem. 250, 1099-1105.
RAPPORT, M. M. & GRAF, L. (1965) Ann. N. Y. Acad. Sci. 122 (part I), 277-279.
VASAN, N. S. & BACHHAWAT, B. K. (1971) J. Neurochem. 18, 1853-1859.
WILLIAMS, E., HAMILTON, J. A. JAIN, M. K. ALLERHAND, A., CORDES, E. H. & OCHS, S. (1973) Science, 181, 869-871.
YOKOYAMA, M., TRAMS, E. G. & BRADY, R. O. (1962) Proc. Soc. Exp. Biol. Med. 111, 350-352.

IV. ESSENTIAL FATTY ACIDS IN NUTRITION

ESSENTIAL FATTY ACIDS IN HUMAN NUTRITION

Ralph T. Holman

The Hormel Institute, University of Minnesota

801 16th Avenue N.E., Austin, Minnesota 55912 USA

STUDIES WITH ANIMALS

Half a century ago, George and Mildred Burr were beginning their studies on the effects of a fat-free diet upon rats (Burr and Burr 1929,1930). Since that time, the story of essential fatty acids (EFA) has progressed to the point that it is now finding practical application in medical nutrition. It will be the purpose of this presentation to trace some of the steps along the way from the first publication in 1929 to the current intense interest in the essential fatty acid deficiency which now is a common occurrence in patients receiving intravenous alimentation. The subject has been reviewed in more detail several times and the reader who wishes more information should turn to those references (Holman, 1971a; Holman, 1971b; Mead, 1971; Burr, 1942; Alfin-Slater & Aftergood, 1976).

By feeding a diet composed of 16% casein, approximately 4% salts and vitamins and the remainder sucrose, Burr and Burr were able to induce a dietary deficiency syndrome which they later traced to the absence of linoleic acid. The principal deficiency symptoms were a diminished growth rate and a scaly dermatitis seen principally on the feet and tail. An example of a fat deficient rat is shown in Figure 1. Most of the organs of the body are affected by the deficiency and aberrations in structure or function of many tissues have been observed. Reproduction is impaired, the liver becomes fatty, kidney function is impaired, the electrocardiogram is abnormal, and the permeability of the skin toward water is much increased. This leads to evaporative loss of water and of heat from the body accounting for an increased food consumption and a diminished caloric efficiency.

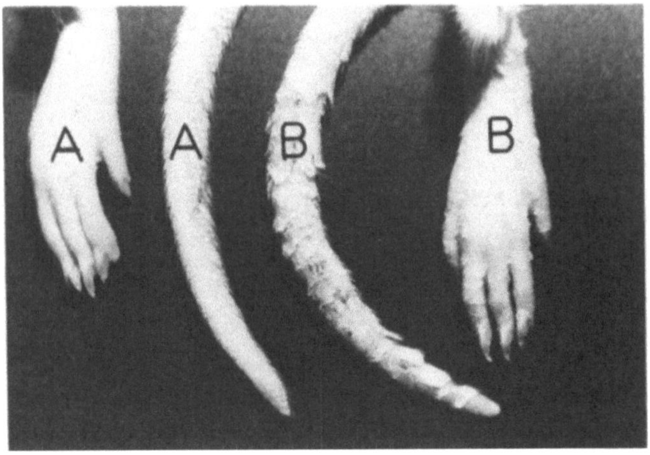

Fig. 1. Comparison of the foot and tail of a normal rat (A) and a rat fed a fat-deficient diet 207 days (B).

Early attempts at description of the changes in lipid composition included a diminished degree of unsaturation in deficient animals, indicated by a decrease of the iodine value of tissue lipids. When alkaline isomerization technique became available for measurement of polyenoic acids, the deficient animal was found to have a diminished content of dienoic acid and tetraenoic acid but an enhanced content of trienoic acid (Rieckehoff *et al.*, 1949). Through supplementation with single fatty acids, it was discovered that linoleic acid induced increases in tetraenoic acid (arachidonic acid) but that linolenic acid induced increases in pentaene and hexaene acids (Widmer & Holman, 1950). These experiments prompted more elegant ones in the laboratories of Mead and Klenk in which the metabolism of linoleic acid to arachidonic acid, linolenic acid to pentaenoic and hexaenoic acids, and oleic to eicosatrienoic acid were described in detail (Mead, 1971).

When gas chromatography became available, it was possible to study in much more detail the changes in individual fatty acid compositions of tissues as a result of deficiency or supplementation. In Figure 2 are shown the changes which occur in fatty acids of liver in response to changes in dietary linoleate, the measurements being made in the fatty acid occurring in the α and β positions of liver lecithin (Pudelkewicz & Holman, 1968). This study confirms that most of the polyunsaturated acids occur in the β position of lecithin and that the content of saturated acids is greater in the α position. In the essential fatty acid deficient state, the content of ω6 fatty acids related to linoleic acid is low in both positions of lecithin. As linoleate is fed in increasing amounts, this family of acids increases abruptly in the β position but less so in the α

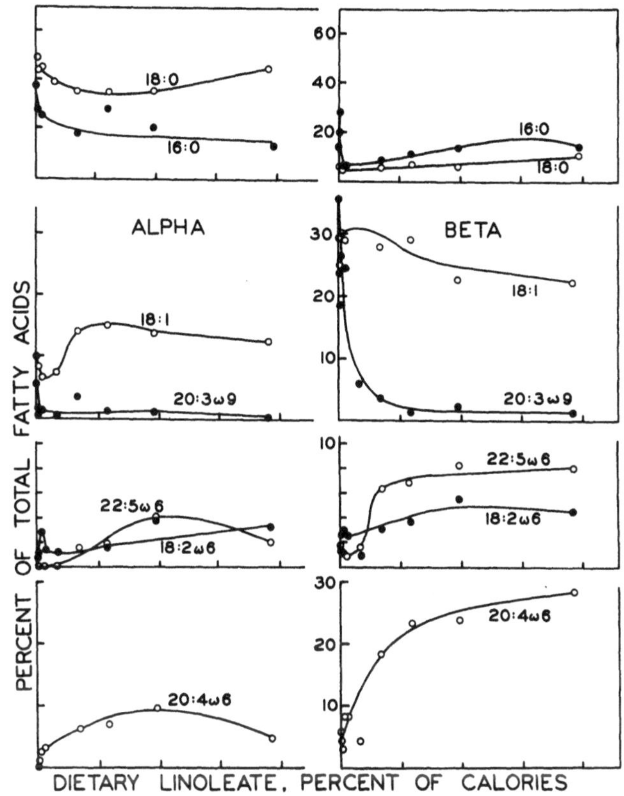

Fig. 2. Changes in the composition of the fatty acids in the α and β positions of liver phosphatidyl choline, induced by different dietary levels of linoleate.

position. In the deficient state, the metabolic product of oleic acid, 20:3ω9, occurs in the β position to an amount greater than 30% of the total acids. As the ω6 acids increase in response to dietary linoleate, this acid is supplanted. These changes in fatty acid composition as a response of essential fatty acid supplementation are really changes in the phospholipids of the membranes of the liver cells and the consequences of these changes are probably changes in physical properties of the phospholipids and the lipoproteins which contain them.

Two of the fatty acids which undergo the most abrupt or dramatic changes are the eicosatrienoic acid, 20:3ω9, derived from oleic acid and arachidonic, 20:4ω6, derived from linoleic acid. The ratio between these two acids has been found convenient for the expression of essential fatty acid nutritive status. When this ratio is plotted against dietary linoleate, a hyperbolic curve is found with its sharp break near 1% of linoleate calories (Holman, 1960). The ratio reaches a low and rather constant level before 2% of cal-

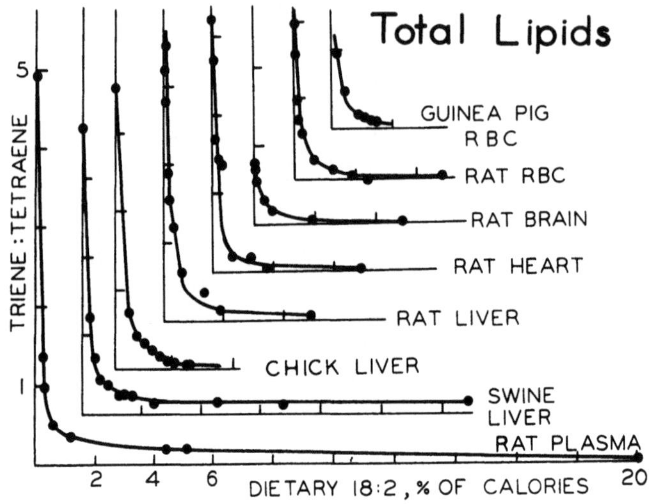

Fig. 3. Curves relating ratio of trienoic acids to tetraenoic acids in tissue lipids from several animals.

ories of linoleate are reached. Triene:tetraene curves for a number of species and several tissues of some species are shown in Figure 3. These curves indicate that the same general biochemical phenomenon occurs in the species and tissues studied (Holman, 1971). It now appears that these changes in fatty acid composition are universal phenomena and that essential fatty acid deficiency probably can be induced in any species of higher animal from whose diet essential fatty acids are deleted.

STUDIES IN HUMANS

1. Infant Eczema

Very soon after the discovery of essential fatty acid phenomena in rats, medical researchers at the University of Minnesota began investigations on humans. The first medical phenomenon related to essential fatty acids was a dermatitis associated with intractable eczema. Hansen and his co-workers chose cases which did not respond to the usual treatments for eczema and gave these patients supplements of lard which contains approximately 10% of linoleic acid and a few percent of arachidonic acid (Hansen, 1937). They found that in the cases of intractable eczema the serum iodine number was low, and that when the diets were supplemented with lard, the iodine value rose to normal and the skin cleared up in 75% of the cases. An example of this disease which responded to essential fatty acids is shown in Figure 4 (Azerad & Grupper, 1949). A study of the histological features of normal and essential fatty acid deficient human skin shown in Figure 5 indicates that in the deficient condition

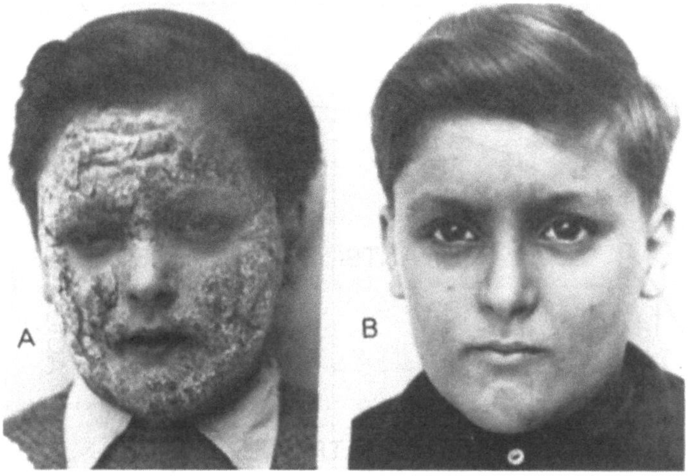

Fig. 4. A case of intractable eczema which responded to dietary supplement with EFA, reported by Azerad and Grupper. A) Before treatment, B) after treatment.

there is a proliferation of the dermis and a loss of squamous epithelium (Hansen, *et al., 1962)*.

2. Changes in Fatty Acid Composition of Serum Lipids

Hansen and his co-workers undertook a long term massive study of essential fatty acid phenomena in infants. In a clinical study of 428 infants, who were fed with five different dietary formulae, they found differences in the dienes, trienes, tetraenes, pentaenes and hexaenes measured by alkaline isomerization. They also found that the content of linoleic acid in the proprietary formulae varied

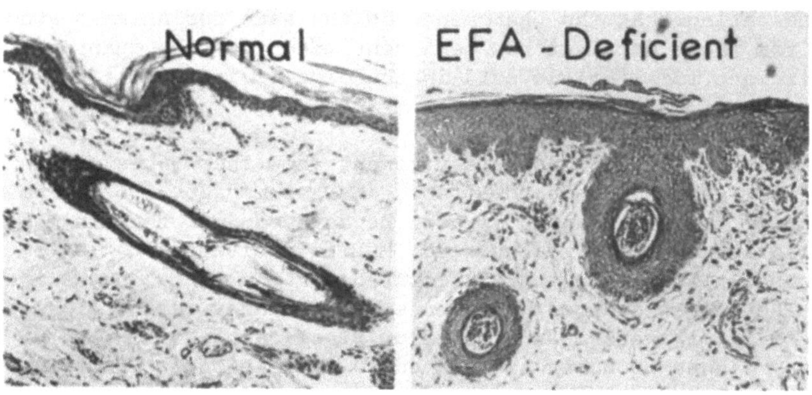

Fig. 5. Histological changes in skin during EFA deficiency.

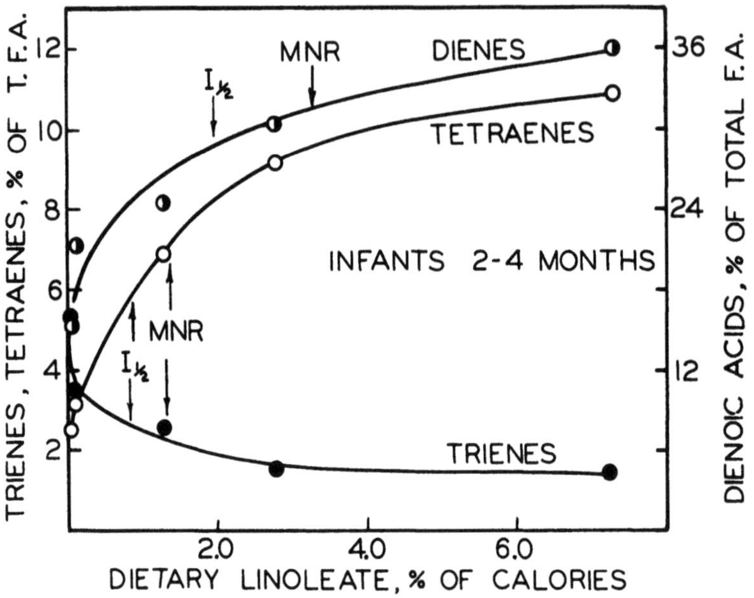

Fig. 6. Effect of dietary linoleate upon dienes, trienes and tetraenes in total fatty acids of lipids of serum of infants.

widely and that one of the formulae actually induced an essential fatty acid deficiency (Hansen *et al.*, 1963). The skim milk-sugar formula which was in current use at that time, also induced an essential fatty acid deficiency with clinical manifestations. Skim milk-sucrose diet is no longer advocated and most infant formulae now contain enhanced amounts of linoleate by addition of vegetable oils. The response of serum dienes, trienes and tetraenes are shown in Figure 6 and the triene:tetraene ratios are plotted against dietary linoleate intake in Figure 7 (Holman *et al.*, 1964). The shape of that curve is very similar to those shown in Figure 3 for animals, clearly indicating that the phenomenon is the same for animals and humans. From these data, it has been concluded that the quantitative requirement for linoleic acid in the diet of infants lies between 1 and 2% of calories.

Estimation of the linoleate requirement from the shapes of the individual curves has been accomplished (Caster *et al.*, 1962). The value for half maximum change, $I_{1/2}$, can be derived and the minimum nutrient requirement (MNR) calculated. MNR is equivalent to 70% of maximum change. These values are indicated for the three curves in Figure 6. The values deduced for trienes and tetraenes relate closely to metabolic parameters and lie below 2% of calories. The value deduced from dienes probably is related to deposition of linoleate and is a much higher value.

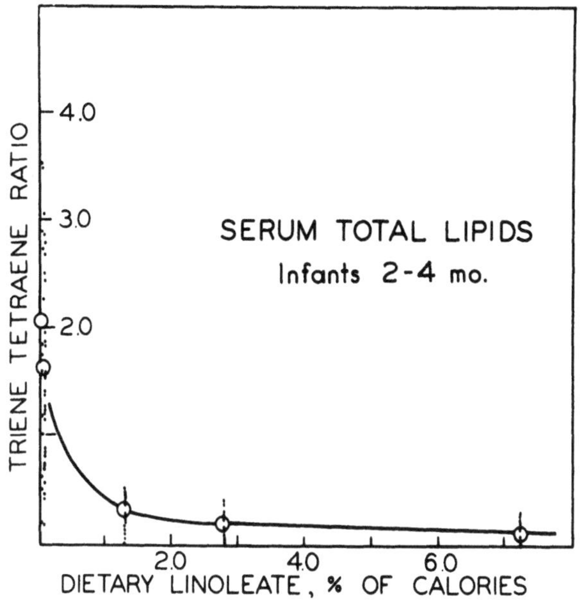

Fig. 7. Effect of dietary linoleate upon triene:tetraene ratio in total fatty acids of serum lipids of infants.

3. Assessment of Nutritive Status

The same data from Hansen's laboratory which produced the quantitative requirement for linoleate, gave also an expression which permits assessment of nutritive status of individuals in comparison with that requirement. In an attempt to find a relationship by which linoleate intake could be measured, linear multiple regression equations were tested for animal data (Caster *et al.*, 1963). These were found to be inadequate and logarithmic multiple regression equations were a considerable improvement (Caster *et al.*, 1963). These latter equations were found to have coefficients for the diene, triene and tetraene terms which were almost equal, and because the pentaenes and hexaenes are not metabolically related directly to linoleate, their inclusion in a multiple regression equation has no direct biochemical meaning. Simplifying the five-term equation to a three-term equation of the form, yielded the following

$$\log \text{dietary linoleate} = a + b\ (\text{diene} - \text{triene} + \text{tetraene}).$$

This simplified equation can be derived graphically from experimental data. Application of this procedure to the data from Hansen's laboratory yielded the relationship shown in Figure 8. The logarithm of dietary linoleate is proportional to the algebraic sum of diene - triene + tetraene. This latter term has found use as a measure of dietary intake of linoleate. For example, Figure 9 shows the re-

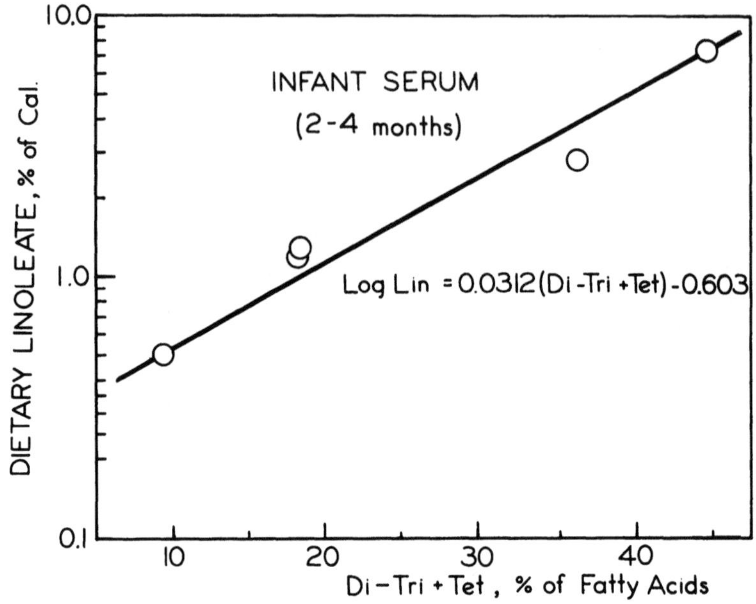

Fig. 8. Relationship between (dienoic - trienoic + tetraenoic) acids in serum lipids of infants and logarithm of dietary linoleate.

lationship of this parameter to time that infants were fed either cows' milk, breast milk or cows' milk supplemented with linoleate (Holman *et al.*, 1965). The amount of diene - triene + tetraene in plasma of the infants which received breast milk rose significantly higher than it did for infants who received only cows' milk formula. A filled milk formula containing linoleate supplement caused responses similar to those found in infants fed breast milk.

4. Diseases with Impaired EFA Metabolism

Some human diseases have been associated with abnormalities in metabolism of polyunsaturated acids. One case of *Acrodermatitis enteropathica* shown in Figure 10 was found to have abnormally low content of arachidonic acid in serum lipids although the linoleic acid content was normal (Cash & Berger, 1969). When the case came to clinic, it was given intravenously fat emulsion containing linoleate. Within a few days, the dermatitis on the face disappeared and the infusion of emulsion was stopped. At this stage, the fatty acid composition of the serum lipids was approaching normal and the previously abundant fatty acids of unknown structure had nearly disappeared. After several days without intravenous emulsion, the condition returned and again the unknown fatty acids increased in amount and arachidonate decreased. This constituted the diagnosis of *Acrodermatitis enteropathica* and the child was fed thereafter a diet con-

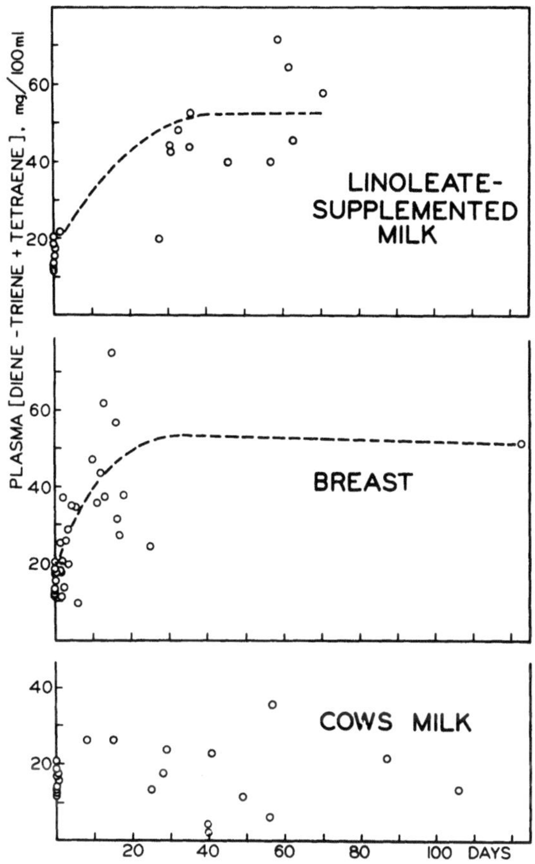

Fig. 9. Variation of the parameter (diene - triene + tetraene) in serum lipid fatty acids of infants with time on one of three diets.

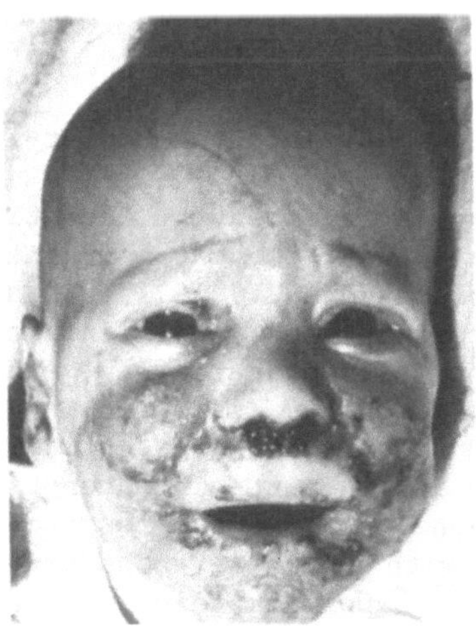

Fig. 10. A case of Achrodermatitis enteropathica.

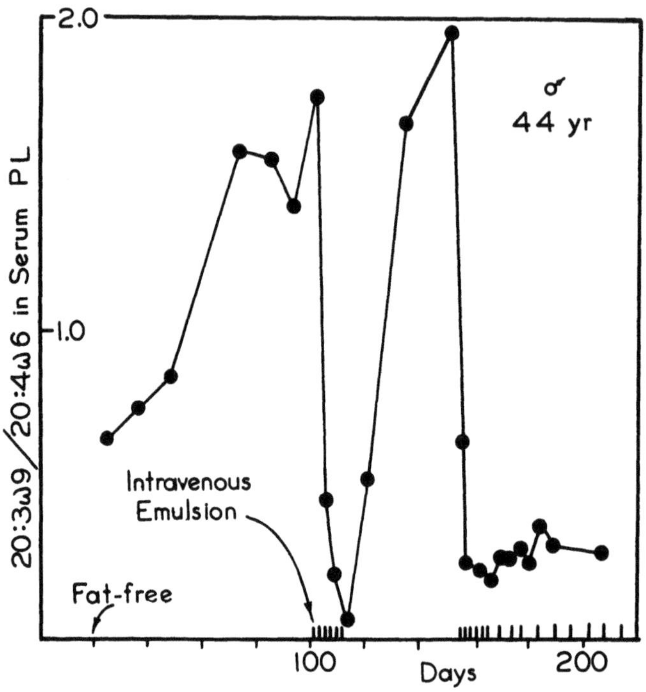

Fig. 11. Changes in triene:tetraene ratio of serum phospholipids in response to fat-free intravenous feeding and intravenous feeding with fat emulsion.

taining no cows' milk and received Diodoquin therapy. Under this treatment, the child has developed normally for several years without exacerbations of the dermatitis. Faulty essential fatty acid metabolism has been postulated as a cause of multiple sclerosis and cystic fibrosis.

5. EFA Deficiency Induced by Intravenous Fat-free Feeding

Collins and his colleagues reported the induction of essential fatty acid deficiency in a 44 year old man fed intravenously with a fat-free preparation (Collins *et al.*, 1971). This individual developed erythema and dermatitis and the triene:tetraene ratio of his serum phospholipids increased dramatically over approximately 100 days on fat-free preparation, as is shown in Figure 11. When an intravenous emulsion was fed providing essential fatty acids, the ratio dropped dramatically and when the intravenous emulsions were stopped, the triene:tetraene ratio again rose to high levels. A second course of treatment with the intravenous emulsions maintained the triene:tetraene ratio for approximately seven weeks. This was the first report of essential fatty acid deficiency in humans caused

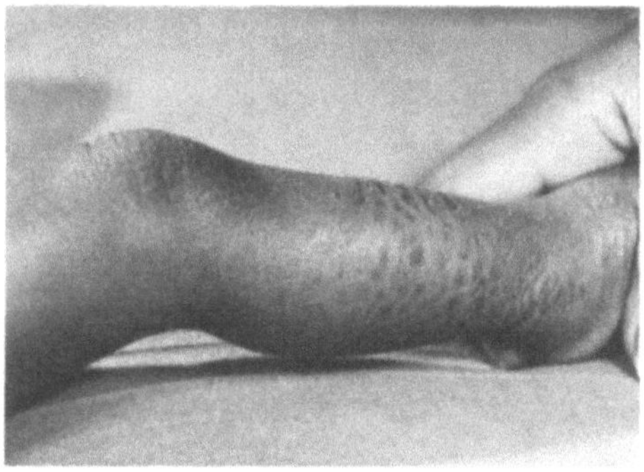

Fig. 12. A case of essential fatty acid deficiency of three months duration in an infant maintained on fat-free intravenous feeding.

by intravenous alimentation with fat-free preparation.

Shortly after the study by Collins *et al.*, our laboratory was asked to make serum lipid analyses on an individual in whom essential fatty acid deficiency was suspected. A child experienced a volvulus at birth and massive resection of the bowel on its first day of life. It was maintained for approximately three months with intravenous feeding with the usual preparation containing glucose, amino acids, vitamins and minerals. At three months of age, it had developed extensive dermatitis shown in Figure 12 (Paulsrud *et al.*, 1972). Analysis of the serum phospholipids revealed abnormally high levels of 20:3ω9 and abnormally low levels of 18:2ω6 and 20:4ω6. The data for this infant are shown in Figure 13 by the double circles indicated by the arrows. The triene:tetraene ratio for this individual in serum phospholipids reached a value of 18 at approximately 100 days, indicating the most severe essential fatty acid deficiency our laboratory had observed in animals or man. When the child succumbed to a systemic infection, autopsy samples of several tissues were subjected to lipid analysis. All tissues were found to exhibit fatty acid patterns typical of essential fatty acid deficiency. The triene:tetraene ratios in the tissue lipids were not as drastic as they had been in the serum lipids. This indicates that analysis of the serum lipids is a useful predictive analysis indicating what may be expected to happen to tissues if the condition is not corrected.

Experience with this case led to a study of several infants who had been given intravenous alimentation for one reason or another. Serial analyses of their serum lipids produced a similar pattern but not as drastic as had been observed in the first case. Data for all of these cases are shown in Figure 13 indicating that as intravenous alimentation without fat proceeded, the condition of essential fatty acid deficiency was induced (Paulsrud *et al.*, 1972).

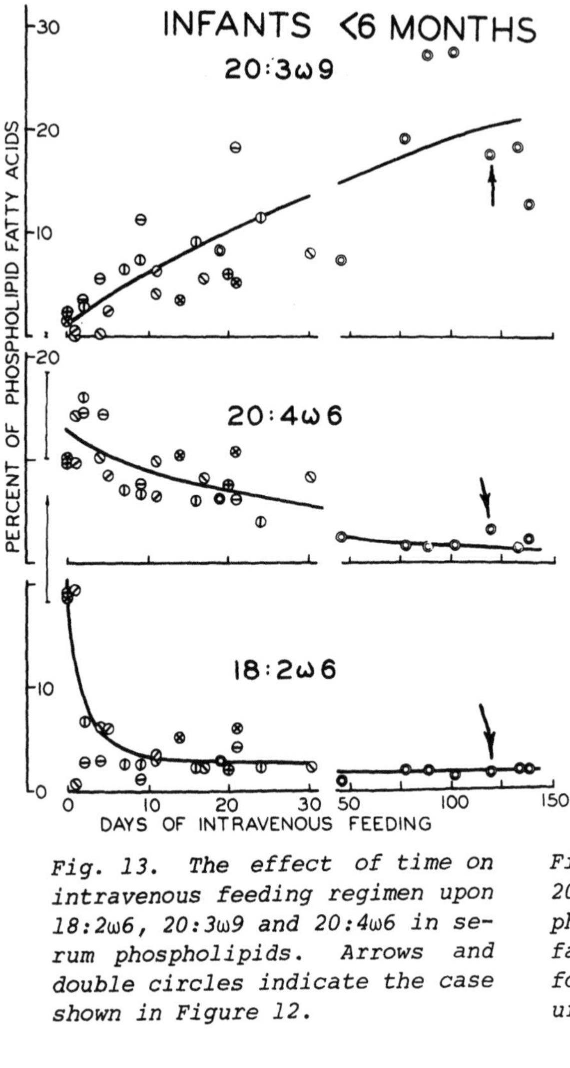

Fig. 13. The effect of time on intravenous feeding regimen upon 18:2ω6, 20:3ω9 and 20:4ω6 in serum phospholipids. Arrows and double circles indicate the case shown in Figure 12.

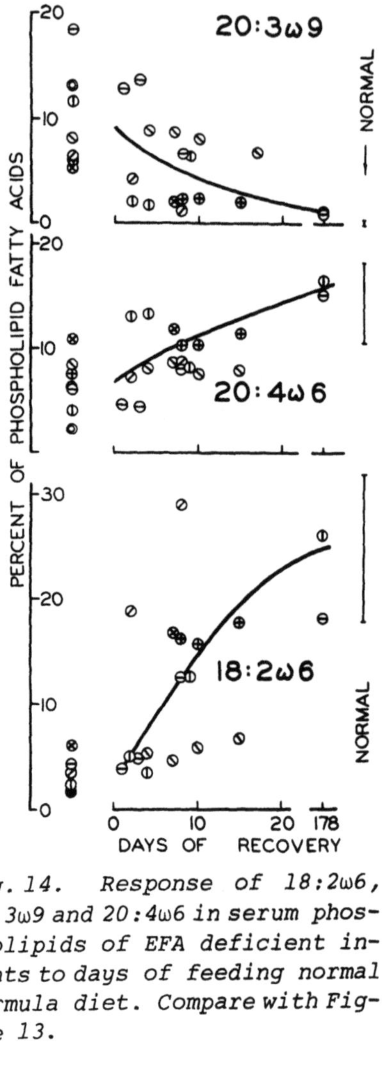

Fig. 14. Response of 18:2ω6, 20:3ω9 and 20:4ω6 in serum phospholipids of EFA deficient infants to days of feeding normal formula diet. Compare with Figure 13.

The magnitude of change for the individuals differed but the direction of the change was the same in each case. When these individuals had gained sufficient weight from the intravenous alimentation, they were later given normal food. During this period, their serum lipids were also analyzed serially for fatty acid content. All of the individuals reverted toward normal, although the magnitude of change again was different for different individuals. These data are shown in Figure 14. The study indicated that although essential fatty acid deficiency can be induced by long term intravenous feeding without fat, recovery from the deficiency is rapid when normal food can be given.

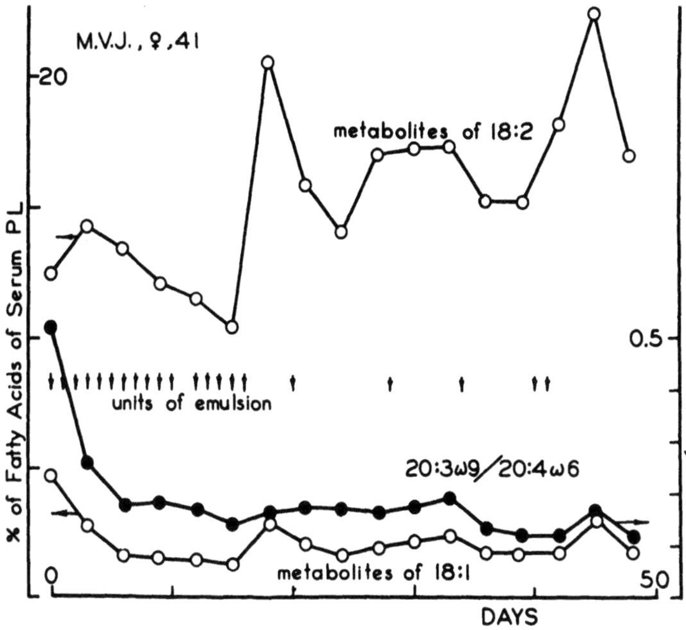

Fig. 15. Effect of intravenous fat emulsion upon serum fatty acids and ratio of 20:3ω9/20:4ω6 in an EFA deficient woman.

6. Recovery from EFA Deficiency by Intravenous Fat Emulsion

When intravenous fat emulsions became available for experimental use in the United States, several individuals were studied during therapy with this emulsion (Holman & Bissen, unpublished data). One such case is shown in Figure 15. After an extensive period of intravenous alimentation without fat, this 41 year old woman had a triene:tetraene ratio greater than 0.5. Serial analysis of serum phospholipids revealed that as the emulsion was administered daily, the triene:tetraene ratio dropped abruptly to the range of 0.1 to 0.2. When the emulsion subsequently was given less frequently at intervals of 4 to 7 days, the triene:tetraene ratio remained low and in that range. During this study, the metabolites of oleic acid which were 20:2ω9 and 20:3ω9 decreased from about 5% to 2%. During the period of daily administration of emulsion, the metabolites of linoleic acid decreased somewhat and when the frequency of administration of emulsion decreased the metabolites increased generally and showed two maxima in the curve. These maxima are 27 days apart and are thought to be related in some way to the menstrual cycle. This will be the subject of future research in our laboratory.

7. Cutaneous Application of EFA Containing Oil

Cutaneous application of sunflower seed oil has been reported

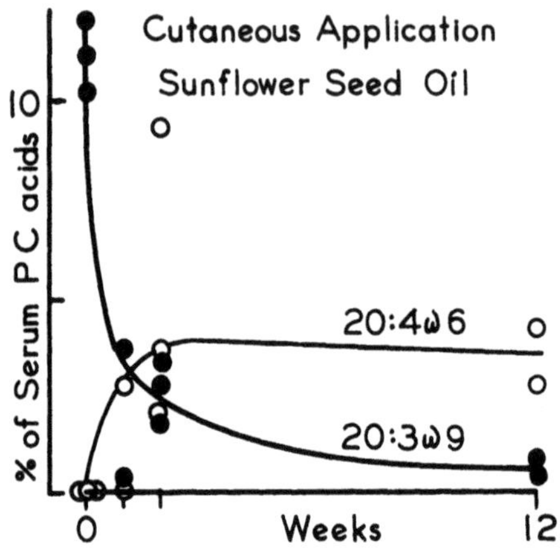

Fig. 16. Effect of cutaneous application of sunflower seed oil upon 20:3ω9 and 20:4ω6 in serum phosphatidyl choline in adults who had been EFA deficient.

(Press *et al.*, 1974) to be effective in correcting essential fatty acid deficiency in humans. Sufficient oil was absorbed through the skin to reverse the pattern of essential fatty acid deficiency in serum phosphatidyl choline. Their data are plotted in Figure 16. At the end of 12 weeks of cutaneous application of sunflower seed oil, the triene:tetraene ratio in serum phosphatidyl choline was in the order of 0.02. The phenomenon has been tested with rats (Böhles *et al.*, 1976) and found that sufficient oil was absorbed through the skin of rats to reverse the pattern of essential fatty acid deficiency. In our laboratory, an attempt to relieve essential fatty acid deficiency in infants was not successful. Cutaneous application of safflower seed oil at approximately the same dose level used by Press *et al.* (1974) caused no real change in triene:tetraene ratio of serum phospholipids. Press *et al.* (1974) have expressed some doubt that the phenomenon they observed was effective over a longer period of time (private communication).

8. Normal Values for Fatty Acid Composition of Serum Lipids

In several of the latter studies quoted, it has been necessary to make comparisons of suspected cases of deficiency against normal humans. In several cases this has been difficult for lack of normal samples with which to compare. Therefore, our laboratory undertook to compile analyses on serum lipids for normal humans of both sexes ranging in age from birth to 90 years (Holman, Smythe & Bissen, previously unpublished). At least ten individuals of each sex from each decade of life were analyzed. The population studied was random patients at the local hospital, and excess serum from diagnostic procedures was made available for analysis for lipids. Patients diagnosed to have metabolic diseases were not included in the study. Total serum lipids were extracted and separated by thin layer chromatography into triglycerides, free fatty acids, cholesteryl esters and phospholipids. The fatty acid compositions of these four general classes and the total lipid mixture were made by gas chromatography of the methyl esters.

Twenty-three fatty acids were found to be present in most samples in more than trace amounts. The mean and standard deviation for each fatty acid was calculated for the entire population of each sex. In addition, the individual values were plotted against age by computer graphic procedure. The slope and intercept of the best fitting straight line for these bodies of data were then calculated. In addition, several parameters which have been used in the past for description of essential fatty acid phenomena were calculated. All of this information is shown in Table I for analysis of serum phospholipids of females. Only a few of the fatty acids measured or parameters calculated vary significantly with age of individual. Palmitic acid tends to increase with age whereas stearic acid tends to decrease. The algebraic sum 18:2ω6 + 20:4ω6 - 20:3ω9 decreases insignificantly with age. The double bond index, which is the number of double bonds per molecule of fatty acid, remains constant as does the triene:tetraene ratio. The total of ω6 acids decreases slightly with age and the total monoenoic acids increases slightly with age in females. It does not appear that the essential fatty acid reserves or metabolism of females change significantly with age.

To illustrate the spread of the data gathered in this population study, the data for double bond index, triene:tetraene ratio, total ω6 acids and the algebraic sum 18:2ω6 + 20:4ω6 - 20:3ω9 are presented in Figure 17. It should be recalled that this population was a free population and there was no dietary control. The constancy of the data is therefore remarkable. The triene:tetraene ratio was found to be approximately 0.11 for the female population and approximately 0.10 for the male population. These values are considerably lower than the 0.4 set previously from studies of animals and infant populations as the limit between normal and abnormal individuals. None of the individuals in the hospitalized population studied had a triene:tetraene ratio as high as 0.4. Therefore, the limit for normalcy for humans must be set at a lower value. At age 0, the intercept is 0.13 and the standard variance is 0.07. Therefore, we suggest that the value 0.2 be used as the upper limit of normalcy. This limit is approximately equal to the mean of the entire population plus 1 standard deviation. The total of ω6 acids which is a useful index of essential fatty acid metabolites, and therefore of essential fatty acid status, has a standard deviation which is approximately 11% of the mean value. With this degree of precision, one can readily distinguish abnormal states of essen-

TABLE I

NORMAL FATTY ACID COMPOSITION OF SERUM

PHOSPHOLIPIDS OF HUMAN FEMALES

FATTY ACID(S)	ALL AGES MEAN	ALL AGES S.D.	FATTY ACID = A(AGE)+B SLOPE A	INTERCEPT B	S.V.
12:0	0.04	0.10	0.0004	0.02	0.10
14:0	0.23	0.25	0.0009	0.20	0.24
14:1	0.15	0.16	0.0012	0.11	0.16
16:0	26.60	4.63	0.0123	26.10	4.61
16:1W7	1.40	0.79	0.0046	1.22	0.78
16:2	0.12	0.19	-0.0000	0.13	0.19
18:0	13.24	2.41	-0.0187	14.00	2.36
18:1W9	13.15	2.44	0.0076	12.84	2.43
18:2W6	18.73	4.58	-0.0043	18.91	4.57
18:3W6	0.27	0.43	-0.0024	0.36	0.43
18:3W3	0.31	0.36	0.0021	0.23	0.36
20:2W9	0.36	0.34	0.0048	0.17	0.31
20:2W6	0.27	0.34	-0.0002	0.28	0.34
20:3W9	1.33	0.95	-0.0094	1.71	0.92
20:3W6	3.54	1.44	-0.0037	3.69	1.44
20:4W6	12.33	2.89	-0.0096	12.72	2.88
20:4W3	0.31	0.47	0.0017	0.24	0.46
20:5W3	1.13	0.70	-0.0057	1.36	0.68
22:4W6	1.73	0.87	-0.0055	1.95	0.85
22:4W3	1.00	1.33	0.0082	0.67	1.31
22:5W6	0.65	0.75	0.0028	0.53	0.74
22:5W3	0.64	0.53	0.0022	0.55	0.52
22:6W3	2.34	1.35	0.0093	1.96	1.33
18:2W6+20:4W6-20:3W9	29.73	4.19	-0.0045	29.92	4.19
DOUBLE BOND INDEX	1.51	0.16	-0.0003	1.52	0.16
20:3W9/20:4W6	0.11	0.07	-0.0005	0.13	0.07
TOTAL W 6 ACIDS	37.52	3.90	-0.0228	38.44	3.85
TOTAL W6 - 18:2	18.78	4.05	-0.0185	19.53	4.02
TOTAL W3 ACIDS	5.74	2.51	0.0179	5.01	2.46
TOTAL W3 - 18:3	5.43	2.43	0.0158	4.79	2.39
TOTAL W9 ACIDS	14.84	2.75	0.0030	14.72	2.75
TOTAL W9 - 18:1	1.69	1.01	-0.0046	1.88	1.01
MONOENE ACIDS	14.71	2.82	0.0134	14.17	2.79
SATURATED ACIDS	40.11	4.25	-0.0051	40.32	4.24

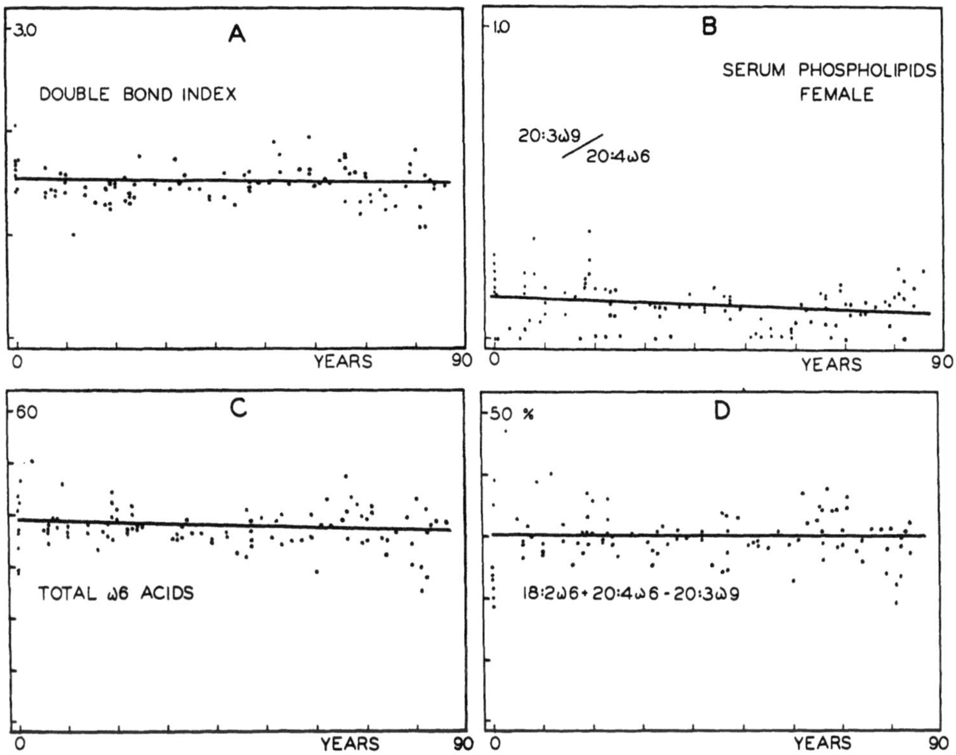

Fig. 17. The noneffect of age of human females upon parameters measured in serum phospholipids: A) double bond index, B) ratio of 20:3ω9/20:4ω6, C) total ω6 acids and D) 18:2ω6 + 20:4ω6 - 20:3ω9.

tial fatty acid status. Standard deviation of the algebraic sum 18:2ω6 + 20:4ω6 - 20:3ω9 is somewhat greater being 14% of the mean value. This parameter is still usable as an index of essential fatty acid status although it is less precise than is total ω6 acids.

The equations relating logarithm of dietary linoleate to tissue fatty acid compositions were derived from data obtained via alkaline isomerization technique. This method is now obsolete and it has been supplanted by a much more precise gas chromatographic method. It is now possible to study variations of more than 20 individual fatty acids as the consequence of a dietary variable. Unfortunately, there is no body of information on humans in which dietary linoleate has been varied systematically and analyses made of serum lipids by means of gas chromatography. With animals, information is available based upon gas chromatography, and the predictive equations have been derived. Logarithm of dietary linoleate bears a linear relationship to the algebraic sum 18:2 + 20:4 - 20:3. Logarithm of dietary linoleate bears a linear relationship to logarithm of total ω6 acids and both can be used as means of estimating intake of linoleate in rats. They should be equally useful with humans once serum lipid analyses have been made using gas chromatography

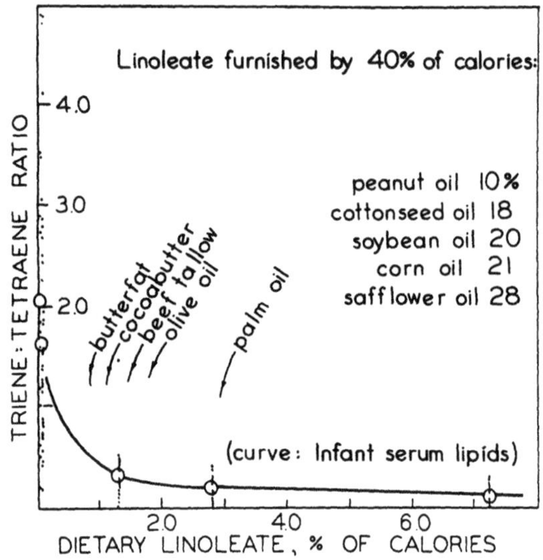

Fig. 18. Level of dietary linoleate provided by 40% of calories of several common diet fats, superimposed over the curve indicating requirement of infants for dietary linoleate.

on samples from individuals whose linoleate intake was known and controlled.

9. Adequacy of Dietary Fats in Providing EFA

The question also remains how well do our eating practices meet the requirement for essential fatty acid? In Figure 18, calculated amount of linoleate provided by 40% of calories of several single dietary fats and oils are presented, superimposed over a triene:tetraene ratio curve derived from infant serum lipids. Butterfat provides seasonably variable amounts of linoleate, averaging to close to 1% of calories. Cocobutter provides approximately 1% of calories, beef tallow and olive oil between 1 and 2% of calories and palm oil approximately 3% of calories. The common salad oils or cooking oils all provide 10% or more of calories of linoleate when consumed singly as 40% of calories. It seems very unlikely that free choice of foods would provide less than the requirement for essential fatty acids. Only by studious avoidance of polyunsaturated oils could an individual arrive at marginal EFA deficiency. Any natural selection of foods following the principles of balanced nutrition, would also provide mixed fats and oils whose linoleate content would lie above the requirement. Conceivably some individuals may have for genetic reasons a metabolic requirement for essential fatty acids which is higher than normal, and for these persons, it may be possible by self-selected diet to encounter a deficiency. It would appear that if the principles of balanced and varied diet are extended to consider the fat and oil contents of the diet, that essential fatty acid deficiency of dietary origin will be rare.

ACKNOWLEDGMENT

Studies reported and reviewed here on the quantitative aspects of the requirement for essential fatty acids were supported in part by NIH grant AM-04524, NIH Program-Project Grant HL-08214, and The Hormel Foundation.

REFERENCES

Alfin-Slater, R. B. & Aftergood, L. (1976) in *Lipid Pharmacology*, vol. II (Paoletti, R. & Glueck, C. J., eds.), pp. 43-82, Academic Press, New York

Azerad, E. & Grupper, C. (1949) *Semaine Hôp. Paris* **25**, 684

Böhles, H., Bieber, M. A. & Heird, W. C. (1976) *Am. J. Clin. Nutr.* **29**, 398-401

Burr, G. O. (1942) *Fed. Proc.* **1**, 224-233

Burr, G. O. & Burr, M. M. (1929) *J. Biol. Chem.* **82**, 345-367

Burr, G. O. & Burr, M. M. (1930) *J. Biol. Chem.* **86**, 587-621

Cash, R. & Berger, C. K. (1969) *J. Pediat.* **74**, 717-729

Caster, W. O., Ahn, P., Hill, E. G., Mohrhauer, H. & Holman, R. T. (1962) *J. Nutr.* **78**, 147-154

Caster, W. O., Hill, E. G. & Holman, R. T. (1963) *J. Anim. Sci.* **22**, 389-392

Collins, F. D., Sinclair, A. J., Royle, J. P., Coats, D. A., Maynard, A. T. & Leonard, R. F. (1971) *Nutr. Metab.* **13**, 150-167

Hansen, A. E. (1937) *Am. J. Dis. Child.* **53**, 933-946

Hansen, A. E., Stewart, R. A., Hughes, G. & Söderhjelm, L. (1962) *Acta Paediat.* Suppl. 137, 1-41

Hansen, A. E., Wiese, H. F., Boelsche, A. N., Haggard, M. E., Adam, D. J. D. & Davis, H. (1963) *Pediat.* **31**, Suppl. 1, part 2, 171-192

Holman, R. T. (1960) *J. Nutr.* **70**, 405-410

Holman, R. T. (1971a) in *Progress in The Chemistry of Fats and Other Lipids*, vol. IX (Holman, R. T., ed.), pp. 275-348, Pergamon Press, Oxford

Holman, R. T. (1971b) in *Progress in The Chemistry of Fats and Other Lipids*, vol. IX (Holman, R. T., ed.), pp. 607-682, Pergamon Press, Oxford

Holman, R. T., Caster, W. O., & Wiese, H. F. (1964) *Am. J. Clin. Nutr.* **14**, 70-75

Holman, R. T., Hayes, H. W., Rinne, A. & Söderhjelm, L. (1965) *Acta Paediat. Scand.* **54**, 573-577

Mead, J. F. (1971) in *Progress in The Chemistry of Fats and Other Lipids*, vol. IX (Holman, R. T., ed.) pp. 159-192, Pergamon Press, Oxford

Paulsrud, J. R., Pensler, L., Whitten, C. F., Stewart, S. & Holman,

R. T. (1972) *Am. J. Clin. Nutr.* **25**, 897-904
Press, M., Hartop, P. J. & Prottey, C. (1974) *Lancet* **1**, 597-598
Pudelkewicz, C., & Holman, R. T. (1968) *Biochim. Biophys. Acta* **152**, 340-345
Rieckehoff, I. G., Holman, R. T. & Burr, G. O. (1949) *Arch. Biochem.* **20**, 331-340
Widmer, C. & Holman, R. T. (1950) *Arch. Biochem.* **25**, 1-12

ESSENTIAL FATTY ACIDS: WHAT LEVEL IN THE DIET IS MOST DESIRABLE?

Kenneth K. Carroll[1]

Department of Biochemistry
University of Western Ontario
London, Ontario, Canada, N6A 5C1

INTRODUCTION

It has been recognized for many years that certain dietary fatty acids, notably linoleic acid, are required to prevent a variety of deficiency symptoms in animals (Burr, 1942; Holman, 1970). A similar need for these fatty acids in human diets has been established more recently. Deficiency symptoms can be demonstrated in children (Söderhjelm et al, 1970) and have also been seen in adults after prolonged parenteral nutrition with fat-deficient preparations (Collins et al, 1971; Riella et al, 1975). It is estimated that these various deficiency symptoms can be prevented by an intake of essential fatty acids in the range of 1 to 2% of total calories (Holman, 1970).

The theme of essential fatty acid deficiency has been discussed in more detail by Holman (1967) and this presentation will therefore be mainly concerned with possible advantages or disadvantages of increasing the intake of essential fatty acids to a level considerably higher than that required to prevent deficiency symptoms.

Essential Fatty Acids in Relation to Coronary Heart Disease

The major incentive for increasing the level of essential fatty acids in the diet has been the belief that this will decrease the incidence of atherosclerosis and coronary heart disease in human

[1]Research Associate of the Medical Research Council of Canada.

populations (American Medical Association Council Statement, 1972). This is based on the knowledge that high levels of serum cholesterol tend to promote the development of atherosclerotic lesions and that a reduction in serum cholesterol levels can be achieved by increasing the dietary intake of polyunsaturated fat. The latter observation was first made by Kinsell and others about 25 years ago, and has been amply confirmed in many laboratories (Kinsell, 1963). In middle-aged human populations, a 15 to 20% reduction in serum cholesterol can be achieved, but a relatively high intake of polyunsaturated fat is required, with an optimum estimated at about 15 to 20% of total calories as linoleic acid (Brown, 1971; Vergroesen, 1972).

A number of clinical trials have been conducted in attempts to determine whether the feeding of polyunsaturated fat has a significant influence on coronary heart disease (Rinzler, 1968; Morris et al, 1968; Dayton et al, 1969; Leren, 1970; Miettinen et al, 1972). These trials have produced some indications of a reduction in mortality from coronary heart disease, but the value of this form of dietary treatment is still in question. A recent report of an Advisory Panel of the British Committee on Medical Aspects of Food Policy (Nutrition) on Diet in Relation to Cardiovascular and Cerebral Vascular Disease (1974) stated "The panel unanimously agree that they cannot recommend an increase in the intake of polyunsaturated fatty acids in the diet as a measure intended to reduce the risk of the development of ischaemic heart disease. In their opinion, the available evidence that such a dietary alteration would reduce that risk in the United Kingdom at the present time is not convincing." Meanwhile there have been expressions of concern about the potential hazards of high dietary levels of polyunsaturated fat (Pinckney, 1973,a,b; Jones, 1974; West & Redgrave, 1974). These include an increased risk of developing cancer, a greater tendency to form gallstones, an increased requirement for fat-soluble vitamins and possible acceleration of phenomena associated with aging.

Essential Fatty Acids in Relation to Carcinogenesis

In analyzing the data from their long-term clinical trial to investigate the effects of a low-cholesterol, high-polyunsaturated fat diet, Pearce and Dayton (1971) noted that in spite of fewer deaths from coronary heart disease, overall mortality in the experimental group was about the same as in the control group. This result was due to a greater number of deaths from cancer in the experimental group. The deaths were caused by various types of cancer, and interpretation of the finding was complicated by the fact that many of the cancer deaths in the experimental group were among those who did not adhere closely to the diet. These

TABLE 1

Effect of Different Dietary Fats and Oils on Incidence and Yield of Mammary Tumors in Rats Given DMBA[1]

Dietary Fat	No. of Rats[2] with Tumors	Total No. of Tumors
20% Sunflower seed oil	26	130
20% Cottonseed oil	28	127
20% Olive oil	26	117
20% Corn oil	27	110
20% Soybean oil	30	103
20% Lard	28	97
20% Butter	26	88
20% Coconut oil	29	73
20% Tallow	24	72
5% Corn oil	23	70
0.5% Corn oil	21	75

[1]Data from Carroll & Khor (1971). Rats were autopsied 4 months after receiving the carcinogen.

[2]Each group consisted of 30 female Sprague-Dawley rats.

observations led to a retrospective analysis of other dietary trials which involved the feeding of increased amounts of polyunsaturated fat, but little evidence of an increased cancer risk was found (Ederer et al, 1971). However, none of these dietary trials were designed to test effects of diet on cancer incidence and this does not rule out the possibility that dietary fat may have an influence on carcinogenesis (West & Redgrave, 1974).

Numerous experiments with mice and rats have shown that animals on high fat diets develop mammary tumors more readily than similarly treated controls on low fat diets (Tannenbaum & Silverstone, 1957; Tannenbaum, 1959; Carroll & Khor, 1975; Carroll, 1975). In our own studies on rats treated with 7,12-dimethylbenz(α)anthracene (DMBA), unsaturated fats increased tumor yields more than saturated fats (Table 1). However, there did not seem to be a direct correlation with essential fatty acid content since lard and olive oil were about as effective as corn oil and soybean oil, although the latter are much richer in linoleic acid (Carroll & Khor, 1971). Dayton and Hashimoto (1976) reached a similar conclusion on the basis of their experiments with an oil rich in oleic acid produced by a mutant safflower. This oil was found to be as effective as normal high-linoleic safflower oil in stimulating development of mammary tumors in rats treated with DMBA.

Epidemiological data on human populations show a strong positive correlation between age-adjusted mortality from breast cancer and dietary fat intake in different countries of the world (Carroll & Khor, 1975; Carroll, 1975) but, as in the experiments with animals, there does not seem to be a correlation with intake of essential fatty acids. More detailed analysis of the data showed a positive correlation with intake of animal fat, but little or no correlation with intake of vegetal fat (Carroll, 1975). Moreover, although breast cancer mortality is about 5 times as high in Americans as in Japanese, the per capita intake of linoleic acid is reported to be about the same in both countries (Insull et al, 1969). In addition, an analysis of the fatty acid composition of adipose tissue showed a level of 16.5% linoleic acid in Japanese compared to 10.2% in Americans (Insull et al, 1969).

Dietary fat also shows a positive correlation with age-adjusted mortality from cancer at certain other sites in the body, including the colon, rectum and prostate (Carroll & Khor, 1975). It should be noted, however, that other dietary variables, such as caloric intake and intake of animal protein, show similar correlations. This simply reflects the fact that fat intake, caloric intake and animal protein intake in different countries each tend to be correlated with one another. One should therefore be wary of assuming that correlations between dietary variables and cancer mortality imply a causative relationship, unless there is other supporting evidence.

The observation that dietary fat stimulates mammary tumorigenesis in animals suggests that this dietary variable should be considered more carefully with respect to breast cancer incidence in human populations. It has also been reported recently that dietary fat enhances yields of intestinal cancer in animals (Reddy et al, 1975). Rose et al (1974) suggested that polyunsaturated fat intake might be implicated in colon cancer, because they noted that blood cholesterol levels were low in colon cancer patients. This proposal has been criticized by Heyden (1975) on the grounds that other factors such as age or the disease itself may have been responsible for the observed difference in blood cholesterol.

Essential Fatty Acids in Relation to Gallstone Formation

Review of autopsy records from the Los Angeles Veterans Administration clinical trial revealed a higher incidence of gallstones in subjects who had eaten the experimental diet containing polyunsaturated fat, in comparison to subjects on the control diet (Sturdevant et al, 1973). The authors concluded that the higher incidence in the experimental group was not due to differences in age or obesity, but after analyzing the data, Heyden (1975) was still of the opinion that obesity may have been a more important factor than dietary fatty acid composition.

The feeding of polyunsaturated fats has also been found to promote the formation of gallstones in squirrel monkeys (Lofland, 1975) and in African green monkeys (Lofland et al, 1975), but Dam (1971) reported that polyunsaturated fat tends to inhibit gallstone formation in guinea pigs. Effects of nutrition on gallstone formation have recently been reviewed by Portman et al (1975).

Essential Fatty Acids in Relation to Lipoperoxidation, Requirements for Vitamin E, and Aging Phenomena

Polyunsaturated fatty acids are subject to autoxidation, but this process is retarded in naturally-occurring oils by vitamin E (Tappel, 1972). Interrelationships between dietary polyunsaturated fatty acids and human requirements for vitamin E have been discussed by Witting (1970, 1972), Horwitt (1974, 1976) and Jager (1975).

A syndrome associated with low plasma levels of vitamin E was reported by Hassan et al (1966) in premature infants receiving formula mixtures with relatively high content of polyunsaturated fatty acids. Symptoms included edema, skin lesions, an elevated platelet count and morphologic changes in erythrocytes. These were relieved or prevented by supplementation with vitamin E.

It is well known that the requirement for vitamin E increases with increasing intake of polyunsaturated fat, but most human diets apparently contain sufficient vitamin E to prevent the appearance of overt symptoms of deficiency in adults (Horwitt, 1974). However, a high intake of polyunsaturated fat may give rise to deleterious effects which are not readily apparent. When tissues contain inadequate amounts of antioxidant, an increase in peroxidation of polyunsaturated fatty acids may occur, the consequences of which are not understood. Harman (1971, 1973) has suggested that degradative changes associated with aging may be due in part to free radical reactions such as those that occur in the peroxidation of fatty acids. He found that increasing the amount and/or degree of unsaturation of dietary fat tended to decrease the mean life span of mice and rats, even when the diet contained appreciable amounts of vitamin E. Anderson (1973) has suggested that a subclinical form of nutritional muscular dystrophy may develop in humans when dietary unsaturated fatty acids are inadequately protected by antioxidants. He has also proposed that this may increase the incidence of myocardial infarction by increasing the vulnerability of the myocardium to atherosclerotic ischemia.

The more saturated fats from both plant and animal sources tend to be deficient in vitamin E, whereas most vegetable oils with a high linoleic acid content contain sufficient tocopherol to meet vitamin E requirements (Jager, 1975). Some of the tocopherol may be lost, however, through oxidation during storage, processing or cooking of the oils. Loss of the protective antioxidants increases the chances of peroxidation of the polyunsaturated fatty acids, and this can give rise to toxic and possibly to carcinogenic products (Artman, 1969; West & Redgrave, 1974).

Other Biological Parameters Influenced by Polyunsaturated Dietary Fat

Jones (1974) compiled a list of parameters influenced by polyunsaturated dietary fat compared to eucaloric amounts of saturated dietary fat. A number of the observed effects were regarded as being beneficial. These include: Reductions in serum cholesterol level, in triglyceride response to dietary carbohydrate, in blood clotting time and in platelet aggregation, as well as enhancement of post-heparin lipolytic activity, clearance of particulate lipid, and platelet survival time. Other parameters, including those already discussed, could have harmful effects.

What is the Most Desirable Dietary Level of Essential Fatty Acids?

In attempting to answer this question, let us first of all consider the extent of variation in the intake of essential fatty acids likely to be encountered in different countries of the world. Data compiled by the Food and Agriculture Organization of the United Nations (1971) indicate that the fat component of human diets may contribute anywhere from just over 10% to more than 40% of the total calories. The various fats and oils that go to make up this dietary fat differ widely in their content of linoleic acid, the major dietary essential fatty acid. The diunsaturated (18:2) fatty acid content of different dietary fats and oils is shown in Table 2. In some cases, this is not all linoleic acid (e.g. butter contains only about 0.8% linoleic acid), but in most cases, the diunsaturated fatty acid can be equated with linoleic acid. Animal fats generally contain less linoleic acid than plant fats, but there is considerable overlap, as seen in Table 2.

Although it is possible that linoleic acid may supply less than 0.5% of total calories in some human diets, this is unlikely because in countries where fat intake is low, the dietary fat is more likely to be derived from plant sources and tends to be more unsaturated than in countries where intake is high. For this reason,

TABLE 2

Diunsaturated (18:2) Fatty Acid Content of Different Dietary Fats and Oils[1]

Animal Fats: Source of Fat	Animal Fats: Percent of Total Fatty Acids	Plant Fats: Source of Fat	Plant Fats: Percent of Total Fatty Acids
Shell fish	0.6-1.2	Coconut oil	2
Salt water fish	1.0-2.2	Cocoa butter	4
Fresh water fish	1.0-7.6	Palm oil	10
Tallow	2	Olive oil	10
Butter	3.4	Rapeseed oil	15-20
Sheep	4	Peanut oil	20-40
Pig	6	Cottonseed oil	51
Goose	10	Soyabean oil	52
Rabbit	8-14	Corn oil	55
Chicken	21	Sunflower seed oil	63
Turkey	23	Safflower seed oil	75

[1]Data from Vergroesen & Gottenbos (1975)

it can also be deduced that there are few natural human diets in which linoleic acid supplies as much as 15% of total calories. These estimates tend to confirm the overall impression that most human diets contain enough essential fatty acids to prevent obvious deficiency symptoms. They also indicate that the calculated intake for optimum lowering of serum cholesterol is at the upper limit of the amount that might be found in natural diets.

Essential Fatty Acid Intake of Children

Infants raised on cow's milk may obtain as little as 1% of their calories from linoleic acid, whereas commercial infant formulas presently in common use normally contain vegetable oils and may provide more than 20% of total calories as linoleic acid (Schubert, 1973). Human milk, which is accepted as the optimal nutritional standard in infancy (Macy & Kelly, 1961; Jelliffe & Jelliffe, 1971) normally provides about 4% of total calories as linoleic acid (Schubert, 1973), but the percentage of linoleic acid in human milk may vary anywhere from 1% to more than 15% (Cuthbertson, 1976). Although various symptoms attributed to dietary deficiencies of essential fatty acids have been observed by a number of investigators, they occur only rarely, and Cuthbertson (1976) considers that the minimum requirement for children is, in fact, less than 0.5% of calories.

There are few documented cases of harmful effects resulting from excessive intake of essential fatty acids in children. The syndrome related to vitamin E deficiency in premature infants described by Hassan et al (1966) was referred to earlier. Presumably, this could be exacerbated by increasing the level of polyunsaturated fat in the diet unless the levels of vitamin E were also increased at the same time. Johnson et al (1974) have provided some evidence that the occurrence of retrolental fibroplasia in premature infants is associated with vitamin E deficiency.

The feeding of formula diets containing high levels of polyunsaturated fat largely prevents the rise in plasma cholesterol that normally occurs in infants raised on mother's milk or on cow's milk (Sweeney et al, 1961; Goalwin & Pomeranze, 1962). This is illustrated in Fig. 1, which also shows that the serum cholesterol rises when the formula containing polyunsaturated fat is supplemented with other foods. The significance of the rise in serum cholesterol that normally occurs after birth is not known, but it is also observed in the young of most other species that have been investigated (Carroll et al, 1973; Carroll & Huff, 1976). It therefore seems possible that a low level of serum cholesterol at this stage of development may have undesirable effects, even though they may not be apparent immediately.

Breast feeding of children should be encouraged for a variety of reasons (Jelliffe & Jelliffe, 1971). When formula feeding is used, however, it would seem desirable from a nutritional point of view to use a formula with a composition patterned as nearly as possible on the normal composition of human milk. This should also apply to the fatty acid composition of the fat used in the formula.

Essential Fatty Acid Intake of Adults

Adults normally contain appreciable stores of essential fatty acids in their tissues and, even when none are supplied in the diet, it takes some time for deficiency symptoms to appear (Collins et al, 1971; Riella et al, 1975). On a high intake of polyunsaturated fat, the fatty acid composition of adipose tissue stores will gradually come to resemble that of the dietary fat, but again, this is a relatively slow process (Hirsch et al, 1960). Thus, in adults perhaps more so than in children, the requirements for dietary essential fatty acids may be considered in terms of long-term trends.

Any decisions on the optimal intake of dietary essential fatty acids involve consideration of a number of different factors, some of which have been discussed above. Individuals at high risk from cardiovascular disease may be justified in increasing their intake to about 15% of total calories in order to achieve maximum lowering

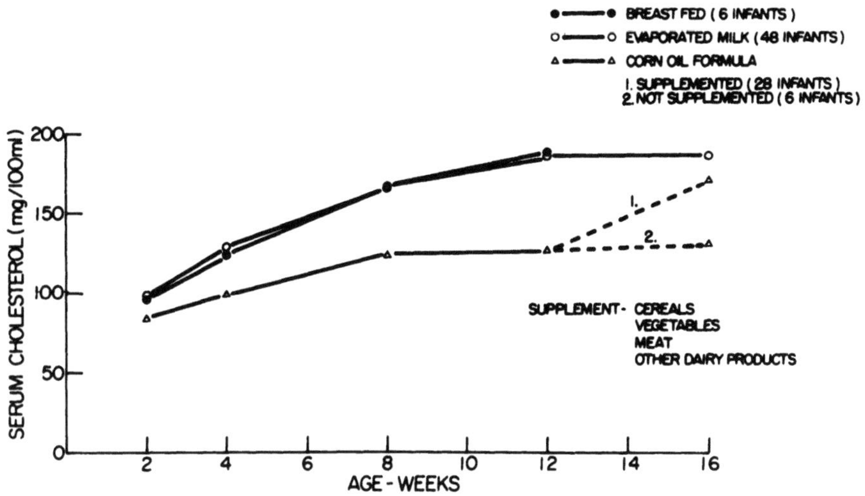

Fig. 1. Effects of diet on serum cholesterol levels in human infants. Based on data of Goalwin & Pomeranze (1962).

of their serum cholesterol levels. In such cases, the possible hazards associated with high intake probably constitute an acceptable risk. For populations as a whole, however, it would seem better in the present state of knowledge to be more conservative and set an upper limit of 10% of total calories for intake of polyunsaturated fatty acids, with 5% as an acceptable level.

Cardiovascular disease continues to be the most frequent cause of death in many industrialized countries, and efforts must be continued to seek a solution to this problem. The relative infrequency of cardiovascular disease in other parts of the world indicates that atherosclerosis is not an inevitable consequence of aging. The fact that dietary manipulation is the most common method of producing atherosclerotic lesions in animals indicates that diet may also be an important factor in the human disease. It seems unlikely, however, that variations in dietary fat intake can explain the geographical distribution of cardiovascular disease, and perhaps there has been too much emphasis on dietary fat in atherosclerosis research. Experiments in animals have provided evidence that non-lipid components of the diet can also influence the level of plasma cholesterol and may play a role in the development of atherosclerosis (Carroll & Hamilton, 1975; Hamilton & Carroll, 1976; Carroll et al, 1976). More attention might profitably be devoted to investigating the possibility that non-lipid components of the diet have an important influence on cardiovascular disease in human populations.

SUMMARY

Most human diets appear to provide at least 1 to 2% of total calories as essential fatty acids, which is enough to prevent deficiency symptoms. In recent years, however, an increase in dietary intake of polyunsaturated fatty acids has been recommended as a means of decreasing serum cholesterol levels and reducing the incidence of coronary heart disease. For maximal lowering of serum cholesterol, an intake of linoleate corresponding to 15 to 20% of total calories is required. This is higher than the levels found in most natural diets, and may have harmful as well as beneficial effects. Concern has been expressed about the possibility of increases in carcinogenesis or the formation of gallstones. Polyunsaturated fatty acids are susceptible to autoxidation and, besides increasing the requirements for vitamin E, this may lead to deleterious effects on tissues. High levels of polyunsaturated fat in infant formulas can keep serum cholesterols low at a time when higher levels might be more beneficial. Human milk normally supplies about 4% of total calories as linoleate, and this is probably a good standard for infant formulas. Adults at high risk from coronary heart disease may be justified in increasing their intake of essential fatty acids to 15% of total calories, but for the general population, 5% of total calories appears to be more than adequate, with 10% as an upper limit.

REFERENCES

American Medical Association Council on Foods & Nutrition and the Food & Nutrition Board, National Academy of Sciences-National Research Council (1972) J. Amer. Med. Ass. 222, 1647.

Anderson, T. W. (1973) Lancet 2, 298-302.

Artman, N. R. (1969) in Adv. Lipid Res. (Paoletti, R. & Kritchevsky, D., eds.), vol. 7, pp. 245-330, Academic Press, New York.

Brown, H. B. (1971) J. Amer. Diet. Ass. 58, 303-311.

Burr, G. O. (1942) Fed. Proc. 1, 224-233.

Carroll, K. K. (1975) Cancer Res. 35, 3374-3383.

Carroll, K. K. & Khor, H. T. (1971) Lipids 6, 415-420.

Carroll, K. K., Hamilton, R.M.G. & MacLeod, G. K. (1973) Lipids 8, 635-640.

Carroll, K. K. & Hamilton, R.M.G. (1975) J. Food Sci. 40, 18-23.

Carroll, K. K. & Khor, H. T. (1975) in Progress in Biochemical Pharmacology, (Carroll, K. K., ed.), vol. 10, Lipids and Tumors, pp. 308-353, Karger, Basel.

Carroll, K. K. & Huff, M. W. (1976) in Proc. Int. Workshop-Conference on Atherosclerosis (Manning, G. W. & Haust, M. D., eds.), Plenum Press, New York, in the press.

Carroll, K. K., Huff, M. W. & Roberts, D.C.K. (1976) in Abst. IV Int. Symp. on Atherosclerosis, p. 55, Tokyo, Japan.

Collins, F. D., Sinclair, A. J., Royle, J. P., Coats, D. A., Maynard, A. T. & Leonard, R. F. (1971) Nutr. Metab. 13, 150-167.
Cuthbertson, W.F.J. (1976) Amer. J. Clin. Nutr. 29, 559-568.
Dam, H. (1971) Amer. J. Med. 51, 596-613.
Dayton, S., Pearce, M. L., Hashimoto, S., Dixon, W. J. & Tomiyasu, U. (1969) Circulation 39-40, Suppl. II, 1-63.
Dayton, S. & Hashimoto, S. (1976) in Abst. IV Int. Symp. on Atherosclerosis, p. 243, Tokyo, Japan.
Ederer, F., Leren, P., Turpeinen, O. & Frantz, I. D. Jr. (1971) Lancet 2, 203-206.
Food and Agriculture Organization of the United Nations, Rome (1971) Food Balance Sheets, 1964-66 Average.
Goalwin, A. & Pomeranze, J. (1962) Arch. Pediat. 79, 58-62.
Hamilton, R.M.G. & Carroll, K. K. (1976) Atherosclerosis 24, 47-62.
Harman, D. (1971) J. Gerontol. 26, 451-457.
Harman, D. (1973) Triangle 12, 153-158.
Hassan, H., Hashim, S. A., Van Itallie, T. B. & Sebrell, W. H. (1966) Amer. J. Clin. Nutr. 19, 147-157.
Heyden, S. (1975) in The Role of Fats in Human Nutrition (Vergroesen, A. J., ed.), Appendix, pp. 96-113, Academic Press, New York.
Hirsch, J., Farquhar, J. W., Ahrens, E. H. Jr., Peterson, M. L. & Stoffel, W. (1960) Amer. J. Clin. Nutr. 8, 499-511.
Holman, R. T. (1970) in Progress in the Chemistry of Fats and Other Lipids (Holman, R. T., ed.), vol. 9, part 5, pp. 607-682, Pergamon Press, Oxford.
Holman, R. T. (1976) This symposium.
Horwitt, M. K. (1974) Amer. J. Clin. Nutr. 27, 1182-1193.
Horwitt, M. K. (1976) Amer. J. Clin. Nutr. 29, 569-578.
Insull, W. Jr., Land, P. D., Hsi, B. P. & Yoshimura, S. (1969) J. Clin. Invest. 48, 1313-1327.
Jager, F. C. (1975) in The Role of Fats in Human Nutrition (Vergroesen, A. J., ed.), pp. 381-432, Academic Press, New York.
Jelliffe, D. B. & Jelliffe, E.F.P. (1971) Amer. J. Clin. Nutr. 24, 1013-1024.
Johnson, L., Schaffer, D. & Boggs, T. R. Jr., (1974) Amer. J. Clin. Nutr. 27, 1158-1173.
Jones, R. J. (1974) J. Amer. Oil Chem. Soc. 51, 251-254.
Kinsell, L. W. (1963) in Progress in the Chemistry of Fats and Other Lipids (Holman, R. T., Lundberg, W. O. & Malkin, T., eds.), vol. 6, pp. 137-170, Pergamon Press, Oxford.
Leren, P. (1970) Circulation 42, 935-942.
Lofland, H. B. (1975) Amer. J. Pathol. 79, 619-622.
Lofland, H. B., Bullock, B. C. & Clarkson, T. B. (1975) Fed. Proc. 34, 862.
Macy, I. G. & Kelly, H. J. (1961) in Milk: The Mammary Gland and its Secretion (Kon, S. K. & Cowie, A. T., eds.), vol. II, pp. 265-304, Academic Press, New York.
Miettinen, M., Turpeinen, O., Karvonen, M. J., Elosuo, R. & Paavilainen, E. (1972) Lancet 2, 835-838.

Morris, J. N. et al (1968) Lancet 2, 693-700.
Pearce, M. L. & Dayton, S. (1971) Lancet 1, 464-467.
Pinckney, E. R. (1973a) Amer. Heart J. 85, 723-726.
Pinckney, E. R. (1973b) Med. Counterpoint 5, 53-71.
Portman, O. W., Osuga, T. & Tanaka, N. (1975) in Adv. Lipid Res. (Paoletti, R. & Kritchevsky, D., eds.), vol. 13, pp. 135-194, Academic Press, New York.
Reddy, B. S., Narisawa, T., Maronpot, R., Weisburger, J. H. & Wynder, E. L. (1975) Cancer Res. 35, 3421-3426.
Report of the Advisory Panel of the British Committee on Medical Aspects of Food Policy (Nutrition) on Diet in relation to Cardiovascular and Cerebrovascular Disease (1974). Her Majesty's Stationery Office, London. Reproduced (1975) in Nutrition Today 10(1), 16-18, 24-27.
Riella, M. C., Broviac, J. W., Wells, M. & Scribner, B. H. (1975) Ann. Intern. Med. 83, 786-789.
Rinzler, S. H. (1968) Bull. N. Y. Acad. Med. 44, 936-949.
Rose, G., Blackburn, H., Keys, A., Taylor, H. L., Kannel, W. B., Paul, O., Reid, D. D. & Stamler, J. (1974) Lancet 1, 181-183.
Schubert, W. K. (1973) Amer. J. Cardiol. 31, 581-587.
Söderhjelm, L., Wiese, H. F. & Holman, R. T. (1970) in Progress in the Chemistry of Fats and Other Lipids (Holman, R. T., ed.), vol. 9, part 4, pp. 555-585, Pergamon Press, Oxford.
Sturdevant, R.A.L., Pearce, M. L., & Dayton, S. (1973) N. Engl. J. Med. 288, 24-27.
Sweeney, M. J., Etteldorf, J. N., Dobbins, W. T., Somervill, B., Fischer, R. & Ferrell, C. (1961) Pediatrics 27, 765-771.
Tannenbaum, A. (1959) in The Physiopathology of Cancer (Homburger, F., ed.), 2nd edn., pp. 517-562, Hoeber-Harper, New York.
Tannenbaum, A. & Silverstone, H. (1957) in Cancer (Raven, R. W., ed.), vol. 1, pp. 306-334, Butterworth, London.
Tappel, A. L. (1972) Ann. N. Y. Acad. Sci. 203, 12-28.
Vergroesen, A. J. (1972) Proc. Nutr. Soc. 31, 323-329.
Vergroesen, A. J. & Gottenbos, J. J. (1975) in The Role of Fats in Human Nutrition (Vergroesen, A. J., ed.), pp. 1-41, Academic Press, New York.
West, C. E. & Redgrave, T. G. (1974) Search 5, 90-94.
Witting, L. A. (1970) in Progress in The Chemistry of Fats and Other Lipids (Holman, R. T., ed.), vol. 9, part 4, pp. 517-553, Pergamon Press, Oxford.
Witting, L. A. (1972) Ann. N. Y. Acad. Sci. 203, 192-198.

THE RELATIONSHIP BETWEEN MEMBRANE FATTY ACIDS AND THE DEVELOPMENT OF THE RAT RETINA

Robert E. Anderson, R. M. Benolken, Margaret B. Jackson, and Maureen B. Maude

Department of Ophthalmology
Baylor College of Medicine
1200 Moursund Ave., Houston, Texas 77030

and

Department of Sensory Sciences
University of Texas Graduate School of Biomedical Sciences
Houston, Texas

INTRODUCTION

The vertebrate rod visual cell is anatomically divided into two compartments, the inner segment and the outer segment (Fig. 1). The rod inner segment contains the cellular organelles responsible for most, if not all, of the biosynthetic activity of the cell. One end of the inner segment makes synaptic contact with horizontal and bipolar cells, while the other is connected to the rod outer segments (ROS) via a short connecting cilium. The ROS are made up of hundreds of membraneous discs. Each is free-floating and apparently does not contact other discs or the plasma membrane.

Recent studies have shown that ROS are constantly renewed (Young, 1967; Young and Droz, 1968). Membrane components are synthesized in the rod inner segments and transported via the connecting cilium to the basal infoldings of the plasma membrane, which pinch off to form single discs. As new discs are added at the base of the outer segment, older discs at the apical tips are shed and phagocytized by the retinal pigment epithelium. Complete renewal of photoreceptor membranes in rats requires 9-10 days (Landis *et al.*, 1973).

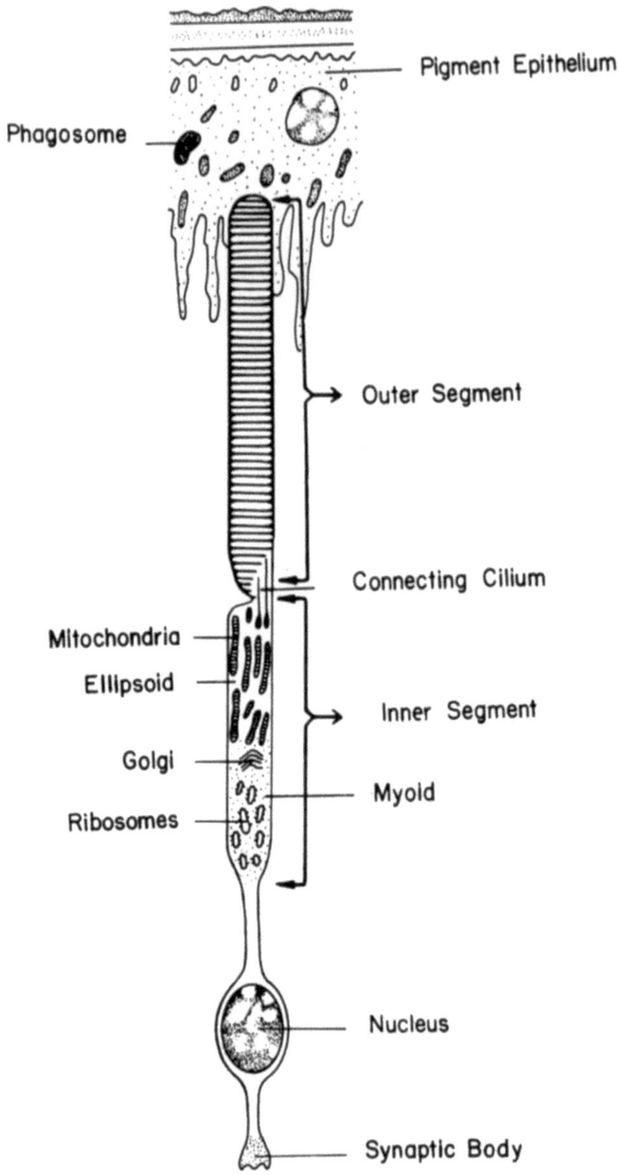

Figure 1. Schematic Diagram of Rod Visual Cell.

Components of ROS are synthesized in the myoid region of the inner segment and transported to the outer segment where they are incorporated into the basal infoldings of the plasma membrane. As new discs are added at the base, older ones are shed at the apical tip and phagocytized by the pigment epithelium.

The chemistry of ROS has been studied by several laboratories, and recently reviewed by Daemen (1973). Over 90% of the protein is rhodopsin, a photosensitive glycoprotein of molecular weight around 35,000, which is imbedded in a lipid bilayer. Phospholipids make up about 96% of the lipids of cattle ROS and cholesterol is the major component of the neutral lipid fraction. Phosphatidyl choline (PC) and phosphatidyl ethanolamine (PE) are the major phospholipids in all species examined, with phosphatidyl serine (PS), phosphatidyl inositol (PI), and sphingomyelin (SPh) present in lesser amounts (Anderson and Maude, 1972). Detailed analysis of the photoreceptor membranes of vertebrate species ranging from frogs to humans have revealed a fairly constant phospholipid class and protein composition (Basinger and Anderson, unpublished).

The fatty acids of ROS phospholipids contain large amounts of polyunsaturated fatty acids (Anderson and Maude, 1972). The most abundant is docosahexaenoic acid (22:6ω3)*, which accounts for almost 50% of the total fatty acids in PE and PS. Several years ago, we attempted to take advantage of the turnover of photoreceptor membranes to replace the long chain polyunsaturates with shorter chain fatty acids by depriving albino rats of essential fatty acids (Anderson and Maude, 1972). To our surprise, after 10 weeks of essential fatty acid deprivation, the composition of the ROS phospholipids remained remarkably similar to that of the animals fed ordinary laboratory chow, although serum fatty acid compositions clearly demonstrated that the animals were deficient in the precursors of long chain polyunsaturated fatty acids. Subsequent experiments demonstrated that the renewal of photoreceptor membranes (Landis *et al.*, 1973) and the electrical response of the retina (Benolken *et al.*, 1973) had been altered by essential fatty acid deficiency.

Experiments in mature albino rats demonstrated that the electroretinographic (ERG) function of the retina could be selectively altered by specific dietary short chain fatty acids (Wheeler *et al.*, 1975). In this report, we present studies on these relationships in rats in the early stages of retinal development. Groups of weanling albino rats were fed either lab chow, fat free, or fat free diets containing individual or mixtures of pure 18:1ω9, 18:2ω6, and 18:3ω3. ERG responses were recorded from each dietary group beginning at an age of 22 days. After 12 weeks on these

---

* Fatty acid nomenclatures as follows: The number before the colon represents the number of carbon atoms. The number after the colon represents the number of *cis* double bonds. The number following the omega represents the position of the first double bond measured from the methyl end of the molecule. It is assumed that all of the double bonds are methylene interrupted.

diets, the fatty acid compositions of the individual phospholipid classes from the ROS of each group were determined as were the compositions of adipose tissue, red blood cells, and plasma.

EXPERIMENTAL PROCEDURES

Dietary Manipulations

Twenty-one day old weanling albino rats (Sprague Dawley from Holtzman, Madison, Wisconsin) were divided into 7 groups and maintained for 12 weeks on one of the following diets: laboratory chow, fat free (Fat Free Test Diet from ICN Pharmaceuticals, Inc., Cleveland, Ohio); fat free plus 0.85% (by weight) 18:1ω9 (all esters were obtained greater than 99% pure from NuChek Prep, Eylsein, Minnesota); fat free plus 0.85% 18:2ω6; fat free plus 0.85% 18:3ω3; fat free plus 0.43% 18:2ω6 plus 0.43% 18:3ω3; or fat free plus 1.7% 18:2ω6. The animals were maintained in cyclic light (12 hours on and 12 hours off) of maximal intensity at the top cage of 4-6 footcandles and had access to food and water *ad libitum*.

Chemical Analyses

The animals were anesthetized with ether and sacrificed by exsanguination via heart puncture. Their eyes were enucleated and the anterior segment removed by dissection. The lens was removed and the retina was striped from the choroid with a pair of smooth forceps. Perirenal fat pads, lungs, kidneys, and liver were removed and stored frozen. Whole blood was collected in evacuated tubes containing sodium EDTA. The plasma was separated from the red blood cells (RBC) by centrifugation and both were lyophilized.

Rod outer segments were prepared from the retinas by the discontinuous sucrose gradient procedure of Papermaster and Dryer (1974) as modified by Basinger, Bok, and Hall (1976). Washed membranes were lyophilized.

Lipids were extracted from plasma, red blood cells, and rod outer segments with chloroform:methanol (2:1). Phospholipids were fractionated by two dimensional thin layer chromatography (Anderson *et al.*, 1969) and methyl esters were prepared with boron trifluoride-methanol (Morrison and Smith, 1964). Aliquots of the total lipid from plasma and of RBC were directly methylated as were small pieces of adipose tissue. Gas liquid chromatography was carried out on a Varian Model 2100 series gas chromatograph (Anderson *et al.*, 1970).

Electrophysiological Procedures

ERGs were measured by procedures similar to those described earlier (Benolken *et al.*, 1973) with a corneal electrode 1 mm in diameter to accommodate the smaller eyes of a 21 day old weanling rat. At 21 days the animals were placed on the dietary regimens, and ERGs were sampled on each eye of 10 anesthetized rats (Nembutal, 45 mgm/kg) from each dietary group at days 22, 24, 27, 31, 36, and 46. The ERGs were measured on dark adapted animals in response to 20 msec flashes of a graded intensity series over a range of 6 log units. Maximum intensity saturated both the a-wave and b-wave of the ERG. No animal was anesthetized more than once.

Adequate sampling proved more of a problem with the weanling animals than it was for our earlier studies on mature animals (Benolken *et al.*, 1973; Wheeler, *et.al.*, 1975). There are two primary reasons for this. First, there was considerable variability in rate of development of the ERGs as a function of time after birth even in normal control populations. Also, contrary to the mature animals, the ERGs from the 2 eyes of a weanling were not statistically independent. When the right eye had developed more rapidly than the average of a sample population, so too had the left eye and to the same extent. Consequently the ERG from the second eye provided no additional information for sampling purposes and, even if other sources of variation had been similar, twice as many weanling animals as mature animals would have been required for equivalent sampling estimates.

With ERGs from 10 weanling animals per group per time point, it was not possible to demonstrate good statistical sampling. These data indicate that good sampling of the populations would be cost prohibitive given the high price of purified fatty acid supplements. However, the data show general trends during the development of retinal function in weanling rats maintained on various fatty acid supplements.

RESULTS

Chemical Results

Rats raised on diets containing less than 1% total lipid exhibit marked changes in the fatty acid patterns in their plasma and RBC lipids (Tables 1 and 2). Short and long chain ω3 and ω6 polyunsaturated fatty acids were decreased in the fat free and 0.85% ω9 groups relative to the laboratory chow group. The decrease in 20:4ω6 in the 0.85% 18:3ω3 group was accompanied by an increase in 20:5ω3. Only those groups receiving 18:3ω3 supple-

Table 1. Fatty Acid Composition of the Total Lipids of Plasma of Rats Raised on Various Diets

Weanling albino rats were fed for 12 weeks on one of the following diets: laboratory chow; fat free; or fat free supplemented with 0.85% 18:1ω9, 0.85% 18:2ω6, 0.85% 18:3ω3, 1.7% 18:2ω6, or 0.43% 18:2ω6 + 0.43% 18:3ω3. The values are reported as relative mole percent. DMA are dimethylacetals derived from methanolysis of plasmalogens. See Experimental Procedures for further details.

Fatty Acid	Diet						
	Lab Chow	Fat Free	0.85% ω3	0.85% ω6	0.85% ω9	1.7% ω6	0.43% ω3 0.43% ω6+
14:0	0.7	0.7	0.1	0.9	--	0.6	0.4
15:0	0.4	0.2	--	0.3	--	--	--
16:0 DMA	--	--	--	0.1	--	--	--
16:0	22.6	18.0	29.4	24.4	25.5	27.8	25.4
16:1ω7	3.3	13.0	8.6	4.1	9.2	7.6	7.7
17:0	--	0.2	--	0.1	0.2	--	--
18:0	9.0	10.5	8.8	14.5	8.4	8.3	10.3
18:1ω9	14.6	30.1	31.6	17.7	41.4	23.3	27.5
18:2ω6	27.3	3.3	1.4	13.0	1.2	15.5	9.9
18:3ω3	0.5	0.4	4.9	0.2	--	--	1.5
20:0	--	1.3	--	--	--	--	--
20:1	--	--	--	--	1.0	0.5	0.5
20:3ω9	0.6	16.3	1.5	0.1	11.0	--	0.9
20:4ω6	15.0	4.1	2.2	19.4	2.2	15.7	9.9
20:5ω3	1.5	0.8	9.1	--	--	--	2.5
22:4ω6	--	--	--	0.4	--	--	--
22:5ω6	--	0.1	--	2.4	--	--	--
22:5ω3	0.9	--	--	1.2	--	0.8	0.3
22:6ω3	3.7	0.3	2.5	0.8	--	--	3.2

ments had significant amounts of 22:6ω3 in their total lipids, but the levels did not exceed 5%. The levels of 20:3ω9, an indicator of essential fatty acid deficiency, were elevated in the groups that did not receive 18:3ω3 or 18:2ω6. It is interesting that 20:3ω9 was not elevated in the 18:3ω3 group, even though the level of linoleic acid was as low in this group as it was in the fat free and 18:1ω9 groups. Evidently both ω3 and ω6 fatty acids must be deficient in order to activate this elongation mechanism.

Table 2. Fatty Acid Composition of the Total Lipids of Red Blood Cells of Rats Raised on Various Diets

Experimental details are given in Table 1 & Experimental Procedures.

Fatty Acid	Diet						
	Lab Chow	Fat Free	0.85% ω3	0.85% ω6	0.85% ω9	1.7% ω6	0.43% ω3 + 0.43% ω6
14:0	0.8	1.6	0.4	0.4	0.4	0.3	0.5
15:0	--	1.2	--	--	0.3	--	--
16:0 DMA	1.9	1.2	1.6	2.2	2.0	1.9	2.1
16:0	23.3	23.6	40.3	35.6	28.9	35.1	43.8
16:1ω7	1.4	5.6	--	--	4.5	--	0.5
17:0	0.8	--	--	--	--	--	--
18:0 DMA	2.2	3.3	1.8	1.7	1.8	1.7	1.6
18:0	17.0	14.8	8.5	13.2	12.1	12.7	14.2
18:1ω9	9.5	21.3	23.2	12.2	23.6	11.8	10.2
18:2ω6	15.2	3.1	0.5	5.9	0.5	4.7	6.7
18:3ω3	0.7	--	1.0	0.2	0.6	--	0.3
20:0	0.9	0.9	--	--	--	0.2	--
20:2ω6	--	2.2	0.1	--	--	0.3	--
20:3ω9	0.1	12.5	1.1	0.4	14.8	0.4	0.3
20:4ω6	20.2	6.6	4.2	22.8	5.6	26.1	9.9
20:5ω3	1.6	1.4	10.7	0.1	1.1	--	3.1
22:4ω6	--	--	--	2.3	0.9	2.1	--
22:5ω6	--	--	--	2.1	2.1	2.2	--
22:5ω3	1.3	--	2.6	0.1	--	--	1.9
22:6ω3	3.0	--	4.6	0.4	0.6	0.4	4.6

The fatty acid compositions of the adipose tissue are predictably affected by the different dietary fatty acids (Table 3). Linoleic acid decreased from 24.7% in the lab chow group to 0.3% and 0.8% respectively in the fat free and 18:1ω9 groups. Small amounts of 18:3ω3 were stored in the fat pads of those animals receiving 18:3ω3 supplements.

The fatty acid compositions of the total lipids and the phospholipid classes of the ROS from the seven groups are given in Tables 4-7. The fat free and 18:1ω9 groups were similar to the lab chow animals, except for a slight decrease in 22:6ω3 and 3-4%

Table 3. Fatty Acid Composition of the Total Lipids of the Perirenal Fat Pad of Rats Raised on Various Diets

Experimental details are given in Table 1 & Experimental Procedures.

Fatty Acid	Diet						
	Lab Chow	Fat Free	0.85% ω3	0.85% ω6	0.85% ω9	1.7% ω6	0.43% ω3 + 0.43% ω6
14:0	2.4	2.7	2.6	2.3	2.4	2.2	2.3
14:1	0.8	0.7	0.6	0.4	0.6	0.5	0.6
15:0	0.1	0.1	0.1	0.1	0.1	0.1	0.1
16:0	28.2	30.4	31.9	30.2	27.4	30.2	32.5
16:1ω7	8.6	16.6	17.1	10.2	17.4	13.4	15.6
17:0	0.7	0.3	0.3	0.3	0.3	0.2	0.2
18:0	3.0	2.4	3.0	3.1	1.9	3.0	2.9
18:1ω9	25.6	45.4	40.1	42.7	47.8	38.0	39.6
18:2ω6	24.7	0.3	0.3	9.1	0.8	11.7	3.1
18:3ω3	2.5	--	3.8	--	--	--	2.6
20:1	--	0.5	--	0.6	0.4	0.4	--
20:3ω9	0.6	0.2	--	--	0.3	0.1	--
20:4ω6	0.6	--	--	0.3	--	0.2	0.1
22:6ω3	1.3	--	--	--	--	--	--

increase in 20:3ω9 in the two EFA deficient groups. Arachidonic acid remained the same in all three groups.

The group fed 18:3ω3 had less 20:4ω6 and more 22:6ω3 than any of the other groups. Those fed 18:2ω6 had significant amounts of 22:5ω6, which replaced some of the 22:6ω3 in each of the ROS phospholipid classes.

The levels of 22:6ω3 were lower in the 18:2ω6 groups than in either of the EFA deficient groups. The ROS of the group receiving both 18:3ω3 and 18:2ω6 fatty acids showed a preference for ω3 fatty acids.

Table 4. Fatty Acid Composition of the Total Lipids of Photoreceptor Membranes of Rats Raised on Various Diets

Experimental details are given in Table 1 & Experimental Procedures.

Fatty Acid	Diet: Lab Chow	Fat Free	0.85% ω3	0.85% ω6	0.85% ω9	1.7% ω6	0.43% ω3 + 0.43% ω6
14:0	0.2	0.2	0.2	0.3	0.6	0.2	0.2
15:0	0.2	0.2	0.2	--	--	--	--
16:0	10.9	15.4	12.7	15.7	16.8	11.4	13.3
16:1ω7	0.1	--	--	--	0.2	--	--
17:0	0.4	0.6	0.4	0.1	0.2	0.2	--
18:0	22.2	15.8	19.4	26.8	6.7	21.0	28.1
18:1ω9	2.0	4.1	2.6	1.6	5.9	6.3	1.5
18:2ω6	0.3	--	--	--	--	0.2	--
18:3ω3	--	--	--	1.6	--	--	--
20:1	0.2	--	--	--	0.3	--	1.3
20:2ω6	0.2	0.4	0.2	2.7	0.2	1.5	2.5
20:3ω9	0.1	3.6	0.2	0.2	2.8	--	0.1
20:4ω6	4.3	5.2	1.9	4.2	4.7	4.5	3.5
20:5ω3	1.4	1.2	0.8	0.5	0.9	0.1	--
22:4ω6	0.6	0.5	0.2	1.9	0.3	1.2	1.0
22:5ω6	0.4	5.5	0.8	12.5	5.7	16.7	0.3
22:6ω3	56.4	46.8	60.4	31.3	54.0	34.8	47.5
UNK	--	--	--	0.6	--	1.3	0.1

Electrophysiological Results

The ERGs in Holtzman rats fed lab chow did not show the advanced development at 22 days of age that one would expect from reported results (Weidman and Kuwabara, 1968) and from our own data (Benolken *et al.*, unpublished) on several litters of 22 day old rats which were raised in our laboratory. We have no explanation for the developmental delays in the normal Holtzman rats used in this study. On the average, by 22 days of age all weanling groups exhibited substantial development of the a-wave of the ERG; none showed substantial development of the b-wave. (The a-wave of the ERG results primarily from the electrical activity of the photo-

Table 5. Fatty Acid Composition of Phosphatidyl Ethanolamine From Photoreceptor Membranes of Rats Raised on Various Diets

Experimental details are given in Table 1 & Experimental Procedures.

Fatty Acid	Diet					
	Lab Chow	Fat Free	0.85% ω3	0.85% ω6	1.7% ω6	0.43% ω3 + 0.43% ω6
14:0	0.2	0.4	0.1	0.3	0.2	0.2
16:0 DMA	0.4	0.8	--	0.4	--	0.2
16:0	5.5	6.7	4.3	8.5	6.6	5.2
16:1ω7	0.3	0.4	0.3	0.1	--	0.3
18:0 DMA	1.2	1.8	--	1.5	--	1.1
18:0	28.1	6.5	27.1	19.1	33.2	25.7
18:1ω9	3.2	7.0	4.6	3.3	--	3.3
18:2ω6	0.6	0.2	--	0.2	--	0.3
18:3ω3	0.4	0.5	--	0.2	--	0.2
20:0	--	0.3	--	--	--	--
20:2ω6	0.8	0.7	0.1	0.5	--	0.4
20:3ω9	--	3.8	--	0.2	--	0.2
20:4ω6	2.9	4.8	0.6	4.7	3.4	1.4
20:5ω3	0.1	0.7	--	0.1	--	0.1
22:4ω6	0.2	0.7	--	0.4	0.6	0.3
22:5ω6	1.1	6.5	--	14.3	19.0	0.8
22:5ω3	--	0.1	--	--	--	--
22:6ω3	54.8	56.5	62.6	44.7	37.1	60.2
UNK	--	--	--	1.0	--	--

receptor cells of the retina and the b-wave results primarily from the electrical activity of other neural cells in the retina (Tomita, 1972).

On the average, by day 24, the lab chow and ω3 plus ω6 groups showed significant b-wave development and by day 31 the b-wave appeared well-developed in the lab chow group. In contrast, the ω3, ω9, and fat free groups did not exhibit significant b-wave function even by day 31. The rate of b-wave development for the two ω6 groups appeared to be somewhat slower than the lab chow group, but faster than the ω3, ω9, and fat free groups. By day 46, substantial b-wave was observed in all groups.

Table 6. Fatty Acid Composition of Phosphatidyl Choline From Photoreceptor Membranes of Rats Raised on Various Diets

Experimental details are given in Table 1 & Experimental Procedures.

Fatty Acid	Diet					
	Lab Chow	Fat Free	0.85% ω3	0.85% ω6	1.7% ω6	0.43% ω3 + 0.43% ω6
14:0	0.9	0.5	0.2	0.4	0.6	0.6
15:0	0.3	--	--	0.1	0.2	--
16:0	29.5	36.6	24.9	25.7	25.7	25.5
16:1ω7	1.0	1.1	--	0.4	0.4	0.4
17:0	0.7	0.5	--	--	--	--
18:0	11.7	4.1	18.1	23.8	18.5	8.8
18:1ω9	8.3	13.6	15.3	7.0	7.5	8.4
18:2ω6	0.8	1.1	--	0.2	0.5	0.5
18:3ω3	0.5	0.3	--	0.1	0.1	0.1
20:0	--	1.9	--	--	0.2	0.3
20:2ω6	0.8	0.7	0.3	0.2	0.4	0.5
20:3ω9	--	2.9	--	--	0.1	--
20:4ω6	3.4	4.7	--	3.3	4.7	2.5
20:5ω3	--	1.3	--	0.1	--	0.1
22:4ω6	--	0.3	--	0.3	0.6	0.2
22:5ω6	3.2	3.1	--	9.5	13.3	1.0
22:5ω3	--	--	--	--	0.1	0.2
22:6ω3	38.4	25.1	41.0	26.7	27.6	49.8
UNK	0.4	1.0	--	1.9	1.8	0.9

DISCUSSION

There is a reciprocal relationship between ω3 and ω6 fatty acids in the ROS that depends on the diet, much like that reported for brain by Galli *et al.* (1971). The groups receiving 0.85% and 1.7% 18:2ω6 showed an accumulation of 22:5ω6 in their ROS phospholipids, which replaced 22:6ω3. However, 22:5ω6 accumulated only when no 18:3ω3 was available in the diet. Given a source of ω3 and ω6 fatty acid precursors, ω3 fatty acids are incorporated into the ROS phospholipids to the almost total exclusion of ω6 fatty acids. An earlier observation supports this notion (Benolken *et al.*, 1973). The amount of 22:6ω3 could be reduced to 19% in the PE of rats only after several generations of ω3 deprivation.

Table 7. Fatty Acid Composition of Phsophatidyl Serine From Photoreceptor Membranes of Rats Raised on Various Diets

Experimental details are given in Table 1 & Experimental Procedures.

Fatty Acid	Diet: Lab Chow	0.85% ω3	0.85% ω6	1.7% ω6	0.43% ω3 + 0.43% ω6
14:0	--	1.5	0.4	0.5	--
15:0	--	--	0.1	--	--
16:0	2.5	--	1.5	1.7	4.8
16:1ω7	--	--	0.3	0.4	0.7
18:0	29.1	--	15.4	14.1	23.8
18:1ω9	1.9	24.7	0.8	1.2	3.7
18:2ω6	0.7	--	0.2	0.4	0.8
18:3ω3	0.7	--	0.1	0.1	0.5
20:0	--	--	--	0.3	--
20:2ω6	0.8	0.4	0.5	0.7	--
20:4ω6	1.9	--	1.4	1.1	1.3
20:5ω3	--	--	0.2	--	--
22:4ω6	--	--	4.9	5.6	--
22:5ω6	--	--	19.9	22.8	--
22:5ω3	--	--	0.2	0.1	--
22:6ω3	61.8	73.5	44.9	41.4	63.8
24:4ω6	--	--	4.9	8.0	--
24:5ω6	--	--	1.9	--	--
UNK	--	--	2.2	--	0.5

Thirty days after returning to a lab chow diet, the levels of 22:6ω3 in the ROS of these animals returned to normal (45%), while the levels of 22:5ω6 decreased from 10% to less than 1%.

The data also indicate that supplementation with 0.43% or 0.85% 18:3ω3 results in an enrichment of 22:6ω3 in photoreceptor membrane phospholipids compared to the other groups. The levels of 22:6ω3 in the lab chow group is a steady state value reflecting the amounts of ω3 fatty acids in that regimen (<0.2%). Up until now, 50% had been assumed to be an upper limit for 22:6ω3 in ROS phospholipids, since this value represents all of 22:6ω3 that can be esterified to the 2-position. This evidently cannot be the case since levels higher than 60% occur in PE and PS.

The ERG data indicate b-wave development is retarded in the absence of either ω3 or ω6 fatty acid precursors. The appearance of the b-wave correlates with the morphological development of the synaptic connections between the nerve cells of the retina (Weidman & Kuwabara, 1968). Apparently both ω3 and ω6 fatty acid precursors are involved in the membrane processes which ultimately determine the wiring diagram of retinal nerve cells.

ACKNOWLEDGEMENTS

The technical assistance of Francis Dohanich and Patricia Lissandrello is gratefully acknowledged. This work was supported by grants from the National Eye Institutes (EY-00871), National Science Foundation (BMS 75-07-7197), Retina Research Foundation (Houston), and Research to Prevent Blindness, Inc.

REFERENCES

Anderson, R. E., Feldman. L. S. and Feldman, G. L. (1970) *Biochim. Biophys. Acta* 202, 367-373
Anderson, R. E. and Maude, M. B. (1972) *Arch. Biochem. Biophys.* 151, 270-276.
Anderson, R. E., Maude, M. B. and Feldman, G. L (1969). *Biochim. Biophys. Acta* 187, 345-353.
Basinger, S. F., Hall, M. O. and Bok, D. (1976) *J. Cell. Biol.* 69, 29-42.
Benolken, R. M., Anderson, R. E. and Wheeler, T. G. (1973) *Science* 182,1253-1254.
Daeman, F. J. M. (1973) *Biochim. Biophys. Acta* 300, 225-288.
Calli, C., Tryeciak, H. J. and Paoletti, R. (1971) *Biochim. Biophys. Acta* 248, 449-454.
Landis, D. J., Dudley, P. A. and Anderson, R. E. (1973) *Science* 182,1144-1146
Morrison, W. R. and Smith, L. M. (1964) *J. Lipid Res.* 5, 600-608.
Papermaster, D. and Dryer, W. J. (1974) *Biochem.* 13, 2438-2444.
Tomita, T. (1972) *Handbook of Sensory Physiology* (M. G. F. Fuortes, ed.), vol. VII, pp. 635-665.
Weidman, T. and Kuwabara, T. (1968) *Arch. Ophthal.* 79, 470-484.
Wheeler, T. G., Benolken, R. M. and Anderson, R. E. (1975) *Science* 188, 1312-1314.
Young, R. W. (1967) *J. Cell. Biol.* 33, 61-72.
Young, R. W. and Droz, B. (1968) *J. Cell. Biol.* 39, 169-184.

DIETARY ESSENTIAL FATTY ACIDS, BRAIN POLYUNSATURATED FATTY ACIDS, AND PROSTAGLANDIN BIOSYNTHESIS

C. Galli, G. Galli*, C. Spagnuolo, E. Bosisio,
L. Tosi, G. C. Folco, and D. Longiave

Institute of Pharmacology and Pharmacognosy
University of Milan, 20129 Milan, Italy
*Laboratory of Applied Biochemistry

INTRODUCTION

The adult brain contains high levels of polyenoic fatty acids deriving from linoleic (18:2 Δ9,12 n - 6) and from linolenic acid (18:3 Δ9,12,15 n - 3). The two major polyunsaturated fatty acids (PUFA) present in brain are arachidonic acid (20:4 Δ5,8,11,14) of the linoleic acid series and especially docosahexaenoic acid (22:6 Δ4,7,10,13,16,19) of the linolenic acid series. The higher levels of 22:6 in brain, in comparison with other tissues, suggest a special function of the n-3 fatty acid family in nervous system. It has been recently found, for instance, that the electrical response of rod outer segments in the retina to illumination is affected by dietary induced changes of 22:6 n-3 levels in this tissue (Wheeler et al., 1970).

The rate of accumulation of PUFA in the developing brain (Sinclair & Crawford, 1972), the formation from the precursors in the CNS (Mead, 1976), the passage from the blood through the blood-brain barrier (Dhopeshwarkar & Mead, 1969; 1970) and the incorporation into brain lipids of both the short-chain unsaturated precursors (linoleic and linolenic acids) and the long-chain polyunsaturated derivatives (arachidonic and docosahexaenoic acids) following oral administration (Sinclair, 1975), have been studied in detail.

Brain polyunsaturated fatty acids, either deriving from "in situ" conversion of linoleic and linolenic acids, or arriving to the brain as preformed compounds, are ultimately found in ester form in the 2-position of phosphoglycerides. The levels of polyunsaturated fatty acids in various phosphoglycerides are quite different, as shown in Table 1.

Levels of both n-6 and n-3 fatty acids are high in EPG, whereas they are low in CPG. SPG contains appreciable levels of 22:6 n-3, and quite lower levels of n-6 fatty acids. The longer chain, more unsaturated members of this fatty acid series predominate. IPG is characterized by a selectively high level of arachidonic acid, whereas other polyunsaturated fatty acids are absent.

The processes reponsible of the different levels of polyenoic fatty acids in brain phosphoglycerides, and their possible differential roles in the various lipid classes are largely unknown. The selectively high levels of arachidonic acid in IPG, in the light of the special metabolic properties of this phospholipid in stimulated tissues, appear of possible significance. IPG has been shown to incorporate very actively intracerebrally injected labelled arachidonic acid (Yau & Sun, 1974). The incorporation of the radioactivity in all lipid classes was constant by 40 minutes after the injection.

The selectivity of incorporation of arachidonic acid in phosphatidyl inositol is quite evident, when values are expressed as relative specific activities (Table 2).

Table 1. Levels of polyunsaturated fatty acids (% of total fatty acids) in the major phosphoglycerides of adult rat brain cortex

Values are the average ± S.E. of determinations carried out on 4 samples.
EPG = ethanolamine phosphoglyceride; CPG = choline phosphoglyceride; SPG = serine phosphoglyceride; IPG = inositol phosphoglyceride.

Fatty acids	EPG	CPG	SPG	IPG
20:4 n-6	19.3±2.6	3.2±0.2	3.8±0.9	32.3±2.9
22:4 n-6	5.2±0.3	tr	1.5±0.4	6.0±0.8
22:5 n-6	3.8±0.9	tr	5.6±1.2	tr
22:6 n-3	18.0±2.7	1.4±0.2	12.7±0.5	tr

Table 2. Incorporation of radioactivity into brain phosphoglycerides one hour after intracerebral injection of [^3H]-arachidonic acid

Lipid classes were separeted by two dimensional thin layer chromatography (Rouser et al., 1970).
Phospholipid composition was determined by the method of Rouser et al. (1970).

$$\text{Relative specific activity} = \frac{\text{\% distribution of radioactivity}}{\text{\% of total phospholipids}}$$

Lipid Classes	% distribution of recovered radioactivity	% of total phospholipids	relative specific activity
CPG	36.4	37.5	0.97
EPG	16.9	38.4	0.44
SPG	7.1	13.0	0.55
IPG	24.1	2.0	12.05
Phosphatidic acid	tr	1.3	--

A correlation between the metabolism of arachidonic acid and phosphatidyl inositol has been observed in platelets, where this phospholipid appears to be the donor of the arachidonic acid which is released after stimulation with aggregating agents (Schoene & Iacono, 1976).

DIETARY ESSENTIAL FATTY ACIDS (EFA) AND BRAIN POLYUNSATURATED FATTY ACIDS

Since tissue PUFA are ultimately derived from dietary EFA precursors, their levels are affected by the intake of linoleic and linolenic acids. This has been shown by a number of investigators to occur in several tissues. Brain appears to be less affected than other tissues by EFA deficiency (Mohrhauer & Holman,

1963). However, when EFA deficient diets are fed in early stages of brain development and for prolonged periods of time, considerable changes are observed in the fatty acids of brain structural phospholipids (Galli et al., 1970), mainly consisting of a rise of the triene/tetraene fatty acid ratio (Holman, 1960). The time course of the changes induced by EFA deficiency in the fatty acids of brain phospholipids (White et al., 1971), their reversibility (White et al., 1971; Sun et al., 1975) and the effects on various brain subcellular membranes (Galli et al., 1972; Sun & Sun, 1974; Karlsson, 1975) have been studied in previous investigations. Also, the influence, on the levels of polyenes in the developing brain, of the maternal dietary history, in respect of EFA intake (Alling et al., 1974) and of the stores of PUFA in peripheral tissues (Galli et al., 1975) of the newborn been considered.

The typical changes in the levels of PUFA of brain ethanolamine phosphoglyceride observed in two-months old EFA deficient rats are shown in Table 3.

Some reduction of tetranoic fatty acids of the n-6 family (20:4 and 22:4) and of 22:6 n-3, and the prominent well known rise of 20:3 n-9 are observed, in the animals raised on the fat free diet.

Relative levels of PUFA of the n-6 and n-3 series in tissues, greatly depend also on the relative balance of linoleic and linolenic acids in the diet. Due to competitive inhibition in the elongation and desaturation of EFA to PUFA (Holman, 1964; Brenner & Peluffo, 1966) levels of arachidonic and docosahexaenoic acids in brain phospholipids are greatly modified by changing the relative ratio of the precursors in the diet (Galli et al., 1971; Alling et al., 1972). The effects of feeding fat free semisynthetic diets supplemented with 1% (w/w) of either an oil rich in n-6 fatty acids (safflower oil) or an oil rich in n-3 fatty acids (fish oil) to growing rats up to three months of age on PUFA levels of brain ethanolamine phosphoglyceride are shown in Table 4.

Changes in the relative levels of PUFA of the n-6 and n-3 series in tissue lipids of animals fed diets with different n-6/n-3 fatty acid ratios, are thus greater than those found in animals fed a fat free diet in comparison with the controls. This could be of practical significance, since linoleic/linolenic acid ratios are quite different in edible fats of common use.

Changes in the fatty acid composition of structural lipids in brain membranes have been reported to modify (Na+K)-ATPase in synaptosomes (Sun & Sun, 1975) and n-6 and n-3 fatty acids are considered important for the activity of certain membrane-bound enzymes (Bernsohn & Spitz, 1974). Modifications in learning

Table 3. Essential fatty acid (EFA) deficiency and polyunsaturated fatty acids (PUFA) in cerebral ethanolamine phosphoglyceride (EPG) in the rat

Two months old rats born from mothers fed either one of the two diets, starting one week before delivery and during lactation. The same diets were fed to the weaning animals up to the day of sacrifice.

Values are the average ± S.E. Number of animals in each group in brackets.

DIET	LEVELS OF DIETARY EFA (% of calories)		PUFA IN BRAIN EPG % of total fatty acids				
	linoleic	linolenic	20:3 n-9	20:4 n-6	22:4 n-6	22:5 n-6	22:6 n-3
Control (6)	3.4	0.85	0.1	12.0±0.5	7.1±0.3	2.6±0.4	17.0±1.2
Fat Free (6)	tr	--	3.7±0.1	10.2±0.2	5.0±0.2	2.8±0.1	15.2±1.7

Table 4. Dietary n-6/n-3 acid ratio and polyunsaturated fatty acids (PUFA) in cerebral ethanolamine phosphoglyceride (EPG) in the rat.

Values are the average ± S.E.

Number of animals in brackets.

DIET		LEVELS OF DIETARY FATTY ACIDS (% of calories)		PUFA IN BRAIN EPG			
		n-6	n-3	20:4 n-6	22:4 n-6	22:5 n-6	22:6 n-3
Safflower oil	(6)	2.1	--	13.2±0.5	7.4±0.3	7.5±0.8	14.2±0.9
Fish oil	(6)	0.01	0.9	8.1±0.3	3.3±0.2	0.6±0.1	27.2±1.6

behavior have also been reported in animals fed EFA deficient diets (Caldwell & Churchill, 1966; Paoletti & Galli, 1972; Galli et al., 1975) or diets with a different linoleic/linolenic acid ratio (Lamptey & Walker, 1976).

BRAIN PUFA AND PROSTAGLANDIN FORMATION

Polyunsaturated fatty acids with 20 carbon atoms deriving from linoleic and linolenic acids are precursors of biologically active substances, the prostaglandins and related compounds. The system which forms prostaglandins from arachidonic acid has been extensively studied in several laboratories. The formation of an endoperoxide intermediate (Hamberg & Samuelsson, 1973) in the synthesis of prostaglandins and thromboxane, a new family of compounds (Hamberg & Samuelsson, 1974) has been described by Samuelsson and coworkers. The rate limiting step in the overall formation of active products from arachidonic acid, is considered the release of the precursor, from phospholipids (Horton, 1975). This step involves activation of a phospholipase A_2 acting on the 2 position of glycerophospholipids. It is not known whether a specific phospholipid is the donor of arachidonic acid for prostaglandin synthesis, although release of arachidonic acid mainly from phosphatidylinositol has been reported in stimulated platelets (Schoene & Iacono, 1976).

Selective increase of free arachidonic acid (FAA) in brain is observed during the ischemia following decapitation (Bazan, 1970) or after administration of convulsant drugs and application of electroshock treatment (Bazan, 1971; Bazan & Rakowski, 1970). An enzymatic process is involved, since sacrifice of animals by means of focussed microwave irradiation of their heads, procedure which rapidly (3 seconds) inactivates brain enzymes (Guidotti et al., 1974), completely prevents the post-mortem changes of brain free fatty acid levels (Cenedella et al., 1975). Thus, the use of a focussed microwave oven for killing small laboratory animals allows a correct determination of the actual levels of brain FAA and prostaglandins, immediately formed from the released precursor, "in vivo". Prostaglandin formation from endogenous precursors in brain tissue has been described "in vitro" (Nicosia & Galli, 1975; Wolfe et al., 1975), whereas negligible conversion of exogenous labelled arachidonic acid to prostaglandins is observed, under the same conditions (Wolfe et al., 1967). Recently conversion of exogenous arachidonic acid to thromboxanes during incubation of brain tissues has been described (Wolfe et al., 1976).

The release of FAA and the formation of prostaglandins in different brain areas during ischemia "in vivo" have been studied in our laboratory (Bosisio et al., 1976). The results are reported in Table 5.

Table 5. Levels of free arachidonic acid (FAA) and of prostaglandins $F_{2\alpha}$ and E_2 in brain cortex and cerebellum of rats sacrificed by microwave irradiation (MW) and by decapitation followed by five minutes of ischemia

Values are the average ± S.E. Number of samples in brackets.

An internal standard of C20:0 was added to the lipid extract and FAA was then measured by GLC of the free fatty acid fraction isolated by TLC and analyzed as methylesters. Prostaglandins were measured by gas-chromatography mass spectrometry using deuterated prostaglandins as internal standards (Nicosia & Galli, 1975).

	BRAIN CORTEX		CEREBELLUM	
	MW	ISCHEMIA	MW	ISCHEMIA
FAA (μg/g)	(8) 10±1	(8) 52±5	(6) 12±2	(6) 17±4
$PGF_{2\alpha}$ (ng/g)	(6) 15±5	(6) 162±22	(8) 17±4	(6) 67±11
PGE_2 (ng/g)	(6) 14±5	(6) 94±5	(8) 14±2	(6) 49±14

Ischemia was induced by decapitation followed by removal of brain from the head kept at 37°after four minutes, and homogenization after five minutes.

A correlation between availability of FAA and formation of prostaglandins is observed. The higher the release of the precursor during ischemia (brain cortex), the higher is the accumulation of prostaglandins (mainly $PGF_{2\alpha}$). Levels of $PGF_{2\alpha}$ are greater than those of PGE_2, as previously reported "in vitro" (Nicosia & Galli, 1975; Wolfe et al., 1975). Greater formation of prostaglandins in brain cortex, in respect of other brain areas "in vitro" has been described by Wolfe et al. (1975).

The study of the time course of the release of FAA and of the formation of prostaglandins in brain cortical tissues "in vitro", shows different trends for the two processes (Table 6).

Active release of FAA occurs in the time interval between decapitation and the beginning of the incubation, followed by a less active release during the incubation assay, whereas

Table 6. Levels of FAA and of $PGF_{2\alpha}$ and PGE_2 in brain cortical slices during incubation

Values are the average ± S.E. obtained in the number of experiments shown in brackets. Determinations were carried out as described in the legend of Table 5.

			Microwave oven		Beginning of incubation		Thirty minutes of incubation
FAA	(μg/g)	(4)	16±2	(4)	129±8	(4)	188±16
$PGF_{2\alpha}$	(ng/g)	(4)	14±2	(4)	116±10	(4)	815±76
PGE_2	(ng/g)	(4)	13±2	(4)	12±2	(4)	288±41

prostaglandins are formed more actively during the period of incubation. It appears from the data that prostaglandins are formed after free arachidonic acid has been actively released. It is also possible that during the time interval between sacrifice and the beginning of the incubation (ischemia), deacylation of phospholipids proceeds very actively in concomitance with low reacylation of the free acids (Bazan, 1976), whereas better conditions for reacylation (temperature, oxygenation and availability of cofactors) are present during incubation. Similarly, formation of prostaglandins could be enhanced in optimalized conditions (incubation).

DIETARY EFA AND THE PROSTAGLANDIN FORMING SYSTEM IN BRAIN

Since tissue levels of arachidonic acid are affected by changes in total and relative supply of the precursor fatty acids, and since PUFA other than arachidonic acid are reported to interact with the cycloxygenase system (Lands et al., 1973), the possible effects of dietary manipulations on the prostaglandin forming system "in vivo" have been studied in our laboratory. Administration of an EFA deficient diet to growing rats reduces the rate of release of total free fatty acids and of FAA during ischemia (Galli & Spagnuolo, 1976). Also the administration to growing rats of semisynthetic equicaloric diets containing 10% (w/w) of fats with different linoleic/linolenic acid ratios results in different release of FAA

Table 7. Free arachidonic acid (FAA) levels (μg/g fresh tissue) in brain cortex and cerebellum after five minutes of ischemia in two-month old rats fed diets containing Linseed Oil (LO) or Sunflower Seed Oil (SSO)

Values are the average ± S.E.

Number of animals in brackets.

Ischemia was obtained as reported in Table 5.

DIETARY FAT	EFA AS % DIETARY CALORIES		FFA	
	linoleic	linolenic	Cortex	Cerebellum
LO	4.8	10.2	40.7±7.4 (6)	18.9±2.4 (6)
SSO	14.4	0.06	76.9±7.7 (6)	27.3±9.2 (6)

from brain cortex during ischemia (Table 7).

Release of FAA during ischemia is lower in brain cortex of animals fed a diet with a high linolenic/linoleic acid ratio. Release is lower in cerebellum than in cortex, as already reported in Table 6 and differences between the two groups are less pronounced in this brain area.

Since release of FAA from phospholipids is correlated with prostaglandin formation, changes of FAA release should result in variations of prostaglandin synthesis. Formation of prostaglandins in brain "in vivo" occurs during ischemia and after administration of convulsant drugs (Folco et al., in press). When convulsant drugs (penthylenetetrazol) are injected in animals fed diets with different EFA contents, differences in brain levels of prostaglandins (only $PGF_{2\alpha}$ has been measured) are observed 90 seconds after the injection (Table 8).

Administration of diets containing saturated fat or high linolenic/linoleic acid ratio, results in reduced formation of prostaglandins in brain, under stimulation, in respect to that observed in the brain of animals fed a more balanced EFA content.

Table 8. Levels of $PGF_{2\alpha}$ in brain cortex of rats fed various dietary fats after administration of penthylenetetrazol

Animals were sacrificed by focussed microwave irradiation 90 seconds after the injection of subconvulsive doses (50 mg/kg i.p.) of penthylenetetrazol.
Animals were fed the semisynthetic diets containing 10% (w/w) of either one of the listed fats, from one week before delivery up to three months of age.
$PGF_{2\alpha}$ was measured by radioimmunoassay (Lindgren et al., 1974).
Proteins were measured following the method of Lowry et al. (1951).

Number of animals in brackets. Values are the average ± S.E.

Dietary Fat		EFA as % of calories		$PGF_{2\alpha}$
		linoleic	linolenic	(ng/mg prot.)
Linseed oil	(8)	4.2	10.2	2.40±0.36**
Olive oil	(8)	1.4	0.2	4.10±0.40
Saturated fat	(8)	0.01	--	3.10±0.04*

Significance of differences from the "olive oil" group" :

* $p < 0.05$

** $p < 0.01$

The administration of a diet rich in linolenic and low in linoleic appears to be more effective than the EFA deficient diet in reducing prostaglandin synthesis.

CONCLUSIONS

It appears from the reported data that manipulations of dietary levels of EFA not only modify the PUFA content of structural phospholipids in brain, but also influence the rate of prostaglandin formation.

The dietary control of prostaglandin and, possibly, of thromboxane formation in tissues appears of general significance, especially in biological systems were a role of these compounds has been elucidated.

ACKNOWLEDGMENTS

This work was supported by CNR - Comitato Tecnologico - Rome, Contract No. 75.00851.11 and by CNR grant No. 75.00886.04.

The excellent technical help of Mr. Claudio Colombo and Miss Milena Blasevich is gratefully appreciated. The Authors wish to acknowledge Mrs Luigina Rossoni for the secretarial help in preparation of their manuscript.

REFERENCES

ALLING, C., BRUCE, Å., KARLSSON, I., SAPIA, O. & SVENNERHOLM, L. (1972), J. Nutr. 102, 773-782.

ALLING, C., BRUCE, Å., KARLSSON, I. & SVENNERHOLM, L. (1972) J. Neurochem. 23, 1263-1270.

BAZAN, N. G. (1970) Biochim. Biophys. Acta 218, 1-10.

BAZAN, N. G. (1971) J. Neurochem. 18, 1379-1385.

BAZAN, N. G. (1976) Chapter in this volume.

BAZAN, N. G. & RAKOWSKI, H. (1970) Life Sci. 9, 501-507.

BERNSOHN, J. & SPITZ, F. J. (1974) Biochem. Biophys. Res. Commun. 57, 293-298.

BOSISIO, E., GALLI, C., GALLI, G., NICOSIA, S., SPAGNUOLO, C. & TOSI, L. (1976) Prostaglandins 11, 773-782.

BRENNER, R. R. & PELUFFO, R. O. (1966) J. Biol. Chem. 241, 5213-5219.

CALDEWELL, D. F. & CHURCHILL, J. A. (1966) Psychol. Rep. 19, 99-102.

CENEDELLA, R. A. J., GALLI, C. & PAOLETTI, R. (1975) Lipids 10, 290-293.

DHOPESHWARKAR, G. A. & MEAD, J. F. (1969) Biochim. Biophys. Acta 187, 461-467.

DHOPESHWARKAR, G. A. & MEAD, J. F. (1970) Biochim. Biophys. Acta 210, 250-256.

FOLCO, G. C., LONGIAVE, D. & BOSISIO, E., Prostaglandins (in press).

GALLI, C., WHITE, H. B. Jr. & PAOLETTI, R. (1970) J. Neurochem. 17, 347-355.

GALLI, C., TRZECIAK, H. I. & PAOLETTI, R. (1971) Biochim. Biophys. Acta 248, 449-454.

GALLI, C., TRZECIAK, H. I. & PAOLETTI, R. (1972) J. Neurochem. 19, 1863-1867.

GALLI, C., AGRADI, E. & PAOLETTI, R. (1975) J. Neurochem. 24, 1187-1191.

GALLI, C., MESSERI, P., OLIVERIO, A. & PAOLETTI, R.(1975) Pharmacol. Res. Commun. 7, 71-80.

GALLI, C & SPAGNUOLO, C.(1976) J. Neurochem 26, 401-404.
GUIDOTTI, A., CHENEY, D. L., TRABUCCHI, M., DOTEUCHI, M., WANG, C. & HAWKINS, R. A. (1974) Neuropharmacol. 13, 1115-1122.
HAMBERG, M. & SAMUELSSON, B. (1973) Proc. Natl. Acad. Sci. USA 70, 899-903.
HAMBERG, M. & SAMUELSSON, B. (1974) Proc. Natl. Acad. Sci. USA 71, 3400-3404.
HOLMAN, R. T. (1960) J. Nutr. 70, 405-410.
HOLMAN, R. T. (1964) Fed. Proc. 23, 1602-1605.
HORTON, E. W. (1975) Scot. Med. J. 20, 155-160.
KARLSSON, I. (1975) J. Neurochem. 25, 101-109.
LAMPTEY, M. S. & WALKER, B. L. (1976) J. Nutr. 106, 86-93.
LANDS. W. E. M., Le TELLIER, P. R., ROME, L. H. & VANDERHOEK, J. Y. (1973) Adv. Biosci. 9, 15-21.
LINDGREN, J.A., KINDAHL, H.& HAMMARSTRÖM, S.(1974) FEBS Letters 48, 22.
LOWRY, H., ROSENBROUGH, N. J., FARR, A.L., RANDALL, R. J. (1951) J. Biol. Chem. 193, 265-275.
MEAD, J. (1976) Chapter in this volume.
MOHRHAUER, H. & HOLMAN, R. T. (1963) J. Neurochem. 10, 523-530.
NICOSIA, S. & GALLI, G. (1975) Prostaglandins 9, 397-403.
PAOLETTI, R. & GALLI, C. (1972) : In "Lipids malnutrition and the developing brain", pp. 121-140, Elsevier-Excerpta Medica-North Holland, Associated Scientific Publisher Co. Amsterdam.
ROUSER, G., FLEISCHER, S. & YAMAMOTO, A. (1970) Lipids 5, 494-496.
SCHOENE, N. W. & IACONO, J. M. (1976) : In " Advances in Prostaglandin and Thromboxane Research", Samuelsson B. & Paoletti, R., eds., vol. 2, pp. 763-776, Raven Press, N.Y.
SINCLAIR, A. J. & CRAWFORD, M. A.(1972) J. Neurochem. 19, 1753-1758.
SINCLAIR, A.J. (1975) Lipids 10, 175-184.
SUN, G. Y., GO, J. & SUN, A. Y. (1974), Lipids 9, 450-454.
SUN, G. Y., WINNICZEK, H., GO, J. & SHENG, S. L. (1975) Lipids 10, 365-373.
SUN, A. Y. & SUN, G. Y. (1976): In "Function and metabolism of phospholipids in the central and peripheral nervous system" (Porcellati, G, Amaducci, L. & Galli, C. eds.) pp. 169-197, Raven Press, N.Y.
WHEELER, T. G., BENOLKEN, R. M. & ANDERSON, R. E. (1975) Science 188, 1312-1314.
WHITE, H. B. Jr., GALLI, C. & PAOLETTI, R. (1971) J. Neurochem. 18, 869-882.
WOLFE, L. S., COCEANI, F. & PACE-ASCIAK, C.(1967):In"Prostaglandins Nobel Symp. 2", (Bergstrom, S. & Samuelsson, B. eds.) pp. 265-275, Almqvist & Wiksell.
WOLFE, L. S., PAPPIUS, H. M & MARION, J. (1975): In "Advances in Prostaglandin and Thromboxane Research" (Samuelsson, B. & Paoletti, R. eds.) vol. 1, pp. 345-355, Raven Press, N. Y.
WOLFE, L. S., ROSTWOROWSKI, K. & MARION, J. (1976), Biochem. Biophys. Res. Commun. 70, 907-913.
YAU, T. M. & SUN, G. Y. (1974) J. Neurochem. 23, 99-104.

ESSENTIAL FATTY ACIDS IN TESTES

J. G. Coniglio,* A. R. Whorton,+ and J. K. Beckman*

*Department of Biochemistry
+Department of Pharmacology
Vanderbilt University
Nashville, Tennessee 37232, USA

Burr & Burr (1929, 1930) originally observed that essential fatty acids are necessary for the maintenance of normal testicular function in rats. Since then the lipid composition of testes and the metabolism of testicular lipids have been studied by many investigators in an attempt to relate these to the development and function of the testis.

Holman & Greenberg (1953) reported the content of polyenoic acids of lamb, swine and beef determined by alkaline isomerization and ultra-violet spectrophotometry. In testis of lamb 15.6% of the fatty acids were hexaenoic and 10.3% were tetraenoic; in testis of hog 1.5% were hexaenoic and 8.8% were tetraenoic; in testis of beef 5.8% were tetranoic acids. Later, indirect evidence was obtained by Cole (1956) suggesting the presence of large amounts of highly unsaturated fatty acids in the testis of the golden hamster.

An increase with age (4 to 21 weeks) in the pentaenoic acid content of rat testis was reported by Aaes-Jorgensen & Holman (1958) and Aaes-Jorgensen (1958). A three-fold increase in pentaenoic acids of rat testis during the age period three weeks to three months was observed by Kirschman & Conoglio (1961). The pentaenoic acid was later shown to be a 22-carbon acid of the linoleic acid family, i.e., docosa-4,7,10,13,16-pentaenoic acid, by structure determination (Davis et al., 1966) and by biosynthesis in testis injected with ^{14}C-linoleate or with ^{14}C-arachidonate (Davis & Coniglio, 1966).

Large concentrations of polyunsaturated fatty acids were found in bovine and porcine testes by Holman & Hofstetter (1965),

Table 1. Polyenoic Fatty Acids of Testes From Various Species*

Fatty Acid	% of Total Fatty Acids							
	Rat	Mouse	Hamster	Rabbit	Dog	Chicken	Guinea Pig	Human
18:2n-6	6.4	5.0	4.4	11.8	4.2	1.4	18.0	5.6
18:3n-3	0.3	0.9	trace	3.1	0.4	0.2	7.9	0.7
20:3n-6	0.8	1.1	0.7	4.9	2.9	0.8	1.2	6.7
20:4n-6	13.7	10.4	11.3	8.4	13.3	11.9	5.0	13.4
22:4n-6	1.7	1.2	trace	1.9	5.7	10.0	0.6	2.1
22:5n-6	13.5	7.3	14.9	10.9	14.0	trace	3.0	0.4
22:6n-3	1.1	4.9	4.4	trace	1.6	trace	2.4	8.5

* Taken from Bieri & Prival (1965)

including two previously unreported fatty acids, docosa-10,13,16-trienoate and tetracosa-9,12,15,18-tetraenoate.

Fatty acids of testicular lipids of various genera were reported by Bieri & Prival (1965). These data are shown in Table 1. Arachidonic acid was present in relatively large quantities in testes of all animals, but the amount of the higher polyenes was more variable. Docosapentaenoic acid was present in large quantities in most species and docosahexaenoic acid was present in large quantities in a few. The testes of mice have significant quantities of each while those of the rooster have only a trace of these two but large quantities of docosatetraenoic acid. In addition to 22:6n-3 and arachidonic acid testis of the human has a large amount of 20:3n-6. The data of Bieri & Prival (1965) for humans are from accidental death victims (presumably adults). Analyses done in our laboratory on testes obtained at orchidectomy (Coniglio,Grogan & Rhamy, 1974) or in the same age group at autopsy (Coniglio et al., 1975) gave values similar to those of Bieri & Prival. However , only minor amounts of 20:3n-6 and 22:6n-3 were found in testes of infants (Coniglio et al., 1975).

Two 24-carbon polyenes, a tetraene and a pentaene, were detected in and isolated from rat testicular tissue by Davis et al.,(1966). The chemical characterization of these as 9,12,15,18-tetracosatetraenoic and 6,9,12,15,18-tetracosapentaenoic acids, respectively, were reported by Bridges & Coniglio (1970a), who also demonstrated their biosynthesis in rat testes injected with ^{14}C-linoleate or ^{14}C-arachidonate. These 24-carbon polyenes have been observed in testes of other animals in small concentrations. Evidence from hydrogenated testicular fatty acids suggests the presence in small amounts of 26-carbon polyenes.

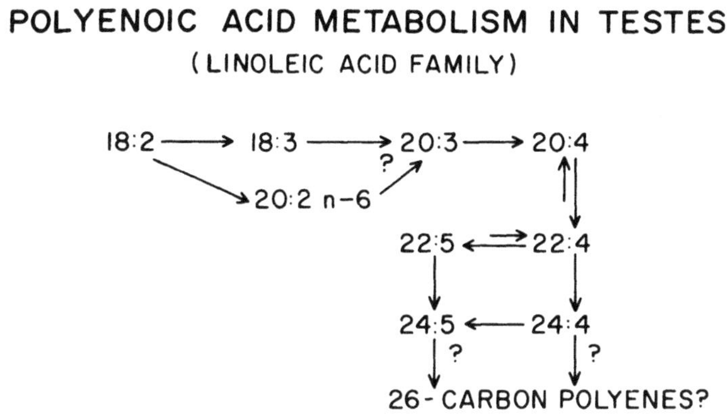

Fig. 1. Pathways of Polyenoic Acid Metabolism in Rat Testicular Tissue

Polyenoic acid biosynthesis in rat testes from linoleate may be summarized as shown in Figure 1. The desaturation of 18:2 to 18:3 and of 20:3 to 20:4 was shown in microsomes of testes by Ayala et al. (1973). It has not yet been established whether or not the pathway 18:2 to 20:2 to 20:3 is active in rat testis. The double arrows from 22:4 to 22:5 and from 20:4 to 22:4 indicate that the process of retroconvertion of 22:5 to 22:4 to 20:4 is active in rat testis (Bridges & Coniglio, 1970b; Ayala et al., 1973).

FACTORS AFFECTING POLYENE METABOLISM IN TESTES

Dietary Factors

The fatty acid composition of rat testes is dependent, at least in part, upon dietary fat. It is well knowns that a deficiency of dietary essential fatty acids leads to accumulation of 20:3n-9 and to decreases in concentrations of 18:2, 20:4 and 22:5 (all n-6) in testes of rats accompanied by sterility. The fatty acid composition of the testis responds to dietary fatty acid if the amount of the fatty acid in question is sufficiently high in the diet fed. However, metabolic derivatives of the fed fatty acid may be accumulated rather than the fed compound. Thus, Bieri & Prival (1965) found that feeding linoleate increased the amount of 22:5n-6 and feeding linolenate increased the amount of 22:6n-3. Rats fed for nine weeks a diet containing cod liver oil as 10% of the diet had increased amounts of 20:5 and 22:6 (n-3) compared to rats fed a diet containing corn oil (Coniglio, Bridges & Ghosal, 1974). The cod liver oil fed had as per cent by weight 9.3% 20:5 and 11.6% 22:6. Testes of rats fed rapeseed oil containing 6.7% 18:3n-3 contained smaller amounts of 22:5n-6 and larger amounts of 22:6n-3 than testes of rats fed corn oil (Coniglio, Grogan & Harris, 1974).

Feeding a fat-free diet to rats results not only in the fatty acid changes cited previously but also in increased synthesis of palmitic acid in rats injected intratesticularly with ^{14}C-acetate or in slices of testes incubated with ^{14}C-acetate. There was relatively less ^{14}C incorporated into the polyenes in the tissue from the deficient rats (Whorton & Coniglio, 1977). It was observed that there was no effect on fatty acid synthetase or on microsomal elongation enzymes, but there was an increase in the acetyl CoA carboxylase activity of testes of the fat-deficient rats.

A deficiency of vitamin A is known to affect lipid metabolism in testes of rats along with degeneration of the germinal tissue. Among the changes found by Krause & Beamer (1974) was a decreased concentration of docosapentaenoic acid in microsomes and in soluble fraction of homogenates, increased concentrations of stearic and oleic acids in the microsomal fraction and of linoleate and arachidonate in the soluble fractions.

Dietary deficiency of vitamin E also leads to degeneration of the seminiferous tubules. Changes in the fatty acid composition of testes of rats receiving a vitamin E-deficient diet have been reported. An increase in arachidonic acid and in 22:4n-6 and a decrease in 22:5n-6 were found by Bieri & Andrews (1964), who suggested that there might be a metabolic block in the conversion of 20:4 to 22:5 in vitamin E-deficiency. Witting et al. (1967) postulated that 22:5n-6 may be destroyed by peroxidation in the testis of the tocopherol-deficient rat, while Carney & Walker (1971)

Table 2. Polyenoic Fatty Acid Composition of Testicular Lipids of Rats Fed a Diet Deficient in or Supplemented with Vitamin E

	% of Total Fatty Acids			
	Group 1*		Group 2*	
Fatty Acid	Supplemented	Deficient	Supplemented	Deficient
18:2	9.9±0.5	10.2±0.5	9.8±1.2	9.1±1.3
20:4	14.0±0.2	13.9±0.4	14.9±0.7	17.9±0.8
22:4	1.8±0.1	1.5±0.1	2.0±0.2	3.6±1.1
22:5	18.7±0.2	16.3±0.4	15.1±0.8	6.6±1.1
> 22:5	3.6±0.2	4.0±0.3	2.9±0.6	2.3±0.9

* Mean ± S.E. Data from 6 rats

attributed the loss of 22:5 to increased turnover of 22:5, including retroconversion to arachidonate, since they found no block in the conversion of linoleate to 22:5 in vitamin E-deficient rats.

We have studied the metabolism of ^{14}C-arachidonate in testes of rats fed a diet containing stripped corn oil (tocopherol-deficient) with or without supplementation with vitamin E. Rats 4 to 6 months old and on the diet since birth were injected intratesticularly with 1-^{14}C-arachidonic acid-albumin complex (40 µl; 38 µg; 7.3 µc). Controls were given 200 mg tocopherol acetate/kg diet. The body weights and testicular weights of the tocopherol-deficient rats were lower than those of the supplemented rats (314 vs. 207; 2.8 vs. 1.7), but the testis weight to body weight ratio was similar. In Table 2 is shown the polyenoic acid composition of testes of rats which were on their respective diets for four to six months. Two groups of animals are presented. In group one there were no significant differences in the fatty acid composition between deficient and supplemented. Histologically, the testes of these deficient rats were almost normal. In the other group there were small but significant increases in testes of deficient animals in arachidonate and in 22:4n-6 and a relatively large decrease in 22:5n-6. These changes were smaller but of the same type reported by Bieri & Andrews (1964). Histologically, the testes of these deficient rats showed severe degeneration of the germinal cells.

Incorporation of ^{14}C from 1-^{14}C-arachidonate into fatty acids, including 22:5, was similar in the deficient and supplemented ani-

Table 3. ^{14}C Distribution in Total Fatty Acids of Testes of Rats Fed a Vitamin E-Supplemented or Deficient Diet and Injected Intratesticularly with 1-^{14}C-arachidonic Acid

	% of Total ^{14}C Recovered in Total Fatty Acids			
	24 Hours (2)		96 Hours (5)	
Fatty Acid	Deficient	Supplemented	Deficient	Supplemented
16:0	6.5±1.8	5.0±0.8	8.0±0.2	9.5±0.2
20:4	65.1±1.8	67.1±0.8	48.1±1.3	49.4±0.5
22:4	12.8±0.0	12.8±0.0	10.1±0.9	8.3±0.3
22:5	8.3±0.1	10.1±1.1	22.0±0.9	21.4±0.6
>22:5	5.8±2.6	4.6±0.9	11.6±0.5	11.4±0.7

* Mean ± S.E.

() Number of rats. The testes of the deficient rats showed minimal or no degeneration when examined histologically.

mals in the group in which there were no fatty acid changes in the deficient rats. These data are shown in Table 3. One group of rats was killed at 24 and the other at 96 hours after intratesticular injection of ^{14}C-arachidonate. The amount of ^{14}C in 20:4 was lower at 96 hours (in both groups) and the amount of ^{14}C in 22:5 and in 24-carbon polyenes was higher at 96 than at 24 hours. The results obtained on the rats showing severe testicular degeneration are shown in Table 4. In most of the experiments more ^{14}C was retained in 20:4 in testes of deficient rats than in testes of the supplemented. This was also true for 22:4 in the 72-hour groups. Smaller amounts of ^{14}C were found in 22:5 and 24-carbon polyenes of testes of the deficient rats than in the supplemented in the 72-hour group. In spite of the degenerated tissue, however, there appeared to be ample conversion of 20:4 to 22:4 and 22:5. Though there may have been a partial inhibition in the conversion of 20:4 to 22:5 in the 72-hour group, it is also possible that there was increased turnover of 22:5 in these testes, including retroconversion to 20:4,

Table 4. ^{14}C Distribution in Total Fatty Acids of Testes of Rats Fed a Vitamin E-Supplemented or Deficient Diet and Interjected Intratesticularly with 1-^{14}C-arachidonic Acid

	% of Total ^{14}C Recovered in Total Fatty Acids*							
	4 Hours (2)		8 Hours (3)		24 Hours (3)		72 Hours (2)	
Fatty Acid	Def.	Supp.	Def.	Supp.	Def.	Supp.	Def.	Supp.
16:0	1.6 ±0.6	4.2 ±0.3	2.2 ±0.02	4.9 ±0.3	5.4 ±1.8	5.4 ±1.1	12.2 ±2.9	9.7 ±0.5
20:4	83.5 ±0.5	72.5 ±2.5	75.1 ±2.5	77.0 ±3.7	70.4 ±2.6	62.0 ±4.7	56.0 ±3.7	51.2 ±2.4
22:4	4.8 ±0.8	8.3 ±5.7	9.7 ±2.4	10.3 ±0.3	14.3 ±1.4	13.3 ±3.3	15.8 ±1.0	10.5 ±0.6
22:5	3.3 ±0.3	3.0 ±1.6	6.4 ±1.1	6.1 ±0.7	6.4 ±1.8	7.7 ±1.8	8.7 ±0.9	15.9 ±0.7
>22:5	-	-	-	-	-	-	5.8 ±1.5	10.4 ±1.9

* Mean ± S.E.

() Number of rats. The testes of the deficient rats all showed gross degeneration when examined histologically.

Table 5. $^{14}CO_2$ Expired by Rats Fed a Vitamin E-deficient or Supplemented Died and Injected Intratesticularly with 1-^{14}C-arachidonic Acid

Time (Hours)	% of Injected Dose[a] Supplemented	% of Injected Dose[a] Deficient	% of Injected Dose/Gm Testis Supplemented	% of Injected Dose/Gm Testis Deficient
4	18.5±0.2[b]	16.9±1.1[b]	4.7±0.0	7.7±0.5
8	32.0±6.3[b]	27.5±2.0[b]	8.3±1.1	10.1±3.7
24	34.6±2.8[c]	35.7±1.2[c]	9.6±0.8	14.5±0.6
96	50.2±2.9[b]	47.2±3.8[b]	35.6±2.9	58.4±3.8

[a] Mean ± S.E.
[b] Two rats per group
[c] Three rats per group

as suggested by Carney & Walker (1971). In some preliminary studies in which biosynthesized ^{14}C-22:5n-6 was injected into testes of vitamin E-deficient and supplemented rats, we obtained results whic suggested increased turnover and increased retroconversion of 22:5 to 20:4 (Coniglio, Bridges & Ghosal, 1974). The amount of $^{14}CO_2$ expired by supplemented and deficient rats injected intratesticularly with 1-^{14}C-arachidonate was determined at several time periods following injection. The data are shown in Table 5. Although some of the expired $^{14}CO_2$ may have been formed by oxidation of ^{14}C compounds by tissues other than testicular tissue, it seems likely that most of it was the result of oxidation in the testis itself-particulary at the earlier time periods. The amount of $^{14}CO_2$ expired per gram testis was greater in all the deficient rats compared to the supplemented. These data support the concept of increased turnover of polyenoic fatty acids in the testis of the deficient rats.

Use of Eicosa-5,8,11,14-tetraynoic Acid (TYA)

Another factor which may affect polyenoic fatty acids in testes is the use of compounds which may inhibit or be competitive in metabolic reactions. Recently, it was reported by Coniglio et al. (1976) that the acetylenic analog of arachidonic acid, eicosa-5,8,11,14-tetraynoic acid (TYA), administered to rats maintained on a fat-free diet, had significant effects on the fatty acid composition of testicular lipids. The amount of arachidonic acid, of 22:5n-6 and of

Table 6. Polyenoic Fatty Acid Composition of Total Lipids From Testes of Rats Receiving Orally TYA in Coconut Oil or Coconut Oil Alone (Controls)

	% of Total Fatty Acids[a]							
	18:2	20:3n-9	20:3n-6	20:4	22:4	22:5n-6	22:6	24:4,24:5
TYA	10.1[c] ±0.18	0.7[b] ±0.16	5.0[c] ±0.67	9.2 ±1.1	0.9 ±0.25	4.6[b] ±0.51	2.4[b] ±0.26	2.1 ±0.41
Control	5.7 ±0.49	0.12 ±0.05	0.34 ±0.07	11.5 ±2.3	0.95 ±0.14	10.4 ±1.9	1.4 ±0.11	1.9 ±0.24

[a] Mean ± S.E. of six rats in each group
[b] P value < 0.01
[c] P value < 0.001

24-carbon polyenes decreased and the amount of oleic, of linoleic and of 20:3n-6 increased in testes of rats receiving TYA. These studies have been extended to rats maintened on a diet containing corn oil. The animals were 20-day old Sprague-Dawley rats maintained on a purified diet containing all essential nutrients and corn oil as 3% of the calories. Rats of one group were given by a stomach tube TYA suspended in 0.75 ml coconut oil twice weekly for a total of eight doses (600 mg in 30 days). The control rats were given by stomach tube 0.75 ml coconut oil twice weekly over the same time period. The fatty acid composition of the testis at this time is shown in Table 6. Values for saturated and monoenoic acids were not different in the two groups. A drastic decrease in 22:5 was noted in the rats receiving TYA along with increases in linoleate, 20:3n-9, 20:3n-6 and 22:6. Similar results were obtained in other organs, such as the liver, heart, kidneys and adrenals. These data are given in Table 7. These data suggest that the TYA might be interfering with the conversion of 18:2 to 20:4 and the further conversion of 20:4 to 22:5. The metabolism of ^{14}C-arachidonic acid was investigated in testes after intratesticular injection. About 25% of the injected dose was retained in lipids of the testes 24 hours after the intratesticular injection, and there was no significant difference between the TYA-fed and the controls. The distribution of the ^{14}C in the fatty acids of the testis is shown in Table 8. More of the ^{14}C was retained in 20-carbon (essentially all arachidonic acid) and less ^{14}C was retained in the 22-carbon fatty acids (22:4n-6 and 22:5n-6) of testicular lipids of the TYA-fed than of the control rats.

Table 7. Polyenoic Fatty Composition of Total Lipids From Organs of Rats Receiving Orally TYA in Coconut Oil or Coconut Oil Alone (Controls)

Fatty Acid	% of Total Fatty Acids[a]							
	Liver		Heart		Kidneys		Adrenals	
	TYA	Control	TYA	Control	TYA	Control	TYA	Control
18:2	17.1±0.3[d]	11.0±1.1	30.3±1.3[d]	20.6±1.6	14.0±0.7[d]	8.3±0.4	6.2±0.5	5.8±2.1
20:3n-9	2.4±0.4[c]	1.0±0.3	1.4±0.3[b]	0.6±0.2	1.2±0.4[b]	0.1±0.0	0.4±0.1[b]	0.1±0.1
20:3n-6	3.8±0.5[d]	1.1±0.2	2.6±0.6[d]	0.7±0.2	2.4±0.4[d]	0.5±0.1	1.1±0.1[c]	0.2±0.1
20:4	9.5±0.3[d]	20.5±0.2	10.1±1.3[c]	18.7±1.8	10.9±2.0[d]	18.1±1.7	20.9±1.8	24.7±6.7
22:4	0.1±0.1	0.2±0.1	0.8±0.4	0.6±0.3	0.8±0.4	1.0±0.4	0.7±0.3[b]	2.7±1.5
22:5n-6	0.5±0.3	1.4±0.4	0.1±0.1[c]	0.9±0.3	0.8±0.3	0.7±0.2	--	-
22:6	2.2±0.4	3.3±0.7	2.4±0.8	2.0±0.8	2.0±0.4	1.4±0.5	0.1±0.1	0.3±0.1

[a] Mean ± S.E. of six rats in each group
[b] P value < 0.05
[c] P value < 0.01
[d] P value < 0.001

These data support the concept that TYA inhibited the conversion of arachidonate to longer chain, more highly unsaturated derivatives.

Almost twice as much ^{14}C was expired as $^{14}CO_2$ by the rats given TYA than the controls (20.7±2.2 vs. 10.6±1.2% of the administered ^{14}C; P<0.01). There was also decreased incorporation of ^{14}C into testicular palmitic acid in these rats. These results suggest that although more ^{14}C-acetyl CoA was produced from ^{14}C-arachidonate in testes of rats given TYA than in controls, use of ^{14}C-acetyl CoA for fatty acid synthesis was inhibited while its oxidation to $^{14}CO_2$ was not. Alternatively, the increased amount of linoleate in testes of TYA-fed rats might be inhibiting fatty acid synthesis (Muto & Gibson, 1970).

Use of Mutants

The availability of rodent mutants is furthering studies of the role of lipids in testicular function. Results of analyses of testicular fatty acids of the quaking mouse indicated no differences in the amount of 22:6n-3 and 22:5n-6 between the mutant and the normal (Coniglio et al., 1975). The mutants are sterile because of faulty spermatid differentiation (Bennett et al., 1971). The fatty acid data are consistent with the hypothesis that the 22-carbon polyenes are associated with the formation of spermatids rather than with their further differentiation into spermatozoa.

Contrasted with this is the mutant called the steel mouse, which has defects in hair pigmentation, hematopoiesis and gametogenesis.

Table 8. Distribution of ^{14}C in Hydrogenated Derivatives of Fatty Acids of Testicular Total Lipid

	% of Total CPM Recovered[a]	
Hydrogenated Derivative	TYA	Control
16:0	1.8±0.6[b]	7.8±0.9
18:0	1.2±1.2	1.7±0.8
20:0	78.0±1.4[c]	63.2±1.8
22:0	9.7±0.6	18.6±2.3
24:0	9.3±2.0	8.5±0.5

[a] Mean ± S.E. of four rats in each group
[b] P value < 0.01
[c] P value < 0.001

Table 9. Fatty Acid Composition of Testes of Steel Mutant and Non-mutant Mice

Fatty Acid	% of Total Fatty Acids	
	Non-mutant	Mutant
14:0	0.6±0.1	1.5±0.3
16:0	30.7±1.0	20.7±0.6
16:1	1.8±0.2	2.8±0.2
18:0	7.2±0.4	9.6±0.7
18:1	13.9±0.5	25.2±1.2
18:2	4.8±0.4	9.9±2.5
20:3	1.2±0.1	0.7±0.1
20:4	12.3±0.7	9.3±1.0
22:4	1.3±0.1	1.5±0.1
22:5	14.6±0.8	6.6±0.7
22:6	7.7±0.4	4.5±0.3
24:?	2.2±0.3	2.1±0.3

* Mean ± S.E. of eight rats in each group.

The absence of germ cells in gonads of mutant embryos was shown by Bennett (1956). In the testis of the adult, spermatogenic cells are absent although normal Sertoli cells are found. Our lipid analyses of testes of the mutant compared with the non-mutant showed a significant elevation of esterified cholesterol in the testis of the mutant. The ratio of esterified cholesterol to unesterified cholesterol was 1.7 for mutants and 0.6 for non-mutants. The fatty acid composition is given in Table 9. The testes of the mutants had more oleic and linoleic and less palmitic, arachidonic, 22:5 and 22:6 acids than testes of the non-mutants. The data support the concept that the 22-carbon polyenes are associated with the germinal cells in the testis. The low value for palmitic acid in the testicular lipids of the mutants is similar to findings in the undeveloped testis which has degenerated germinal epithelium such as the surgical crytorchidic testis and may indicate that palmitic acid and the 22-carbon polyenes may be esterified in the same phospholipid molecule.

The final study to be described concerns the fatty acid composition of testicular lipids of a rat mutant-the restricted color (H^{re}) rat. The germinal epithelium of the testis of this mutant develops more slowly than and not as extensively as that of the non-mutant control (Gumbreck et al., 1972). These investigators reported

Table 10. Fatty Acid Composition of Testes of the H^{re} Mutant Rats and of the Non-Mutant Controls of Various Ages

Fatty Acid	% of Total Fatty Acids									
Age	41 Days		60 Days		80 Days		100 Days		239-269 Days	
	C(2)[a]	H^{re}(2)	C(2)	H^{re}(2)	C(2)	H^{re}(2)	C(3)	H^{re}(3)	C(5)	H^{re}(4)
16:0	26.7±2.1	29.7±1.6	34.8±3.0	33.6±0.8	34.8±1.4	44.8±5.6	34.1±2.3	33.4±1.1	38.1±2.0	40.6±3.5
16:1	5.9±1.7	3.9±1.3	3.0±0.1	3.7±1.2	2.0±0.3	2.7±1.2	2.1±0.2	2.2±0.6	1.2±0.2	0.9±0.1
18:0	7.9±0.3	9.1±2.0	6.8±0.8	7.1±0.8	8.4±2.1	6.9±1.0	4.2±0.3	5.5±0.3	6.1±0.5	10.3±1.0
18:1	21.5±0.3	23.1±1.1	11.1±0.1	14.2±2.0	12.8±0.5	15.9±2.5	11.5±1.1	13.4±0.6	13.1±0.9	17.1±0.7
18:2	6.8±2.1	16.7±2.4	5.0±0.6	7.2±1.6	7.3±0.5	7.3±1.3	5.5±0.2	6.8±1.5	4.9±1.2	4.2±0.4
20:3	0.5±0.1	(n.d.)[b]	1.4±0.7	0.3±0.3	1.8±0.5	0.5±0.1	0.9±0.2	0.8±0.2	1.0±0.2	0.5±0.1
20:4	19.0±4.0	8.2±0.6	16.5±0.2	13.0±1.9	16.3±0.2	11.9±0.7	17.2±0.9	14.5±1.1	14.8±0.9	16.8±1.9
22:4	0.9±0.2	0.8±0.3	1.3±0.2	1.2±0.2	1.8±0.2	0.7±0.5	1.7±0.1	1.1±0.2	1.3±0.4	1.0±0.5
22:5	7.6±3.3	4.2±1.0	16.3±0.3	13.0±4.2	15.3±0.2	10.2±1.7	18.2±1.0	17.8±1.1	16.2±0.7	8.4±2.1
24:4 24:5	1.5±0.7	1.5±0.3	1.5±0.6	2.8±0.6	1.2±0.3	1.6±0.2	1.6±0.2	1.8±0.5	1.7±0.2	0.9±0.2

a Mean ± S.E. Number in parenthesis is number of rats in the group. C is non-mutant

b n.d. is not detectable

that spermatozoa were observed at 45 days of age in both mutants and non-mutants but were less numerous in mutants. When spermatogenesis was present in H^{re} rats, it was greatest between 45 and 90 days and was reduced or absent in older animals. We have analyzed the fatty acid composition of the testes of H^{re} and non-mutant control rats of several ages and the results are shown in Table 10. In the 41-day sample there was hitologically a denser population of germinal cells in the non-mutants. The concentrations of 20:3n-6, 20:4 and 22:5 were lower in the testes of the mutants than in those of the controls, while the concentration of linoleic acid was much higher. Histological studies were not made on the 60-day animals, but the concentrations of the polyenoic acids of the two groups were more similar than at 41 days. In the 80-day group one mutant had noticeably less sperm in the testes than did its normal control, but the second mutant had an almost normal pattern. The fatty acid analyses showed differences similar to the 41-day group except for linoleic acid, the concentration of which was not different between the two groups. At 100 days there were essentially no differences in the concentrations of the polyenoic acids of testes of mutants and non-mutants. Histologically, there was very little difference in the population of germinal cells in the two groups. In the last group (239-269 days) the main difference in the two groups was the smaller content of 22:5 (8.4 vs. 16.2) in the testis of the mutant. Histologically, the germinal epithelium of one H^{re} rat was severely degenerated while the testes of the other two H^{re} rats were noticeably less populated by germinal cells than were the testes of the controls. Thus, the concentration of 22:5n-6 and the presence or absence of germinal cells correlate well in this mutant.

SUMMARY

The polyenoic acids derived from linoleic or α-linolenic acid are prominent components of testicular tissue of various animals. In the rat the 22-carbon pentaenoic acid, derived from linoleic via arachidonic acid and 22:4, triples in concentration during sexual maturation and appears to be associated with the appearance and development of spermatids. Further testicular metabolism of the 22:5 include its oxidation to CO_2, its conversion to 24-carbon polyenes and its retroconversion to arachidonic acid.

Dietary factors affecting the fatty acid composition of and metabolism in rat testes include: a fat-free diet, which results in testicular accumulation of 20:3n-9 and loss of 18:2, 20:4 and 22:5 (all n-6) and increased activity of acetyl CoA carboxylase; various dietary oils, such as cod liver oil and rapeseed oil, which result in testicular accumulation of the n-3 polyenoic acid derivatives of α-linolenic acid; feeding a vitamin E-deficient diet, which results in a decreased concentration of 22:5n-6 and increased concentrations of 20:4 and 22:4n-6 in those testes which show severe degeneration of the germinal epithelium. The decreased concentration of 22:5n-6 may be due to increased turnover, including retroconversion to 20:4 rather than to a block in the conversion of 20:4 to 22:5; feeding eicosa-5,8,11,14-tetraynoic acid (TYA), the acetylenic analog of arachidonic acid, which results in decreased concentrations of 20:4 and 22:5 and increased concentrations of 18:2 in testes and several other organs of the rat. TYA interfered with the conversion of intratesticularly injected 1-^{14}C 20:4n-6 to 22:5n-6, indicating that the decreased concentration of 22:5 was due to inhibition of the conversion of arachidonic acid to the docosapentaenoic acid.

The apparent relationship of the 22-carbon polyenoic acids to the development of the germinal cells in the rodent was studied further by determining the fatty acid patterns of testicular lipids of mutant mice and rats,which are entirely or relatively sterile. The quaking mouse, which is sterile because of a defect in the differentiation of the spermatids to spermatozoa, has similar fatty acid patterns in testes of the mutant and the non-mutant. Therefore, the spermatids have a normal complement of 22:6n-3 and 22:5n-6 at that stage of development. The steel mouse, a sterile mutant which has testicular tissue largely devoid of germinal cells but has a normal complement of Sertoli cells, has greatly decreased amounts of 22:6n-3 and 22:5n-6. The H^{re} rat, a mutant which has a slower development of germinal cells than the normal non-mutant, which is fertile only for a short time and which becomes sterile in its later adult life, shows changes in the testicular concentration of 22:5n-6 which parallel the presence or absence of significant numbers of germinal cells. The presence of the 22-carbon polyenoic acids in testes of these rodents is apparently important to the development of germinal cells from the spermatogonia or spermatocyte stage to the spermatid stage.

ACKNOWLEDGEMENTS

These studies have been supported, in part, by Research Grants Nos. HD-06070, HD-07694, and AM-06483 from the United States Public Health Service. We are grateful to Hoffman-La Roche, Nutley, New Jersey for the TYA.

REFERENCES

AAES-JORGENSEN, E. (1958) J. Nutr. 66, 465-483.
AAES-JORGENSEN, E. & HOLMAN, R. T. (1958) J. Nutr. 65, 633-641.
AYALA, S., GASPAR, G., BRENNER, R. R., PELUFFO, R. O. & KUNAU, W. (1973) J. Lipid Res. 14, 296-305.
BENNETT, D. (1956) J. Morphol. 98, 199-233.
BENNET, W. I., GALL, A. M., SOUTHARD, J. L. & SIDMAN, R. L. (1971)

Biol. Reprod. 5, 30-58.
BIERI, J. G. & ANDREWS, E. L. (1964) Biochem. Biophys. Res. Commun. 17, 115-119.
BIERI, J. G. & PRIVAL, E. L. (1965) Comp. Biochem. Physiol. 15, 275-282.
BRIDGES, R. B. & CONIGLIO, J. G. (1970a) J. Biol. Chem. 245, 46-49.
BRIDGES, R. B. & CONIGLIO, J. G. (1970b) Biochem. Biophys. Acta 218, 29-35.
BURR, G. O. & BURR, M. M. (1929) J. Biol. Chem. 82, 345-367.
BURR, G. O. & BURR, M. M. (1930) J. Biol. Chem. 86, 587-621.
CARNEY, J. A. & WALKER, B. L. (1971) Nutr. Reports Int. 4, 103-108.
COLE, B. T. (1956) Proc. Soc. Expt'l. Biol. Med. 93, 290-294.
CONIGLIO, J. G., BRIDGES, R. B. & GHOSAL, J. (1974) La Rivista Italiana Delle Sostanze Grasse LI, 312-314.
CONIGLIO, J. G., BUCH, D. & GROGAN, W. M.,Jr. (1976) Lipids 11, 143-147.
CONIGLIO, J. G., GROGAN, W. M.,Jr. & HARRIS, D. G. (1974) Proc. Soc. Expt'l. Biol. Med. 146, 738-741.
CONIGLIO, J. G., GROGAN, W. M.,Jr., HARRIS, D. G. & FITZHUGH, M. L. (1975) Lipids 10, 109-112.
CONIGLIO, J. G., GROGAN, W. M.,Jr. & RHAMY, R. K. (1974) J. Reprod. Fert. 41, 67-73.
CONIGLIO, J. G., GROGAN, W. M.,Jr. & RHAMY, R. K. (1975) Biol. Reprod. 12, 255-259.
DAVIS, J. T. & CONIGLIO, J. G. (1966) J. Biol. Chem. 241, 610-612.
DAVIS, J. T., BRIDGES, R. B. & CONIGLIO, J. G. (1966) Biochem. J. 98, 342-346.
GUMBRECK, L. G., STANLEY, A. J., ALLISON, J. E. & EASLEY, R. B. (1972) J. Expt'l. Zoology 180, 333-350.
HOLMAN, R. T. & GREENBERG, S. I. (1953) J. Am. Oil Chem. Soc. 30, 600-601.
HOLMAN, R. T. & HOFSTETTER, H. H. (1965) J. Am. Oil Chem. Soc. 42, 540-544.
KIRSCHMAN, J. C. & CONIGLIO, J. G. (1961) Arch. Biochem. Biophys. 93, 297-301.
KRAUSE, R. H. & BEAMER, K. C. (1974) J. Nutr. 104, 629-637.
MUTO, Y. & GIBSON, D. M. (1970) Biochem. Biophys. Res. Commun. 38, 9-15.
WHORTON, A. R. & CONIGLIO, J. G. (1977) J. Nutr. (In Press).
WITTING, L. A., LIKHITE, V. N. & HORWITT, M. K. (1967) Lipids 2, 103-108.

MEMBRANE STRUCTURE AND FUNCTION WITH DIFFERENT LIPID-SUPPLEMENTED DIETS

R. N. Farías* and R. E. Trucco**

*Instituto de Química Biológica, Facultad de Bioquímica, Química y Farmacia, Universidad Nacional de Tucumán, San Miguel de Tucumán, Tucumán, Argentina
**Centro de Investigaciones de Tecnología Pesquera, Buenos Aires and Mar del Plata, Argentina

INTRODUCTION

In 1966 we made the observation that the (Na^+, K^+)-ATPase and (Mg^{2+})-ATPase from rat erythrocyte of animals fed with a commercial diet showed a lesser degree of cooperativity as compared with animals fed with the same diet supplemented either with lard or cod liver oil (Farías, 1967). The work performed by our group in the last ten years confirmed our early hypothesis that changes in the lipid composition of the cell membranes modify the regulatory activity of the membrane-associated enzymes (Farías et al., 1975). Conclusive evidence has been accumulated indicating that allosteric transitions and modification of other kinetic parameters of a given enzyme could be correlated with changes in the "conformation" or "state" of the protein involved. The determination of these kinetic parameters in membrane-associated enzymes may give some clues as to how these enzymes are regulated by the membrane. On the other side, changes in the behavior of enzymes may indicate if some modifications in membrane structure take place under special situations such as changes in lipid composition or action of hormones (Massa et al., 1975; Moreno & Farías, 1976; De Mendoza et al., 1977) or other chemical agents (Domenech et al., 1976).

The modification of the regulatory activity of the membranes on its bound enzymes has the particularity of being highly specific for each particular enzyme since it depends on the alteration of the lipid environment surrounding the enzyme. The modification of the lipid environment through feeding conditions led to some particular protein-lipid interaction in each membrane for each enzyme obtained from rats fed a different lipid supplemented diet. The experimental evidence supporting the hypothesis described will be presented.

RESULTS

Kinetic changes of erythrocyte membrane-bound acetylcholinesterase from rats fed fat-free diet

The mechanism for the modification of the kinetic parameters under the influence of the dietary lipids will be easily understood by describing the studies performed with the acetylcholinesterase from rat erythrocytes. The allosteric behavior (Hill plots) and temperature-dependent activity (Arrhenius plots) of the enzyme from rats fed a fat-free diet will be discussed in detail in the first part of this presentation.

a) Hill plots

An early observation indicated that the Hill coefficient for the inhibition by F^- of the erythrocyte acetylcholinesterase had a value of 1.6 for the animal fed corn-oil supplemented diet and of 1.0 for those fed with fat-free diet (Fig. 1). When the acetylcholinesterase from both groups of animals were solubilized with Triton X-100, it was observed that while the value of n remained

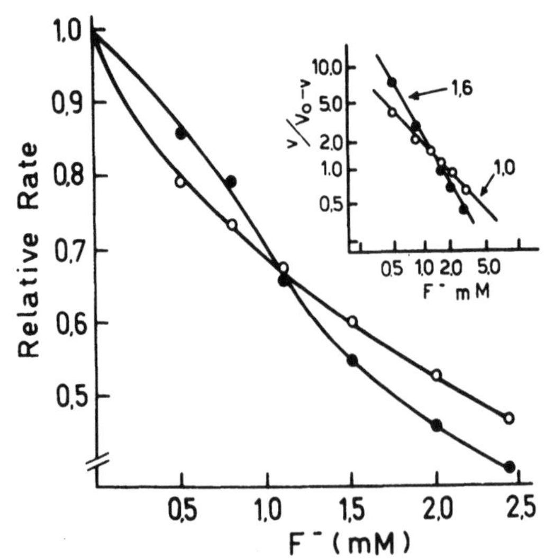

Fig. 1. Effect of F^- on the acetylcholinesterase from rats fed a fat-sufficient (•-•) and a fat-deficient (o-o) diet. Insert: Hill plot.

Table 1. Rf values of soluble acetylcholinesterase from rats fed fat-free and fat supplemented diets

Rf values were calculated by taking as reference the distance that the tracking dye (bromophenol blue) moved. The results are given as the mean ± S. E. The number of cases are in parentheses.

Solubilization treatment		Diet[a] Fat-sufficient		Fat-deficient	
Triton X-100	0.04 %[b]	0.15 ± 0.01	(4)	0.14 ± 0.01	(3)
	0.12 %	0.13 ± 0.01 0.19 ± 0.02	(10)	0.15 ± 0.02 0.22 ± 0.01	(6)
	0.4 %	0.24 ± 0.01	(7)	0.27 ± 0.02	(3)
Deoxycholate	10 mM	0.40 ± 0.01	(10)	0.41 ± 0.01	(12)
Ultrasound		0.21		0.20	

a. The Rf values from rats fed a fat-sufficient and a fat-deficient diet are not significantly different in any of the solubilization treatments.

b. The Rf values for 0.04 % and 0.4 % Triton X-100 concentrations are significantly different ($p < 0.001$)

(Adapted from Martínez de Melián et al., 1976)

constant for the enzyme from the fat sufficient animals, it increased to 1.6 in the case of the soluble enzyme from the fat deficient animals (Morero et al., 1973).

The electrophorectic patterns of acetylcholinesterase (solubilized by different agents such as Triton X-100 at different concentrations, sodium deoxycholate or ultrasonic irradiation) indicated the presence of multiple enzymatic forms, the chemical nature of which are only partially understood. However, the multiple forms for either the enzyme from the sufficient or deficient animals, respectively, were not significantly different (Martínez de Melián et al., 1976). These results are presented in Table 1. A single band of activity in polyacrylamide-gel, independent upon the method employed, can be obtained under controlled and standard conditions.

These facts indicated that only one isoenzymatic form of acetylcholinesterase is present in the rat erythrocyte membrane independently on the diet they were fed with.

The reported experiments seem to indicate that the membrane integrity may be responsible for the manifestation of the effect of the lipid composition on the modification of the value of n. By dialysis of the 100,000 g supernatant of the Triton solubilized membrane against a phosphate buffer in the presence of $MgCl_2$ and $CaCl_2$, a reaggregation material was obtained, which could be pelleted by centrifugation at 30,000 g, resuspended in sodium phosphate and assayed for acetylcholinesterase (Morero et al., 1973). In Table 2, the values of n are presented for the intact membrane, solubilized membrane, reaggregated material and Triton X-100 solubilized reaggregated material from animals fed corn oil supplemented diet and fat-free diet, respectively. As it can be observed, the value of n is around 1.6 for all the enzyme preparations from the fat sufficient animals. For the preparations from the fat deficient animals instead it is in the order of 1.0 in the case of intact membrane and reaggregation membrane like material, and around 1.6 for the solubilized preparations. Similar results were obtained with the other solubilizing agents (Martínez de Melián et al., 1976). Further evidence on the role played by the membrane was obtained when the acetylcholinesterase band from the gel electrophoresis of solubilized acetylcholinesterase from fat sufficient animals was eluted, mixed with lipid extracted from red cell membranes of rat fed a fat-free diet, and diffused against buffer. A value of n = 1.0, which corresponds to that of the intact membrane from the deficient

Table 2. Effect of solubilization and formation of membrane-like material on the values of n

Enzyme preparation	Diet	
	Fat-sufficient	Fat-deficient
Erythrocyte membrane	1.7	1.0
Supernatant of erythrocyte membrane solubilized with Triton X-100	1.6	1.6
Material reaggregated by dialysis	1.6	1.0
Supernatant at the treatment of re-aggregated material with Triton X-100	1.6	1.7

(Adapted from Morero et al., 1972)

animals, was obtained for the reaggregated material, while, in the case of the soluble enzyme eluted from the polyacrylamide gel, a value n = 1.6 was found (Martínez de Melián et al., 1976).

b) Arrhenius plots

Another kinetic parameter of the erythrocyte membrane-bound acetylcholinesterase, the Arrhenius plot, was affected in fat deficient animals (Bloj et al., 1974). It was found that this plot of membrane-bound acetylcholinesterase from rats fed corn-oil diet has a breaking point at about 20 C at pH 8.0 with lower activation energy at higher temperatures. The enzyme from rat fed fat-free diet exhibited a breaking point at about 28 C, the activation energies being lower than that of the enzyme from fat-sufficient animals. With the Triton solubilized enzyme from the fat-deficient animals, a shift in the breaking point and an increase in the activation energies were observed and no changes were detected with preparations from fat-sufficient animals. Thus, the soluble form of the enzyme from both groups become identical. (Table 3 summarizes these results). After reconstitution of membrane-like material from the soluble fat-deficient preparation, the characteristic enzymatic behavior was restored (Fig. 2).

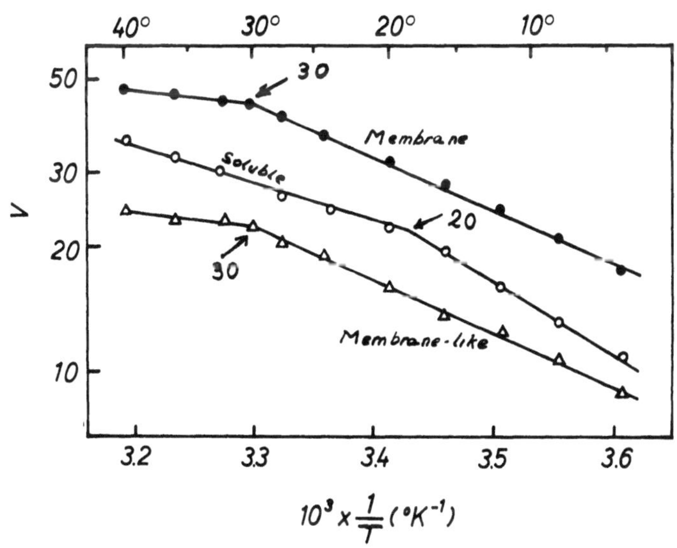

Fig. 2. Effect of membrane reconstitution on the Arrhenius plot of acetylcholinesterase from rats fed fat-free diet. Enzyme in the original membrane (•); in the solubilized state (o) and in the reconstituted vesicles (Δ). Initial velocity was expressed in an arbitrary scale.

Table 3. Breakpoint temperatures and activation energies of acetylcholinesterase from rat fed fat-sufficient and fat-deficient diets[a]

Diet	Breakpoint temperature (°C)		Activation energy in the temperature interval (°C)					
			4 - 20		20 - 28		28 - 40	
Fat-sufficient	19.9	1.2	7.5	0.3	4.1	0.4	4.1	0.4
Fat-deficient	28.8	0.8	6.0	0.2	6.0	0.2	1.9	0.4

[a]. Mean values in each of the different columns differ significantly ($p < 0.001$) and are the average of at last nine animals.

(Adapted from Boj et al., 1974)

Influence of fat-free diet on cooperativity of several membrane-bound enzymes

The values of the Hill coefficients for several membrane-bound enzymes from different sources (tissues) under the action of different types of effectors have been found to differ when the animals are fed either with an essential fatty acid supplemented diet or with a fat-free diet, respectively. The results which have been presented elsewhere (Farías et al., 1975) are summarized in Table 4.

Relationship between membrane lipid fluidity and enzyme cooperativity

a) Changes in fatty acid composition

The mixed fatty acid composition of rat membrane erythrocyte was dependent on the nature of the lipid supplement of the diet which particulary influence the total double bonds in the membrane (Bloj et al., 1973; Rahm & Holman, 1964; van Golde & van Deenen, 1966).

The Hill coefficient for the inhibition by F^- of the erythrocyte acetylcholinesterase obtained from seven groups of rats (five fed with diets supplemented with hydrogenated fat, lard, linseed oil, olive oil and corn oil, respectively, one fed with fat-free diet, and another fed with commercial standard diet) was determined (Bloj et al., 1973). Values ranging from 0.9 in the case of lard-supplemented diet to 1.6 in the case of corn-oil supplemented diet were obtained. Similar studies were performed for the inhibition by $F^-$of the ATPase

Table 4. Values of n for several membrane bound enzymes of different organs

Enzymes	Source	Effector		n values[a]	
		Type	Action	Fat-suf.	Fat-free
(Na^+, K^+)-ATPase	Red cell	Na^+	Activator	2.9	2.0
		K^+	Activator	2.1	1.4
		F^-	Inhibitor	2.8	1.6
(Mg^{2+})-ATPase	Red cell	F^-	Inhibitor	2.1	1.4
	Heart	F^-	Inhibitor	1.9	1.0
	Kidney	F^-	Inhibitor	1.9	1.0
	Brain	F^-	Inhibitor	1.6	1.8
(Mg^{2+}) p-nitrophenyl phosphatase	Red cell	F^-	Inhibitor	1.7	1.3
(Mg^{2+}, K^+) p-nitrophenyl phosphatase	Red cell	K^+	Activator	1.9	1.0
Acetylcholinesterase	Red cell	F^-	Inhibitor	1.6[b]	1.0
(Mg^{2+}, Ca^{2+}) ATPase	Red cell	Mg^{2+}	Activator	2.0[b]	1.3

[a] The values for the two groups are significantly different ($p < 0.01$ at least) except for brain microsomes which are not statistically different.

[b] Corn oil was used as EFA-supplement. In the other cases lard was used.

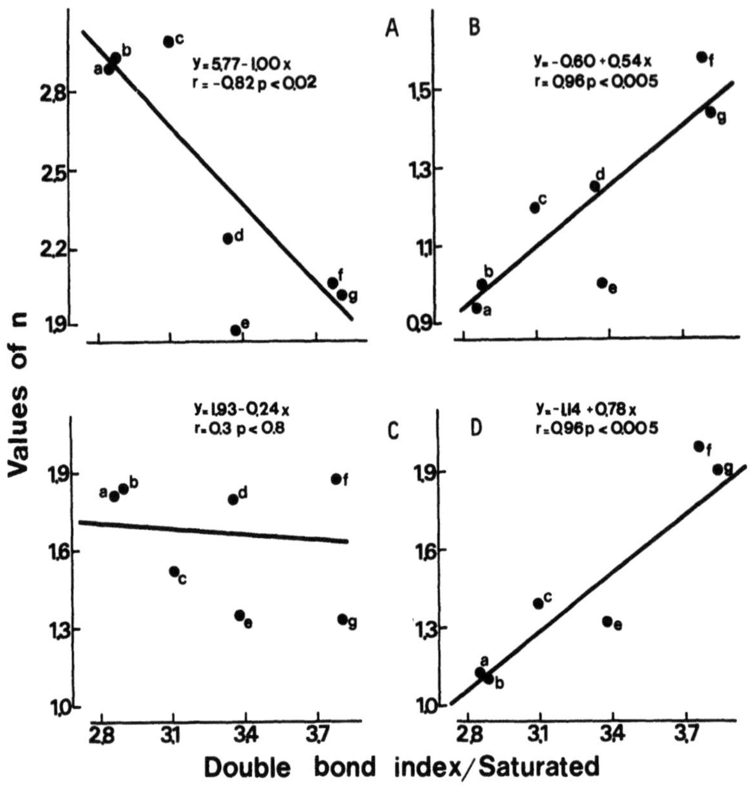

Fig. 3. Scattergram of the correlation between the values of n (Hill coefficient) and the ratio double bond index per saturated fatty acids from membrane erythrocytes lipid. The equation of the regression line and overall correlation coefficient r with its significance are included A, (Na^+, K^+)-ATPase; B, acetylcholinesterase; C, (Mg^{2+})-ATPase; D, (Ca^{2+}, Mg^{2+})-ATPase. Diet supplements: a, hydrogenated fat; b, lard; c, linseed oil; d, olive oil; e, fat-free; f, corn oil and g, standard diet. (Adapted from Bloj et al., 1973a and Galo et al., 1975).

of the erythrocyte membrane and, in the case of the (Na^+, K^+)-ATPase, values ranging form 2.1 for rat fed with corn oil supplemented diet to 3.0 for those fed with linseed oil supplemented diet were obtained.

The possibility that the modification of the cooperativity was due to direct action of some fatty acid family was excluded since no correlation could be established between the value of n and the amount of the members of any of the fatty acid families (n - 3), (n - 6), (n - 7) or (n - 9). In addition, different allosteric behavior for the inhibition by F^- of the (Mg^{2+})-ATPase and (Na^+, K^+)-ATPase of membrane erythrocyte from rats fed a fat-free diet or hydrogenated fat supplemented diet respectively, were obtained (Bloj et al., 1973). This latter fact rules out the possibility that the changes of the Hill coefficient in the seven groups of rats studied were dependent only on the essential fatty acid deficiency.

The possibility was then considered that the dietary lipid effect could be produced through changes in the membrane fluidity. Since the fluidity of a lipid may be more directly dependent on the total number of double bond index/saturated fatty acid ratio from membrane fatty acid ratio composition of each group was calculated and plotted against the value of n for the inhibition by F^- of the enzymes. As it can be seen in Fig. 3, in the case of animals with a higher double bond index/saturated fatty acid ratio, the inhibition by F^- of the (Na^+, K^+)-ATPase exhibited lower values of n (Fig. 3A). This relationship was inverse for the acetylcholinesterase (Fig. 3B). (Mg^{2+})-ATPase did not display any significant correlation (Fig. 3C). Fig. 3D shows that in the case of activation by Mg^{2+} of the erythrocyte (Ca^{2+}, Mg^{2+})-ATPase in the presence of Ca^{2+}, the values of n increased with the double bond index/saturated ratio (Galo et al., 1975). The assumption that the double bond lipid index/saturation ratio could be taken as an indicator of membrane fluidity was supported by physical studies on membranes (Engleman, 1971; Esfahani et al., 1971; Raison et al., 1971; Williams et al., 1972). Saturated fatty acids tend to render to membrane less fluid, while unsaturated fatty acids have the opposite effect.

b) <u>Changes in cholesterol content</u>

It is known that cholesterol interacts with erythrocyte phospholipids reducing its molecular area (Demel et al., 1967) and, as a consequence, a decrease in the local fluidity of the lipid matrix of the erythrocyte membrane occurs, as it was determined by electron spin resonance studies (Kroes et al., 1972). The results obtained for the inhibition by F^- of the erythrocyte membrane acetylcholinesterase and (Na^+, K^+)-ATPase from rats fed corn oil and corn oil-plus-cholesterol supplemented diet, respectively, are presented in Table 5. In the case of acetylcholinesterase, the values of n change from 1.5 to 1.0 because of cholesterol dietary effect. Consequently, in the (Na^+, K^+)-ATPase the values of n vary in an inverse manner (from 2.0 to 3.6). (Bloj et al., 1973b).

Since the composition of fatty acid was identical in the membranes from the animals fed the cholesterol-free and cholesterol-supplemented diets (Bloj et al., 1976a) and, since the cholesterol effect was not present in soluble acetylcholinesterase (Bloj et al., 1973b), it can be assumed that the effect of cholesterol on the

Table 5. Values of n of membrane-bound enzymes and cholesterol feedings

Diet	N[b]	n[a]		
		Acetylcholinesterase	(Na^+, K^+)-ATPase	(Mg^{2+})-ATPase
Corn oil	5	1.5 ± 0.04	2.07 ± 0.14	1.67 ± 0.02
Corn oil + cholesterol	5	1.0 ± 0.04	3.60 ± 0.18	1.32 ± 0.14
		$p < 0.001$	$p < 0.001$	$p < 0.05$

[a] Expressed as the mean ± S.E.

[b] Number of animals

(Adapted from Bloj et al., 1973b)

membrane occurred through direct incorporation in their lipid phase. Furthermore, the "in vitro" experiments clearly showed that additional cholesterol modified the kinetic parameters of the enzymes in the presence of the same membrane fatty acid composition (Bloj et al., 1973b). The changes of the values of n for the two enzymes in response to cholesterol "condensing effect" were in the same direction as those observed in the case of the modification of membrane fatty acid composition (Fig. 3A, B). In other words, changes in fluidity produced similar effects on the cooperative behavior of the mammalian membrane enzymes independently of the dietary procedure used to modify the membrane lipid fluidity.

Influence of the fatty acid composition of the diet on the effect of cholesterol feeding

The experiments described above on the effect of cholesterol feeding were performed with an EFA (essential fatty acid) sufficient diet using corn oil as fat supplement and were interpreted as a consequence of the modification of the membrane fluidity due to cholesterol "condensing effect". The effect of cholesterol feeding on the kinetic parameters of membrane-associated enzymes was further investigated using different fat supplemented diets.

a) Hill plots

Table 6 displays the results obtained for the F^- inhibition of

Table 6. Effect of cholesterol feeding on n values of acetylcholinesterase and (Ca^{2+}, Mg^{2+}) ATPase from rats fed diets containing different fats[a]

Diet	Acetylcholinesterase		(Ca^{2+}, Mg^{2+}) ATPase	
		plus cholesterol		plus cholesterol
Corn oil	1.5 ± 0.04	1.0 ± 0.04 ±	1.90 ± 0.08	1.3 ± 0.02
	$p < 0.001$		$p < 0.02$	
Standard	1.38 ± 0.04	0.97 ± 0.09		
	$p < 0.001$			
Lard	0.90 ± 0.05	0.95 ± 0.03	1.08 ± 0.03	1.10 ± 0.02
	n.s.		n.s.	
Hydrogenated fat	1.00 ± 0.05	0.92 ± 0.03		
	n.s.			
Fat-free	1.00 ± 0.03	1.02 ± 0.3		
	n.s.			

[a] The results are expressed as mean ± S.E. and are the average of at least three animals.

(Adapted from Bloj et al. 1976a)

the acetylcholinesterase. The results show that the value of n was not modified when cholesterol was fed with fat-free diet, lard or hydrogenated fat supplemented diets, all of which induced membranes with low a fluidity parameter. The values of n decrease from 1.4 to 1.0 in rats fed a standard diet.

The kinetic changes by feeding corn oil supplemented diet with cholesterol are noticeable only after 24 hours (Bloj et al 1973b). When the rats receiving the diet with cholesterol for 11 days were fed the same diet without cholesterol, the kinetic parameters recovered their original values within 4 to 11 days (Bloj et al., 1976a). Values of $n = 1.6$ and lower K 0.5 values were found for the acetylcholinesterase after membrane solubilization irrespective of the kind of dietary lipid used.

Table 6 presents also the results of an experiment showing the effect of cholesterol on the Hill coefficient for the activation by Mg^{2+} in the presence of Ca^{2+} of the membrane (Ca^{2+}, Mg^{2+})-ATPase from two groups of animals which had been fed with corn oil supplemented diet and lard supplemented diet, respectively. As in the acetylcholinesterase system, the cholesterol modified the value of n only in corn oil-fed animals (Bloj et al., 1976b). Both enzymatic systems present a positive correlation between the values of n and the fluidity parameter (Bloj et al., 1973a; Galo et al., 1975). Similar results were obtained by feeding fat supplemented diets with cholesterol. Finally there was no further modification in the kinetic parameter when membranes from rats fed with the corn oil diet in the "in vivo" cholesterol experiment were loaded with cholesterol "in vitro". Cholesterol feeding presaturated all the membrane cholesterol sites which are related to this phenomenon (Bloj et al., 1976a).

b) Arrhenius plot

When the effect of cholesterol feeding on the Arrhenius plot was investigated, it was found (Bloj et al., 1976b) that, in the case of the enzyme from rats fed corn oil-plus-cholesterol supplemented diet, a curve with two breaking points, one at 20°C and another at 29°C, was obtained (Fig. 4). After solubilization of

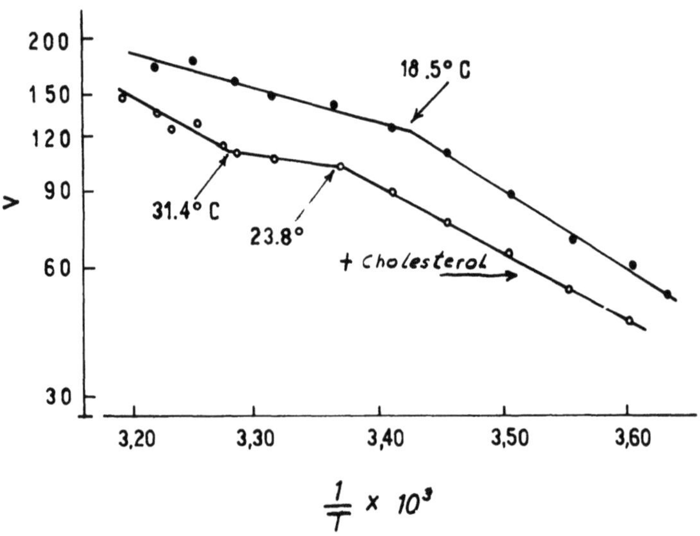

Fig. 4. Arrhenius plot of acetylcholinesterase from rats fed corn oil (•-•) and corn oil + 1% cholesterol (o-o) diet.

the membrane, the enzyme recovered its original temperature activity dependence. No differences were detected when the animals were fed either the standard diet or the lard supplemented diet plus cholesterol. A breaking point around 20°C was obtained, similar to rats fed a corn oil diet. By feeding cholesterol with fat-free diet or hydrogenated fat supplemented diet a shift from to 20°C in the breaking point could be observed. After solubilization of the membrane, the breaking point remained at 20°C. The cholesterol effect on the kinetic parameters of the membrane-bound enzymes can be obtained by cholesterol loading "in vitro" (Bloj et al., 1976b).

DISCUSSION

Singer (1972) has divided the membrane proteins in two major classes according to their location on the membrane structure. "Peripheral" are those protein molecules which are associated only with the membrane surface and, therefore, are easily released from the membrane by mild treatment. "Integral" are those protein molecules that penetrate into the membrane and require a much more drastic treatment to be separated from the membrane. Acetylcholinesterase may be considered as a peripheral protein because it can be dissociated from erythrocyte ghosts at high ionic strength, whereas the (Na^+, K^+)-ATPase the (Ca^{2+}, Mg^{2+})-ATPase from erythrocytes are integral proteins (Coleman, 1973; Zwaal et al., 1973).

Another difference in the characteristic of the integral ATPases and peripheral acetylcholinesterase is the dependence on lipids for enzymatic activity. It has been shown that, on one hand, the (Na^+, K^+)-ATPase and (Ca^{2+}, M^{2+})-ATPase were deactivated by phospholipase treatment and reactivated by phospholipids and, on the other hand, the activity of the acetylcholinesterase was not affected (Coleman, 1973, Zwaal et al., 1973).

The results of our work present a new situation on the effect of lipid composition of the membrane on the role played by the membrane and on the regulation of the membrane-associated enzymes. The lipid composition of the membrane depends on the lipid composition of the diet which is primarily responsible for the regulatory characteristics affecting its associated-enzymes, independently of their locations and lipid dependency.

A very distintive characteristic of the regulatory phenomenon already described is that the observed changes of the kinetic parameters require the integrity of the membrane, which means that the modification of the fatty acid composition change the behavior of the enzyme when membrane lipids are structuraly organized. This work shows that complementary information could be obtained from the studies of allosteric behavior and temperature-dependent activity of the membrane-bound enzymes. It is interesting to note that to cause a shift in the position of the breaking point in the Arrhenius plot, changes in the interaction energy between the membrane and the enzyme of the order of 3 Kcal/mol are necessary, while just a variation of 0.7 to 0.8 Kcal/mol will be enough to produce modifications of the allosteric behavior (Siñeriz et al., 1975).

Some selected results of the modification of the kinetic parameters studied here for three membrane-associated enzymes with different diets are summarized in Table 7.

The values for the temperature breaking point of the acetylcholinesterase are 20°C and 28°C for the EFA-sufficient and EFA-deficient diets, respectively. When any of the EFA-deficient, the fat-free or the fat-hydrogenated supplemented diets is supplemented with cholesterol, the breaking point shifts to 20°C. Similar values of n for F^- inhibition, breaking point and breaking point in cholesterol supplementation were obtained with any of both EFA-deficient diets. In other words, the behavior of the acetylcholinesterase is independent on the kind of the EFA-deficient diet used for feeding. However, the values of n for F^- inhibition of the (Na^+, K^+)-ATPase and for Mg^{2+} activation in presence of Ca^{2+} of the (Ca^{2+}, Mg^{2+})-ATPase indicated that a different lipid-protein interaction in the membrane is produced with the different EFA-deficient diets. A similar pattern can be observed in EFA-sufficient animals as it can be seen in Table 7. In the case of the EFA-sufficient diets, the values of n for the inhibition by $F^-$of the (Na^+, K^+)-ATPase and Mg^{2+} activation of the (Ca^{2+}, Mg^{2+})-ATPase are 2.0 and for the inhibition by F^- of the acetylcholinesterase is 1.5 when animals are fed either the corn oil supplemented or standard diet. With lard supplemented diet, the value of n increases to 3.0 in the (Na^+, K^+)-ATPase and decreases to 1.0 in the (Ca^{2+}, Mg^{2+})-ATPase and acetylcholinesterase, respectively. Then, whithin the EFA-sufficient animal the cooperative behavior of rat fed lard diet was clearly different from rat fed with corn oil and standard diet. The breaking point temperature is 20°C with any of the three EFA-sufficient diets. By supplementation with cholesterol the corn oil supplemented diet differentiates from the other two diets since an Arrhenius curve with two points one at 22°C and another at 28°C, was obtained. Thus, this latter alteration shows a difference in the membrane from rats fed corn oil and standard diet. The findings presented here provide conclusive evidence that feeding conditions are able to modify one kinetic parameter of the membrane-associated enzymes which in turn may function as a good indicator of structural modifications of the membrane caused by change in the lipid composition of the diet. The results obtained with the different experimental approaches used in our studies pointed that the mechanism for the modifications of the kinetic parameters do no respond to a single effector such as EFA, membrane fluidity, enzyme

Table 7. Hill coefficient and transition temperatures of erythrocyte membrane-bound enzymes from rats fed different lipid supplemented diets

Diet	(Na^+, K^+) ATPase[a]	(Ca^{2+}, Mg^{2+}) ATPase[b]	Acetyl-cholin-esterase[a]	Acetylcholinesterase[c, d]	Acetylcholinesterase[c, d] plus cholesterol in diet
	n values			Breakpoint (°C)	
EFA-sufficient					
Corn oil	2.0*	2.0*	1.5*	20*	22-28*
Standard	2.0*	2.0*	1.4*	20*	20*
Lard	3.0**	1.1**	0.9**	20*	20*
EFA-deficient					
Fat-free	2.0*	1.3***	1.0**	28**	20*
Hydrogenated fat	3.0**	1.1**	1.0**	28**	20*

*, **, ***. The values in each column differ significantly when not followed by the same symbol ($p < 0.05$) at least.

Adapted from a) Bloj et al., (1973a); b) Galo et al., 1975; c) Bloj et al., 1974 and d) Bloj et al., (1976b)

localization or lipid dependent activity. This is a very complex phenomenon by which every particular enzyme responds in a specific unpredictable manner to changes of the lipid environment while they are attached to the membrane.

Another remarkable characteristic of the dietary lipid effect is the interaction among administered lipids such as the case of cholesterol which influences the kinetic parameters according to the fatty acid composition of the diet.

CONCLUSIONS

In conclusion, with each particular diet a specific regulatory pattern for the membrane-associated enzyme was obtained which changed with the diet. From this picture, two questions still remain unanswered at the moment: which is the "normal" regulatory pattern and which are the physiological implications of the described effect.

ACKNOWLEDGEMENTS

We acknowledge the excellent secretarial assistance of Miss Susana Bustos. Special thanks are due to Dr. Eugenio Valentinuzzi for his assistance in the preparations of English manuscript. This investigation was supported by Consejo Nacional de Investigaciones Científicas y Técnicas and Secretaría de Ciencia y Técnica de la Universidad Nacional de Tucumán (República Argentina).

REFERENCES

BLOJ B., MORERO R. D., FARIAS R. N. & TRUCCO R. E. (1973a) Biochim. Biophys. Acta 311, 67-69

BLOJ B., MORERO R. D. & FARIAS R. N. (1973b) FEBS Lett. 38, 101-105

BLOJ B., MORERO R. D. & FARIAS R. N. (1974) J. Nutrition 104, 1265-1272

BLOJ B., GALO M. G., MORERO R. D. & FARIAS R. N. (1976a) J. Nutrition 106, 1827-1834.

BLOJ B., MORERO R. D. & FARIAS R. N. (1976b) Paper in preparation .

COLEMAN R. (1973) Biochim. Biophys. Acta 300, 1-30

DE MENDOZA D., MASSA E. M., MORERO R. D. & FARIAS R. N. (1976) Fed. Proc. 35 N° 3 2139

DEMEL R. A., van DEENEN L. L. M. & PETHICA B. A. (1967) Biochim. Biophys. Acta 135, 11-19

DOMENECH C. E., MACHADO de DOMENECH E. E., BALEGNO H. F., DE MENDOZA D. & FARIAS R. N. (1977) FEBS Lett. 74, 243-246.

ENGLEMAN D. M. (1971) J. Mol. Biol. 58, 153-165

ESFAHANI M., LIMBRICH A. R., KNOTTON S., OKA T. & WAKIL S. J.

(1971) Proc. Natl. Acad. Sci. U. S. A. 68, 3180-3184
FARIAS R. N. (1967) Doctoral dissertation. Facultad de Ciencias Exactas y Naturales, Universidad de Buenos Aires (Argentina)
FARIAS R. N., BLOJ B., MORERO R. D., SIÑERIZ F. & TRUCCO R. E. (1975) Biochim. Biophys. Acta 415, 231-251
GALO M. G., BLOJ B. & FARIAS R. N. (1975) J. Biol. Chem. 250, 6204-6207
KROES J., OSTWALD R. & KEITH A. D. (1972) Biochim. Biophys. Acta 274, 71-74
MASSA E. M., MORERO R. D., BLOJ B. & FARIAS R. N. (1975) Biochem. Biophys. Res. Commun. 66, 115-122
MARTINEZ de MELIAN E. R., MORERO R. D. & FARIAS R. N. (1976) Biochim. Biophys. Acta 422, 127-137
MORENO H. & FARIAS R. N. (1976) Biochem. Biophys. Res. Commun. 72, 74-80
MORERO R. D., BLOJ B., FARIAS R. N. & TRUCCO R. E. (1972) Biochim. Biophys. Acta 282, 157-165
RAISON J. K., LYONS J. M., MELHORN R. J. & KEITH A. D. (1971) J. Biol. Chem. 246, 4036-4040
RAHM J. J. & HOLMAN R. T. (1964) J. Lipid Res. 5, 169-176
SINGER S. J. (1971) in Structure and Function of Biological Membrane (Rothfield L. I. ed.) pp 146-223. Academic Press. London
SIÑERIZ F., FARIAS R. N. & TRUCCO R. E. (1975) J. Theor. Biol. 52, 113-120
VAN GOLDE L. M. G. & VAN DEENEN L. L. M. (1966) Biochim. Biophys. Acta 125, 496-509
WILLIAMS M. A., STANCLIFF R. C., PACKER L. & KEITH A. D. (1972) Biochim. Biophys. Acta 267, 444-456
ZWAAL R. F. A, ROELOFSEN B. & COLLEY C. M. (1973) Biochim. Biophys. Acta 300, 159-182

EFFECT OF THYROXINE ON Δ6 AND Δ9 DESATURATION ACTIVITY

I.N.T. de Gómez Dumm,* M.J.T. de Alaniz,* and R.R. Brenner*

Instituto de Fisiología, Facultad de Ciencias Médicas
Universidad Nacional de La Plata
La Plata, Argentina

INTRODUCTION

Many studies have revealed that thyroid hormones markedly affect lipid metabolism in man and in several species of animals. Concerning fatty acid biosynthesis it was demonstrated that the administration of thyroxine stimulates the incorporation of $|1\text{-}^{14}C|$ acetate into fatty acids in rats and mice (Dayton et al, 1960) (Gompertz and Greenbaum, 1966) (Marchi and Mayer, 1959). According to Gompertz and Greenbaum (1966) these observations appear to be associated with an increase of stearyl-CoA desaturase activity. Moreover Myant and Iliffe (1963) found that rats treated with thyroxine showed an inhibition of acetate incorporation, but not of malonate incorporation, into fatty acids by mitochondria free, subcellular liver preparations. Other authors have shown that the thyrotoxic state was accompanied by an increased incorporation of acetyl-CoA to fatty acid and a rise in the activity of fatty acid synthetase in rat livers (Diamant et al, 1972) (Roncari and Murthy, 1975). However, in vitro studies of fatty acid synthesis in which liver supernatant of 105,000 xg and microsomal preparations were incubated with the hormone showed that thyroxine inhibits de novo synthesis of palmitate and stimulates the desaturation reactions (Faas et al, 1972).

Despite these findings and as far as we know, the influence of thyroid hormones on the biosynthesis of polyunsaturated fatty acids

*The authors are members of the Carrera del Investigador Científico of the Consejo Nacional de Investigaciones Científicas y Técnicas.

of the essential series is not established. Therefore, the purpose of the experiments reported in this paper was to study the effect of thyroxine and a hypothyroid agent upon linoleic acid desaturation and elongation. Comparison with effect of the hormone on Δ9 desaturation activity was also investigated.

MATERIAL AND METHODS

Adult female Wistar rats, weighing 180-220 g and maintained on standard Purina chow were used. One group of animals was administered L-thyroxine as sodium salt dissolved in 0.005 N NaOH. Once a day each animal was injected with 40 mg/kg body weight for 12 days. Another group of rats was treated with propylthiouracil (gift from G. Ramon Lab. Argentina) dissolved in 0.005 N NaOH at a single dose of 50 mg/kg body weight per day, for 18 days. Control animals were injected with 0.005 N NaOH solution only. All compounds were injected intraperitoneally.

Rats were killed by decapitation and samples of blood were collected for glucose determination and fatty acid composition. Samples of liver were taken to measure cyclic AMP concentration. The rest of the liver was excised and immediately homogenized (Castuma et al, 1972). Samples were taken to measure protein and glycogen content and total fatty acid composition. Liver microsomes were separated by differential centrifugation at 100,000 xg for 60 min in a Spinco ultracentrifuge and the pellets were resuspended in the homogenizing solution.

The fatty acid desaturation of liver microsomes was measured by estimation of the percentage conversion of $|1\text{-}^{14}C|$ linoleic acid (61 mC/mmole, 99 % radiochemical purity, purchased from New England Nuclear Corp., Boston, Mass) to γ-linolenic acid and $|1\text{-}^{14}C|$ palmitic acid (55.5 mC/mmole, 99 % radiochemical purity, purchased from the Radiochemical Centre, Amersham, England) to palmitoleic acid. Three nmoles of labeled acid and 97 nmoles of unlabeled acid were incubated aerobically with 5 mg microsomal protein and the necessary cofactors at 35°C during 20 min, according to the procedure described previously (Gómez Dumm et al, 1975).

The elongation of linoleic acid by liver microsomal preparation was measured by estimating the conversion of $|1\text{-}^{14}C|$ linoleic acid to eicosa-11, 14-dienoic acid. Five nmoles labeled linoleic acid and 135 nmoles unlabeled acid and malonyl CoA were incubated anaerobically with 5 mg microsomal protein at 35°C for 10 min with the cofactors described in a previous paper (Gómez Dumm et al, 1976).

After incubation fatty acids were recovered and esterified. The distribution of the radioactivity between substrate and product was determined by gas liquid radiochromatography. The labeled methyl esters were identified by equivalent chain length determination and comparison with standards.

Lipids of plasma and liver homogenate were extracted by the procedure of Folch et al (1957). The fatty acid composition was determined using gas liquid chromatography. Blood glucose was measured by the o-toluidine method (Dubowsky, 1962) and liver glycogen by the method of Van Handel (1965). Proteins were determined by the biuret method of Gornall et al (1949) and liver cyclic AMP by the method of Mato and Serrano Rios (1973).

RESULTS AND DISCUSSION

The influence of thyroid hormones upon fatty acid desaturation reaction has been studied in vivo and in vitro, but in most cases indirect evidence was given. Thus, Ellefson and Mason (1964) analyzing plasma and liver fatty acid composition of rats treated with thyroxine suggested that under the influence of the hormone there was an enhancement in the conversion of linoleic acid to arachidonic acid. In vivo effect of thyroxine was also reported by Gompertz and Greenbaum (1966), who studied the desaturation reaction in rat liver microsomal preparation. Under these conditions it was found that treatment of the rats with thyroxine greatly stimulated the desaturation of stearic acid to oleic acid. Faas et al (1972) provide additional data in this respect adding thyroxine to subcellular fractions of rat liver, and studying de novo synthesis of fatty acids. Regarding the proportion of saturated fatty acids formed in the presence of thyroxine, these authors indicate that the hormone stimulates the microsomal desaturation reactions but has no effect on the chain elongation reactions. The present experiment gives direct evidence that thyroxine administration to normal rats produces a significant increase of palmitoyl $\Delta 9$ desaturation activity of microsomal preparation as it is shown in Fig. 1. This observation agrees with those of Gompertz and Greenbaum (1966) for stearic acid desaturation activity. In contrast to the marked increase of palmitoyl $\Delta 9$ desaturation activity, linoleyl $\Delta 6$ desaturation activity showed a significant decrease (Fig. 1). These results are in conflict with those reported by other authors (Ellefson and Mason, 1964; Faas et al, 1972), but in the present experiment the desaturation reaction was measured through the conversion of labeled linoleic acid to γ-linolenic acid in the microsomal fraction, whereas in the others only indirect conclusions were obtained.

The simultaneous effects of thyroxine and propylthiouracil on other parameters (plasma glucose, liver glycogen, and liver cyclic AMP levels) are summarized in Table 1. The concentration of serum

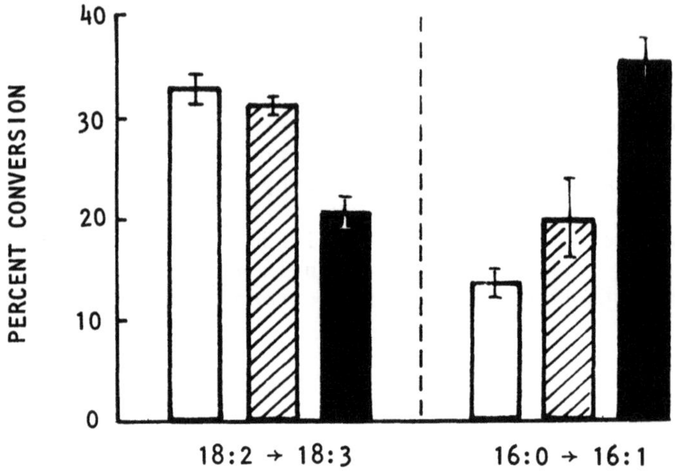

Fig. 1. Effect of propylthiouracil ▨ and thyroxine ■ administration compared to controls □ on the oxidative desaturation of |1-^{14}C| linoleic acid to γ-linolenic acid (18:2→18:3) and palmitic acid to palmitoleic acid (16:0→16:1). Results are means of analysis of 5 animals (each analysis was performed in duplicate). Vertical lines represents 1 SEM. Results corresponding to thyroxine-treated rats are significantly different from the controls ($p < 0.001$).

glucose was maintained within the normal control range after the treatment of the animals with either thyroxine or propylthiouracil. In contrast to the glucose levels, hepatic glycogen concentration tended to decrease in the hyperthyroid group and increased significantly after propylthiouracil treatment. The modifications observed in liver glycogen content are in accordance with those obtained by other authors (Myant and Iliffe, 1963; Freeland, 1965), who showed that liver glycogen levels were increased by thyroidectomy and markedly decreased when thyroid hormones were administered to normal rats. The diminished liver glycogen in the thyroxine-treated rats could reflect an enhanced glycogenolysis as a consequence of increased adenylate cyclase activity. In this respect it was demonstrated that phosphorylase _a_ activity increased significantly following thyroxine administration to thyroidectomized rats (Takahashi and Suzuki, 1975). This effect was associated with an increase of cyclic AMP in liver. However, after 12 days of thyroxine treatment no increase of cyclic AMP was detected in these experiments (Table 1). This result may be related to the long period of thyroxine administration. Glucagon and epinephrine produce a rise in cyclic AMP followed by enhanced glycogenolysis, and depressed Δ6

Table 1. Effect of the administration of thyroxine and propylthiouracil to intact rats

Averages of the analyses of 5 rats ±one SEM. * Results significantly different from the control.

CONDITION	Glycemia mg%	Liver glycogen µg/mg protein	Liver cAMP pmoles/mg protein
control	120 ± 4	0.123 ± 0.004	4.1 ± 0.8
propylthiouracil	123 ± 5	0.194* ± 0.026	5.0 ± 0.9
thyroxine	124 ± 3	0.104* ± 0.005	2.9 ± 0.5

desaturation activity. This last effect was interpreted to be a consequence of the rise in intracellular levels of cyclic AMP (Gómez Dumm et al, 1975, 1976). Therefore it is reasonable to think that the effect of thyroxine on Δ6 desaturation activity could also be produced through an enhancement of the intracellular levels of cyclic AMP or, indirectly, through an increase of glucose metabolism, since glucose oxidation and removal might be increased in hyperthyroidism (Rabinowitz and Myerson, 1967). This last possibility is consistent with the report that shows that a glucose diet depressed the Δ6 desaturation of fatty acids (Peluffo et al,

Table 2. Effect of thyroxine and propylthiouracil on the elongation of |1-^{14}C| linoleic acid

Conditions of incubation are detailed in the experimental part. Means of five animals analysed in duplicate ± one SEM.

CONDITION	20:2 (11, 14) nmoles formed/mg protein/min
control	0.37 ± 0.04
propylthiouracil	0.38 ± 0.03
thyroxine	0.44 ± 0.03

Table 3. Effect of thyroxine and propylthiouracil in the percent distribution of total fatty acids in plasma and liver

PPT= propythiouracil, L-T_4=thyroxine. Mean values of five observations
* Results significantly different from the controls.

Fatty acids	Liver			Plasma		
	control	PPT	L-T_4	control	PPT	L-T_4
14:0	0.3	0.2	0.3	1.4	0.8	0.8
16:0	25.9	26.0	24.4	26.1	25.7	25.8
16:1	1.9	1.3	1.4	2.1	2.5	2.4
18:0	22.9	24.9	26.8*	16.5	14.9	16.7
18:1	12.2	11.8	12.7	15.6	16.1	18.2
18:2	16.3	13.8*	13.2*	21.3	21.3	20.6
20:4	20.7	22.1	21.1	17.4	18.8	15.5

1971). Moreover, no direct effect of cyclic AMP on the Δ6 desaturation activity of rat liver microsomes could yet be demonstrated.* On the other hand, thyroxine stimulated Δ9 desaturation activity (Fig. 1). This effect could also be produced by a stimulation of glucose metabolism. The work of Mercuri et al (1974) gives support to this suggestion since the administration of glycerol, glucose, fructose, or stearic acid to normal rats increases the microsomal specific activity of the Δ9 desaturase. These results are also consistent with the observations that Δ9 desaturase is different from Δ6 desaturase (Brenner, 1971). However, while Δ9 desaturase seemed to be insensitive to glucagon or epinephrine (Gómez Dumm et al, 1975, 1976) activity increased significantly under thyroxine treatment.

In contrast to the marked effect of thyroxine on the desaturation reactions, the elongation reactions were not affected by the administration of the hormone (Table 2). Previously it was demonstrated that neither glucagon nor epinephrine modified the elongation reactions (Gómez Dumm et al, in press). Therefore, the present results corroborate the assumption that the microsomal elongating system is insensitive to several hormones in the conditions of our experiments.

Although the doses of propylthiouracil (a hypothyroid agent) given to the rats seem to be sufficient to produce hypothyroidism

* Gómez Dumm et al. Unpublished observations.

(Nelson and Cornatzer, 1965), no modifications of the desaturating or elongating reactions compared to the controls was observed. (Fig. 1, Table 2).

The fatty acid composition of serum and liver lipid fractions in the thyroxine, propylthiouracil, and control groups of rats is given in Table 3. Quantitatively, the most important change found by thyroxine administration was an increase of stearic acid and a decrease of linoleic acid in liver lipids. These results agree with those reported by Mitchel and Truchot (1962). Besides, the data in Table 3 show that arachidonic acid does not appear to be modified in the same way by thyroxine as linoleic acid, as it was reported previously (Mitchel and Truchot, 1962).

Propylthiouracil-treated animals also showed a decrease in the percent of linoleic acid distribution in liver. In spite of the results obtained by different authors (Ellefson and Mason, 1964; Mitchel and Truchot, 1962; Kirkeby, 1972), the changes in the percent distribution of total fatty acids in plasma are small and unimportant (Table 3). The probable explanation for the low linoleic acid concentration in liver could be that the supply of essential fatty acids from the diet does not parallel the increase of lipid degradation (Kirkeby, 1972), thereby giving a relative linoleic acid deficiency despite the food intake. The changes on Δ6 and Δ9 desaturation activities are not apparently related with the modifications observed in the percent distribution of total fatty acids in liver. This fact could be explained considering that the composition of liver and plasma lipids is a result of a balance between dietary fat, lipid, and carbohydrate metabolism, lipolysis of tissue lipids, and endocrinological factors.

Although controversy still exists concerning the effect of thyroxine on lipid metabolism, there is a consensus that Δ9 desaturation activity is enhanced under the influence of the hormone, which is corroborated in this paper. However, the present investigation demonstrates that in vivo, Δ6 desaturation activity is modified in an opposite direction to palmitic desaturase.

CONCLUSIONS

The administration of thyroxine to normal rats for several days produced an increase of Δ9 desaturation activity of liver microsomal preparations. Conversely, Δ6 desaturation activity decreased significantly under the hormone treatment. These results were attributed to an increase in glucose metabolism. The elongating enzyme system of linoleic acid seems to be insensitive to thyroxine. Neither the desaturation nor the elongation reactions were modified under propylthiouracil treatment of rats.

REFERENCES

BRENNER R. R. (1971) Lipids 6, 567-575
CASTUMA J. C., CATALA A. & BRENNER R. R. (1972) J. Lipid Res. 13, 783-789.
DAYTON S., DAYTON J., DRIMMER F. & KENDALL F. E. (1960) Am. J. Physiol. 199, 71-76
DIAMANT S., GORIN E. & SHAFRIR E. (1972) Eur. J. Biochem., 26, 553-559
DUBOWSKY K. M. (1962) Clin. Chem., 8, 215-235
ELLEFSON R. D. & MASON H. L. (1964) Endocrinology 75, 179-186
FAAS F. H., CARTER W. J. & WYNN J. (1972) Endocrinology 91, 1481-1492
FOLCH J., LEES M. & SLOANE-STANLEY G. H. (1957) J. Biol. Chem. 226, 497-509
FREELAND R. A. (1965) Endocrinology 77, 19-27
GOMEZ DUMM I. N. T. de, ALANIZ M. J. T. de & BRENNER R. R. (1975) J. Lipid Res. 16, 264-268
GOMEZ DUMM I. N. T. de, ALANIZ M. J. T. de & BRENNER R. R. (1976) J. Lipid Res. 17, 616-621
GOMEZ DUMM I. N. T. de, ALANIZ M. J. T. de & BRENNER R. R. Lipids in press
GOMPERTZ D. & GREENBAUM A. L. (1966) Biochem. Biophys. Acta 116, 441-459
GORNALL A. G., BARDAWILL C. J. & DAVID M. M. (1949) J. Biol. Chem. 117, 751-766
KIRKEBY K. (1972) Acta Endocrinol. 71, 62-72
MARCHI P. & MAYER J. (1959) Experientia 15, 359-361
MATO J. M. & SERRANO RIOS M. (1973) Rev. Esp. Fisiol. 29, 233-238
MERCURI O., PELUFFO R. O. & DE TOMAS M. E. (1974) Biochim. Biophys. Acta 369, 264-268
MITCHEL R. & TRUCHOT R. (1962) Bull. Soc. Chim. Biol. 44, 1141-1152
MYANT N. B. & ILIFFE J. (1963) in "The control of lipid metabolism" (J. K. Grant, ed.) p. 145-154. Academic Press, London and New York
NELSON D. R. & CORNATZER W. E. (1965) Endocrinology 77, 37-44
PELUFFO R. O., GOMEZ DUMM I. N. T. de, ALANIZ M. J. T. de & BRENNER R. R. (1971) J. Nutr. 101, 1075-1084
RABINOWITZ J. L. & MYERSON R. M. (1967) Metab. Clin. Exp. 16, 68-75
RONCARI D. & MURTHY V. K. (1975) J. Biol. Chem. 250, 4134-4138
TAKAHASHI T. & SUZUKI M. (1975) Endocrinol. Jap. 22, 187-194
VAN HANDEL E. (1965) Anal. Biochem. 11, 256-265

BIOSYNTHESIS OF POLYUNSATURATED FATTY ACIDS FROM THE LINOLEIC ACID FAMILY IN CULTURED CELLS

M.J.T. de Alaniz,* I.N.T. de Gómez Dumm,* and R.R. Brenner*

Cátedra de Bioquímica, Instituto de Fisiología, Facultad de Ciencias Médicas, Universidad Nacional de La Plata, La Plata, Argentina

INTRODUCTION

HTC cells (designated HTC for hepatoma tissue culture) were derived from the ascites form of a rat-carried Morris hepatoma 7288 C (Thompson et al, 1966). Previous studies have revealed that cells of this kind are able to desaturate and elongate fatty acids. In this respect it was demonstrated that culture HTC cells preserved the ability to desaturate stearic to oleic acid (Δ9 desaturase), α-linolenic acid to octdeca-6,9,12,15-tetraenoic acid (Δ6 desaturase), and eicosa-8,11,14-trienoic acid to arachidonic acid (Δ5 desaturase) (Alaniz et al, 1975). They are also able to convert α-linolenic acid to higher homologs with 5 and 6 double bonds by desaturation and elongation reactions. These results also proved the existence of Δ4 desaturase activity (Alaniz et al, 1975). However, it was shown that the cell cultured in Swim's medium supplemented with serum possessed a very low capacity to convert labeled linoleic acid of the medium to arachidonic acid. Nevertheless, these tumor cells readily converted eicosa-8,11,14-trienoic acid to arachidonic acid (Alaniz et al, 1975; Gaspar et al, 1975). Therefore, the difficulty of these cells to synthesize arachidonic acid from linoleic acid may reside in a step previous to the Δ5 desaturation of eicosa-8, 11, 14-trienoic acid. This step could be a Δ6 desaturation of linoleic acid. The discrepancy between the biosynthesis of linoleic and α-linolenic acid series is difficult to explain considering that the same enzyme desaturates

---

* The authors are members of the Carrera del Investigador Científico of the Consejo Nacional de Investigaciones Científicas y Técnicas.

linoleic and α-linolenic acids in Δ6 position (Brenner and Peluffo, 1966) (Brenner, 1971; Brenner, 1974) (Ninno et al, 1974).

For this reason it was important to investigate the routes of arachidonic acid synthesis from labeled linoleic acid and the possible incorporation of labeled acids in cell lipids.

MATERIAL AND METHODS

HTC 7288 C cells were maintained and grown at 37°C in confluent layers attached to glass on Swim's 77 medium supplemented with 10% calf serum (Thompson et al, 1966) (Alaniz et al, 1975). After 48 h, when the bottles contained approximately 5×10^6 cells, the medium was changed for 10 ml of S 77 medium without serum, to which 5 nmoles of $|1\text{-}^{14}C|$ linoleate (61 mC/mmole, 99 % radiochemical purity) per bottle was added, and the cells were maintained in this medium for 24 h. The acid was added as sodium salt bound to defatted albumin (Goodman, 1957) according to Spector et al (1965). After the end of the incubation period the attached cells were washed with 0.85 % NaCl, removed, and suspended in the same saline solution. An aliquot of the suspension was used to determine the amount of cellular protein by the method of Lowry et al (1951) and the rest was centrifuged. The lipids were extracted by the procedure of Folch et al (1957) and separated by thin layer chromatography (TLC). The solvent mixture used was chloroform-methanol-water 65:25:4 v/v/v. An aliquot of the corresponding fraction of the samples was counted in a Packard Tricarb Scintillation counter. Other aliquots were esterified. The distribution of the radioactivity between the fatty acids was determined by gas liquid radiochromatography (Alaniz et al, 1976). The labeled methyl esters were identified by equivalent chain length determination and comparison with authentic standards. The fatty acid composition of serum, HTC cells and lipid fractions of culture cells was analyzed by gas liquid chromatography. The specific radioactivities for linoleic and arachidonic acids were calculated with those data after measuring the radioactivity in an aliquot in which the mass distribution of the fatty acids had previously been determined by gas liquid chromatography in the presence of an internal standard of eicosaenoic acid.

RESULTS AND DISCUSSION

The fatty acid composition of serum and different lipid fractions of HTC cells is shown in Table 1. The pattern is similar to that of the serum used in the preparation of the medium, except for changes in oleic, linoleic and α-linolenic acids. Whereas the relative amount of oleic acid was higher in the cells than in the

Table 1. Fatty acid composition of serum and different lipid fractions of HTC cells.

Cells grown for 2 days with Swim's medium and calf serum were incubated with |1-^{14}C| linoleic acid for 24 h in a new Swim's medium without serum. Means of three bottles ± 1 SEM. Total fatty acids (TFA), phosphatidyl choline (PC), phosphatidyl ethanolamine (PE), free acids (FA) and neutral lipids (NL) fractions.

Fatty acids	Serum %	Lipid fractions TFA %	PC %	PE %	FA %	NL %
14:0	1.3 ±0.1	1.9 ±0.2	3.6 ±0.2	0.2 ±0.03	1.9 ±0.4	3.3 ±0.1
16:0	27.8 ±1.2	25.0 ±1.5	46.2 ±1.3	12.7 ±1.4	23.6 ±4.2	26.9 ±2.3
16:1	3.7 ±0.4	4.5 ±0.6	3.3 ±0.1	2.1 ±0.3	4.9 ±0.6	3.9 ±1.3
18:0	21.9 ±1.1	17.6 ±0.3	6.8 ±0.4	27.0 ±1.9	22.1 ±3.5	10.9 ±0.6
18:1	22.6 ±1.5	41.8 ±2.2	36.1 ±0.05	46.7 ±1.0	35.9 ±3.7	45.8 ±2.8
18:2	15.8 ±0.5	5.9 ±0.3	2.8 ±1.4	4.9 ±1.1	9.5 ±1.9	7.9 ±0.8
18:3	4.4 ±0.2	tr.	tr.	tr.	tr.	tr.
20:4	2.5 ±0.4	3.3 ±0.2	1.0 ±0.1	6.5 ±0.5	2.1 ±0.8	1.3 ±0.03
Ratio 20:4/ 18:2	0.16	0.56	0.36	1.33	0.22	0.16

serum the percentages of linoleic and α-linolenic acid were lower. The low levels of linoleic and arachidonic acid in HTC cells could result from a difficulty of the cells to absorb linoleic acid readily, a very active catabolism of the acid, or an absence of low activity of any enzyme involved in the synthesis of polyunsaturated acids. The ready absorption of linoleic acid was proved by Gaspar

Table 2. Distribution of radioactivity in the lipids of HTC cells after incubation with [$1-^{14}C$] linoeic acid.

Results are the means of 3 bottles expressed as percentage of total radioactivity on the plate ± 1 SEM. Radioactivity on the rest of the plate makes up to 100 %. Experimental conditions as in Table 1.

Lipid fractions	Labeling distribution %
Phosphatidyl choline	33.0 ± 0.5
Phosphatidyl ethanolamine	22.8 ± 1.0
Free acids	14.5 ± 0.3
Neutral lipids	19.1 ± 0.9

et al* measuring the incorporation of labeled linoleic acid from the medium to the cell. The incorporation increased with time of incubation and the amount of acid in the medium. The same authors showed that the saturation curves of incorporation of linoleic, oleic and α-linolenic acids were similar. However, the plateau of the curve was lower for oleic and linoleic acids than for α-linolenic acid. Besides, linoleic acid is catabolized very little by the cells, since Gaspar et al* showed that labeled acid in the medium is preferentially incorporated rather than converted to CO_2.

The amount of arachidonic acid in all cell lipids is low. The highest proportion is found in the phosphatidyl ethanolamine fraction. Moreover, the arachidonic:linoleic acid ratio is the highest in the phosphatidyl ethanolamine fraction. The lowest ratios correspond to free and neutral lipid fractions. These results are similar to those reported previously (Wood, 1973; Wood and Falch, 1973). Therefore it is possible to admit that the fatty acid composition of HTC cells is regulated in an active way and the low content of arachidonic acid is not necessarily the only consequence of low linoleic acid uptake from the medium. Besides, the low content of linoleic and arachidonic acids is not apparently due to a preferential oxidation to CO_2.

The incorporation of labeled linoleic acid of the medium in

* Gaspar G., Alaniz M. J. T. de and Brenner R. R. Mol. and Cell Biochem. (sent for publication).

Table 3. Pattern of labeled fatty acids in HTC cells incubated in the presence of [1-^{14}C] linoleic acid

Results are the percentages of total measurable radioactivity. They are the means of three bottles ± 1 SEM. Experimental conditions and symbols as in Table 1.

Fatty acids	Lipid fractions % radioactivity distributions				
	TFA	PC	PE	FA	NL
18:2 (Δ9, 12)	84.3 ±1.7	79.0 ±4.3	83.9 ±0.9	100.0	91.0 ±0.6
20:2 (Δ11, 14)	5.8 ±0.6	9.1 ±2.2	4.8 ±0.7		9.0 ±0.6
20:3 (Δ5, 11, 14)	2.8 ±0.3	4.2 ±0.9	2.9 ±0.3		
20:3 (Δ8, 11, 14)	2.9 ±0.3	3.8 ±0.7	3.3 ±0.2		
20:4 (Δ5, 8, 11, 14)	4.2 ±0.8	3.9 ±0.9	5.1 ±0.3		

cell lipids is shown in Table 2. It shows that phospholipids incorporated the highest percentage of the radioactivity. This selective concentration of unsaturated acids in the phospholipids helped to investigate the products of linoleic acid conversion by analysis of the fatty acid composition of the labeled products. In Table 3 it is possible to recognize that the free acid fraction of the cell lipids contained only labeled linoleic acid. In neutral lipids the elongation product of linoleic acid: 20:2 (Δ11, 14) was also found. In phosphatidyl choline and phosphatidyl ethanolamine, not only were linoleic acid, 20:2 (Δ11, 14) and arachidonic acid incorporated, but also two 20:3 acids. These 20:3 acids were identified as 20:3 (Δ5, 11, 14) and 20:3 (Δ8, 11, 14) acids by comparison of the retention times with standards and with the data published by Ullman and Sprecher (1971).

The acids detected suggest the existence of two metabolic pathways in the conversion of linoleic acid to higher homologs. These pathways are equivalents to the routes used by HTC cells to desaturate and elongate α-linolenic acid (Alaniz et al, 1976). One of the routes is initiated by a Δ6 desaturation and the other by an

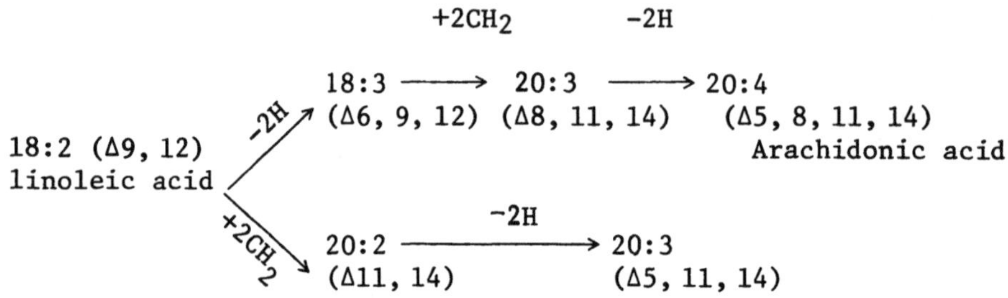

elongation. In the first route, through a Δ6 desaturation linoleic acid is converted to γ-linolenic acid which is rapidly elongated to 20:3 (8, 11, 14) and then desaturated in Δ5 to arachidonic acid. The elongation of linoleic acid in the second route leads to 20:2 (11, 14). This acid would not be desaturated to 20:3 (8, 11, 14) acid since it has been reported that a Δ8 desaturase is absent in liver (Alaniz et al, 1976; Sprecher et al 1975). However, 20:2 (11, 14) acid would be a substrate for the Δ5 desaturase and 20:3 (5, 11, 14) acid is so formed. This last pathway has been described in normal rat liver to occur in very specific conditions (Ullman and Sprecher, 1971).

The crescent importance of the elongation route that converts linoleic acid to 20:2 (Δ11, 14) and 20:3 (Δ5, 11, 14) acid and also converts α-linolenic acid to 20:3 (Δ11, 14, 17) and 20:4 (Δ5, 11, 14, 17) in HTC cells compared to normal tissue may be easily explained by a decrease of the activity of the Δ6 desaturase/elongating enzyme ratio in the transformed cells and, besides, by a relatively high Δ5 desaturation activity.

The Δ6 desaturase is a microsomal enzyme but the elongating enzymes may be found in microsomes and mitochondria. Therefore it is possible that the microsomes of HTC cells have been altered. It is suggestive that we have been unable as yet to obtain active microsomes from the cells in spite of using different mild procedures.

Experiments carried out by Dunbar and Bailey (1975) suggest that generally heteroploid or transformed cell lines lose Δ6 desaturase by dilation. HTC cells cannot be included in this generalization since they possess a Δ6 desaturase. However, Δ6 desaturation activity for linoleic acid is extremely low compared to α-linolenic acid.

Unsaturated acids are incorporated in a normal way in HTC cell phospholipids (Tables 1 and 2). Arachidonic acid would be preferentially incorporated in phosphatidyl choline and phosphatidyl ethanolamine. The measurement of the specific radioactivity of

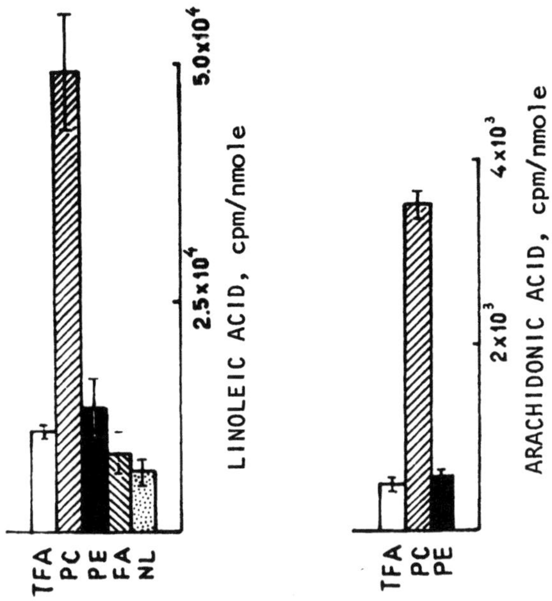

Fig. 1. Specific radioactivities of linoleic and arachidonic acids in different lipid fractions of HTC cells incubated in the presence of labeled linoleic acid.

Results are the means of three incubations. Vertical lines represent 1 SEM. Details of experimental conditions described in material and methods. Symbols as in Table 1.

linoleic and arachidonic acids in the different lipid fractions (Fig. 1) indicates that both acids present the highest values in phosphatidyl choline. However, the highest percentage of arachidonic acid was found in phosphatidyl ethanolamine (Table 1). Therefore it is reasonable to assume that labeled arachidonic acid is diluted in phosphatidyl ethanolamine by unlabeled endogenous acid. Both labeled acids would be incorporated in both phospholipids preferentially by an interchange reaction. Therefore, these results do not help to explain the specifically low desaturation activity for linoleic acid.

CONCLUSIONS

HTC cells incubated in Swim's 77 medium with 10 % calf serum contain small amounts of linoleic and arachidonic acids. Labeled linoleic acid of the medium is incorporated in phospholipids and neutral lipids. It is converted to higher homologs following two routes. One route is 18:2 (Δ9, 12) ⟶ 18:3 (Δ6, 9, 12) ⟶ 20:3 (Δ8, 11, 14) ⟶ 20:4 (Δ5, 8, 11, 14). The other route is 18:2 (Δ9, 12) ⟶ 20:2 (Δ11, 14) ⟶ 20:3 (Δ5, 11, 14). These products are also incorporated in phosphatidyl choline and phosphatidyl ethanolamine. Linoleic acid Δ6 desaturation is very low compared to α-linolenic acid, whereas Δ5 desaturation of 20:3 (Δ8, 11, 14) to arachidonic acid is very high.

REFERENCES

ALANIZ M. J. T. de, PONZ G. & BRENNER R. R. (1975) Acta Physiol. Latinoam. 25, 1-11

ALANIZ M. J. T. de, GOMEZ DUMM I. N. T. de & BRENNER R. R. (1976) Mol. and Cellular Biochem. 12, 3-8

BRENNER R. R. (1971) Lipids. 6, 567-575

BRENNER R. R. (1974) Mol. and Cellular Biochem. 3, 41-52

BRENNER R. R. & PELUFFO R. O. (1966) J. Biol. Chem. 241, 5213-5219

DUNBAR L. M. & BAILEY J. M. (1975) J. Biol. Chem. 250, 1152-1153

FOLCH J., LEES M. & SLOANE-STANLEY H. (1957) J. Biol. Chem. 226, 497-509

GASPAR G., ALANIZ M. J. T. de & BRENNER R. R. (1975) Lipids. 10, 726-731

GOODMAN D. S. (1957) Science. 125, 1296-1297

LOWRY O. H., ROSEBROUGH M. J., FARR A. L. & RANDALL R. J. (1951) J. Biol. Chem. 193, 265-275

NINNO R. E., TORRENGO M. A. P. de, CASTUMA J. C. & BRENNER R. R. (1974) Biochim. Biophys. Acta. 360, 124-133

SPECTOR A. A., STEINBERG D. & TANAKA A. (1965) J. Biol. Chem. 240, 1032-1041

SPRECHER H. & LEE Ch. L. (1975) Biochim. Biophys. Acta. 388, 113-125

THOMPSON E. B., TOMKINS G. M. & CURRAN J. F. (1966) Proc. Nat. Acad. Sci. 56, 296-303

ULLMAN D. & SPRECHER H. (1971) Biochim. Biophys.Acta 248, 186-197

WOOD R. (1973) Lipids. 8, 690-701

WOOD R. & FALCH J. (1973) Lipids. 8, 702-710

SUBJECT INDEX

MIX
Papier aus verantwortungsvollen Quellen
Paper from responsible sources
FSC® C105338

If you have any concerns about our products, you can contact us on
ProductSafety@springernature.com

In case Publisher is established outside the EU, the EU authorized representative is:
Springer Nature Customer Service Center GmbH
Europaplatz 3, 69115 Heidelberg, Germany

Printed by Libri Plureos GmbH
in Hamburg, Germany